ENGINEERING RESINS

ENGINEERING RESINS

An Industrial Guide

by

Ernest W. Flick

np
NOYES PUBLICATIONS
Park Ridge, New Jersey, U.S.A.

Library of Congress Catalog Card Number: 88-25249
ISBN: 0-8155-1172-8
Printed in the United States

Published in the United States of America by
Noyes Publications
Mill Road, Park Ridge, New Jersey 07656

10 9 8 7 6 5 4 3 2 1

Library of Congress Cataloging-in-Publication Data

Flick, Ernest W.
Engineering resins.

Includes indexes.
1. Gums and resins, Synthetic. I. Title.
TP978.F54 1988 668'.37 88-25249
ISBN 0-8155-1172-8

To
The Previous Flick Generation
Albert, Clementine, Louis, Ernest E., Blanche, René, George

Preface

This volume describes more than 2,500 engineering resins which are currently available for industrial usage. The definition of engineering resins differs depending on whom one asks. The book contains those products which are termed "engineering plastics" as well as "upgraded commodity resins." The book will be of value to industrial technical and managerial personnel involved in the specification and use of these products. It has been compiled from information received from numerous industrial companies and other organizations.

The data included represent selections from manufacturers' descriptions made at no cost to, nor influence from, the makers or distributors of the materials.

Only the most recent information has been included. Products covered include blends, composites, alloys, and resins with superior impact resistance, heat or cold resistance, and/or stiffness. The products have been manufactured to meet high performance requirements in major end-use industries—automotive, packaging, electronics, computers and business equipment, aerospace and medical supplies. Many can be tailored to specific end-uses on request. It is believed that all of the products listed here are currently available, which will be of interest to readers concerned with product discontinuances.

Products are presented by company, and the companies are listed alphabetically. The table of contents is organized in such a way as to serve as a company index to the book. Also included is a Trade Name Index, for easy and rapid location of products by the reader. In addition, another section, which will be useful, contains the Suppliers' Addresses. It can be found immediately following the Product Information Section.

The book lists the following product information, as available, in the manufacturer's own words:

(1) Company name and product category.

(2) Trade name and product numbers.

(3) Product Description: a description of the product, as presented by the supplier.

My fullest appreciation is expressed to the companies and organizations who supplied the data included in this book.

Newburyport, Massachusetts
August 1988

Ernest W. Flick

NOTICE

To the best of our knowledge the information in this publication is accurate; however, the Publisher does not assume any responsibility for the accuracy or completeness of, or consequences arising from, such information. This Industrial Guide does not purport to contain detailed user instructions, and by its range and scope could not possibly do so. Mention of trade names or commercial products does not constitute endorsement or recommendation for use by the Publisher.

Some engineering resins could be toxic, and therefore due caution should always be exercised in the use of potentially hazardous materials. Final determination of the suitability of any information or product for use contemplated by any user, and the manner of that use, is the sole responsibility of the user. We strongly recommend that users seek and adhere to a manufacturer's or supplier's current instructions for handling each material they use.

The Author and Publisher have used their best efforts to include only the most recent data available. The reader is cautioned to consult the supplier in case of questions regarding current availability.

Contents

Product Information

ALLIED-SIGNAL INC.: CAPRON Nylon Copolymers for Molding and Extrusion:

CAPRON Nylon is based on type 6 polyamide resin, the most versatile and performance-proven of engineering thermoplastics.

CAPRON Copolymers:

Copolymerization of CAPRON Nylon results in a polymer structure exhibiting varying levels of toughness and flexibility, combined with the excellent thermal and chemical resistance properties provided by the polyamide backbone.

8253:
Injection molding and extrusion grade copolymer developed for applications requiring improved dry as molded toughness over nylon homopolymers, while maintaining excellent strength and stiffness.

8254:
Extrusion grade copolymer developed for tubing applications requiring a balance of performance properties including high flexibility, toughness, and abrasion resistance (Developed as CPN 1220)

8255:
Injection molding grade copolymer developed for applications requiring a high degree of flexibility and toughness. (Developed as CPN 1235)

8259:
Injection molding grade copolymer developed for fastener applications requiring resistance to abusive impact loading, combined with a high level of pullout strength (Developed as CPN 909A)

8350:
Injection molding and extrusion grade resin exhibiting the highest impact performance within the copolymer product line combined with an excellent balance of strength and stiffness. (Developed as CPN 1450)

Applications:
* Plugs, receptacles, flexible connector covers, weed trimmer components, clips, fasteners, flanges, key housings, and other high performance molding applications.

* Convoluted and emissions tubing, cable jacketing, automotive vacuum tubing, hydraulic and air conditioning hose, and other tubing and jacketing applications requiring heat, chemical and abrasion resistance.

ALLIED-SIGNAL INC.: CAPRON Nylon Copolymers for Molding and Extrusion (Continued):

Properties of CAPRON Copolymers (73F, Dry as Molded):

8253:

Tensile Strength at Yield: psi: 9,500
Ultimate Elongation: %: 150
Flexural Strength: psi: 12,000
Flexural Modulus: psi: 320,000

8254:

Tensile Strength at Yield: psi: 5,300
Ultimate Elongation: %: 240
Flexural Strength: psi: 4,200
Flexural Modulus: psi: 110,000

8255:

Tensile Strength at Yield: psi: 5,000
Ultimate Elongation: %: 230
Flexural Strength: psi: 4,000
Flexural Modulus: psi: 120,000

8259:

Tensile Strength at Yield: psi: 8,500
Ultimate Elongation: %: 150
Flexural Strength: psi: 11,000
Flexural Modulus: psi: 300,000

8350:

Tensile Strength at Yield: psi: 8,000
Ultimate Elongation: %: 260
Flexural Strength: psi: 9,500
Flexural Modulus: psi: 260,000

ALLIED-SIGNAL INC.: CAPRON Nylon Homopolymers for Molding and Extrusion:

CAPRON Nylon is based on type 6 polyamide resin, the most versatile and performance-proven of engineering thermoplastics.

CAPRON Homopolymers:

CAPRON homopolymers are standard nylon 6 grades, used in a wide variety of molding and extrusion applications. These resins exhibit the combination of strength, stiffness, and toughness that is associated with nylon 6 as well as excellent chemical and abrasion resistance. There are three categories of CAPRON homopolymers: Standard, Alpha and Wire Jacketing Grades.

A. Standard Grades:

8200:
Medium viscosity injection molding grade exhibiting resistance to sink-mark formations in thick sections and improved toughness over conventional lower viscosity grades. Gears, mechanical components, fittings, handles, and wiring devices are produced from this grade.

8202:
Low viscosity injection molding grade for fast fill and thin sections. Clips and fasteners, fuel filters, electrical receptacles, connectors, and wire ties are typical end-uses.

8202L:
8202 type with excellent flow and mold release characteristics. Meets UL flammability requirements for 94 V-2 in sections as thin as 1/32 inch. UL regulated electrical applications include switches, plugs, and receptacles.

B. Alpha Grades:

Alpha grades differ from conventional type 6 materials in crystalline form. As a result, they offer an increase in strength, stiffness, and heat distortion temperature while improving the excellent processing properties normally associated with CAPRON Nylon. Alpha grades are denoted by the suffix "C", and include:

8202C:
Low viscosity, highly crystalline resin: fastest molding grade available. Gears, furniture casters, and electrical devices are among the numerous end-uses for this grade.

ALLIED-SIGNAL INC.: CAPRON Nylon Homopolymers for Molding and Extrusion(Continued):

B. Alpha Grades(Continued):

8202CL:
8202C type: meets UL flammability requirements for 94 V-2 in sections as thin as 1/32 inch.

8203C:
Intermediate viscosity, highly crystalline resin: exhibits fast set up for extrusion of tubing and cable liners.

C. Wire Jacketing Grades:

8220HS:
Medium viscosity heat stabilized extrusion grade for THHN, THWN building wire, shipboard cable, and non-metallic, sheathed cable primary jacketing.

8221HS:
Medium viscosity heat stabilized, modified 8220. Utilized in large diameter wire applications where increased flexibility is required.

Properties of CAPRON Homopolymers(73F, Dry as Molded):

Standard Grades:

8200:
Tensile Strength at Yield: psi: 12,000
Ultimate Elongation: %: 150
Flexural Strength: psi: 16,000
Flexural Modulus: psi: 410,000

8202:
Tensile Strength at Yield: psi: 11,500
Ultimate Elongation: %: 70
Flexural Strength: psi: 15,700
Flexural Modulus: psi: 410,000

8202L:
Tensile Strength at Yield: psi: 11,500
Ultimate Elongation: %: 50
Flexural Strength: psi: 15,700
Flexural Modulus: psi: 410,000

ALLIED-SIGNAL INC.: CAPRON Nylon Homopolymers for Molding and Extrusion(Continued):

Properties of CAPRON Homopolymers (73F, Dry as Molded) (Continued):

Alpha Grades:

8202C:

Tensile Strength at Yield: psi: 13,000
Ultimate Elongation: %: 15
Flexural Strength: psi: 16,000
Flexural Modulus: 460,000

8202CL:

Tensile Strength at Yield: psi: 12,500
Ultimate Elongation: %: 10
Flexural Strength: psi: 17,000
Flexural Modulus: psi: 440,000

8203C:

Tensile Strength at Yield: psi: 12,000
Ultimate Elongation: %: 200
Flexural Strength: psi: 17,000
Flexural Modulus: psi: 460,000

Wire Jacketing Grades:

8220 HS:

Tensile Strength at Yield: psi: 12,000
Ultimate Elongation: %: 80
Flexural Strength: psi: 15,500
Flexural Modulus: psi: 410,000

8221 HS:

Tensile Strength at Yield: psi: 8,500
Ultimate Elongation: %: 200
Flexural Strength: psi: 10,000
Flexural Modulus: psi: 290,000

ALLIED-SIGNAL INC.: Mineral Reinforced CAPRON Nylon for Injection Molding:

CAPRON Nylon is based on type 6 polyamide resin, the most versatile and performance-proven of engineering thermoplastics.

Mineral Reinforced CAPRON Nylon:
The addition of mineral reinforcement enhances the performance characteristics of CAPRON Nylon molding compounds. Mineral reinforced CAPRON products exhibit high stiffness, dimensional stability, and performance at elevated temperatures combined with low warpage and resistance to sink mark formation.

8260:
40% mineral reinforced: excellent balance of engineering properties with the unique ability to be pigmented, painted, or chrome plated.

8360:
34% mineral reinforced: improved toughness and ductility combined with excellent procesability and surface aesthetics.

Applications:
* Marine hardware, brackets, fittings, bobbins, office furniture, appliance components, power tool housings.
* Automotive fuel filler pockets, door and window hardware, wheel hubs and covers, mirror housings, and interior lighting components.

Decorating Techniques:
* Colorability
* Painting
* Plating

Properties of CAPRON Mineral Reinforced Grades (73F, Dry as Molded):

8260:
Tensile Strength at Yield: psi: 13,000
Ultimate Elongation: %: 10
Flexural Strength: psi: 20,000
Flexural Modulus: psi: 800,000

8360:
Tensile Strength at Yield: psi: 13,000
Ultimate Elongation: %: 15
Flexural Strength: psi: 19,000
Flexural Modulus: psi: 725,000

ALLIED-SIGNAL INC.: Reinforced CAPRON Nylon for Injection Molding:

CAPRON Nylon is based on type 6 polyamide resin, the most versatile and performance-proven of engineering thermoplastics.

Reinforced CAPRON Nylon:

The addition of glass fiber or a combination of glass fiber and selected mineral reinforcement enhances the performance characteristics of CAPRON Nylon molding compounds. Glass reinforced CAPRON products exhibit high strength, stiffness, and performance at elevated temperatures. Mineral/Glass reinforced products offer a balance of engineering properties combined with low warpage and resistance to sink-mark formation.

A. Glass Reinforced:

8230G HS:

6% glass reinforced: improved stiffness and dimensional stability in thin sections.

8231G HS:

14% glass reinforced: improved mechanical properties with excellent processability and surface aesthetics.

8233G HS:

33% glass reinforced: high strength and rigidity combined with excellent dimensional stability, creep resistance and surface appearance.

8234G HS:

44% glass reinforced: the highest level of strength, rigidity, and mechanical performance.

B. Mineral/Glass Reinforced:

8266G HS:

40% mineral/glass reinforced: an excellent balance of mechanical properties combined with good sink mark and warp resistance.

8267G HS:

40% mineral/glass reinforced: similar to 8266G HS but specifically formulated to improve surface appearance when pigmentation is required.

ALLIED-SIGNAL INC.: Reinforced CAPRON Nylon for Injection Molding (Continued):

Applications:

* Handles and housings for power tools, chain saws, lawn and garden equipment.
* Automotive under-hood components including cooling fans and shrouds, timing chain covers, and vacuum reservoirs.
* Painted exterior automotive body parts, door and window hardware, gears, connectors, and many other similar applications.

Properties of CAPRON Reinforced Grades (73F, Dry as Molded):

Glass-Reinforced Grades:

8230G HS:
Tensile Strength at Yield: psi: 12,000
Ultimate Elongation: %: 5.0
Flexural Strength: psi: 18,000
Flexural Modulus: psi: 580,000

8231G HS:
Tensile Strength at Yield: psi: 18,000
Ultimate Elongation: %: 3.5
Flexural Strength: psi: 25,000
Flexural Modulus: psi: 800,000

8233G HS:
Tensile Strength at Yield: psi: 29,000
Ultimate Elongation: %: 3.0
Flexural Strength: psi: 40,000
Flexural Modulus: psi: 1,360,000

8234G HS:
Tensile Strength at Yield: psi: 33,000
Ultimate Elongation: %: 2.0
Flexural Strength: psi: 41,000
Flexural Modulus: psi: 1,720,000

Mineral/Glass Grades:

8266G HS:
Tensile Strength at Yield: psi: 19,000
Ultimate Elongation: %: 4.0
Flexural Strength: psi: 27,500
Flexural Modulus: psi: 1,330,000

8267G HS:
Tensile Strength at Yield: psi: 20,000
Ultimate Elongation: %: 4.0
Flexural Strength: psi: 30,000
Flexural Modulus: psi: 1,110,000

ALLIED-SIGNAL, INC.: NYPEL Nylon Injection Molding Compounds:

NYPEL Nylon injection molding compounds are based on selected type 6 nylon feedstocks and are specifically formulated to combine quality, performance, and processability at a lower cost.

NYPEL Homopolymers:

NYPEL unreinforced resins are standard nylon 6 homopolymers used in a wide variety of molding applications. These resins possess an overall balance of strength, stiffness and toughness combined with excellent chemical and abrasion resistance.

2314:

Low viscosity injection molding grade for fast fill and thin sections. Fittings, cord reels, connectors, gears, pulleys, and washers are typical end uses.

2314 FCAT:

Low viscosity injection molding compound possessing a modified crystalline structure for increased performance and faster cycles. Bushings, fittings, wall plugs and tension rollers are among the numerous end uses.

NYPEL Reinforced Grades:

Glass fiber, mineral reinforcement, or a combination of glass fiber and mineral reinforcement, strengthen the performance characteristics of NYPEL molding compounds. Glass reinforced NYPEL products combine high strength, stiffness and performance at elevated temperatures. Mineral reinforced products possess rigidity and dimensional stability. Mineral/glass reinforced products offer a combination of engineering properties with low warpage and resistance to sink mark formation.

2360:

40% mineral reinforced: high stiffness, dimensional stability, and heat resistance combined with excellent processability and the ability to be painted. Applications include fuel filler pockets, wheel covers and supports, and housings. (Developed as NPN 1519)

2365G:

36% mineral and glass reinforced: offers excellent mechanical properties with good sink mark and warp resistance. Fans, shrouds, brackets and wheel hubs are some of the numerous applications.

6033G:

33% glass reinforced: high strength and rigidity combined with excellent creep resistance and dimensional stability. Bolts, switch components, gears and pressure regulators are typical applications.

ALLIED-SIGNAL INC.: NYPEL Nylon Injection Molding Compounds (Continued):

Properties of NYPEL Nylon Resins(73F, Dry as Molded):

Homopolymers:

2314:

Tensile Strength at Yield: psi: 11,500
Ultimate Elongation: %: 30
Flexural Strength: psi: 15,700
Flexural Modulus: psi: 410,000

2314 FCAT:

Tensile Strength at Yield: psi: 12,000
Ultimate Elongation: %: 12
Flexural Strength: psi: 16,000
Flexural Modulus: psi: 460,000

Reinforced:

2360:

Tensile Strength at Yield: psi: 13,000
Ultimate Elongation: %: 10
Flexural Strength: psi: 20,000
Flexural Modulus: psi: 800,000

2365G:

Tensile Strength at Yield: psi: 17,000
Ultimate Elongation: %: 5
Flexural Strength: psi: 25,000
Flexural Modulus: psi: 1,200,000

6033G:

Tensile Strength at Yield: psi: 24,000
Ultimate Elongation: %: 3
Flexural Strength: psi: 35,000
Flexural Modulus: psi: 1,360,000

ALLIED-SIGNAL INC.: Reinforced PETRA Thermoplastic Polyesters for Injection Molding:

PETRA thermoplastic polyester is based on polyethylene terephthalate (PET) and comprises a family of injection molding resins which offer a balance of engineering properties combined with high temperature performance, chemical resistance, electrical properties, and ease of processability.

PETRA Compounds:

The PETRA product line is formulated from PET polymer reinforced with glass fibers or a combination of glass fibers and selected minerals. Glass reinforced grades exhibit strength, stiffness, dimensional stability and performance at elevated temperatures. Mineral/glass reinforced grades offer a balance of engineering properties combined with low warpage and resistance to sink mark formation. Glass reinforced flame retarded products offer performance combined with an all color UL 94 V-O flammability rating.

PETRA 130:

30% glass reinforced, strength and rigidity combined with excellent dimensional stability, creep resistance and high temperature performance.

PETRA 130 FR:

30% glass reinforced, flame retarded, excellent balance of engineering properties including an all color UL 94 V-O rating in sections as thin as 1/32 inch and a 155C continuous use temperature.

PETRA 140:

45% glass reinforced for increased strength, stiffness and engineering property performance.

PETRA 149:

45% glass reinforced, modified to improve flexural properties in large structural components at elevated temperatures.

PETRA 150:

55% glass reinforced for the highest level of strength, stiffness and engineering performance.

PETRA 230:

35% mineral/glass reinforced, low warp and high resistance to sink mark formation.

Applications:

* Automotive ignition and carburetor components
* Electrical connectors, bobbins, relays and switches
* Structural and mechanical components that maintain dimensional stability under load, or performance at temperatures above 150C, or in presence of moisture.
* Parts with thin sections or that are mechanically joined by snap fit, force fit or threading methods

ALLIED SIGNAL INC.: Reinforced PETRA Thermoplastic Polyesters for Injection Molding(Continued):

Properties of PETRA Reinforced Grades(73F, Dry as Molded):

PETRA 130:

Tensile Strength at Yield: psi: 22,000
Ultimate Elongation: %: 2.2
Flexural Strength: psi: 34,000
Flexural Modulus: psi: 1,300,000

PETRA 130 FR:

Tensile Strength at Yield: psi: 19,000
Ultimate Elongation: %: 2.0
Flexural Strength: psi: 27,500
Flexural Modulus: psi: 1,440,000

PETRA 140:

Tensile Strength at Yield: psi: 26,000
Ultimate Elongation: %: 1.6
Flexural Strength: psi: 41,500
Flexural Modulus: psi: 2,000,000

PETRA 149:

Tensile Strength at Yield: psi: 27,000
Ultimate Elongation: %: 1.6
Flexural Strength: psi: 42,500
Flexural Modulus: psi: 2,000,000

PETRA 150:

Tensile Strength at Yield: psi: 29,000
Ultimate Elongation: %: 1.4
Flexural Strength: psi: 44,500
Flexural Modulus: psi: 2,500,000

PETRA 230:

Tensile Strength at Yield: psi: 15,000
Ultimate Elongation: %: 1.5
Flexural Strength: psi: 25,000
Flexural Modulus: psi: 1,500,000

AMOCO PERFORMANCE PRODUCTS, INC.: MINDEL B-322 Engineering Resin:

MINDEL B-322 Resin is a glass reinforced thermoplastic resin developed specifically for molded electrical connector applications requiring excellent dimensional stability and resistance to warpage. This material is characterized by a high heat distortion temperature, good creep resistance and the outstanding rigidity, tensile and flexural strengths expected of an engineering polymer.

MINDEL B-322 provides a unique combination of the best properties of amorphous and crystalline polymers, such as providing good stress cracking resistance to common solvents and maintaining outstanding resistance to shrinkage and warpage during processing.

Warpage is significantly less than that experienced with glass reinforced thermoplastic polyesters.

MINDEL B-322 Resin has outstanding electrical properties even at high frequencies and temperatures and is rated V-O in the UL-94 flammability standard to 1/32" or .031" (0.79 mm) in all colors.

Humidity resistance is another strong point. MINDEL B-322 unlike polyesters retains a high level of mechanical and electrical properties even at elevated temperatures making it a suitable replacement for high performance thermosets.

MINDEL B-322 flows well and can be molded into thin sections which exhibit good resilience and low warpage. It can be fabricated on conventional equipment.

MINDEL B-322 is a creamy white translucent material in its natural form and may be readily colored.

INJECTION MOLDING:

MINDEL B-322 Resin can be readily molded in most screw injection molding machines.

TYPICAL PHYSICAL PROPERTIES OF MINDEL B-322 RESIN:

General

Properties	ASTM Method	Value
Specific Gravity	D-1505	1.47
Hardness, Rockwell	D-785	M80
Water Absorption, % in 24 hrs.	D-570	0.14

AMOCO PERFORMANCE PRODUCTS, INC.: MINDEL B-340 Engineering Resin:

MINDEL B-340 Resin is a glass reinforced thermoplastic resin developed for electrical/electronic applications requiring excellent dimensional stability, resistance to warpage, cleaning chemical resistance and high stiffness.

The material, like other MINDEL B Resins, is characterized by a high heat distortion temperature, good creep resistance and outstanding rigidity expected of an engineering polymer.

MINDEL B-340 offers designers an opportunity to obtain the high stiffness (1.6 MM psi modulus) and thermal index of thermosets with processing economics of a thermoplastic.

MINDEL B-340 Resin's warpage resistance resulting from its low isotropic shrinkage makes it uniquely suited for demanding connector, switch, relay, bobbin and electronic housing designs. The resin's regrind stability and heat staking ability are additional processing advantages.

MINDEL B-340, unlike polyesters, retains a high level of mechanical and electrical properties even at elevated temperatures and humidities.

The resin is a light tan opaque material in its natural form and may be readily colored via color concentrates. A variety of colors is also available.

INJECTION MOLDING:

The molding characteristics of MINDEL B-340 are similar to those of GF thermoplastic polyesters. Melt temperatures of 520-550F and injection pressures of 15-20,000 psi will serve as a good starting point.

Properties	ASTM Method	Values
General		
Specific Gravity	D-1505	1.66
Hardness, Rockwell	D-785	M82
Water Absorption,% in 24 hrs	D-570	0.07

AMOCO PERFORMANCE PRODUCTS, INC.: TORLON Engineering Polymer 4203L:

ENGINEERING PROPERTIES	ASTM Method	Units	4203L
Tensile strength	D1708	10 3 psi	
-321F			31.5
73F			27.8
275F			16.9
450F			9.5
Tensile elongation	D1708	%	
-321F			6
73F			15
275F			21
450F			22
Tensile modulus	D1708	10 5 psi	
73F			7.0
Flexural strength	D790	10 3 psi	
-321F			41.0
73F			34.9
275F			24.8
450F			17.1
Flexural modulus	D790	10 5 psi	
-321F			11.4
73F			7.3
275F			5.6
450F			5.2
Compressive strength	D695	10 3 psi	32.1
Shear strength	D732	10 3 psi	
73F			18.5
Izod impact strength (1/8 inch)	D256	ft.lb/in	
notched			2.7
unnotched			20.0
Poisson's ratio			0.38
GENERAL PROPERTIES			
Specific gravity	D792		1.42
Hardness, Rockwell E	D785		86
Water absorption	D570	%	0.33

AMOCO PERFORMANCE PRODUCTS, INC.: TORLON Engineering Polymer 4275:

ENGINEERING PROPERTIES	ASTM Method	Units	4275
Tensile strength	D1708	10 3 psi	
-321F			18.8
73F			22.0
275F			16.3
450F			8.1
Tensile elongation	D1708	%	
-321F			3
73F			7
275F			15
450F			17
Tensile modulus	D1708	10 5 psi	
73F			11.3
Flexural strength	D790	10 3 psi	
-321F			29.0
73F			30.2
275F			22.4
450F			15.8
Flexural modulus	D790	10 3 psi	
-321F			13.9
73F			10.6
275F			8.1
450F			7.4
Compressive strength	D695	10 3 psi	17.8
Shear strength 73F	D732	10 3 psi	11.1
Izod impact strength (1/8 in)	D256	ft.lb/in	
notched			1.6
unnotched			4.7
Poisson's ratio			0.30

GENERAL PROPERTIES			
Specific gravity	D792		1.51
Hardness, Rockwell E	D785		70
Water absorption	D570		0.33

AMOCO PERFORMANCE PRODUCTS, INC.: TORLON Engineering Polymer 4301:

ENGINEERING PROPERTIES	ASTM Method	Units	4301
Tensile strength	D1708	10 3 psi	
73F			23.7
275F			16.3
450F			10.6
Tensile elongation	D1708	%	
73F			7
275F			20
450F			17
Tensile Modulus	D1708	10 5 psi	
73F			9.5
Flexural strength	D790	10 3 psi	
73F			31.2
275F			23.5
450F			16.2
Flexural modulus	D790	10 5 psi	
73F			10.0
275F			7.9
450F			7.2
Compressive strength	D695	10 3 psi	24.1
Shear strength	D732	10 3 psi	
73F			16.1
Izod impact strength (1/8 in)	D256	ft.lb/in	
notched			1.2
unnotched			7.6
Poisson's ratio			0.33
GENERAL PROPERTIES			
Specific gravity	D792		1.46
Hardness, Rockwell E	D785		72
Water absorption	D570	%	0.28

AMOCO PERFORMANCE PRODUCTS, INC.: TORLON Engineering Polymer 4347:

ENGINEERING PROPERTIES	ASTM Method	Units	4347
Tensile strength	D1708	10 3 psi	
73F			17.8
275F			15.1
450F			7.8
Tensile elongation	D1708	%	
73F			9
275F			21
450F			15
Tensile modulus	D1708	10 5 psi	8.7
Flexural strength	D790	10 3 psi	
73F			27.0
275F			20.5
450F			14.3
Flexural modulus	D790	10 5 psi	
73F			9.1
275F			6.4
450F			6.2
Compressive strength	D695	10 3 psi	18.3
Shear strength	D732	10 3 psi	
73F			11.5
Izod impact strength (1/8 in)	D256	ft.lb/in	
notched			1.3
Poisson's ratio			0.33

GENERAL PROPERTIES			
Specific gravity	D792		1.50
Hardness, Rockwell E	D785		66
Water absorption	D570		0.17

AMOCO PERFORMANCE PRODUCTS, INC.: TORLON Engineering Polymer 5030:

ENGINEERING PROPERTIES	ASTM Method	Units	5030
Tensile strength	D1708	10^3 psi	
-321F			29.5
73F			29.7
275F			23.1
450F			16.3
Tensile elongation	D1708	%	
-321F			4
73F			7
275F			15
450F			12
Tensile modulus 73F	D1708	10^5 psi	15.6
Flexural strength	D790	10^3 psi	
-321F			54.4
73F			48.3
275F			35.9
450F			26.2
Flexural modulus	D790	10^5 psi	
-321F			20.4
73F			17.0
275F			15.5
450F			14.3
Compressive strength	D695	10^3 psi	38.3
Shear strength 73F	D732	10^3 psi	20.1
Izod impact strength (1/8 in)	D256	ft.lb/in	
notched			1.5
unnotched			9.5
Poisson's ratio			0.40

GENERAL PROPERTIES			
Specific Gravity	D792		1.61
Hardness, Rockwell E	D785		94
Water absorption	D570		0.24

AMOCO PERFORMANCE PRODUCTS, INC.: TORLON Engineering Polymer 7130:

ENGINEERING PROPERTIES	ASTM Method	Units	7130
Tensile strength	D1708	10 3 psi	
-321F			22.8
73F			29.4
275F			22.8
450F			15.7
Tensile elongation	D1708	%	
-321F			3
73F			6
275F			14
450F			11
Tensile modulus	D1708	10 5 psi	32.2
Flexural strength	D790	10 3 psi	
-321F			45.0
73F			50.7
275F			37.6
450F			25.2
Flexural modulus	D790	10 5 psi	
-321F			35.7
73F			28.8
275F			27.2
450F			22.8
Compressive strength	D695	10 3 psi	36.9
Shear strength	D732	10 3 psi	
73F			17.3
Izod impact strength (1/8 in)	D256	ft.lb/in	
notched			0.9
unnotched			6.4
Poisson's ratio			0.37

GENERAL PROPERTIES			
Specific gravity	D792		1.48
Hardness, Rockwell E	D785		94
Water absorption	D570		0.26

AMOCO ENGINEERING PRODUCTS, INC.: TORLON Engineering Polymer 9040:

ENGINEERING PROPERTIES	ASTM Method	Units	9040
Tensile Strength	D1708	10 3 psi	
73F			31.8
275F			25.0
450F			19.9
Tensile Elongation	D1708	%	
73F			7
275F			6
450F			8
Tensile modulus 73F	D1708	10 5 psi	20.2
Flexural strength	D790	10 3 psi	
73F			52.0
275F			43.6
450F			38.9
Flexural modulus	D790	10 5 psi	
73F			21.0
275F			20.0
450F			20.0
Compressive strength	D695	10 3 psi	46.7
Shear strength 73F	D732	10 3 psi	22.7
Izod impact strength (1/8 in)	D256	ft.lb/in	
notched			1.5
unnotched			8.1
Poisson's ratio			0.33

GENERAL PROPERTIES			
Specific gravity	D792		1.68
Hardness, Rockwell E	D785		107
Water absorption	D570		0.21

AMOCO PERFORMANCE PRODUCTS, INC.: TORLON Engineering Polymer:

Grade	Application
4203L	Connectors, switches, relays, thrust washers, spline liners, valve seats, poppets, mechanical linkages, bushings, wear rings, insulators, cams, picker fingers, ball bearings, rollers, thermal insulators.
4301,4275,4347	Bearings, thrust washers, wear pads, strips, piston rings and seals, vanes, valve seats.
5030	Burn-in sockets, gears, valve plates, fairings, tube clamps, impellers, rotors, housings, back-up rings, terminal strips, insulators, brackets.
7130	Metal replacement, housings, mechanical linkages, gears, fasteners, spline liners, cargo rollers, brackets, valves, labyrinth seals, fairings, tube clamps, standoffs, impellers, shrouds. Potential use for EMI shielding.
9040	Switches, relays, terminal strips, wear bands, back-up rings, housings, impellers, brackets, thermal insulators.

Use TORLON Polymers...

Reduce weight
Reduce unit cost
Improve performance
Obtain design flexibility
Consolidate parts
Implement new design concepts
Eliminate galling
Reduce frictional forces
Eliminate corrosion
Prevent galvanic corrosion

AMOCO PERFORMANCE PRODUCTS, INC.: TORLON Polyamide-imide Engineering Plastic:

TORLON is a unique high performance plastic that can handle the toughest jobs. If your application requires strength at high temperatures, resistance to creep and wear, and the economies of injection molding, TORLON engineering plastic can solve your design problem.

For strength and stiffness up to 500F

TORLON engineering resin can be used at temperatures well beyond the limits of most other high performance plastics. At high temperatures, TORLON plastic is stiffer and stronger than many high performance materials are at room temperature.

For resistance to wear in high-friction use

TORLON performance plastic is ideal for moving parts because it offers low wear, excellent creep resistance, and self-lubricity.

For low flammability and smoke generation

In tests of flammability, smoke density, and toxic gas emission, TORLON polyamide-imide meets stringent aerospace requirements, and is used extensively in applications where flammability is a critical consideration.

For thermal and dimensional stability

TORLON parts are reliable for long-term use at high temperatures. TORLON parts hold their dimensions under loads and at high temperatures. Low coefficients of linear thermal expansion (CLTE) facilitate integration of TORLON components with metal.

For resistance to harsh chemicals

TORLON polyamide-imide is virtually unaffected by aliphatic and aromatic hydrocarbons, chlorinated and fluorinated hydrocarbons, and most acids at moderate temperature. It is attacked by saturated steam, strong bases, and some acids at high temperature.

For impact resistance

TORLON engineering plastic can absorb impact energy better than most other high modulus materials.

AMOCO PERFORMANCE PRODUCTS, INC.: UDEL Polysulfone:

An Outstanding Engineering Polymer for Molding and Extrusion

* UL Thermal Index 320F (160C)
* Steam Resistant - Hydrolytically Stable
* Long-Term Resistance to Creep
* Excellent Mechanical and Electrical Properties
* Easily Molded to Close Tolerances
* Transparent, Opaque and Filled Grades
* Resistant to Acids and Alkalis, Even at Elevated Temperatures
* Complies with Food Contact Regulations

UDEL Polysulfone is a tough, rigid, high strength thermoplastic which maintains its properties over a wide temperature range from -150F(-101C) to above 300F(149C). Extended time thermal aging at 300F(149C) has little effect on physical and electrical properties of polysulfone.

APPLICATIONS:

The range of applications for UDEL Polysulfone is unusually wide.

ELECTRICAL/ELECTRONIC

Printed circuit boards. Integrated circuit carriers. Coil bobbins. Housings. Connectors, bushings, and terminal blocks. TV and stereo components. Capacitor film. Brush holders. Alkaline battery cases and cells. Business machine parts.

AUTOMOTIVE/AEROSPACE

Under-the-hood components. Lamp bezels. Electrical gear. Battery caps. Fuses. Electronic ignition components. Sensors. Aircraft interior and exterior components. Astronaut's outer face mask shields.

CONSUMER PRODUCTS

Microwave cookware. Coffee brewers. Humidifiers. Hair dryers. Clothes steamers. Hot lather dispensers. Popcorn poppers, egg cookers and other specialized cooking appliances. Digital watch cases. Camera cases. Projector components. Beverage and food dispensers.

MEDICAL

Surgical tool trays. Nebulizers and humidifiers. Flow controls. Instrument housings. Dental and surgical instruments. Fluid containers. Heart valve cases. Pacemakers. Respirators. Lab equipment.

OTHER

Process and sanitary pipe. Steam table pans. Plumbing components. Water heater dip tubes. Milking machine parts. Pollution control equipment. Filtration membranes and housings.

AMOCO PERFORMANCE PRODUCTS, INC.: UDEL Polysulfone(Continued):

Product Name:

P-1700:

Appearance in Natural Form: Natural 11 transparent light amber. Available in opaque and transparent colors.
Application: General purpose, transparent, FDA

General:

Melt Flow, at 650F(343.3C), 44 psi(0.030MPa), g/10 minutes: ASTM D1238: 6.5
Density, Mg/cu m: ASTM D1505: 1.24
Mold Shrinkage, in/in or mm/mm: ASTM D955: 0.007
Water Absorption, % in 24 hours: ASTM D570: 0.3

P-1700FR:

Appearance in Natural Form: Natural, translucent
Application: UL94 V-O grade

General:

Melt Flow, at 650F(343.3C), 44 psi(0.030 MPa), g/10 minutes: ASTM D1238: 6.5
Density, Mg/cu m: ASTM D1505: 1.24
Mold Shrinkage, in/in or mm/mm: ASTM D955: 0.007
Water Absorption, % in 24 hours: ASTM D 570: 0.3

P-1710:

Appearance in Natural Form: Natural 15 light beige opaque.
Application: General purpose, opaque, FDA

General:

Melt Flow, at 650F(343.3C), 44 psi(0.030 MPa), g/10 minutes: ASTM D1238: 6.5
Density, Mg/cu m: ASTM D1505: 1.25
Mold Shrinkage, in/in or mm/mm: ASTM D955: 0.007
Water Absorption, % in 24 hours: ASTM D570: 0.3

AMOCO PERFORMANCE PRODUCTS, INC.: UDEL Polysulfone(Continued):

Product Name:

P-1720:

Appearance in Natural Form: Natural 13 light beige translucent.
Application: UL94 V-O grade

General:

Melt Flow, at 650F(343.3C), 44 psi(0.30 MPa), g/10 minutes: ASTM D1238: 6.5
Density, Mg/cu m: ASTM D1505: 1.25
Mold Shrinkage, in/in or mm/mm: ASTM D955: 0.007
Water Absorption, % in 24 hours: 0.3

P-3500:

Appearance in Natural Form: Natural 11 transparent light amber. Available in opaque and transparent colors.
Application: Extrusion

General:

Melt Flow, at 650F(343.3C), 44 psi(0.30 MPa), g/10 minutes: ASTM D1238: 3.5
Density, Mg/cu m: 1.24
Mold Shrinkage, in/in or mm/mm: ASTM D955: 0.007
Water Absorption, % in 24 hours: ASTM D570: 0.3

APPLIED COMPOSITES: Sheet Molding Compounds:

Distinctive Features:

Product Name:

8079 Communications/Electronics:

- Low moisture absorption
- Paintable
- Flexible design parameters
- Thick cross sections
- High flow

8041 Standard Business Machine:

- Meets UL 94 V-O
- Radiant panel < 15 flame spread (ASTM E 162)
- Paintable, low profile
- Low sink and shrink

8659 Automotive:

- Improved flow
- Low profile, paintable
- Low sink and shrink
- Increased surface smoothness
- Supplied in slabs, ease of handling
- Reduced porosity
- Ideal for injection molding

8050 Pigmentable Business Machine:

- Radiant panel
- Meets UL 94V-O
- Low shrink
- Pigmentable, available in most colors

8038 Low Smoke and Fire:

- Construction and mass transportion
- Oxygen Index 80+
- E84-61 Tunnel:
  - --fuel contribution, 5
  - --smoke developed, 120
  - --flame spread index, 20
- Pigmentable

APPLIED COMPOSITES: Sheet Molding Compounds(Continued):

Distinctive Features:

Product Name:

8085 Low Profile, Fast Cure:

Automotive applications
High strength in bosses and ribs
Low profile, paintable
No shrink and sink
Dimensionally stable
Excellent flow

8057 High Modulus Graphite:

Aerospace type fiber
High stiffness/modulus
Available in hybrid forms with fiberglass
Compression moldable

8072 Automotive:

Superior Class "A" finish
Zero shrink
Exterior body panels
Improved impact
Accepts mold coatings

8049 High Stress:

Developed for high stress, torsion, vibration and abrasion considerations
Low coefficient of friction
Improved molding characteristics, dimensional stability with minimal flow lines, over most other SMC materials
Low shrink and sink
Low profile
Pigmentable or paintable

8031 Molded-in-Color

Pigmentable. Uniform pigmentation on all surfaces
High gloss
Low shrink
Low porosity characteristics
Patented formulation

APPLIED COMPOSITES: Sheet Molding Compounds(Continued):

Distinctive Features:

Product Name:

8020 General Purpose:

Pigmentable
Ease of molding (forgiving compound)
Low cost

8007 Waste Water Treatment:

Pigmentable
Low water absorption
Good corrosion resistance in waste water environment
Weathering resistance

8022 Underground and Corrosive Environments:

Ideal for buried applications
High corrosion resistance in waste water treatment and water control systems
Resistance to bacteria and fungus
Excellent physical strength
Dimensional stability

8061 High Strength:

High physical properties
Low shrink
High impact resistance
Pigmentable and paintable
Industrial applications

8016 Military/Defense:

High strength--specially developed
Consistent and more predictable
Low shrink
Self-extinguishing

ARCO CHEMICAL CO.: DYLARK Engineering Resins:

These unique engineering polymers offer an optimum ratio of high heat resistance and low cost per cubic inch in many applications. DYLARK resins are characterized by:

* High quality and product consistency
* Excellent heat resistance
* Easily processed in injection molding and extrusion equipment
* Outstanding dimensional stability

Grades of DYLARK Resins:

* Transparent Grades combine excellent water-white clarity with high heat resistance and structural rigidity.
* Impact Grades afford moderate to high levels of impact strength combined with high heat resistance and processability.
* Glass Reinforced Grades exhibit the highest heat resistance, improved dimensional stability, improved tensile and flexural properties, and outstanding processability in precompounded (P) and dry blended (F) form.

Glass Reinforced Impact Grades Technical Data:

Outstanding Physical Properties

* Very high modulus
* High heat resistance
* Impact
* Adhesion to urethane
* Excellent affinity for glass fiber

Typical Applications

Automotive:

* Instrument panel substrates
* Consoles
* Cluster housings
* Interior trim

Commercial/Industrial

* Appliance components

DYLARK Resin Glass Reinforced Impact Grades:

<u>Dylark</u> <u>Grade/Typical</u> <u>Values</u>

238P20:

Specific Gravity, 23/23C: D-792: 1.22
Melt Flow, g/10 min. Condition "L": D-1238: 0.9
Glass Fiber Content: 20

ARCO CHEMICAL CO.: DYLARK Engineering Resins(Continued):

DYLARK Resin Glass Reinforced Impact Grades

DYLARK Grade/Typical Values:

Pre-Compounds

250P10:
Specific Gravity, 23/23C: 1.13
Melt Flow, g/10 min. Condition "L": D-1238: 0.6
Glass Fiber Content: 10

250P20:
Specific Gravity, 23/23C: 1.20
Melt Flow, g/10 min. Condition "L": 0.35
Glass Fiber Content: 20

378P20:
Specific Gravity, 23/23C: 1.21
Melt Flow, g/10 min. Condition "L": 0.4
Glass Fiber Content: 20

Dry Blends:

238F20:
Specific Gravity, 23/23C: 1.23
Melt Flow, g/10 min. Condition "L": NA
Glass Fiber Content: 20

378F20:
Specific Gravity, 23/23C: 1.22
Melt Flow, g/10 min. Condition "L": NA
Glass Fiber Content: 20

Impact Grades Technical Data:

Outstanding Physical Properties
* High heat resistance
* High impact strength
* Good processability

Typical Applications

Automotive
* Consoles
* Trim parts
* Glove box doors
* Structural foam

Non-Automotive
* Air conditioner grills
* Appliance/houseware components

ARCO CHEMICAL CO.: DYLARK Engineering Resins(Continued):

DYLARK Resin Impact Grades:
DYLARK Grades/Typical Values

238:
Specific Gravity, 23/23C: D-792: 1.08
Melt Flow, g/10 min., Condition "L": D-1238: 1.8

250:
Specific Gravity, 23/23C: D-792: 1.06
Melt Flow, g/10 min., Condition "L": D-1238: 0.9

378:
Specific Gravity, 23/23C: D-792: 1.06
Melt Flow, g/10 min., Condition "L": D-1238: 1.0

700:
Specific Gravity, 23/23C: D-792: 1.06
Melt Flow, g/10 min., Condition "L": D-1238: 0.5

Transparent Grades Technical Data

Outstanding Physical Properties
* High heat resistance
* Sparkling water-white clarity
* Good chemical resistance
* High modulus resulting in superior shear and torque resistance

Typical Applications
* Dishwasher-safe tumblers
* Faucet handles
* Tape cassette boxes
* Small lens
* Heat resistant appliance covers

DYLARK Resin Transparent Grades
DYLARK Grade/Typical Values

132:
Specific Gravity, 23/23C: D-792: 1.08
Melt Flow, g/10 mm Condition "L": D-1238: 1.5
Melt Flow, g/10 mm Condition "N": D-1238: 1.9

232:
Specific Gravity, 23/23C: D-792: 1.08
Melt Flow, g/10 mm Condition "L": D-1238: 1.7
Melt Flow, g/10 mm Condition "N": D-1238: 1.8

332:
Specific Gravity, 23/23C: D-792: 1.10
Melt Flow, g/10 mm Condition "L": D-1238: 1.9
Melt Flow, g/10 mm Condition "N": D-1238: NA

ARCO Chemical Co.: ARLOY Thermoplastic Resins:

New High-Performance Options for Molding and Extrusion

ARLOY Thermoplastic Resins are blends of DYLARK Styrene Copolymers and polycarbonate. They have an excellent balance of physical properties, superior toughness, high heat resistance and outstanding processability. ARLOY resins offer exceptional performance and value for the most demanding applications.

ARLOY resins 1000, 1100 and 1200 are opaque compositions, suitable for applications where toughness and ductility are important:

* One-piece automotive instrument panels
* Scuff plates
* Seat belt retractor housings
* Speaker grills
* Tool housings
* Surgical appliances
* Hospital feeding trays
* Camera components

ARLOY resin 2000 is a crystal-clear product which is FDA-regulated for food contact applications. It is ideal for the clear components of the following items:

* Cookware
* Food processors
* Microwaveable deli trays
* Dishware
* Food containers
* Hair curler covers

Properties of ARLOY Thermoplastic Resins:

ARLOY Resin Grades:

1000:
Melt Flow, g/10 min. Condition "L": D1238: 0.7
Specific Gravity, 23/23C: D-792: 1.13

1100:
Melt Flow, g/10 min. Condition "L": D-1238: 0.9
Specific Gravity, 23/23C: D-792: 1.14

1200:
Melt Flow, g/10 min. Condition "L": D-1238: 1.3
Specific Gravity, 23/23C: D-792: 1.15

2000:
Melt Flow, g/10 min. Condition "L": D-1238: 3.2
Specific Gravity, 23/23C: D-792: 1.13

ARCO CHEMICAL CO.: Family of Foams for Design Engineering:

Performance Characteristics/Application Areas

Product:

ARPRO:
- Expanded Polypropylene Beads
- Performance Features:
  - High energy absorption
  - Excellent cushioning over broad loading range
  - High service temperature
- Application Areas:
  - Protective packaging of high value, highly fragile electronics
  - Heat resistant packaging
  - Energy absorbing automotive bumper systems

ARPAK:
- Expanded Polyethylene Beads
- Performance Features:
  - Resilient--very low permanent set
  - Excellent multiple drop cushioning
  - High solvent resistance
- Application Areas:
  - Protective packaging of high value, fragile items
  - Solvent resistant packaging
  - Furniture--bedding and seating

ARCEL:
- Moldable Polyethylene Copolymers
- Performance Features:
  - A balance of flexibility and strength
  - Tough--puncture resistant
  - Very good solvent resistance
- Application Areas:
  - Reusable packaging and packaging of heavy weight items
  - Bulk pack trays and tote bins
  - Water sports equipment

ARSAN:
- Moldable Resilient Resin
- Performance Features:
  - Good cushioning performance
  - Low dynamic set
  - Good compressive creep resistance
- Application Areas:
  - Packaging of moderately fragile items
  - Packaging exposed to solvents
  - Boat flotation

ARCO CHEMICAL CO.: Family of Foams for Design Engineering (Continued):

Performance Characteristics/Application Areas

Product:

DYLITE:
- Expandable Polystyrene
- Performance Features:
  - Good strength
  - Good 1st drop cushioning performance
  - Excellent insulation value
- Application Areas:
  - Packaging for business machines, appliances, toys
  - Food service--cups, coolers
  - Insulation--roof insulation, sheathing board

DYTHERM:
- Expandable Copolymers
- Performance Features:
  - High strength characteristics
  - Moldable in very high densities, up to 30 pct
  - High service temperature, up to 250F
- Application Areas:
  - Interior automotive components
  - Building panels and solar panels
  - Cabinetry

Typical Physical Properties of Foams Molded from ARCO Chemical's Family of Foams:

ARPRO Resins:
- Typical Density range (pcf): 1.3-4.0
- Density of Foam Tested (pcf): 1.3 (1.9)

ARPAK Resins:
- Typical Density Range (pcf): 2.2
- Density of Foam Tested (pcf): 2.2

ARCEL Resins:
- Typical Density Range (pcf): 1.5-4.0
- Density of Foam Tested (pcf): 2.0

ARSAN Resins:
- Typical Density Range (pcf): 0.8-3.0
- Density of Foam Tested (pcf): 1.0

DYLITE Resins:
- Typical Density Range (pcf): 0.9-5.0
- Density of Foam Tested (pcf): 1.25

DYTHERM Resins:
- Typical Density Range (pcf): 1.0-30.0
- Density of Foam Tested (pcf): 6.0

ASHLEY POLYMERS, INC.: Nylon Resins for Molding and Extrusion:

Nylon, a relatively high-cost polymer, is used for applications which require particular properties that the lower priced so-called "commodity resins" cannot provide. Since these properties are of critical importance, users take great care in deciding upon the particular nylon to be used for each application.

Injection Molding Grade Resins:

Nylon 66:

<u>Product Number and Characteristics:</u>

528:

General Purpose unlubricated basic 66 nylon.

528L:

General Purpose lubricated for improved machine feed and mold release characteristics.

528E:

An externally lubricated version of 528L.

528LW:

General Purpose lubricated with the UL-94V2 rating.

525LD:

Modified for superior moldability with extra toughness.

527LD:

A specialized nylon with very high impact resistance.

528MS:

Molybdenum Disulfide modified, self-lubricating type, with excellent wear resistance.

528GF:

Graphite modified 66 polymer.

ASHLEY POLYMERS, INC.: Nylon Resins for Molding and Extrusion (Continued):

Injection Molding Grade Resins(Continued):

Nylon 6:

Product Number and Characteristics:

830:

Basic Nylon 6 Medium Viscosity. Unlubricated.

830L:

Same as 830 but lubricated for improved machine feed and mold release.

830E:

Same as 830L but externally lubricated.

735D:

A plasticized copolymer with improved impact resistance.

734D:

A plasticized copolymer with high impact resistance.

738D:

A plasticized copolymer with maximum impact resistance.

Flame Retardant:

528BR-WO:

Rust colored nylon 66 flame retardant resin with V-O rating by U. L.

ASHLEY POLYMERS, INC.: Nylon Resins for Molding and Extrusion (Continued):

Injection Molding Grade Resins:

Reinforced Nylon 66:

Product Number and Characteristics:

528L-13G:
528L-25G:
528L-33G:

General purpose glass fiber reinforced 66 designed for parts requiring excellent thermal and dimensional stability and higher strength than conventional resins. Available with 13, 25 and 33% glass. Lubricated for improved machine feed and mold release.

525LD-13G:

High impact. Most popular with 13% glass but available with any percentage of glass. Lubricated.

527LD-13G:

Very high impact with 13% glass. Lubricated.

Nylon 6:

830L-13G:
830L-33G:

General Purpose reinforced with 13 and 30% glass fibers. Lubricated. Available with any percentage of glass.

735LD-13G:

High Impact with 13% glass fiber. Lubricated.

Ashley also has mineral and carbon fiber reinforced nylons and custom compounds other specialty nylons with glass, mineral, carbon fiber, etc.

ASHLEY POLYMERS, INC.: Nylon Resins for Molding and Extrusion (Continued):

Economy Injection Molding Grade Resins:

In plastic design, it is important to understand the application thoroughly. Many applications with wider tolerances can take advantage of lower priced resins. Ashley's line includes a series of nylon resins which are priced to compete with reprocessed nylons in most cases, yet provide physical properties and processing characteristics close to prime virgin resin.

Appearance is better than generally available from reprocessed nylon and Ashley provides good lot to lot uniformity.

NYLON 66:

Product Name and Characteristics:

520:

General Purpose basic 66 resin, slightly lubricated.

521:

General purpose containing approximately 0.2% TiO_2 giving it a neutral color.

522:

Same as 521 but containing approximately 1% TiO_2 giving it a white color.

520B:

General purpose basic 66 in black.

520BU:

Utility grade based on reprocessed resin.

525:

Modified for superior moldability with extra toughness.

ASHLEY POLYMERS, INC.: Nylon Resins for Molding and Extrusion (Continued):

Product Name and Characteristics:

Nylon 66(Continued):

527:

A specialized nylon with very high impact resistance.

520MS:

Molybdenum Disulfide modified, self-lubricating type, with excellent wear resistance.

520GF:

Graphite modified 66 polymer.

Nylon 6:

630:

General purpose basic 6 resin. Slightly lubricated.

630B:

General purpose basic 6 in black.

630BU:

Utility black based on reprocessed resin.

735:

A plasticized copolymer with improved impact resistance.

734:

A plasticized copolymer with high impact resistance.

738:

A plasticized copolymer with maximum impact resistance.

ASHLEY POLYMERS, INC.: Nylon Resins for Molding and Extrusion (Continued):

Reinforced Nylon 66:

Product Name and Characteristics:

520-13G:
520-25G:
520-33G:

General purpose glass fiber reinforced 66 designed for parts requiring excellent thermal and dimensional stability and higher strength than conventional resins. Available with 13, 25 and 33% glass. Lubricated for improved machine feed and mold release.

525-13G:

High impact with 13% glass. Lubricated.

527-13G:

Very high impact with 13% glass. Lubricated.

Nylon 6:

630-13G:
630-33G:

General purpose reinforced with 13% and 30% glass fibers. Lubricated.

735-13G:

High impact with 13% glass. Lubricated.

Ashley also has mineral and carbon fiber reinforced nylons and custom compounds other specialty nylons with glass, mineral, carbon fiber, etc.

ASHLEY POLYMERS, INC.: Nylon Resins for Molding and Extrusion (Continued):

ASHLENE Extrusion Grade Nylon Resins:

Ashley Polymers offers various ASHLENE nylon resins specifically designed for extrusion. Although any of the unreinforced and unmodified ASHLENE nylon resin such as 528 and 830 may be extruded, there are many practical limitations due to the relatively low viscosity of Ashley's basic general purpose 6 and 66 resins.

Nylon 66:

538:

Description: High Viscosity
Characteristics & Uses: General purpose for thin wall tubing, rods and profile.

541:

Description: Very High Viscosity
Characteristics & Uses: For thick slab material, large rod stock and large complex profiles, pipes and slabs.

541S:

Description: Heat Stabilized, High Viscosity
Characteristics & Uses: Sheet, film and blown film extrusion.

Nylon 6:

840:

Description: High Viscosity
Characteristics & Uses: General purpose for film, rigid pipe, large profile, and thick slab extrusions.

741:

Description: Highly Plasticized
Characteristics & Uses: For applications requiring excellent flexibility, high strength and good abrasion resistance. For flexible tubing, monofilament, fishing line.

737:

Description: Flexible Copolymer
Characteristics & Uses: A specialized highly flexible copolymer for corrugated tubing, wire and cable jacketing.

ATOCHEM,INC.: FORAFLON PVDF:

FORAFLON polyvinylidene fluoride has a molecule based upon ethylene and containing two atoms of fluorine. This gives it an excellent combination of properties which are rarely found in plastics:

- excellent resistance to UV ageing
- high strength and good abrasion resistance
- very low permeability to gases and liquids
- good resistance to nearly all chemical products
- ease of processing on conventional thermoplastics machinery

Range:
FORAFLON granules are available in a range of different base grades having different melt flow characteristics suitable for conversion using the following processes:

- extrusion and coextrusion
- injection moulding
- compression moulding

Chemical Resistance:

Heat Resistance/Flammability:
Since FORAFLON retains its excellent mechanical properties at temperatures very close to its melt point, it can be used continuously at temperatures between -50C and +150C.

- UL 94: VO at 0.8mm thickness
- ASTM D 635-68: 'non burning' at 3 mm thickness
- ASTM D 2863/70: limiting oxygen index 44%

Ageing:
FORAFLON is already on many of the world's oceans.

- resists UV ageing
- resists corrosion
- has good long term mechanical properties

ATOCHEM,INC.: VOLTALEF PCTFE:

VOLTALEF polychlorotrifluoroethylene is a thermoplastic material with a high fluorine content, giving it exceptionally high performance:

- outstanding chemical resistance (practically inert to most chemical reagents)
- excellent fire resistance
- operating temperature range -250C to +160C/200C
- very high compression resistance

Cyogenics: Widely used in the cryogenics industry

Safety: Non flammable

Range: Grades are available in powder or granule form for compression moulding or extrusion.

ATOCHEM, INC.: ORGAMIDE Nylon 6:

The generally high technical specification of nylon 6 makes it the most widely sold engineering polymer in the world.

In the field of nylon 6, ORGAMIDE is particularly well known for its ease of conversion, good mechanical properties, toughness, excellent surface finish and above all for close tolerance moulding with economy.

Mechanical Performance

The fine, regular crystalline structure of ORGAMIDE gives good tensile and impact strength over a wide range of temperatures.

Abrasion

ORGAMIDE'S resistance to abrasion and its low coefficient of friction are used to great effect in gear wheels, bearings and many industrial components.

Chemical Resistance

Electrical Insulation

Its electrical properties are good compared with other materials when taking into account its cost and mechanical performance.

Range

The ORGAMIDE range developed by ATOCHEM contains grades suitable for injection moulding, tube extrusion, cast film and profile extrusion.

Property:

RMNOCD:

Specific Gravity: 1.13
Melting Point: 218

RMNO P40 CD:

Specific Gravity: 1.13
Melting Point: 218

RZM 15/30:

Specific Gravity: 1.20/1.35
Melting Point: 218

ATOCHEM, INC.: ORGATER PBT:

ORGATER Polybutylene Terephthalate is a thermoplastic polyester having high regidity, good long term heat stability, high dielectric strength, and outstanding chemical resistance.

It is easy to process, has good tensile strength wear and creep resistance and low moisture absorption giving good stability of all requirements in a finished component.

Surface Finish:
The fine, high quality hard finish which can be obtained with ORGATER makes it an ideal material for mouldings requiring a high quality appearance.

Dielectric Properties:
ORGATER is an excellent insulator.

Heat Resistance/Rigidity:
Because of its lack of distortion at continuous elevated temperatures combined with good surface finish, automotive headlamp manufacturers are now specifying ORGATER to mould reflector bodies.

- high service temperatures;
- dimensional stability giving perfect light beam properties;
- high rigidity;
- good surface finish ready for vacuum metallisation without a primer lacquer;
- reduction in finished part cost.

Range:

- glass fibre and bead reinforced;
- flame retardant additives;
- special compounds.

Property:

TMNO:
Specific Gravity: 1.30
Melting Point: C: 225

TZM 20:
Specific Gravity: 1.45
Melting Point: C: 225

TZM 30 W8:
Specific Gravity: 1.68
Melting Point: C: 225

ATOCHEM, INC.: PEBAX Polyether Block Amides:

The PEBAX range forms a new family of polymers Polyether Block Amides (PEBA). The structure of these products consists of a linear and regular chain of rigid polyamide segments and flexible polyether segments.

Principal characteristics:

Because of their bi-phasic structure the various PEBAX grades offer many original characteristics:

* very wide range of flexibilty - absence of any plasticiser - excellent impact resistance, even at low temperatures.
* Little change in flexibility between -40C and +80C.
* Low density - very high elastic memory - excellent mechanical properties - good chemical resistance.

All these properties make it possible to use PEBAX in many sectors. PEBAX can be easily converted using the normal polyamide techniques

Anti-static grades:	5533 SN 70 Black
	4033 SN 70 Black
	2533 SN 70 Black

Properties

PEBAX grades 5533 SN 70 Black, 4033 SN 70 Black and 2533 SN 70 Black compounded with special fillers to make them anti-static.

These 3 grades offer a relatively low surface resistivity (10 3 to 10 4 Ω/cm) and increased mechanical properties.

Applications

The areas of use for these grades are very varied:

- car radio antenna
- hair brushes
- paint spray hose
- items for the textile industry
- explosion proof lamp housings
- castors.

ATOCHEM, INC.: PEBAX Polyether Block Amides(Continued):

Glass fibre reinforced grades MX 1500
MX 1501
MX 1502
MX 1503

Properties

The new PEBAX grades MX 1500, MX 1501, MX 1502 and MX 1503 are loaded with 20% glass fibre.

These 4 grades exhibit the following performance characteristics:

- Flexural modulus ranging from 900 to 2,000 MPa
- Improved heat distortion temperature under high load.
- Good impact resistance
- Exceptionally good elongation at break for reinforced grades (MX 1502 and MX 1503)
- Good dimensional stability.

These grades are easily processed by both injection moulding and extrusion and allows the production of finished parts with a good surface finish.

Applications

The use of these grades is envisaged in various business sectors due to their unique properties:

- electrical (connection boxes)
- sports (football boot studs)
- materials handling.

Improved chemical resistance grades MX 1057
Improved abrasion resistance MX 1058
Improved thermal resistance MX 1059
MX 1060

Properties

PEBAX grades MX 1057, MX 1058, MX 1059 and MX 1060 are modified grades which offer improved properties to the 33 series, with an equivalent hardness:

- improved chemical resistance (to oils and greases),
- improved heat resistance (higher Vicat point),
- Improved abrasion resistance.

All the other properties are little or not affected.

Applications

Due to the enhanced chemical and thermal properties of these 4 new grades, they are suitable for all the static or dynamic applications of traditional PEBAX. They are particularly suitable for applications in the mechanical and automotive industries:

- gaiters (for constant velocity joints, etc).
- bellows.

Processing

- moulding - extrusion -extrusion-blow moulding

ATOCHEM, INC.: PEBAX Polyether Block Amides(Continued):

Powders for coatings with elastomeric properties

PEBAX is an engineering thermoplastic elastomer used in a wide variety of applications. Four special powder grades for use in anti-corrosion and anti-slip coating of substrates are available: PEBAX 2533-3533-5562-6312 natural and black colours.

GRADES 2533 and 3533 can be applied using the fluidized bed technique. Grades 5562 and 6312 can be applied using either fluidized bed (T) or electrostatic spraying (ES).

Properties

PEBAX coatings are used in wide range of applications where the following properties are required:

- corrosion resistance
- resistance to abrasion, wear and impact
- vibration and noise reduction
- electrical insulation
- anti-slip characteristics
- flexibility even at low temperatures
- easy processing
- low powder fusion temperature allows coating of substrates (e.g. tempered metals) which do not resist elevated preheat temperature.

Properties

Grade 5562:
Melting Point (ASTM 2117): 120C
Density (ASTM D792): 1,06

Grade 3533:
Melting Point (ASTM 2117): 152C
Density (ASTM D792): 1,01

Grade 2533:
Melting Point (ASTM 2117): 148C
Density (ASTM D792): 1,01

Grade 6312:
Melting Point (ASTM 2117): 205C
Density (ASTM D792): 1,11

ATOCHEM, Inc.: RILSAN Resins:

A unique combination of properties:
Outstanding dimensional stability.
Excellent chemical resistance.
Flexibility and elastic memory.
Easy to mold.
Easy to extrude.
High impact and abrasion resistance.

RILSAN Nylon 11 Injection Molding Grades:

BMNO TL
Specific Gravity (D792): 1.03
Melting Point (D789) (C) (F): 186 (367)
Water Absorption (D570) (%) (24 Hr. Immersion): 0.3
Tensile Strength, Ultimate (D638) (psi x 10 3) (176F): 9.1
Tensile Elongation, Ultimate (D638) (%): 420
Flexural Modulus, Tangent (D790) (psi x 10 5) (176F): 0.43
Impact Strength, Notched Izod (D256) (Ft-Lb/In) (73F): 1.4
Rockwell Hardness R Scale @73F & 50% R.H.: 108
Deflection Temp. Unannealed @264 psi (D648) F: 115

BMVO P10:
Specific Gravity (D792): 1.04
Melting Point (D789) (C) (F): 186 (367)
Water Absorption (24 Hr. Immersion) (D570) (%): 0.3
Tensile Strength, Ultimate (176F) (D638) (psi x 10 3): 8.9
Tensile Elongation, Ultimate (176F) (D638) (psi x 10 5): 0.32
Flexural Modulus, Tangent (176F) (D790) (psi x 10 5): 0.32
Impact Strength, Notched Izod (73F) (D256) (Ft-Lb/In): 2.0
Rockwell Hardness R Scale @73F & 50% R.H. (D785): 97
Deflection Temp. Unannealed @ 264 psi (D648) (F): 1.08

BMNO P20:
Specific Gravity (D792): 1.05
Melting Point (D789) (C) (F): 186 (367)
Water Absorption (24 Hr. Immersion) (D570) (%): 0.3
Tensile Strength, Ultimate (D638) (psi x 10 3) (%): 8.8
Tensile Elongation, Ultimate (D638) (%): 470
Flexural Modulus, Tangent (D790) (psi x 10 5) (176F): 0.28
Impact Strength, Notched Izod (D256) (73F) (Ft-Lb/In): 3.4
Rockwell Hardness R Scale @ 73F & 50% R.H. (D785): 87
Deflection Temp. Unannealed @ 264 psi (D648) (F): 108

BMNO P40:
Specific Gravity (D792): 1.05
Melting Point (D789) (C) (F): 186 (367)
Water Absorption (24 Hr. Immersion) (D570) (%): 0.3
Technical Strength, Ultimate (176F) (D638) (%): 7.8
Tensile Elongation, Ultimate (176F) (%): 490
Flexural Modulus, Tangent (73F) (D790) (Ft-Lb/In): 6.0
Rockwell Hardness, R Scale @ 73F & 50% R.H. (D785): 78
Deflection Temp. Unannealed @ 264 psi (D648) (F): 106

ATOCHEM, INC.: RILSAN Resins(Continued):

RILSAN Nylon 11 Injection Molding Grades(Continued):

BMNO F15:
- Specific Gravity (D792): 1.05
- Melting Point (D789) (C) (F): 186 (364)
- Water Absorption (24 Hr. Immersion) (D570) (%): 0.3
- Tensile Strength, Ultimate (176F) (D638) (psi x 10 3): 4.5
- Tensile Elongation, Ultimate (176F) (D638) (%): 400
- Flexural Modulus, Tangent (176F) (D790) (psi x 10 5): 0.13
- Impact Strength, Notched Izod (-40F) (D256) (Ft-Lb/In): 0.4
- Rockwell Hardness R Scale @ 73F & 50% R.H. (D785): 53

RILSAN Nylon 11 Cable Grades:

BECNO TL:
- Specific Gravity (D792): 1.03
- Melting Point (D789) C (F): 186 (367)
- Water Absorption (24 Hr. Immersion) (D570) (%): 0.3
- Tensile Strength, Ultimate (176F) (D638) (psi x 10 3): 9.1
- Tensile Elongation, Ultimate (176F) (D638) (%): 420
- Flexural Modulus, Tangent (176F) (D790) (psi x 10 5): 0.35
- Impact Strength, Notched Izod (73F, 1/4" section) (D256) (Ft-Lb/In): 1.4
- Rockwell Hardness R Scale @73F & 50% R.H. (D785): 108
- Deflection Temp., Unannealed @ 264 psi (D648) (F): 113

BECVO P40 TL:
- Specific Gravity (D792): 1.05
- Melting Point (D789) C (F): 186 (367)
- Water Absoprtion (24 Hr. Immersion) (D570) (%): 0.3
- Tensile Strength, Ultimate (176F) (D638) (%): 440
- Flexural Modulus, Tangent (176F) (D790) (psi x 10 5): 0.24
- Impact Strength, Notched Izod (73F, 1/4" section) (D256) (Ft-Lb/In): 8
- Rockwell Hardness R Scale @ 73F & 50% R.H. (D785): 78
- Deflection Temp. Unannealed @ 264 psi (D648) (F): 106

RILSAN Nylon 11 Extrusion Grades:

BESNO TL:
- Specific Gravity (D792): 1.03
- Melting Point (D789) C (F): 186 (367)
- Water Absorption (24 Hr. Immersion) (D570) (%): 0.3
- Tensile Strength, Ultimate (176F) (D638) (psi x 10 3): 9.5
- Tensile Elongation, Ultimate (176F) (D638) (%): 420
- Flexural Modulus, Tangent (176F) (D790) (psi x 10 5): 0.37
- Impact Strength, Notched, Izod (73F, 1/4" section) (73F) (D256) (Ft-Lb/In): 1.8
- Rockwell Hardness R Scale @ 73F & 50% R.H. (D785): 108
- Deflection Temp. Unannealed @ 264 psi (D648) (F): 117

ATOCHEM, INC.: RILSAN Resins(Continued):

RILSAN Nylon 11 Extrusion Grades:

BESHVO:
Specific Gravity (D792): 1.03
Melting Point (D789) (C) (F): 186 (367)
Water Absorption (24 Hr. Immersion) (D570) (%): 0.3
Tensile Strength, Ultimate (176F) (D638) (psi x 10 3): 9.5
Tensile Elongation, Ultimate (176F) (D638) (%): 400
Flexural Modulus, Tangent (176F) (D790) (psi x10 5): 0.39
Impact Strength, Notched Izod (73F, 1/4" section) (D256) (Ft-Lb/In): 1.8
Rockwell Hardness R Scale @ 73F & 50% R.H. (D785): 108
Deflection Temp. Unannealed @ 264 psi (D648) (F): 117

BESNO P10 TL:
Specific Gravity (D792): 1.04
Melting Point (D789) (C) (F): 186 (367)
Water Absorption (24 Hr. Immersion) (D570) (%): 0.3
Tensile Strength, Ultimate (176F) (D638) (psi x 10 3): 9.4
Tensile Elongation, Ultimate (176F) (D638) (%): 410
Flexural Modulus, Tangent (176F) (D790) (psi x10 5): 0.33
Impact Strength, Notched Izod (73F, 1/4" section): 2.9
Rockwell Hardness, R Scale @ 73F & 50% R.H. (D785): 97
Deflection Temp. Unannealed @ 264 psi (D648) (F): 113

BESNO P20 TL:
Specific Gravity (D792): 1.04
Melting Point (D789) (C) (F): 186 (367)
Water Absorption (D570) (24 Hr. Immersion) (%): 0.3
Tensile Strength, Ultimate (176F) (D638) (%): 9.3
Tensile Elongation, Ultimate (176F) (D638) (%): 440
Flexural Modulus, Tangent (176F) (D790) (psi x 10 5): 0.27
Impact Strength, Notched Izod (73F, 1/4" section) (D256) (Ft-Lb/In): 4.0
Rockwell Hardness R Scale @ 73F & 50% R.H. (D785): 87
Deflection Temp. Unannealed @ 264 psi (D648) (F): 111

BESNO P40 TL:
Specific Gravity (D792): 1.05
Melting Point (D789) C (F): 186 (367)
Water Absorption (24 Hr. Immersion) (D570) (%): 0.3
Tensile Strength, Ultimate (176F) (D638) (%): 420
Tensile Elongation, Ultimate (176F) (D638) (%): 420
Flexural Modulus, Tangent (176F) (D790) (psi x10 5): 0.23
Impact Strength, Notched Izod (73F, 1/4" section) (D256) (Ft-Lb/In): NB
Rockwell Hardness R Scale @ 73F & 50% R.H. (D785): 78
Deflection Temp. Unannealed @ 264 psi (D648) (F): 110

ATOCHEM, INC.: RILSAN Resins(Continued):

RILSAN Nylon 11:

The Ultimate Coating For Metal

Characteristics:

A significant characteristic of RILSAN Nylon 11 powder is its ease of application. The formulation provides wide latitude in processing and resistance to film defects. The melt viscosity is high enough to prevent sagging and running, yet fluid enough to prevent sagging and running, yet fluid enough to achieve proper leveling, flow-out and wetting of the substrate.

Properties

RDP15-10 Natural
Fluid Bed:
60 Gloss-air cooled: 36
Coverage, Ft 2 - mil/Pound: 173

Electrostatic:
60 Gloss - air cooled: 41
Coverage, Ft 2 - mil/Pound: 173

RDP21 White
Fluid Bed:
60 Gloss-air cooled: 68
Coverage, Ft 2 - mil/Pound: 163

Electrostatic:
60 Gloss - air cooled: 68
Coverage, Ft 2 - mil/Pound: 162

MAC Colors
60 Gloss - air cooled: 45-85
Coverage, Ft 2 - mil/Pound: 160

ATOCHEM, INC.: RILSAN Resins(Continued):

RILSAN Nylon 12 Molding Grades:

AMNO:
Specific Gravity (D792): 1.02
Melting Point (D789) C (F): 174 (345)
Moisture Absorption (24 hr. immersion) (D570) (%): 0.3
Tensile Strength, Ultimate (176F) (D638) (psi x 10 3): 8.6
Tensile Elongation, Ultimate (176F) (D638) (%): 430
Flexural Modulus, Tangent (176F) (D790) (psi x 10 5): 0.45
Impact Strength, Notched, Izod (176F, 1/4" section) (D256) (Ft-Lb/in): 0.8
Deflection Temp., Unannealed @ 264 psi (D648) (F): 122
Rockwell Hardness, R Scale (D785): 106

AZM-23-G9:
Specific Gravity (D792): 1.20
Melting Point (D789) C (F): 174 (345)
Moisture Absorption (24 hr. immersion) (D570) (%): 0.2
Tensile Strength, Ultimate (176F) (D638) (psi x10 3): 7.4
Tensile Elongation, Ultimate (176F) (D638) (%): 7
Flexural Modulus, Tangent (176F) (D790) (psi x 10 5): 4.0
Impact Strength, Notched, Izod (176 F, 1/4" section) (D256) (Ft-Lb/In): 4.1
Deflection Temp., Unannealed @ 264 psi (D648) (F): 320
Rockwell Hardness, R Scale (D785): 109

AZMO 30:
Specific Gravity (D792): 1.23
Melting Point (D789): C (F): 174 (345)
Moisture Absorption (24 hr. immersion) (D570) (%): 0.2
Tensile Strength, Ultimate (176F) (D638) (psi x 10 3): 8.8
Tensile Elongation, Ultimate (176F) (D638) (psi x 10 5): 17
Flexural Modulus, Tangent (176F) (D790) (psi x 10 5): 4.4
Impact Strength, Notched, Izod (176, 1/4" section) (D256) (Ft-Lb/In): 8.0
Deflection Temp. Unannealed @ 264 psi (D648) (F): 322
Rockwell Hardness, R Scale (D785): 111

AZM-43-G9:
Specific Gravity (D792): 1.44
Melting Point (D789) C (F): 174 (345)
Moisture Absorption (24 hr. immersion) (D570) (%): 0.2
Tensile Strength, Ultimate (176F) (D638) (psi x 10 3): 11
Tensile Elongation, Ultimate (176F) (D638) (%): 9
Flexural Modulus, Tangent (176F) (D790) (psi x 10 5): 7.8
Impact Strength, Notched, Izod (176F, 1/4" section) (D256) (Ft-Lb/In): 4.0
Deflection Temp., Unannealed @ 264 psi (D648) (F): 329
Rockwell Hardness, R Scale (D785): 111

BASF CORP.: ULTRAFORM (Polyoxymethylene POM):

Commercial and development products:

Unreinforced Ultraform:

Z 2320-003:
Z 2330-003:
Physical form:
Standard shades: Only Black 120
Application: Extremely easy-flow products for moldings with very thin walls and adverse ratios of runner length to wall thickness.
Permit very rapid cycling.
Z 2330 X-003 contains an antistatic.

W 2320-003:
W 2330-003:
Physical Form:
Standard shades: Only black 120
Application: Very easy flow; rapidly solidifying products to meet extreme demands on processing characteristics but only moderate demands on mechanical performance.
W 2330-003 contains an antistatic

S 2320:
S 2320-003:
Application: Rapidly solidifying products with easy flow; specially intended for injection moldings that are difficult to produce, e.g., with thin walls and adverse ratios of runner length to wall thickness.
Contains demolding aid.

N 2320:
N 2320-003:
Application: Rapidly solidifying, general-purpose injection molding products. Contains demolding aid.

H 2320:
H 2320-004:
H 2320-006:
Application: Product with high molecular weight for extrusion:
for thick stock for machining
for thin tubes and panels
suitable for thick injection moldings

BASF CORP.: ULTRAFORM (Polyoxymethylene POM) (Continued):

Commercial and development products:

ULTRAFORM with special additives:

N 2211 PVX:
Application: Injection molding compound with special lubricant that allows a very low coefficient of friction and considerably reduced wear yet has no adverse effect on the mechanical properties.

N 2320 black
11005 MO:
Application: Injection molding compound modified by molybdenum disulphide for the production of friction bearings.

N 2320 black
11001 UV:
Application: Injection molding compound UV-stabilized for outdoor applications.

Impact-Modified ULTRAFORM:

N 2540 X:
Physical form:
Standard shades: Black 120
Application: Appreciably toughened products for injection molding

N 2640 X:
Physical form:
Standard shades: Black 120
Application: For injection moldings that have to meet the severest demands on impact resistance

Reinforced ULTRAFORM:

N 2200 G5:
Physical form:
Standard shades: Only Black 5120
Application: Glass-reinforced ULTRAFORM for injection molding. Features very great strength and rigidity.

BORG WARNER CHEMICALS: Automotive Polymers:

CYCOLAC ABS:

CYCOLAC ABS is a broad family of ABS resins available with a wide variety of properties and processability. CYCOLAC ABS engineering polymers are especially suited where toughness and rigidity are necessary over a range of temperatures.

PREVEX Polymers:

PREVEX polymers are a family of phenylene ether copolymer (PEC) alloys combining impact strength with excellent heat resistance. They provide good overall mechanical properties, outstanding creep resistance and aging performance in higher humidities and use temperatures.

PROLOY Polymers:

PROLOY Polymers are a family of thermoplastic alloys that combine functional heat resistance with superior impact strength and wide processing latitude. They offer excellent thermal stability during processing and provide a cost-effective choice for applications requiring greater heat resistance and longer term property retention than ABS or polystyrene.

BORG WARNER CHEMICALS: Automotive Polymers (Continued):

GENERAL PURPOSE GRADES:

Product Name:

CYCOLAC AR:

Offers a good overall balance of toughness, rigidity and flow. Recommended for applications requiring aesthetic qualities, scratch resistance and good dimensional stability, such as interior trim, exterior painted grilles, tail light and side marker housings, air conditioner/heater ducts and outlets

CYCOLAC TA:

Provides higher flow than AR for complex shapes and thin-wall, difficult-to-fill molds. Used primarily in interior trim parts, quarter panels and consoles.

LOW GLOSS GRADES:

CYCOLAC ZA4:

Similar to AR, but developed specifically to meet the automotive industries' requirements for low gloss part appearance.

CYCOLAC ZA7:

Similar to X11, but developed specifically to meet the automotive industries' requirements for a more heat-resistant, low-gloss type material.

IMPACT-RESISTANT GRADES:

CYCOLAC ZA2:

Offers a good balance of properties with emphasis on good modulus, excellent heat deflection and high impact strength.

CYCOLAC HM:

Provides high impact resistance and toughness in applications such as glove compartments and kick plates. Offers higher flow than CYCOLAC L with slightly less toughness at low temperatures.

BORG WARNER CHEMICALS: Automotive Polymers(Continued):

IMPACT RESISTANT GRADES(Continued):

Product Name:

CYCOLAC L:

Offers maximum low temperature ductility and impact strength with better processability than other high-impact engineering materials for applications such as instrument panels.

CYCOLAC LXB:

Offers consistently good low temperature impact resistance. Recommended for automotive bumpers where a balance of low temperature abuse resistance, processabilty and chemical resistance, and ease of fabrication are required.

HEAT-RESISTANT GRADES:

CYCOLAC X11:

Toughest of the high heat-resistant grades, it provides better processabilty than X17. Applications include interior trim and exterior painted radiator grilles.

CYCOLAC X17:

Offers higher heat resistance than X11, but lower impact strength. X17 is recommended where impact resistance is secondary to heat resistance. Applications include upper rear quarter trim panels, tail light and parking lamp housings.

CYCOLAC X15:

This improved high heat-resistant ABS offers almost twice the impact resistance of X17 with higher heat deflection and improved processability.

CYCOLAC Z48:

Offers the highest heat resistance in the ABS product line. Provides good moldability and excellent dimensional stability at high use temperatures. Used for interior trim and high temperature requirement parts.

BORG WARNER CHEMICALS: Automotive Polymers(Continued):

HEAT-RESISTANT GRADES(Continued):

Product Name:

PREVEX W30:

Provides high flow, good impact strength and a heat deflection temperature of 235F. Suggested applications include instrument panels and interior trim components.

PREVEX W40:

Provides higher impact strength with a heat deflection temperature of 235F. Suggested applications include instrument panels and other components requiring both heat and impact resistance.

PREVEX W50:

Offers good dimensional stability and a heat deflection temperature of 245F. Suggested applications include wheel covers, defroster and speaker grilles.

PREVEX W70:

Offers good impact resistance with a heat deflection temperature of 265F.

IMPACT-RESISTANT, WARP-RESISTANT GRADES:

PROLOY HC2 and HC3:

These high performance products provide excellent impact strength up to 8.4 and 10.5 ft.-lbs./in. respectively. Molded parts have endured oven exposures up to 250F without significant warpage or distortion. PROLOY grades present a cost-effective alternative to metals and other engineering plastics such as polycarbonates, nylons, thermoplastic polyesters and thermosets.

FLAME-RETARDANT GRADES:

CYCOLAC KJB:

A medium impact-resistant, high flow, flame-retardant grade. Can be used in speaker grilles and other critical flame-retardant applications.

BORG WARNER CHEMICALS: Automotive Polymers(Continued):

PLATING GRADES:

Product Name:

CYCOLAC EP-3510:

An electroplating product which provides high flow for molding complex, stress-free parts. Applications include large, difficult-to-mold exterior grilles and bezels.

CYCOLAC EPB-3570:

Provides maximum thermocyling performance in plated applications. This material should be considered for applications which must maintain high performance over a wide temperature range with a low coefficient of thermal expansion/contraction. Used for exterior automotive grilles and trim.

CYCOLAC ETS-3540:

Offers increased versatility because it is suitable for both electroplating and sputter-coating applications. Excellent processability and thermal cycle performance.

HEAT RESISTANT, PLATABLE GRADE:

CYCOLOY EHA:

An ABS/polycarbonate alloy. Provides superior thermal cycle performance and resistance to warpage. Suggested applications include wheel covers and side mirror housings. Can be electroplated, painted or sputtercoated.

BORG WARNER CHEMICALS: Extrusion Polymers:

CYCOLAC ABS:

CYCOLAC ABS is a broad family of ABS resins available with a wide variety of properties and processability. CYCOLAC ABS engineering polymers are especially suitable where toughness and rigidity are necessary over a range of temperatures. Flame retardant grades are also available in both high gloss and satin-finish grades.

CYCOVIN Alloys:

CYCOVIN alloys combine the inherent flame retardant properties of polyvinyl chloride (PVC) with the good processability of ABS. This family of products is also available in both high gloss and satin-finish grades.

PREVEX Polymers:

PREVEX polymers are a family of phenylene ether copolymers (PEC) alloys combining impact strength with excellent heat resistance. They provide good overall mechanical properties and performance in higher humidities and use temperatures than most other extrusion polymers. Flame-retardant grades are available which do not sacrifice impact strength or heat resistance.

BORG WARNER CHEMICALS: Extrusion Polymers(Continued):

HIGH-IMPACT GRADES:

Product Name:

CYCOLAC GSE:

CYCOLAC GSE is a high gloss ABS material combining high impact strength with rigidity while still offering exceptional extrusion and formability characteristics. The aesthetics of CYCOLAC GSE parts make it ideally suited for such applications as ice chests, recreational vehicle parts, and other applications requiring high gloss and rigidity.

CYCOLAC CRS:

CYCOLAC CRS is also a high impact grade of ABS, with impact values greater than CYCOLAC GSE and similar rigidity. The product offers the best balance of properties with good processing characteristics. CYCOLAC CRS has been widely used as a coextrusion substrate for weatherable polymers in outdoor applications such as spas, pool steps, boats, and recreational vehicle parts.

CYCOLAC LS:

CYCOLAC LS offers the greatest toughness at room temperature and below. The unique balance of properties and processing characteristics of this ABS grade has established it as the "industry standard" for high abuse applications such as luggage, agricultural and automotive component parts.

BORG WARNER CHEMICALS: Extrusion Polymers(Continued):

FLAME-RETARDANT GRADES:

Product Name:

CYCOLAC KCS:

CYCOLAC KCS is a high gloss ABS grade with the best overall U.L. Flame Class Rating of 94 V-O at 0.060 in. and 94-5V at 0.125 in. thicknesses. This product doesn't contain PVC, thus it has superior thermal stability compared to many competitive flame retardant polymers. The product has demonstrated excellent processing and thermoforming characteristics in both vacuum and pressure formed applications.

CYCOLAC KLG:

CYCOLAC KLG is a low gloss flame retardant ABS grade which provides the thermal stability properties of CYCOLAC KCS. It contains no PVC and it exhibits low gloss (18%) with minimal gloss-up during the forming operation. CYCOLAC KLG has a U.L. Flame Class Rating of 94 V-O at 0.090 in. thickness.

CYCOVIN K25:

CYCOVIN K25 is an ABS/PVC alloy with a low gloss and a U.L. Flame Class Rating of 94 V-O at 0.100 in. and 94-5V at 0.125 thicknesses in all colors. This alloy has high impact strength, with low notch sensitivity and gloss levels which make it an ideal choice for transportation and electrical/electronic housing applications at an economical price.

PREVEX VF1:

PREVEX VF1 is a phenylene ether copolymer alloy for sheet and profile extrusions requiring both flame retardance and high impact strength. The 185F heat deflection temperature, low specific gravity (1.06), U.L. Flame Class Rating of 94 V-O at 0.62 in. thickness and excellent regrind stability makes it a good choice for applications such as thermoformed electronic enclosures.

PREVEX VQ1:

PREVEX VQ1 is a PEC alloy with a heat deflection temperature of 265F. VQ1 has U.L. Flame Class Rating of 94 V-1 at 0.062 in. thickness. The balance of properties make this extrusion polymer an ideal choice for thermoformed or profile applications requiring both flame retardance and exposure to higher temperatures.

BORG WARNER CHEMICALS: Extrusion Polymers(Continued):

HEAT-RESISTANT GRADES:

Product Name:

CYCOLAC ZA5:

CYCOLAC ZA5 replaces CYCOLAC Z36. This ABS grade offers improvements in both impact strength and heat deflection temperatures. The high impact and rigidity of ZA5 have proven appropriate for automotive related applications such as shrouds and duct work.

PREVEX PQ1:

PREVEX PQ1 is a PEC alloy with a heat deflection temperature of 265F. PQ1 has low moisture absorption and excellent thermal stability which makes it ideal for sheet and profile applications subjected to the hot, humid environments.

BORG WARNER CHEMICALS: Extrusion Polymers(Continued):

SPECIALTY GRADES:

Controlled Gloss:

CYCOLAC CGA:

CYCOLAC CGA is a low gloss (16% Gardner Gloss) ABS grade with good impact strength and ductility. The low gloss is retained through the extrusion and thermoforming process. This product has been successful in many applications requiring non-glare appearance in markets such as automotive, transportation, agriculture and luggage.

Semi-flexible:

CYCOLAC SF:

CYCOLAC SF is a semi-flexible, low modulus ABS grade specifically designed not to stress whiten when flexed severely. The excellent low temperature impact properties make this polymer especially suitable for profile extrusion applications normally specified in semi-rigid PVC.

Profile:

CYCOLAC ASC:

CYCOLAC ASC is a medium impact, high modulus ABS grade designed for profile extrusion. The low coefficient of thermal expansion, low die swell and high heat deflection temperature make it suitable for general purpose profile extrusion applications.

CYCOVIN K29:

CYCOVIN K29 is an ABS/PVC alloy with a balance of properties showing improved warp resistance and low specific gravity compared to standard PVC polymers. CYCOVIN K29 offers high impact strength and gloss values with a U.L. Flame Class Rating of 94 V-1 at 0.125 in. thickness. The low die swell and ease of processability make it the choice for profile extrusion applications requiring flame retardance.

BORG WARNER CHEMICALS: Injection Molding Polymers:

IMPACT GRADES:

T:

CYCOLAC T is the most widely used grade offered in the general purpose product line. It is recognized as the standard for the ABS industry. CYCOLAC T offers an optimum balance of physical properties with high impact and good flow at an economical price. Successful applications include telephone housings, office equipment, kitchen appliances, toys and furniture components.

GSM:

Toughest of the general purpose molding grades. GSM offers high impact and tensile properties; however, the material does not flow as well as CYCOLAC T. It offers a Relative Temperature Index of 70C. Applications are in areas where additional abuse resistance is required, such as golf club heads, vacuum cleaner components, power tools housings, lawn and garden equipment.

L:

CYCOLAC L offers maximum low temperature toughness. The viscosity of L makes it ideal for thick-wall applications. This material exhibits excellent toughness and superior impact resistance. Current applications include electrical connectors, small appliances, video tape housings and snowmobile components.

DFA-R:

Our most economical molding grade, DFA-R has a high modulus with medium impact resistance and high flow. It is now being used for housewares, appliances, decorative office and home furnishings, hardware and toys.

BORG WARNER CHEMICALS: Injection Molding Polymers(Continued):

HIGH HEAT GRADES:

DH:

The most economical high heat, high modulus and tensile, medium impact grade in the high heat product line. It has a better flow than X-11 or X-17.

X-11:

Tougnest of the high heat grades. Compared to CYCOLAC T, products made from X-11 exhibit better heat resistance and equivalent abuse resistance. It has better processability than X-17. Applications include building products and interior automotive trim.

X-17:

Higher heat resistance than X-11, but less impact strength, X-17 should be used where impact resistance is secondary to heat resistance. Current examples include automotive parts, small appliances, and power tools.

X-37:

One of the best heat resistant materials in the ABS product line, surpassing X-17 in both heat resistance and impact strength. It was formulated to withstand temperatures up to 240F (115C) extending the use temperature range of CYCOLAC ABS materials by approximately 20F (10C).

BORG WARNER CHEMICALS: Injection Molding Polymers(Continued):

FLAME RETARDANT CYCOLAC ABS GRADES:

Product Name:

KJB:

Offers a U.L. flame class rating of 94V-O at 0.058"(1.47 mm) and 94-5V at 0.120"(3.05 mm) with a Relative Temperature Index of 80C. The physical property balance is similar to a medium impact ABS with high melt flow. KJB is used primarily in electrical applications requiring regulatory agency approval, such as business machines and computer components, electrical appliances, and power tools. KJB should be used where U.V. resistance to color change is important in indoor applications.

KJW:

Offers a U.L. flame class rating of 94V-O at 0.058"(1.47mm) and 94-5V at 0.120"(3.05mm) with a Relative Temperature Index of 60C. It offers good impact strength which makes it suitable for many applications requiring good abuse resistance. Suggested applications are computer components, electrical appliances, power tools and bsuiness machines.

KJT:

The physical property data and U.L. ratings of this product are similar to KJB. Priced below KJB, it offers a flame class rating of 94V-O at 0.120" (3.05mm). Recommended applications include business machine housings and components, electrical appliances, power tools, and television parts.

KJU:

Offers a flame class rating of 94V-O at 0.058"(1.47mm). It has superior U.V. stability resulting in reduced color change for indoor applications. Recommended for computer components and consumer electronic applications.

BORG WARNER CHEMICALS: Injection Molding Polymers(Continued):

FLAME RETARDANT CYCOLAC ABS GRADES(Continued):

Product Name:

Z-86:

Has a U.L. flame class rating of 94V-O at 0.125"(3.17mm). It was designed for injection molded applications and features a good balance of physical properties. Recommended for applications in consumer and industrial electronic components, television, VCR and video disc cabinets and computer disc paks.

KJM:

Offers a U.L. flame class rating of 94V-O at 0.125"(3.17 mm). It can provide a cost savings for many applications previously designed for flame retardant structural foam. Recommended wall thickness is 0.150"(3.81mm) with generous radii.

BORG WARNER CHEMICALS: Injection Molding Polymers(Continued):

PLATING GRADES:

Product Name:

EPB-3570:

EPB-3570 is designed to provide maximum thermocycling performance in plating applications. It is used for automotive exterior trim, large and small appliances, plumbing fixtures, and marine equipment. This material should be considered for applications which must maintain high performance over a wide temperature range in varied environments.

EP-3510:

EP-3510 is an electroplating product with high melt flow designed to mold complex, stress-free parts. Applications include small appliance housings, interior and exterior automotive components, cosmetic cases, appliance knobs, etc.

CYCOLOY EHA:

An ABS/polycarbonate alloy designed with emphasis on impact strength, superior thermal cycle performance and resistance to warpage. Suggested applications include plated exterior automotive trim, plumbing and marine equipment.

TRANSPARENT GRADES:

CTB:

A transparent ABS featuring high modulus and medium impact resistance. Typical applications are industrial and cosmetic packaging, refrigerator crisper trays, and toys, where the highest clarity and hardest surface finish are desired.

BORG WARNER CHEMICALS: Injection Molding Polymers (Continued):

FLAME RETARDANT CYCOVIN ABS/PVC ALLOYS:

Product Name:

K20:

A satin finished product featuring excellent impact and good processability. It offers a U.L. flame class rating of 94V-0 @ 0.100". Applications include appliances, business machines and the communications market.

K29:

K29 is a versatile alloy which combines excellent processability with a U.L. flame class rating of 94V-1 at 0.125". Application areas include appliances, business machines and telecommumications.

BORG WARNER CHEMICALS: PREVEX Polymers:

Flame Retardants:

Product Name:

PREVEX VFA:

PREVEX VFA offers a UL 94 V-2 Flame Class Rating along with high impact strength and good dimensional stability. VFA is designed for use in business machine applications, internal parts and other enclosures.

PREVEX VFS:

PREVEX VFS exhibits good mechanical property performance. VFS is designed for use in business machines and is particularly suited for applications requiring UV stability.

PREVEX VGA:

PREVEX VGA combines flame retardance with excellent impact strength and good dimensional stability. VGA is designed for use in business machine applications, internal parts and other enclosures.

PREVEX VHA:

PREVEX VHA meets property performance requirements for many business machine enclosures, internal parts and other enclosure applications.

PREVEX VJA:

PREVEX VJA provides a good combination of impact and heat resistance in a flame-retardant product. VJA is classified V-1 (at 0.062 inch) according to UL 94. VJA has a 95C RTI and is designed for business machine housings and telecommunications equipment.

PREVEX VJB:

PREVEX VJB offers good dimensional stability and heat resistance in a flame-retardant grade. VJB is classified V-1 (at 0.125 inch) according to UL Standard 94 and is suitable for snap-fit applications in housings and telecommunications equipment.

BORG WARNER CHEMICALS: PREVEX Polymers(Continued):

Flame Retardants(Continued):

Product Name:

PREVEX VKA:

PREVEX VKA is a grade with a 94 V-O classification (at 0.062 inch) for thin wall designs that also offers excellent electrical properties. VKA, with a 95C RTI, is suitable for electronic enclosures, outlet boxes, solenoids, and many appliance components

PREVEX VMA:

PREVEX VMA exhibits good overall mechanical property performance and is classified V-1 according to UL 94. VMA has a 105C RTI rating.

PREVEX VQA:

PREVEX VQA meets performance requirements for many electrical devices, enclosures, industrial instruments and connectors with its 94 V-1 classification. VQA provides good dimensional stability, durability and high heat resistance.

GENERAL PURPOSE:

PREVEX PMA:

PREVEX PMA provides good heat resistance (DTUL of 245F) and dimensional stability. This property balance can be used in small appliances, lawn care tools, power tools and many industrial devices.

PREVEX PQA:

PREVEX PQA is ideal for portable appliances, power tools, hot water appliances and industrial products. PQA provides outstanding dimensional stability and toughness with a DTUL of 265F.

BORG WARNER CHEMICALS: PREVEX Polymers(Continued):

Reinforced/Flame Retardant:

<u>Product Name:</u>

PREVEX V2A:

PREVEX V2A is a 20% glass-reinforced, flame-retardant grade that combines superior modulus with flame-retardant properties required for connectors, business equipment, motor housings, terminal blocks and electrical applications.

PREVEX V3A:

PREVEX V3A is a 30% glass-reinforced flame-retardant grade with the highest flexural modulus of all PREVEX flame-retardant grades (1.0 x 10 6) psi). V3A is an outstanding choice for applications in electrical markets and industrial control products.

REINFORCED/GENERAL PURPOSE:

PREVEX P2A:

PREVEX P2A is a 20% glass-reinforced grade that features high flexural modulus with low mold shrinkage. P2A is an attractive replacement for die cast metal parts used in pumps, fluid meters and other plumbing applications.

PREVEX P3A:

PREVEX P3A is a 30% glass-reinforced grade with the highest flexural modulus of all PREVEX general-purpose grades (1.0 x 10 6 psi). P3A is especially suited as a die cast replacement in more demanding pump, fluid meter and plumbing applications.

BORG WARNER CHEMICALS: PREVEX Polymers(Structural Foam Grades):

Flame Retardants:

Product Name:

PREVEX BJA:

PREVEX BJA is a phenylene ether polymer alloy which exhibits good overall mechanical property performance at higher use temperatures than ABS or polystyrene. PREVEX BJA, a flame retardant foam, exhibits a heat deflection temperature of 170F at 10% density reduction with a UL Flame Class Rating of V-O at 0.125 inch and 5V at 0.175 inch and is widely accepted for thin-wall applications from 0.150 inch to 0.250 inch conventional applications.

PREVEX BJB:

PREVEX BJB is a general-purpose, flame-retardant structural foam with a UL recognition of 94 V-1 and 5V at 0.225 inch. PREVEX BJB provides the performance necessary for most business-machine applications while maintaining its economic competitiveness when compared to low cost, lower performance engineering materials.

PREVEX BJF:

PREVEX BJF is a high-productivity phenylene ether polymer alloy grade which allows faster cycle times and offers smoother surfaces than other foamable materials. PREVEX BJF is especially effective in thick wall sections where coring is not practical, and where a UL 94 5V (0.250"), a V-O (0.250") or a V-1 (0.150") flame class rating are needed.

BORG WARNER CHEMICALS: PREVEX Polymers(Structural Foam Grades) (Continued):

Counterpressure:

Product Name:

PREVEX BJC:

PREVEX BJC is a high flow, flame-retardant foam grade specifically designed for use in gas counterpressure molding. PREVEX BJC exhibits good overall mechanical properties, with a UL Flame Class rating of V-O and 5V at 0.200 inch, and is suitable for many high performance applications.

PREVEX BJU:

PREVEX BJU, a high flow, flame-retardant foam for use in gas counterpressure molding, offers improved UV stabilization to help prevent color shift in unpainted gas counterpressure applications. PREVEX BJU exhibits good overall mechanical properties, with a UL Flame Class Rating of V-O and 5V at 0.200 inch, and is suitable for many high performance applications.

Reinforced:

PREVEX B1A:

PREVEX B1A is a glass-reinforced phenylene ether polymer alloy grade that combines high strength, stiffness and good impact resistance with a 190F DTUL. PREVEX B1A has the highest DTUL among the PREVEX structural foam grades, a UL Flame Class rating of V-1 at 0.150" and has been used effectively in load-bearing applications such as printers and card cages.

THE BUDD CO.: Sheet Molding (SMC) Compounds:

Compound Number:

DSM-780:

Reinforcement: Glass

Typical Applications: Parts requiring high strength, dimensional stability, excellent surface smoothness without sink marks. Parts to be painted required automotive type finish. Molded parts are more resilient than parts molded from BUDD DSM-730, etc.

Exterior automobile body parts, such as front upper grille panels, fender skirts, hoods, trunk lids, etc.

Tail lamp housing	Tractor parts
Blower housing	Camper body panels
Snowmobile hoods	Instrument panels
Entrance doors	Space frame housing

DSM-795:

(HMC-Low Density Black)

Reinforcement: Glass

Typical Applications: Parts requiring high strength and not smooth surface. Non Class A parts such as radiator support housing.

DSM-929:

(Diesel Grade SMC)

Reinforcement: Glass

Typical Applications: Parts requiring high strength at room temperature, good retention of properties at high temperature and corrosion resistant such as diesel engine external body parts.

DSM-929N:

(Diesel Grade SMC - Natural)

Reinforcement: Glass

Typical Applications: Parts requiring high strength at room temperature, good retention of properties at high temperature and corrosion resistant such as diesel engine external body parts.

DSM-930:

(Low Profile)

Reinforcement: Glass

Typical Applications: Parts requiring high strength, dimensional stability, excellent surface smoothness without sink marks. Parts to be painted requiring an automotive finish.

Exterior automobile body parts such as front upper grille panels, fender skirts, hoods, trunk lids, etc.

Tail lamp housings	Tractor parts
Snowmobile hoods	Camper body panels
Space frame housings	Instrument panels
Entrance doors	Blower housings

THE BUDD CO.: Sheet Molding Compounds (SMC) (Continued):

Compound Number:

DSM-933:
(Grey Black)
(High Strength Low Shrink)
Reinforcement: Glass
Typical Applications: Parts requiring high strength and dimensional stability. Parts to be painted requiring an automotive finish.
Underbody automobile body parts, such as floor panels.

DSM 934:
(C30R20 Diesel Reinforcement SMC)
Reinforcement: Glass
Specific Gravity: D-792: 1.87
Water Absorption (24 Hrs. @ 23C): D-570: <0.35%

DSM 935:
Bumper Grade C30R20 Natural
Reinforcement: Glass
Specific Gravity: D-792: 1.85
Barcol Hardness: D-2583: 50 - 55

DSM-936:
Bumper Grade R60-HMC, Natural DSM-936
Reinforcement: Glass
Specific Gravity: D-792-70: 1.71
Barcol Hardness: 45 - 50

DSM 938:
(Diesel Grade SMC - Vinyl Ester - Blue)
Reinforcement: Glass
Typical Applications: Parts requiring high strength at room temperature. Good retention of properties at high temperature and corrosion resistant such as diesel engine body parts.

DSM 942:
(Diesel Grade SMC - Vinyl Ester - Natural)
Reinforcement: Glass
Typical Applications: Parts requiring high strength at room temperature and corrosion resistant such as diesel engine external body parts.

DSM 943:
Bumper Grade C30R20 Black, Exterior
Reinforcement: Glass
Specific Gravity: D-792: 1.89
Barcol Hardness: D-2583: 45 - 55

THE BUDD CO.: Sheet Molding (SMC) Compounds(Continued):

Compound Number:

DSM 950:
(Smooth Surface - Class A)
Reinforcement: Glass
Typical Applications: Parts requiring high strength, dimensional stability, excellent surface smoothness without sink marks. Parts to be painted requiring automotive type finish.
Exterior automotive body parts for large class body panels such as door hoods, fenders, deck lids, etc.

DSM 9501:
(Lower Density SMC)
Reinforcement: Glass
Typical Applications: Parts requiring lower specific gravity, dimensional stability, not class A surface.
Non Class A automotive body parts such as inner reinforcement panels for large class A body panels (Doors, hoods, fenders, etc.)

MX 7387:
Structural Grade C50R10 Black
Reinforcement: Glass
Specific Gravity: D-792: 1.86
Barcol Hardness: D-2583: 45 - 55

CELANESE ENGINEERING RESINS, INC.: CELANEX Thermoplastic Polyester:

CELANEX thermoplastic polyester resins are highly crystalline and exhibit a unique combination of properties including high strength, rigidity and toughness, low creep even at elevated temperatures, minimal moisture absorption, exceptional dimensional stability, resistance to a wide range of chemical attack, excellent electrical properties and outstanding processing characteristics.

Available in a wide range of formulations, including unreinforced, glass and mineral reinforced, polymer blends, and impact modified, as well as flame retarded to meet the requirements of Underwriters' Laboratories Bulletin UL 94V-O, CELANEX resins have demonstrated a versatility of performance at economical overall finished part cost, leading to their use in an unusually wide range of applications.

The series listed below, are designed to provide the design engineer and end user a spectrum of performance to enable a formulation choice to be made that closely tailors both cost and performance to specific application needs, in the economic replacement of metals, thermosets and competitive thermoplastics.

CELANEX COMPOSITIONS:

2000 Unreinforced Series:

* Chemical Resistance
* Wear Resistance and Lubricity
* Easy Flow

The CELANEX 2000 series offers a unique combination of chemical and wear resistance, low coefficient of friction, minimal moisture absorption, excellent colorability, and surface gloss. These properties combined with outstanding processing characteristics meke the CELANEX series ideal for such diverse applications as typewriter and terminal keytops, cable liners, switches, paint brush bristles, stator insulation, and motor end caps.

There are three unreinforced grades available in the CELANEX 2000 series: CELANEX 2000, noted for its easy flow, CELANEX 2002, a tougher version of 2000; and CELANEX 2012, a flame retarded version of 2002 recognized as 94V-O down to 0.028 inch minimum thickness.

CELANESE ENGINEERING RESINS, INC.: CELANEX Thermoplastic Polyester(Continued):

CELANEX Compositions(Continued):

3000 General Purpose Series:

* High Strength and Rigidity
* Excellent Electrical Properties
* High Temperature Performance

The CELANEX 3000 general purpose series consists of glass reinforced injection moldable grades that offer an unusual combination of excellent mechanical, electrical and thermal properties, together with outstanding processability. In high energy auto ignition systems, the unique properties of the CELANEX 3000 series enabled it to replace performance thermosets, providing greatly improved dielectric strength and arc resistance, in both wet and dry environments at temperatures as high as 300F. The high performance characteristics of the CELANEX 3000 series have made it ideal for use in automotive distributor caps, coil bobbins, coil cases, rotors and other ignition components, and in a seemingly endless array of electrical, telecommunications, appliance, and industrial applications.

The 3000 general purpose series is available in three levels of glass reinforcement (15%, 30%, and 40%) to offer the end user the maximum in design flexibility in terms of rigidity and ductility.

Grade:	Description
3200:	15% glass reinforced general purpose resin.
3300:	30% glass reinforced general purpose resin.
3400:	40% glass reinforced general purpose resin.

CELANESE ENGINEERING RESINS, INC.: CELANEX Thermoplastic Polyester(Continued):

CELANEX Compositions(Continued):

3000 Flame Retarded Series:

* High Strength and Rigidity
* Excellent Electrical Properties
* High Temperature Performance
* UL 94V-O

Like its general purpose counterpart, the CELANEX 3000 flame retarded series consists of glass reinforced inmjection moldable grades that maintain excellent mechanical and electrical properties throughout a wide range of aggressive end-use environments, while exhibiting superior thermal stability and exceptional processing characteristics. The excellent electrical profile, dimensional stability, and chemical resistance of the 3000 flame retarded series make it an outstanding candidate for a multitide of connector applications; DIP sockets and switches; motor components, terminal boards, stator insulation, brush holders, and motor control devices; relays and protectors; all types of switches; bobbins for use in solenoids and transformers; and relay cases and covers.

The 3000 flame retarded series is available in from 12% to 30% glass reinforcement levels offering the maximum design flexibility and all grades are recognized by UL as 94V-O in all recommended colors. Of particular note, CELANEX 3211 and 3311 meet UL's requirements for the Sole Support of current carrying components and CELANEX 3310 and 3311 have electrical temperature index ratings of 140C. All other grades in this series have UL electrical temperature index ratings of 130C.

Grade:	Description
3112:	12% glass reinforced resin recognized by UL as 94V-O
3210:	18% glass reinforced resin recognized by UL as 94V-O and 94-5V
3211:	19% glass reinforced resin recognized by UL as 94V-O and 94-5V. Meets UL's requirements for Sole Support of current carrying components.
3310:	30% glass reinforced resin recognized as 94V-O and 94-5V.
3311:	30% glass reinforced resin recognized by UL as 94V-O and 94-5V. Meets UL's requirements for sole support of current carrying components.

CELANESE ENGINEERING RESINS, INC.: CELANEX Thermoplastic Polyester (Continued):

CELANEX Compositions (Continued):

4000 Higher Impact Series:

* Toughness
* Flexibility and Ductility
* Surface Appearance and Colorability

Through the use of impact modifiers and higher molecular weight polymers, the new CELANEX 4000 series offers glass reinforced grades that combine excellent mechanical strength with improved impact performance and ductility. The 4000 series is an ideal candidate for a wide array of applications including large and small industrial pump housings and impellers; gears, bearings and other load bearing components; mechanical winches; chain saw housings; power tool components, and generally any greater impact strength and toughness is required.

The CELANEX 4000 series consists of two 30% glass reinforced grades: 4330 which offers the greatest impact strength in the CELANEX product line; and 4300 which provides impact strength intermediate between CELANEX 3300 and 4330. Both grades offer outstanding colorability with CELANEX 4330 providing exceptional surface gloss as well.

Grade:	Description
4300:	30% glass reinforced resin with greater imnpact strength than 3300
4330:	30% glass reinforced resin offering the greatest impact strength in the CELANEX product line

CELANESE ENGINEERING RESINS, INC.: CELANEX Thermoplastic Polyester (Continued);

CELANEX Compositions (Continued):

5000 Polyester Blend Series:

* Surface Appearance and Gloss
* Resistance to Warpage
* Good Mechanical Profile

In the CELANEX 5000 series, unique glass reinforced polymer blends offer exceptional surface appearance and gloss with resistance to warpage. The 5000 series is ideally suited to such various applications as fire extinguisher cartridges; automotive door handles, bases, and housings for such small appliances as toasters, cookers and fryers; and generally any CELANEX 3000 type application where improved surface appearance and some reduction in warpage is needed.

Mechanical, thermal and electrical properties of CELANEX 5000 series are similar to corresponding general purpose grades in the 3000 series

Grade:	Description
5300:	30% glass reinforced resin exhibiting excellent surface gloss and appearance

CELANESE ENGINEERING RESINS, INC.: CELANEX Thermoplastic Polyester (Continued):

CELANEX Compositions (Continued)

6000 Mineral/Glass General Purpose Series:

* Excellent Warp Resistance
* High Rigidity
* Low Cost

For applications requiring flatness and extra rigidity the new CELANEX 6000 series has mineral/glass general purpose formulations that offer maximum warp resistance at a low cost. Typical applications for the 6000 series include: fan blades and shrouds; headlamp actuators and housings; painted exterior body panels, and generally any CELANEX 3000 type application where flatness, stiffness and surface gloss are required.

The thermal and electrical properties of CELANEX 6400 are similar to its glass reinforced counterpart, while it offers an increase in rigidity of over 30% compared to CELANEX 3300.

It is the most cost effective general purpose reinforced polyester material, ideally suited for applications where strength and ductility are not critical.

Grade:	Description
6400:	40% mineral/glass reinforced general purpose resin providing excellent warp resistance.

CELANESE ENGINEERING RESINS, INC.: CELANEX Thermoplastic Polyester(Continued):

CELANEX Compositions(Continued):

7000 Mineral/Glass Flame Retarded Series:

* Excellent Electrical Properties
* Low Warpage
* Low Cost

Like its general purpose counterpart, the CELANEX 7000 mineral/glass flame retarded series offers excellent warpage control, stiffness, and low cost combined with superior electrical properties. CELANEX 7000 series is particularly suited for high voltage applications such as television components, switches, bobbins, circuit breakers, relays, connectors, and housings for timers and other electrical controls.

CELANEX 7700 has UL flammability of 94V-O at 0.028 inch thickness and 94-5V at 0.115 inch thickness. In addition, CELANEX 7700 has a UL electrical temperature index rating of 140C.

The properties of CELANEX 7700 are similar to its general purpose counterpart, CELANEX 6400.

CELANEX 7700 is the material of choice in applications where rigidity, low warpage, dimensional stability and tracking resistance are critical.

It is the most cost effective, reinforced, flame retarded material and should be specified in applications where impact strength and other "toughness" properties are not critical.

Grade:	Description
7700:	A mineral/glass reinforced resin recognized by UL as 94V-O and 94-5V.

CELANESE ENGINEERING RESINS, INC.: CELANEX K-Series:

CELANEX K-Series thermoplastic resins have been designed specifically to meet the exacting quality of leading keytop manufacturers. By combining new process technology and rigid quality control measures, Celanese Engineering Resins provides superior materials for printed polyester keytops.

Keytops molded from CELANEX K-Series resins deliver outstanding printability, abrasion resistance, UV stability and resistance to oils and solvents. For the ultimate in design flexibility, from high toughness integral plungers to thin wall sections, there's a CELANEX K-Series product to meet your needs. Developed in conjunction with industry leaders, CELANEX K-Series resins in both natural and colored grades meet the most stringent end-user specifications for black specks and color variability, as well as providing ultrafast molding cycles for maximum productivity.

CUSTOM KEYTOP COLORS

CELANEX K-Series resins offer a broad spectrum of keytop colors and the capacity to match your DIN standard earthtone colors accurately and quickly.

For those who prefer to compound or let down their own special colors, the aesthetics and color uniformity of the CELANEX base resins offer a proven rate to superior final parts.

OUTSTANDING PROCESSABILITY

Celanese Engineering Resins has employed its extensive experience in thermoplastic polyesters to produce CELANEX K-Series resins which demonstrate excellent mold flow and quick set-up leading to fast cycle times, clean molding and low reject rates.

WORLDWIDE CAPABILITIES

CELANEX K-Series resins are marketed worldwide.

CELANESE ENGINEERING RESINS, INC.: CELCON Acetal Copolymer:

CELCON is the registered trademark of Celanese Corporation used for its acetal copolymer. CELCON acetal copolymer is a high strength, crystalline, thermoplastic engineering resin whose unique balance of properties and ease of processability offer performance and cost advantages over many other materials.

CELCON is widely accepted by engineers to design parts where thermosets, laminates, wood, die-cast and stamped metals such as magnesium, zinc, brass, iron and steel were formerly used. The superior performance of CELCON results from its high mechanical strength, stiffness, toughness, practical impact strength and, most importantly, its ability to maintain these properties over a broad range of temperatures and environments.

In addition to the versatility of design which CELCON offers engineers, it can also be processed by many conventional means including injection molding, blow molding, extrusion and rotational casting. Rod and slab stock from CELCON can also be easily machined.

Because of its unique properties and its ease of processing CELCON is used in an extraordinarily wide range of applications in a variety of end uses.

* high tensile strength and stiffness
* exceptional dynamic fatigue strength and dimensional stability
* practical toughness and resilience
* minimal moisture absorption
* low friction and wear properties making it one of the few natural bearing materials
* hard, high gloss surface
* superior property retention under extended exposure at temperatures up to 220F in air and 180F in water.
* exceptional resistance to a wide range of chemicals, oils, greases and solvents
* easy to process and fabricate

In addition, CELCON is available in a wide range of melt flows, specialty grades and colors to match functional requirements.

Further contributing to the broad acceptance of CELCON are numerous agency code approvals. Although not a complete listing, end use approvals and/or specifications have been obtained for various CELCON grades from ASTM, FDA, NSF, USDA, UL, as well as plumbing code bodies such as IAPMO and BOCA.

CELANESE ENGINEERING RESINS, INC.: CELCON Acetal Copolymer (Continued):

STANDARD CELCON Grades:

Series:

U10:

Melt Index: 1.0
Description: A terpolymer with excellent melt strength and processability in extrusion, blow molding and compression molding
Typical Applications: Aerosol containers, industrial tanks and floats, rod, tube, slab and profiles.

M25:

Melt Index: 2.5
Description: High molecular weight grade primarily used for extrusion and selected injection molding applications in easy-to-fill molds.
Typical Applications: Wire coatings, rod, tube, sheet and slab as well as injection molded items requiring extra toughness and elongation such as chain links, plumbing fittings and ski bindings.

M50:

Melt Index: 5.0
Description: An extrusion and injection molding grade with an intermediate molecular weight tailored for selected applications in extrusion and injection molding.
Typical Applications: Extruded rod, tube and slab as well as selected molded items requiring additional toughness.

M90:

Melt Index: 9.0
Description: A general purpose injection molding grade with a molecular weight designed for excellent moldability and optimum properties in demanding applications.
Typical Applications: A wide range of injection molded items where toughness and dimensional tolerances are important such as cams, gears, springs, knobs, check valves and drapery supports.

M140:

Melt Index: 14.0
Description: A general purpose injection molding grade with slightly higher flow for use in hard-to-fill molds.
Typical Applications: Injection molded items with less critical demands on toughness and dimensional tolerance.

CELANESE ENGINEERING RESINS, INC.: CELCON Acetal Copolymer (Continued):

Standard CELCON Grades:

Series:

M270:

Melt Index: 27.0
Description: A lower molecular weight, high-flow grade designed for superior moldability in multi-cavity, intricate or hard-to-fill applications.
Typical Applications: Parts molded on short cycle times which are less demanding on toughness than for an M90 material. Ideal for thin wall parts or long flow paths such as combs, marking pen bodies and housings.

M450:

Melt Index: 45.0
Description: An extra high flow grade with reduced toughness designed for optimal cycle performance.
Typical Applications: Dimensionally stable thin walled, low-load bearing parts such as audio/video tape hubs and micro gears.

CELANESE ENGINEERING RESINS, INC.: CELCON Acetal Copolymer (Continued):

Specialty CELCON Grades:

Special Property: U.V. Stabilized
Product: UV90
UV25
Description: Stabilized for use where U.V. degradation is a problem. Available in natural and custom matched colors.
Application: Automotive interiors, and recreational items where exposure to sunlight causes discoloration and property loss, i.e., knobs, buttons, toys, cams and levers.

Special Property: Weather Resistant
Product: WR90 BLack
WR25 Black
Description: Stabilized for use where maximum U.V. and outdoor weathering resistance is required.
Applications: Automotive, irrigation and recreational items.

Special Property: Impact Modified
Product: C-400
C-401
Description: M90 grade modified for use where enhanced impact is required.
Application: Automotive and industrial applications.

Special Property: Glass Coupled
Product: GC-25A
Description: A glass coupled formulation containing 25% fiber glass reinforcement. This grade offers an excellent balance of physical properties for applications requiring increased strength, stiffness, a reduced thermal expansion coefficient and increased heat distortion temperature.
For specialized applications lower levels can be achieved by blending with M90.
Application: Gears, machine housings, pressure vessels, valve bodies, marine hardware.

CELANESE ENGINEERING RESINS, INC.: CELCON Acetal Copolymer (Continued):

Specialty CELCON Grades(Continued):

Special Property: Low Wear
Product: LW 90
Description: This is an M90 grade formulated for high speed, low load wear applications against metal.

Product: LW90-S2
LWGC-S2
LW90-SC
Description: The S2 suffix denotes a 2% silicone modification for use in low speed, high load applications. LWGC is 25% glass coupled grade. SC is a 20% silicone concentrate for use with all CELCON resins.
Application: Bearings, side plates, bushings, wear surfaces and conveyor links or plates.

Special Property: Anti-Static
Product: AS270
AS450
Description: M270, M450 grades formulated to reduce static build-up on molded parts.
Application: Audio or video cassette components, medical appliances and electrical items.

Special Property: Mineral Coupled
Product: MC90
MC90-HM
MC270
MC270-HM
Description: M90 and M270 grades formulated to provide increased stiffness, better dimensional stability and lower warpage. The HM suffix denotes highest modulus.
Application: Gears, cams and gear pump covers where maximum stiffness and dimensional stability are required.

Special Property: Electroplatable
Product: EP90
Description: An M90 grade designed to be electroplated using conventional electroless/electrolytic ABS processes.
Application: High strength chrome plated parts to reduce plated non-ferrous metals, i.e., faucets, automotive door knobs, window cranks, plumbing valves and appliance and consuner applications.

CELANESE ENGINEERING RESINS, INC.: CELCON Acetal Copolymer (Continued):

In addition to the standard grades of CELCON, Celanese Engineering Resins has developed a series of CELCON specialty grades offering distinct performance characteristics.

Grade:

PROPERTY	ASTM	UNITS	
Antistatic:			
AS270:			
STRENGTH:			
Tensile Strength 73F(23C)	D-638	psi	8,800
Tensile Elongation 73F(23C)	D-638	%	50
STIFFNESS AND CREEP:			
Flexural Strength 73F(23C)	D-790	psi	13,000
Flexural Modulus 73F(23C)	D-790	psi	375,000
Heat Deflection-Temperature			
@ 264 psi	D-648	F(C)	230(110)
@ 66 psi	D-648	F(C)	316(158)
TOUGHNESS:			
Notched Izod Impact Strength 73F(23C)	D-256	Ft-lb/in of notch	1.1(6.0)
Shear Strength 73F(23C)	D-732	psi	7,700
AS450:			
STRENGTH:			
Tensile Strength 73F(23C)	D-638	psi	8,800
Tensile Elongation 73F(23C)	D-638	%	25
STIFFNESS AND CREEP:			
Flexural Strength 73F(23C)	D-790	psi	13,000
Flexural Modulus 73F(23C)	D-790	psi	375,000
Heat Deflection-Temperature			
@ 264 psi	D-648	F(C)	230(110)
@ 66 psi		F(C)	316(158)
TOUGHNESS:			
Notched Izod Impact Strength 73F (23C)	D-256	Ft-lb/in	0.9(4.9)

CELANESE ENGINEERING RESINS, INC.: CELCON Acetal Copolymer (Continued):

Grade:

PROPERTY	ASTM	UNITS	
Mineral Coupled:			
MC90:			
STRENGTH:			
Tensile Strength 73F(23C)	D-638	psi	7,700
Tensile Elongation 73F(23C)	D-638	%	55
STIFFNESS AND CREEP:			
Flexural Strength 73F(23C)	D-790	psi	13,000
Flexural Modulus 73F(23C)	D-790	psi	430,000
Heat Deflection-Temperature	D-648		
@ 264 psi		F(C)	200(93)
@ 66 psi		F(C)	310(154)
TOUGHNESS:			
Notched Izod Impact Strength 73F(23C)	D-256	Ft-lb/in of notch	1.2(6.5)
Shear Strength 73F(23C)	D-732	psi	7,200
MC90-HM:			
STRENGTH:			
Tensile Strength 73F(23C)	D-638	psi	6,400
Tensile Elongation 73F(23C)	D-638	%	50
STIFFNESS AND CREEP:			
Flexural Strength 73F(23C)	D-790	psi	12,500
Flexural Modulus 73F(23C)	D-790	psi	500,000
Heat Deflection-Temperature	D-648		
@ 264 psi		F(C)	207(97)
@ 66 psi		F(C)	302(150)
TOUGHNESS:			
Notched Izod Impact Strength 73F(23C)	D-256	Ft-lb/in of notch	1.2(6.5)
Shear Strength 73F(23C)	D-732	psi	6,500

CELANESE ENGINEERING RESINS, INC.: CELCON Acetal Copolymer (Continued):

Grade:

PROPERTY	ASTM	UNITS	
Mineral Coupled:			
MC270:			
STRENGTH:			
Tensile Strength 73F(23C)	D-638	psi	7,700
Tensile Elongation 73F(23C)	D-638	%	40
STIFFNESS AND CREEP:			
Flexural Strength 73F(23C)	D-790	psi	13,000
Flexural Modulus 73F(23C)	D-790	psi	430,000
Heat Deflection Temperature	D-648	F(C)	
@ 264 psi			200(93)
@ 66 psi			310(154)
TOUGHNESS:			
Notched Izod Impact Strength 73F(23C)	D-256	ft-lb/in of notch	1.0(5.4)
Shear Strength 73F(23C)	D-732	psi	7,200
MC270-HM:			
STRENGTH:			
Tensile Strength 73F(23C)	D-638	psi	6,400
Tensile Elongation 73F(23C)	D-638	%	35
STIFFNESS AND CREEP:			
Flexural Strength 73F(23C)	D-790	psi	12,500
Flexural Modulus 73F(23C)	D-790	psi	500,000
Heat Deflection-Temperature	D-648	F(C)	
@ 264 psi			207(97)
@ 66 psi			302(150)
TOUGHNESS:			
Notched Izod Impact Strength 73F(23C)	D-256	Ft-lb/in of notch	0.9(4.9)
Shear Strength 73F(23C)	D-732	psi	6,500(457)

CELANESE ENGINEERING RESINS, INC.: CELCON Acetal Copolymer (Continued):

Grade:

PROPERTY	ASTM	UNITS	
Low Wear:			
LW90:			
STRENGTH:			
Tensile Strength 73F(23C)	D-638	psi	8,500
Tensile Elongation 73F(23C)	D-638	%	50
STIFFNESS AND CREEP:			
Flexural Strength 73F(23C)	D-790	psi	12,300
Flexural Modulus 73F(23C)	D-790	psi	370,000
Heat Deflection Temperature	D-648	F(C)	
@ 264 psi			207(97)
TOUGHNESS:			
Notched Izod Impact Strength 73F(23C)	D-256	Ft-lb/in of notch	1.2
LW90-S2:			
STRENGTH:			
Tensile Strength 73F(23C)	D-638	psi	7,400
Tensile Elongation 73F(23C)	D-638	%	60
STIFFNESS AND CREEP:			
Flexural Strength 73F(23C)	D-790	psi	12,000
Flexural Modulus 73F(23C)	D-790	psi	350,000
Heat Deflection Temperature	D-648	F(C)	
@ 264 psi			230(110)
@ 66 psi			316(158)
TOUGHNESS:			
Notched Izod Impact Strength 73F(23C)	D-256	Ft-lb/in of notch	1.1(7.5)

CELANESE ENGINEERING RESINS, INC.: CELCON Acetal Copolymer (Continued):

Grade:

PROPERTY	ASTM	UNITS	
Low Wear (Continued):			
LW GC-S2:			
STRENGTH:			
Tensile Strength 73F(23C)	D-638	psi	14,000
Tensile Elongation 73F(23C)	D-638	%	2.5
STIFFNESS AND CREEP:			
Flexural Strength 73F(23C)	D-790	psi	25,000
Flexural Modulus 73F(23C)	D-790	psi	930,000
Heat Deflection-Temperature	D-648	F(C)	
@ 264 psi			293(145)
TOUGHNESS:			
Notched Izod Impact Strength 73F(23C)	D-256	Ft-lb/in of notch	1.1(6.0)
TOUGH X:			
TX-90:			
STRENGTH:			
Tensile Strength 73F(23C)	D-638	psi	7,200
Tensile Elongation 73F(23C)	D-638	%	100+
STIFFNESS AND CREEP:			
Flexural Strength 73F(23C)	D790	psi	9,500
Flexural Modulus 73F(23C)	D-790	psi	280,000
Heat Deflection-Temperature	D-648	F(C)	
@ 264 psi			214(101)
@ 66 psi			302(150)
TOUGHNESS:			
Notched Izod Impact Strength 73F(23C)	D-256	Ft-lb/in of notch	1.8(9.8)

CELANESE ENGINEERING RESINS, INC.: CELCON Acetal Copolymer (Continued):

Grade:

PROPERTY	ASTM	UNITS	
TOUGH X:			
TX-90 Plus:			
STRENGTH:			
Tensile Strength 73F(23C)	D-638	psi	6,000
Tensile Elongation 73F(23C)	D-638	%	200+
STIFFNESS AND CREEP:			
Flexural Strength 73F(23C)	D-790	psi	7,100
Flexural Modulus 73F(23C)	D-790	psi	220,000
Heat Deflection-Temperature	D-648	F(C)	
@ 264 psi			176(80)
@ 66 psi			275(135)
TOUGHNESS:			
Notched Izod Impact Strength 73F(23C)	D-256	Ft-lb/in of notch	2.5(13.6)
Conductive:			
EC90 Plus:			
STRENGTH:			
Tensile Strength 73F(23C)	D-638	psi	5,300
Tensile Elongation 73F(23C)	D-638	%	30
STIFFNESS AND CREEP:			
Flexural Strength 73F(23C)	D-790	psi	7,600
Flexural Modulus 73F(23C)	D-790	psi	250,000
Heat Deflection-Temperature	D-648	F(C)	
@ 264 psi			217(103)
TOUGHNESS:			
Notched Izod Impact Strength 73F(23C)	D-256	Ft-lb/in of notch	1.0(5.5)

CELANESE ENGINEERING RESINS, INC.: CELCON Acetal Copolymer (Continued):

Grade:

PROPERTY	ASTM	UNITS	
ELECTROPLATE EP90:			
Unplated:			
STRENGTH:			
Tensile Strength 73F(23C)	D-638	psi	8,200
Tensile Elongation 73F(23C)	D-638	%	70
STIFFNESS AND CREEP:			
Flexural Strength 73F(23C)	D-790	psi	12,000
Flexural Modulus 73F(23C)	D-790	psi	350,000
Heat Deflection-Temperature @ 264 psi	D-648	F(C)	217(103)
TOUGHNESS:			
Notched Izod Impact Strength @ 264 psi	D-256	Ft-lb/in of notch	1.1(6.0)
Plated:			
STRENGTH:			
Tensile Strength 73F(23C)	D-638	psi	16,000
Tensile Elongation 73F(23C)	D-638	%	925,000
STIFFNESS AND CREEP:			
Flexural Strength 73F(23C)	D-790	psi	16,000
Flexural Modulus 73F(23C)	D-790	psi	925,000
Heat Deflection-Temperature @ 264 psi	D-648	F(C)	324(162)
TOUGHNESS:			
Notched Izod Impact Strength 73F(23C)	D-256	Ft-lb/in of notch	4.8(26.1)

CELANESE ENGINEERING RESINS, INC.: CELCON Acetal Copolymer (Continued):

Grade:

PROPERTY	ASTM	UNITS	
U/V:			
UV90:			
STRENGTH:			
Tensile Strength 73F(23C)	D-638	psi	8,800
Tensile Elongation 73F(23C)	D-638	%	60
STIFFNESS AND CREEP:			
Flexural Strength 73F(23C)	D-790	psi	13,000
Flexural Modulus 73F(23C)	D-790	psi	375,000
Heat Deflection-Temperature	D-648	F(C)	
@ 264 psi			230(110)
@ 66 psi			316(158)
TOUGHNESS:			
Notched Izod Impact Strength 73F(23C)	D-256	Ft-lb/in of notch	1.3(7.0)
Shear Strength 73F(23C)	D-732	psi	7,700(540)
Weather Resistant:			
WR90:			
STRENGTH:			
Tensile Strength 73F(23C)	D-638	psi	8,800
Tensile Elongation 73F(23C)	D-638	%	40
STIFFNESS AND CREEP:			
Flexural Strength 73F(23C)	D-790	psi	13,000
Flexural Modulus 73F(23C)	D-790	psi	375,000
Heat Deflection-Temperature	D-648	F(C)	
@ 264 psi			230(110)
@ 66 psi			316(158)
TOUGHNESS:			
Notched Izod Impact Strength 73F(23C)	D-256	Ft-lb/in of notch	1.2(6.5)
Shear Strength 73F(23C)	D-732	psi	7,700

CELANESE ENGINEERING RESINS, INC.: CELCON Acetal Copolymer (Continued):

Grade:

PROPERTY	ASTM	UNITS	
Appliance Grade M90-34:			
STRENGTH:			
Tensile Strength 73F(23C)	D-638	psi	8,800
Tensile Elongation 73F(23C)	D-638	%	60
STIFFNESS AND CREEP:			
Flexural Strength 73F(23C)	D-790	psi	13,000
Flexural Modulus 73F(23C)	D-790	psi	375,000
Heat Deflection-Temperature	D-648	F(C)	
@ 264 psi			230(110)
@ 66 psi			316(158)
TOUGHNESS:			
Notched Izod Impact Strength 73F(23C)	D-256	Ft-lb/in of notch	1.3(7.0)
Shear Strength 73F(23C)	D-732	psi	7,700(540)

CELANESE ENGINEERING RESINS, INC.: DUREL Polyarylate:

DUREL polyarylate is a new, tough, clear engineering thermoplastic. Advanced technology builds flame resistance and weatherability right into the thermoplastic molecule. And with 340F heat resistance DUREL polyarylate handles applications previously reserved for other more exotic materials.

Select clear or opaque versions of the DUREL polyarylate family. Either delivers a unique combination of properties that allow the design versatility and durability you need.

The combination of properties that makes DUREL polyarylate truly versatile is:

* Optical clarity
* High heat-distortion temperature
* Inherent flame resistance
* Inherent UV stability
* High impact strength
* Excellent surface hardness
* Excellent flexural recovery
* Good electrical properties

OUTSTANDING PROPERTY HIGHLIGHTS

* Transparency	>85%
* Toughness	Outstanding Impact Strength under a Variety of End-Use Conditions
* High Heat Deflection Temperature	340F
* Inherent Flame Retardance	Inherently V-O
* Excellent Weatherability	Outstanding Outdoor Property Retention
* Rockwell Hardness	M87-M95

TYPICAL POTENTIAL APPLICATIONS

* Automotive * Headlamps * Reflector * High Warning Taillight
* Support Structures/Housings(Low Warpage) * Body Parts and Panels * Glazing * Electrical Connectors
* High Intensity Discharge Lighting
* Microwave Cookware
* Specialty Film and Sheet * Glazing/Solar Panels
* Aircraft * Transparencies * Glazing * Interior Panels * Lighting and Lenses
* Appliance
* Electrical/Electronics
* Businesss Machines

CELANESE ENGINEERING RESINS, INC.: DUREL Polyarylate (Continued):

Properties of DUREL 400:
General Purpose, Unfilled

GENERAL	ASTM	ENGLISH
Specific Gravity	D792	1.21
Specific Volume	----	22.9 cu in/lb.
Mold Shrinkage	D955	0.006 in/in
Water Absorption, 73F (23C) 24 Hrs	D570	0.26%
Equilibrium		0.50%
Transmittance, 1/8"	D1003	84-88%
Haze, 1/8"	D1003	1-2%
Refractive Index	-----	1.61
THERMAL		
Heat Deflection Temperature		
66 psi (Annealed)	D648	355F
264 psi (Unannealed)	D648	316F
264 psi (Annealed)	D648	340F
Coefficient of Thermal Expansion	D696	3.50 X 10 -5 in/in/F
Thermal Conductivity	----	1.25 BTU/hr/ sq ft/F/in
Flammability UL94, 1/8"	----	V-0
Limiting Oxygen Index	D2863	38%
MECHANICAL		
Tensile Strength @ Yield	D638	10,000 psi
Elongation at Break	D638	50%
Flexural Strength	D790	14,500 psi
Flexural Modulus	D790	3.3 X 10 5 psi
Compressive Strength	D695	12,000 psi
Izod Impact Strength (1/8"-3.2 mm)	D256	(ft-lb/in)
Notched 73F(23C)		5.5
Notched -40F(-40C)		4.0
Unnotched 73F(23C)		No break
Unnotched -40F(-40C)		No break
Direct Impact (1/8" plaque, 3/4" dart)		>40 ft-lb
Taber Abrasion Resistance	D1044	
mg loss/1000 cycles		4
Rockwell Hardness	D785	M87
Rockwell Hardness (Annealed)	D785	M95

CELANESE ENGINEERING RESINS, INC.: FORTRON Polyphenylene Sulfide:

FORTRON polyphenylene sulfide molding resin is the newest addition to the broad range of engineering plastics offered by Celanese Engineering Resins, Inc. It is a high performance crystalline material that offers design engineers and manufacturers an excellent combination of good processability and superior thermal and mechanical properties. Applications that require high temperature stability, chemical resistance, inherent flame resistance and good moldability will find FORTRON PPS the material of choice.

FORTRON PPS can be ordered in a variety of formulations, including glass-fiber reinforced and mineral/glass combinations. The unfilled resin is available in low or medium viscosity grades.

ADVANTAGES:

FORTRON is an improved PPS resin. It has a lower level of ionic impurities, which means it is a purer, more consistent product. Its natural color is a light tan, making it easy to color. Product consistency in viscosity is assured through tight quality control, which means you will have less waste and greater productivity. FORTRON PPS is tougher than conventional PPS which means it is less brittle in thin wall sections.

APPLICATIONS:

FORTRON PPS's good physical properties combined with its excellent thermal and electrical characteristics make it an ideal material for connectors, switches, coil bobbins and vapor phase soldering applications.

Uses in the automotive industry include engine compartment components, fuel line systems, and other applications where solvent resistance and high continuous use temperatures are required.

The mechanical handling industry will also find FORTRON PPS the right choice for high temperature and corrosive environments. Applications include caps, gears, fittings, and other industrial parts.

AGENCY STATUS:

FORTRON PPS has received a rating of 94 V-O from the Underwriters Laboratories on the 40% glass-reinforced grade down to 1/64 of an inch

FORTRON PPS has also been granted permission to test under Military Specification M-24519.

CELANESE ENGINEERING RESINS, INC.: FORTRON Polyphenylene Sulfide(Continued):

TYPICAL PROPERTIES:

Product Name:

0214P1:

PHYSICAL:	ASTM	UNITS	GRADE
Density	D-1505	9/cc	1.35
Water Absorption 24 Hr. Imms.	D-570	%	0.01
MECHANICAL @ 73F.:			
Tensile Strength	D-638	psi	12,400
Elongation	D-638	%	3.0
Flex Strength	D-790	psi	21,300
Flex Modulus	D-790	10 6 psi	0.54
Rockwell Hardness	D-785	M-Scale	93
Izod Impact Strength 73F	D-256		
Notched	-----	ft.lb./in	0.5
Unnotched	-----	ft.lb./in	11.6
THERMAL:			
Heat Deflection Temp.	D-648		
@ 264 psi	-----	F	221
ELECTRICAL:			
Dielectric Strength	D-149	V/mil	520
Dielectric Constant @ 73F	D-150		
1 KHz	-----	-----	3.0
100 KHz	-----	-----	3.0
Dissipation Factor @ 73F	D-150		
1 KHz	-----	-----	0.0013
100 KHz	-----	-----	0.0010
Volume Resistivity	D-257	ohm-cm	1 x 10 16
Arc Resistance	D-495	secs.	124

FORTRON Grade Characteristics

0214P1: Medium viscosity, unfilled pellets.

CELANESE ENGINEERING RESINS, INC. FORTRON Polyphenylene Sulfide(Continued):

TYPICAL PROPERTIES:

Product Name:

0205P4:

PHYSICAL:	ASTM	UNITS	GRADE
Density	D-1505	g/cc	1.35
Water Absorption, 24 hrs. Imms.	D-570	%	0.01
MECHANICAL @ 73F:			
Tensile Strength	D-638	psi	12,800
Elongation	D-638	%	4.1
Flex Strength	D-790	psi	21,200
Flex Modulus	D-790	10^6 psi	0.55
Rockwell Hardness	D-785	M-Scale	93
Izod Impact Strength 73F	D-256		
Notched	-----	ft.lb./in	0.5
Unnotched	-----	ft.lb./in	10.7
THERMAL:			
Heat Deflection Temp.	D-648		
@ 264 psi		F	221
ELECTRICAL:			
Dielectric Strength	D-149	V/mil	410
Dielectric Constant @ 73F	D-150		
1 KHz	-----	-----	2.9
100 KHz	-----	-----	2.9
Dissipation Factor @ 73F	D-150		
1 KHz	-----	-----	0.0013
100 KHz	-----	-----	0.0011
Volume Resistivity	D-257	ohm-cm	1×10^{16}
Arc Resistance	D-495	secs.	120

FORTRON GRADE CHARACTERISTICS:

0205P4 Low viscosity, unfilled pellets

CELANESE ENGINEERING RESINS, INC.: FORTRON Polyphenylene Sulfide(Continued):

TYPICAL PROPERTIES:

Product Name:

1140B4:

PHYSICAL:	ASTM	UNITS	GRADE
Density	D-1505	9/cc	1.64
Water Absorption 24 Hr. Imms.	D-570	%	0.015
Mold Shrinkage Flow Direct.	-----	mils/in	1-3
Transverse Direct	-----	mils/in	3-6
MECHANICAL @ 73F:			
Tensile Strength	D-638	psi	23,000
Elongation	D-638	%	1.5
Flex Strength	D-790	psi	32,000
Flex Modulus	D-790	10 6 psi	2.0
Rockwell Hardness	D-785	M-Scale	98
Izod Impact Strength 73F	D-256	-----	
Notched	-----	ft.lb./in	1.4
Unnotched	-----	ft.lb./in	7.1
THERMAL:			
Heat Deflection Temp.	D-648		
@ 264 PSI	-----	F	502
ELECTRICAL			
Dielectric Strength	D-149	V/mil	500
Dielectric Constant @ 73F	D-150	-----	
1 KHz	-----	-----	3.5
100 KHz	-----	-----	3.5
Dissipation Factor @ 73F	D-150	-----	
1 KHz	-----	-----	0.0013
100 KHz	-----	-----	0.0011
Volume Resistivity	D-257	ohm-cm	1 x 10 16
Arc Resistance	D-495	secs.	134

FORTRAN GRADE CHARACTERISTICS

1140B4 Low viscosity, 40% fiberglass

CELANESE ENGINEERING RESINS, INC.: FORTRON Polyphenylene Sulfide (Continued):

TYPICAL PROPERTIES:

Product Name:

1140B1:

PHYSICAL:	ASTM	UNITS	GRADE
Density	D-1505	g/cc	1.64
Water Absorption 24 Hr. Imms.	D-570	%	0.015
Mold Shrinkage Flow Direct	----	mils/in	1-3
Transverse Direct	----	mils/in	3-6
MECHANICAL @ 73F:			
Tensile Strength	D-638	psi	24,000
Elongation	D-638	%	1.5
Flex Strength	D-790	psi	35,000
Flex Modulus	D-790	10^6 psi	1.9
Rockwell Hardness	D-785	M Scale	100
Izod Impact Strength 73F	D-256	-----	
Notched	-----	ft.lb./in	1.6
Unnotched	-----	ft.lb./in	10.7
THERMAL:			
Heat Deflection Temp.	D-648	-----	
@ 264 psi	-----	F	505
Flammability			
Underwriters Lab Rating @ 1/64"	-----	(UL94)	V-O
ELECTRICAL:			
Dielectric Strength	D-149	V/mil	400
Dielectric Constant @ 73F	D-150	-----	
1 KHz	-----	-----	3.6
100 KHz	-----	-----	3.5
Dissipation Factor @ 73F	D-150	-----	
1 KHz	-----	-----	0.0020
100 KHz	-----	-----	0.0010
Volume Resistivity	D-257	ohm-cm	1×10^{16}
Arc Resistance	D-495	secs.	136

FORTRON GRADE CHARACTERISTICS

1140B1 Medium viscosity, 40% fiberglass, improved toughness.

CELANESE ENGINEERING RESINS, INC.: FORTRON Polyphenylene Sulfide(Continued):

TYPICAL PROPERTIES:

Product Name:

6165B4:

PHYSICAL:	ASTM	UNITS	GRADE
Density	D-1505	9/cc	2.08
MECHANICAL @ 73F:			
Tensile Strength	D-638	psi	16,500
Elongation	D-638	%	1.0
Flex Strength	D-790	psi	26,300
Flex Modulus	D-790	10 6 psi	2.4
Rockwell Hardness	D-785	M-Scale	98
Izod Impact Strength 73F	D-256	-----	
Notched	-----	ft.lb./in	0.8
Unnotched	-----	ft.lb./in	4.6
THERMAL:			
Heat Deflection Temp.	D-648	-----	
@ 264 psi	-----	F	498
ELECTRICAL			
Dielectric Strength	D-149	V/mil	420
Dielectric Constant @ 73F	D-150	-----	
1 KHz	-----	-----	4.5
100 KHz	-----	-----	4.5
Dissipation Factor @ 73F	D-150	-----	
1 KHz	-----	-----	0.0010
100 KHz	-----	-----	0.0009
Volume Resistivity	D-257	ohm-cm	1 x 10 16
Arc Resistance	D-485	secs.	182

FORTRON Grade Characteristics

6165B4 Low viscosity, 65% CaCO3/fiberglass.

CELANESE ENGINEERING RESINS, INC.: FORTRON Polyphenylene Sulfide(Continued):

TYPICAL PROPERTIES:

<u>Product Name:</u>

6265B4:

PHYSICAL:	ASTM	UNITS	GRADE
Density	D-1505	9/cc	2.08
Water Absorption 24 Hr. Imms.	D-570	%	0.01
MECHANICAL @ 73F:			
Tensile Strength	D-638	psi	14,100
Elongation	D-638	%	0.7
Flex Strength	D-790	psi	21,900
Flex Modulus	D-790	10 6 psi	2.8
Rockwell Hardness	D-785	M-Scale	85
Izod Impact Strength 73F	D-256	-----	
Notched	-----	ft.lb./in	0.7
Unnotched	-----	ft.lb./in	2.8
THERMAL:			
Heat Deflection Temp.	D-648	-----	
@ 264 psi	-----	F	514
ELECTRICAL:			
Dielectric Strength	D-149	V/mil	450
Dielectric Constant @ 73F	D-150	-----	
1 KHz	-----	-----	4.0
100 KHz	-----	-----	3.9
Dissipation Factor @ 73F	D-150	-----	
1 KHz	-----	-----	0.0060
100 KHz	-----	-----	0.0040
Volume Resistivity	D-257	ohm-cm	1x10 16
Arc Resistance	D-495	secs.	195

FORTRON Grade Chacteristics

6265B4 Low viscosity, 65% talc/fiberglass.

CELANESE ENGINEERING RESINS, INC.: Glass Reinforced Nylon:

Celanese Nylon 1500 and 1503, were first introduced to the market in 1966. From their initial introduction, these resins had oustanding properties. Since that time, the 1400 and 1600 series have been introduced.

The 1500 series are injection molding grade 6/6 nylons with 33% glass fiber reinforcement. The 1600 series contains 43% glass fiber reinforcement. Both series are available in heat stabilized grades (1503, 1603). Other glass fiber reinforced grades are available in general purpose and heat stabilized formulations. These include products with 13% glass fiber reinforcement. (1400 series)

Glass fiber reinforced nylons combine the inherently good properties of nylon with the high strength characteristics of glass fibers. The presence of specially treated glass fibers increases the properties of nylon significantly because of the efficient chemical bonding of the glass fibers to the nylon. Some of the factors which have made possible these improvements in Celanese Nylon 1400, 1500 and 1600 are:

(1) The base nylon resin used for these products has a high and stable molecular weight which is retained throughout the process of molding into finished parts, and any subsequent regrinding and reworking.

(2) Glass fibers of proper diameter and rigidity have been carefully chosen and coated with an effective coupling agent to bond them to the nylon. The length to diameter ratio of the glass fibers was selected to yield the best overall balance in strength, rigidity, and moldability.

(3) The coated glass fibers are well dispersed throughout the nylon resin, and are bonded to the resin under optimum conditions of temperature and pressure.

CELANESE ENGINEERING RESINS, INC.: Glass Reinforced Nylon (Continued):

Typical properties of Celanese Nylon glass fiber reinforced grades 1400 series, 1500 series and 1600 series:

English Units - Nylon 1400

PROPERTY	ASTM	UNITS	DAM
Specific Gravity	D792	-----	1.23
Density	D792	lbs/cu in	0.0444
Tensile Strength	D638	psi	17,000
Elongation at Break	D638	%	2-4
Flexural Strength	D790	psi	27,500
Flexural Modulus	D790	psi	750,000
Izod Impact (Notched)	D256	ft-lb/in	0.95
Tensile Impact	D1822	ft-lb/ sq in	35
Melting Point	D789	F	495
Temperature (.46 MPa) 264 psi	D648	F	470
Coefficient of Linear Thermal Expansion	D696	in/in/F	1.7x10 5
Dielectric Strength	D149	v/mil	400
Hardness	D785	Rockwell	M95
Flammability	UL94	-----	HB

CELANESE ENGINEERING RESINS,INC.: Glass Reinforced Nylon (Continued):

Typical properties of Celanese Nylon glass fiber reinforced grades 1400 series, 1500 series and 1600 series:

English Units - Nylon 1500

PROPERTY	ASTM	UNITS	DAM
Specific Gravity	D792	-----	1.38
Density	D792	lbs/cu in	0.0498
Water Absorption			
(24 hrs., 73F/23C)	D570	%	1.0
Tensile Strength	D638	psi	28,000
Elongation at Break	D638	%	2-4
Shear Strength	D732	psi	13,000
Compressive Strength	D695	psi	28,000
Flexural Strength	D790	psi	40,000
Flexural Modulus	D790	psi	1.2×10^{6}
Izod Impact (Notched)	D256	ft-lb/in	2.2
Tensile Impact	D1822	ft-lb/ sq in	70
Melting Point	D789	F	495
Temperature (.46 MPa)			
66 psi	D648	F	495
264 psi	D648	F	485
Specific Heat (1.82 MPa)	----	----	0.3
Thermal Conductivity	----	BTU/m/ sq ft/in	1.5
Coefficient of Linear Thermal Expansion	D696	in/in/F	1.6×10^{5}
Dielectric Strength	D149	v/mil	400
Dielectric Constant			
@ 60 Hertz	D150	-----	4.1
@ 1000 Hertz	D150	-----	4.0
Volume Resistivity	D257	ohm-om	10^{14}
Power Factor			
@ 60 Hertz	D150	-----	0.006
@ 1000 Hertz	D150	-----	0.009
Hardness	D785	Rockwell	M100
Taber Abrasion	D1044	mg/1000 cycles	20
Dynamic			
vs. itself	----	-----	0.23
vs. steel	----	-----	0.15
vs. aluminum	----	-----	0.12
vs. brass	----	-----	0.11
Flammability	D635	in/min	1.1
	UL94	-----	HB

CELANESE ENGINEERING RESINS, INC.: Glass Reinforced Nylon (Continued):

Typical properties of Celanese Nylon glass fiber reinforced grades 1400 series, 1500 series and 1600 series:

English Units - Nylon 1600

PROPERTY	ASTM	UNITS	DAM
Specific Gravity	D792	-----	1.47
Density	D792	lbs/cu in	0.0531
Water Absorption			
(24 hrs., 73F/23C)	D570	%	0.9
Tensile Strength	D638	psi	32,000
Elongation at Break	D638	%	2-4
Shear Strength	D732	psi	13,000
Flexural Strength	D790	psi	49,000
Flexural Modulus	D790	psi	1.6x10 6
Izod Impact (Notched)	D256	ft-lb/in	2.6
Melting Point	D789	F	495
Temperature (.46 MPa)			
66psi	D648	F	495
264psi	D648	F	488
Coefficient of Linear			
Thermal Expansion	D696	in/in/F	1.4X10 5
Dielectric Strength	D149	v/mil	400
Dielectric Constant			
@ 100 Hertz	D150	----	4.4
@ 1000 Hertz	D150	----	4.4
Volume Resistivity	D257	ohm-om	10 14
Power Factor			
@ 100 Hertz	D150	----	0.009
@ 1000 Hertz	D150	----	0.012
Hardness	D785	Rockwell	M100
Dynamic			
vs. itself	----	----	0.23
vs. steel	----	----	0.15
vs. aluminum	----	----	0.12
vs. brass	----	----	0.11
Flammability	UL94	----	HB

CELANESE ENGINEERING RESINS, INC.: NYLON Resins:

CELANESE Nylon 1000 Series resins are general purpose and heat-stabilized injection molding grade 6/6 type polyamides, produced in a fully continuous process from a diamine salt and a dibasic acid. Of the most common nylon resin types currently available (nylon 6/6, nylon 6, nylon 6/12), nylon 6/6 resins have the highest melting point, the highest tensile strength, the highest flexural modulus, and the best creep resistance. The very high strength levels of these resins, combined with their unusual toughness, abrasion resistance, and resistance to chemical attack, make them ideal materials in many demanding engineering applications. They have found widespread use in the replacement of metals, providing reduced weight, corrosion resistance, self-lubricating qualities and colorability at substantially lower costs, frequently performing tasks beyond the scope of metals.

Celanese Nylon 1300 and 1310 are unique nylon 6/6 molding materials. Celanese Nylon 1300 Series Resins offer fast cycling in injection molding machines with toughness not normally characteristic of fast cycling nylon 6/6 resins. They are slightly stronger and stiffer than general purpose nylon 6/6 resins in the "dry, as molded condition."

Celanese Nylon 1300 Series resins are specifically designed to provide molders and end users with a unique combination of properties not available before in nylon resins. More precisely, these properties include:

* Faster Cycling
* Improved Processability (Flow and Mold release)
* Toughness
* Increased Surface Hardness

Celanese Nylon 1300 Series resins are produces by a patented controlled crystallization process which provides a unique product. Significantly shorter cycles and toughness properties characteristic of unmodified general purpose nylon 6/6 are realized. As a result, an additional water immersion operation is usually unnecessary.

Celanese Nylon 1300 Series crystallization controlled resins offer improved processing over Celanese Nylon 1000 Series resins, while offering almost equivalnet physical properties.

In most cases, Celanese Nylon 1300 Series resins are interchangeble with general purpose nylon 6/6 resins.

CELANESE ENGINEERING RESINS, INC.: NYLON Resins(Continued):

Designation:

Celanese Nylon 1000 Series:

1000-1:
Description: General purpose for injection molding and extrusion.

1000-2:
Description: Surface-lubricated form of 1000-1.
Applications: Mechanical parts, gears, bearings, hardware, etc.

1000-4:
Description: General purpose resin for easy mold release.

1003-1:
Description: Heat-stabilized form of 1000-1.
Applications: High temperature resistant applications, e.g. under-the-hood automotive parts.

1003-2:
Description: Surface-lubricated form of 1003-1.

Celanese Nylon 1200 Series:

1200-1:
Description: High viscosity extrusion grade.
Applications: Extruded profiles, film, sheet and injection molded parts with improved toughness.

Celanese Nylon 1300 Series:

1310-1:
Description: Controlled crystallization resin for fast cycling in injection molding with general purpose product toughness.

1310-2:
Description: Surface-lubricated form 1310-1 for easier processing and mold release.

1310-4:
Description: Controlled crystallization resin for easy mold release.

Applications: Parts similar to 1000 series applications which can benefit from higher production rates due to faster cycles, easy mold release, etc.

CELANESE ENGINEERING RESINS, INC.: NYLON Resins(Continued):

Designation:

Celanese Nylon 1500 Series:

Designation:

1500-1:
Description: Glass fiber (33%) reinforced nylon 6/6, for use in screw injection molding machines.

1500-2:
Description: Surface-lubricated form of 1500-1, for use in screw or ram type injection molding machines.

Applications: Injection molded parts with excellent impact, high tensile strength and greatest stiffness.

1503-1:
Description: Heat-stabilized form of 1500-1, for use in screw injection molding machines.

1503-2:
Description: Surface-lubricated form of 1503-1. Most widely used as general purpose glass fiber reinforced grade in both screw and ram type injection molding machines.

Applications: Strong, highest strength parts requiring temperature resistance up to 350F (177C).

Celanese Nylon 1600 Series:

1600-1:
Description: Glass fiber 40% reinforced nylon 6/6 injection molding for use in screw machines.
Applications: Injection molded parts with exceptionally high impact, high tensile strength and superior stiffness.

1600-2:
Description: Surface-lubricated form of 1600-1.

1603-1:
Description: Heat stabilized form of 1600-1 for use in screw injection molding machines.

1603-2:
Description: Surface lubricated form of 1603-1.

Applications: Parts requiring extra high strength and rigidity and temperature resistance up to 350F(177C).

CELANESE ENGINEERING RESINS, INC.: NYLON Resins(Continued):

Characterization of properties of Celanese Nylon compounds (typical) - English units

General Purpose Grades: Nylon 1000/Nylon 1003:

PROPERTY	ASTM	UNITS	AS MOLDED
Moisture Content	----	%	<0.3
STRENGTH: at 73F:			
Tensile Strength	D638	psi	12,000
Elongation (break)	D638	%	40-80
Yield Strength	D638	psi	12,000
Shear Strength	D732	psi	9,500
RIGIDITY: at 73F:			
Flexural Strength	D790	psi	17,000
Flexural Modulus	D790	psi	420,000
IMPACT & TOUGHNESS: at 73F			
Izod Impact: 1/8" bar	D256	ft.lb./ in.notch	1.0
THERMAL			
Melting Point	D789	F	495
Specific Heat	Cenco	----	0.4
Coefficient of Expansion	Fitch D696	in./in./ F	5 x 10 -5
Heat Distortion @ 66 psi	D648	F	470
Heat Distortion @ 264 psi	D648	F	170
MISCELLANEOUS			
Density	D792	lbs/cu in	0.0412
Water Absorption (24 hr)	D570	%	1.5
Hardness, Rockwell	D785	----	M79-M86
ELECTRICAL: at 73			
Dielectric Strength			
Short Time	D149	v/mil	600
Step by Step	D149	v/mil	290
Dielectric Constant			
100 Hz	D150	-----	5.5
1000 Hz	D150	-----	5.3
Power Factor			
60 Hz	D150	-----	0.020
1000 Hz	D150	-----	0.035

CELANESE ENGINEERING RESINS, INC.: NYLON Resins (Continued):

Characterization of properties of Celanese Nylon compounds (typical) - English units

Nylon 1300:

Controlled Crystallization Grades (Rapid Cycle):

PROPERTY	ASTM	UNITS	AS MOLDED
Moisture Content	----	%	<0.3
STRENGTH: at 73F			
Tensile Strength	D638	psi	14,000
Elongation (break)	D638	%	25-40
Yield Strength	D638	psi	14,000
Shear Strength	D732	psi	9,400
RIGIDITY: at 73F			
Flexural Strength	D790	psi	19,000
Flexural Modulus	D790	psi	500,000
IMPACT & TOUGHNESS at 73F			
Izod Impact: 1/8" bar	D256	ft.lb./ in.notch	0.8-0.9
THERMAL			
Melting Point	D789	F	495
Specific Heat	Cenco	----	0.4
Heat Distortion @ 66 psi	D648	F	465
Heat Distortion @ 264 psi	D648	F	174
MISCELLANEOUS			
Density	D792	lbs/cu in	0.0412
Water Absorption (24 hr)	D570	%	1.4
Hardness, Rockwell	D785	----	M86-M89
ELECTRICAL: at 73			
Dielectric Strength			
Short Time	D149	v/mil	600
Dielectric Constant			
100 Hz	D150	----	5.4
1000 Hz	D150	----	5.2
Power Factor			
100 Hz	D150	----	.037
1000 Hz	D150	----	.027

CELANESE ENGINEERING RESINS, INC.: NYLON Resins(Continued):

Characterization of properties of Celanese Nylon compounds (typical) - English units

Nylon 1310:

Controlled Crystallization Grades (Rapid Cycle):

PROPERTY	ASTM	UNITS	AS MOLDED
Moisture Content	----	%	<0.3
STRENGTH: at 73F			
Tensile Strength	D638	psi	13,600
Elongation (break)	D638	%	25-40
Yield Strength	D638	psi	13,600
Shear Strength	D732	psi	9,400
RIGIDITY: at 73F			
Flexural Strength	D790	psi	17,900
Flexural Modulus	D790	psi	470,000
IMPACT & TOUGHNESS: at 73F			
Izod Impact: 1/8" bar	D256	ft.lb./ in. notch	0.9
THERMAL			
Melting Point	D789	F	495
Specific Heat	Cenco	----	0.4
Heat Distortion @ 66 psi	D648	F	425
Heat Distortion @ 264 psi	D648	F	171
MISCELLANEOUS			
Density	D792	lbs/cu in	0.0412
Water Absorption (24 hr)	D570	%	1.4
Hardness, Rockwell	D785	----	M84-M88
ELECTRICAL: at 73			
Dielectric Strength			
Short Time	D149	v/mil	600
Dielectric Constant			
100 Hz	D150	----	5.4
1100 Hz	D150	----	5.2
Power Factor			
100 Hz	D150	----	.037
1000 Hz	D150	----	.027

CELANESE ENGINEERING RESINS, INC.: VECTRA High-Performance Resins Electrical:

The VECTRA family of self-reinforcing liquid crystal polymer (LCP) materials delever a performance profile ideally suited to the needs of the expanding electrical/electronics marketplace.

With VECTRA resins, you get an unmatched array of benefits:

* Inherent flammability resistance, with no unstable or corrosive additives.
* Superior high temperature performance.
* Exceptional dimensional stability and rigidity.
* Outstanding insulation and arc resistance properties.
* Remarkable chemical resistance, even under stress and at elevated temperatures.
* Excellent impact strength.
* Very low warpage and very low mold shrinkage.
* Fast molding cycles, superb processability in intricate, thin-walled components.

VECTRA Resin Selector:

A130:-General purpose electrical grade with outstanding balance of properties, suitable for many surface mount devices.

C130:-Best performer at elevated temperatures and mechanical stresses common to surface mount applications.

A410:-Most rigid VECTRA variant, allowing thin wall designs with extra measure of stiffness. Best arc resistance properties.

A515:-Highest flow grade, with excellent impact and best ductility, ideal for applications where toughness is critical.

Suggested Application Areas:

Surface Mount Components	HDI Connectors
IC Encapsulation	D-Subminiature Connectors
DIP/SIP Sockets	DIP Switches
ZIF Connectors/Sockets	Solenoid Bobbins
Plastic-Leaded Chip Carriers	Relay Cases and Bobbins
Pin Grid Arrays	Backplane Connectors
Burn-in/Test Sockets	Card Edge Connectors
SOIC Sockets	Post Headers/Receptacles

CELANESE ENGINEERING RESINS, INC.: VECTRA High-Performance Resins Electrical(Continued):

PROPERTIES Of VECTRA E/E Primary Grades:

Physical/Mechanical/Thermal Properties:

PROPERTY	ASTM	VECTRA A130
Specific Gravity	D792	1.57
Moisture Absorption, % @ equil.	D570	0.01-0.02
Mold Shrinkage, in/in(mm/mm)	D955	0-0.001
Tensile Strength, psi (MPa)	D638	29,000(202)
Tensile Modulus, psi (GPa)	D638	2.5x10^6(17.5)
Elongation @ Break, %	D638	2.2
Flex Strength, psi (MPa)	D790	36,100(252)
Flex Modulus, psi (GPa)	D790	2.1x10^6(14.7)
Notched Izod Impact, ft-lb/inch(J/m)	D256	2.5(133)
Heat Defl. Temp., F @ 264 psi, (C@ 1.8 MPa)	D648	445(229)

Electrical/Flammability Properties:

Dielectric Constant, 1KHz/1MHz/1GHz	D150	4.1/3.7/3.7
Dissipation Factor, 1 GHz	D150	0.006
Arc Resistance, seconds	D495	137
Dielectric Strength, volts/mil (KV/mm)	D149	1100(43)
Volume Resistivity, ohm-cm	D257	10^{15}
CTI, volts	D3638	175
Flammability Rating	UL 94	V-O
Limiting Oxygen Index, %	D2863	37

CELANESE ENGINEERING RESINS, INC.: VECTRA High-Performance Resins Electrical (Continued):

Properties of VECTRA E/E Primary Grades:

Physical/Mechanical/Thermal Properties:

PROPERTY	ASTM	VECTRA C130
Specific Gravity	D792	1.57
Moisture Absorption, % @ equil.	D570	0.01-0.02
Mold Shrinkage, in/in (mm/mm)	D955	0-0.001
Tensile Strength, psi (MPa)	D638	24,000(168)
Tensile Modulus, psi (GPa)	D638	2.3×10^{6}(16.1)
Elongation @ Break, %	D638	1.9
Flex Strength, psi (MPa)	D790	31,800(223)
Flex Modulus, psi (GPa)	D790	2.1×10^{6}(14.7)
Notched Izod Impact, ft-lb/inch (J/m)	D256	2.2(117)
Heat Defl. Temp., F @ 264 psi, (C @ 1.8 MPa)	D648	465(241)

ELECTRICAL/FLAMMABILITY PROPERTIES:

Dielectric Constant, 1 KHz/1MHz/1Ghz	D150	4.1/3.7/3.8
Dissipation Factor, 1 GHz	D150	0.004
Arc Resistance, seconds	D495	137
Dielectric Strength, volts/mil (KV/mm)	D149	1100(43)
Volume Resistivity, ohm-cm	D257	10^{15}
CTI, volts	D3638	175
Flammability Rating	UL 94	V-O
Limiting Oxygen Index, %	D2863	37

CELANESE ENGINEERING RESINS, INC.: VECTRA High-Performance Resins Electrical (Continued):

Properties of VECTRA E/E Primary Grades:

Physical/Mechanical/Thermal Properties:

PROPERTY	ASTM	VECTRA A410
Specific Gravity	D792	1.78
Moisture Absorption, % @ equil.	D570	0.01-0.02
Mold Shrinkage, in/in(mm/mm)	D955	0-0.001
Tensile Strength, psi (MPa)	D638	24,400(170)
Tensile Modulus, psi (GPa)	D638	2.9x10 6(20.3)
Elongation @ Break, %	D638	1.8
Flex Strength, psi (MPa)	D790	32,000(224)
Flex Modulus, psi (GPa)	D790	2.7x10 6(18.9)
Notched Izod Impact, ft-lb/inch(J/m)	D256	1.7(90)

Electrical/Flammability Properties:

Dielectric Constant, 1KHz/1MHz/1GHz	D150	4.5/4.2/4.1
Dissipation Factor, 1 GHz	D150	0.004
Arc Resistance, seconds	D495	163
Dielectric Strength, volts/mil (KV/mm)	D149	850(33)
Volume Resistivity, ohm-cm	D257	10 15
CTI, volts	D3638	175
Flammability Rating	UL 94	V-O
Limiting Oxygen Index, %	D2863	37

CELANESE ENGINEERING RESINS, INC.: VECTRA High-Performance Resins Electrical (Continued):

Properties of VECTRA E/E Primary Grades:

Physical/Mechanical/Thermal Properties:

PROPERTY	ASTM	VECTRA A515
Specific Gravity	D792	1.48
Moisture Absorption, % @ equil.	D570	0.02-0.03
Mold Shrinkage, in/in (mm/mm)	D955	0.001-0.004
Tensile Strength, psi (MPa)	D638	26,300(184)
Tensile Modulus, psi (GPa)	D638	$1.7\text{x}10^{6}$(11.9)
Elongation @ Break, %	D638	4.4
Flex Strength, psi (MPa)	D790	23,100(162)
Flex Modulus, psi (GPa)	D790	$1.5\text{x}10^{6}$(10.5)
Notched Izod Impact, ft-lb/inch(J/m)	D256	7.0(371)
Heat Defl. Temp., F @ 264 psi, (C @ 1.8 MPa)	D648	370(188)

Electrical/Flammability Properties:

PROPERTY	ASTM	VECTRA A515
Dielectric Constant, 1KHz/1MHz/1GHz	D150	4.2/3.8/3.8
Dissipation Factor, 1 GHz	D150	0.004
Arc Resistance, seconds	D495	137
Dielectric Strength, volts/mil (KV/mm)	D149	890(35)
Volume Resistivity, ohm-cm	D257	10^{15}
CTI, volts	D3638	175
Flammability Rating	UL 94	V-O
Limiting Oxygen Index, %	D2863	33

COLORITE PLASTICS CO.: UNICHEM Compounds: Clear Non-Toxic Compounds for Intravenous and Blood Tubing:

A series of vinyl compounds have been developed specifically for use in the critical applications of intravenous and blood tubing. These materials are produced to the highest standards of performance to comply with all USP and FDA requirements.

Complete traceability and lot identification of all components is maintained.

Products available are as follows:

Compound:

7811A-02:

Hardness: 78
Specific Gravity: 1.22

7877-02:

Hardness: 78
Specific Gravity: 1.24

8011A-20:

Hardness: 80
Specific Gravity: 1.23

COLORITE PLASTICS CO.: UNICHEM Compounds: Clear Non-Toxic Flexible Compounds:

Product Number:

Extrusion: 4011:
Molding: 4012:

Type: Flexible Clear
Specific Gravity: 1.16
Hardness: 40 (A)

Extrusion: 4511:
Molding: 4512:

Type: Flexible Clear
Specific Gravity: 1.16
Hardness: 45 (A)

Extrusion: 5011:
Molding: 5012:

Type: Flexible Clear
Specific Gravity: 1.16
Hardness: 50 (A)

Extrusion: 5511:
Molding: 5512:

Type: Flexible Clear
Specific Gravity: 1.17
Hardness: 55 (A)

Extrusion: 6011:
Molding: 6012:

Type: Flexible Clear
Specific Gravity: 1.18
Hardness: 60 (A)

Extrusion: 6511:
Molding: 6512:

Type: Flexible Clear
Specific Gravity: 1.19
Hardness: 65 (A)

COLORITE PLASTICS CO.: UNICHEM Compounds: Clear Non-Toxic Flexible Compounds (Continued):

Product Number:

Extrusion: 7011:
Molding: 7012:

Type: Flexible Clear
Specific Gravity: 1.21
Hardness: 70 (A)

Extrusion: 7511:
Molding: 7512:

Type: Flexible Clear
Specific Gravity: 1.22
Hardness: 75 (A)

Extrusion: 7811:
Molding: 7812:

Type: Flexible Clear
Specific Gravity: 1.23
Hardness: 78 (A)

Extrusion: 8011:
Molding: 8012:

Type: Flexible Clear
Specific Gravity: 1.23
Hardness: 80 (A)

Extrusion: 8511:
Molding: 8512:

Type: Flexible Clear
Specific Gravity: 1.25
Hardness: 85 (A)

Extrusion: 9011:
Molding: 9012:

Type: Flexible Clear
Specific Gravity: 1.27
Hardness: 90 (A)

COLORITE PLASTICS CO.: UNICHEM Compounds: Clear Non-Toxic Flexible Compounds(Continued):

Product Number:

Extrusion: 9511:
Molding: 9512:

Type: Flexible Clear
Specific Gravity: 1.28
Hardness: 95 (A)

Extrusion: 9611:
Molding: 9612:

Type: Semirigid Clear
Specific Gravity: 1.29
Hardness: 52 (D)

Extrusion: 9811:
Molding: 9812:

Type: Semirigid Clear
Specific Gravity: 1.31
Hardness: 66 (D)

Extrusion: 9904:
Molding: 9912:

Type: Semirigid Clear
Specific Gravity: 1.33
Hardness: 69 (D)

COLORITE PLASTICS CO.: UNICHEM Compounds: Electrical Compounds:

Building Wires & Appliance Wiring Materials:

<u>Compound:</u>

4060:

Application: T, TW, UF, NMC, 60C & 80C - AWM
Specific Gravity: 1.33
Duro.: 80

4060A:

Application: T, TW, UF, NMC, 60C & 80C - AWM
Specific Gravity: 1.38
Duro.: 87

4075:

Application: 90C THW/MTW/AWM/Thinwall
Specific Gravity: 1.36
Duro.: 88

4075H:

Application: 90C THW/MTW/AWM
Specific Gravity: 1.39
Duro.: 90

40-105:

Application: 105C THHN, THWN/AWN
Specific Gravity: 1.36
Duro.: 97

4114:

Application: 105C THHN, THWN/90C - 1000 volt, AWM
Specific Gravity: 1.36
Duro.: 95

COLORITE PLASTICS CO.: UNICHEM Compounds: Electrical Compounds (Continued):

Building Wires & Appliance Wiring Materials(Continued):

Compound:

4195:

Application: T 60C Semi Rigid
Specific Gravity: 1.34
Duro.: 95

5433:

Application: T, TW, 60C oil, 80C AWM, UF, NMC, 90C - 1000 UV
Specific Gravity: 1.37
Duro.: 89

COLORITE PLASTICS CO.: UNICHEM Compounds: Electrical Compounds (Continued):

GENERAL INSULATION TYPES:

Compound:

4160:

Application: 60C SPT, SJT, SVT, TF, XT, CXT
Specific Gravity: 1.40
Duro.: 80

4160C:

Application: 60C SPT
Specific Gravity: 1.23
Duro.: 80

4160LT/FR-1:

Application: 80C SJT, SVT, Jackets
Specific Gravity: 1.39
Duro.: 80

4260-75:

Application: Cord Sets
Specific Gravity: 1.34
Duro.: 75

4260-80:

Application: Cord Sets
Specific Gravity: 1.38
Duro.: 80

4260-90:

Application: Cord Sets
Specific Gravity: 1.42
Duro.: 90

COLORITE PLASTICS CO.: UNICHEM Compounds: Electrical Compounds (Continued):

GENERAL INSULATION TYPES(Continued):

Compound:

8005:

Application: 105C SPT, SJT, SVT, Jackets
Specific Gravity: 1.37
Duro.: 80

8005C:

Application: 105C SPT
Specific Gravity: 1.23
Duro.: 80

8006MG:

Application: 105C Plug Compound
Specific Gravity: 1.38
Duro.: 74

8060:

Application: 60C SPT, SJT, SVT, TF, XT, CXT
Specific Gravity: 1.45
Duro.: 80

8560:

Application: 60C SPT, SJT, SVT, TF, XT, CXT
Specific Gravity: 1.45
Duro.: 85

8580:

Application: 80C SJT, SVT, Jackets
Specific Gravity: 1.45
Duro.: 85

COLORITE PLASTICS CO.: UNICHEM Compounds: Electrical Compounds (Continued):

GENERAL INSULATION TYPES(Continued):

<u>Compound:</u>

8085:

Application: 105C Non-Migrating & Oil Resistant
Specific Gravity: 1.38
Duro.: 80

8190:

Application: 90 Jackets, SJT, SPT, CSA Primary Wire
Specific Gravity: 1.38
Duro.: 80

8205:

Application: 105C SPT
Specific Gravity: 1.46
Duro.: 83

8705:

Application: 105C Style 1015 AWM-32
Specific Gravity: 1.36
Duro.: 87

9005:

Application: 105C Mil-W-16878, low odor
Specific Gravity: 1.36
Duro.: 90

9805:

Application: Semi-Rigid Business Machine, Mil W-16878
Specific Gravity: 1.38
Duro.: 97

COLORITE PLASTICS CO.: UNICHEM Compounds: Extraction Resistant Low Temperature Compounds:

Colorite Plastics is developing a series of modified vinyl compounds designed to provide essentially complete resistance to extraction by solvents along with exceptionally good low temperature impact properties. The unusual performance of this class of compounds suggests its use in applications previously unsuited for flexible PVC.

Examples of such uses are as follows:

A) Gasoline Tubing

B) Paintable Flexible Vinyl

C) Various Automotive Applications Requiring Solvent Resistance and Extreme Low Temperature Impact Strength.

This series of materials will be available in a range of hardnesses as well as with varying degrees of gloss.

The first product available for evaluation is UNICHEM 7555-80G which is a high gloss product.

Typical Properties of UNICHEM 7555-80G are as Follows:

Specific Gravity	1.14
Shore A Durometer	75
Tensile Strength	2270 psi
100% Modulus	880 psi
Brittle Temperature (ASTM D-746)	-65C

Extraction Resistance:

14 Days Gasoline	1.70% Weight Loss
14 Days Kerosene	0.47% Weight Loss
14 Days Vegetable Oil	0.46% Weight Loss

COLORITE PLASTICS CO.: UNICHEM Compounds: General Purpose Extrusion and Molding Compounds:

General Purpose Crystal Clears:

<u>Compound</u> <u>#:</u>

7000:

Specific Gravity	1.20
Hardness	70

7500:

Specific Gravity	1.22
Hardness	75

8000:

Specific Gravity	1.23
Hardness	80

8500:

Specific Gravity	1.25
Hardness	85

9000:

Specific Gravity	1.27
Hardness	90

9500:

Specific Gravity	1.29
Hardness	95

COLORITE PLASTICS CO.: UNICHEM Compounds: General Purpose Extrusion and Molding Compounds (Continued):

General Purpose Molding & Profile Extrusion Compounds:

Compound #:

4543:

Specific Gravity: 1.31
Hardness: 45

4875:

Specific Gravity: 1.38
Hardness: 48

5543:

Specific Gravity: 1.25
Hardness: 55

5545:

Specific Gravity: 1.32
Hardness: 55

5790:

Specific Gravity: 1.34
Hardness: 58

6575:

Specific Gravity: 1.37
Hardness: 66

7075:

Specific Gravity: 1.39
Hardness: 70

COLORITE PLASTICS CO.: UNICHEM Compounds: General Purpose Extrusion and Molding Compounds (Continued):

General Purpose Molding & Profile Extrusion Compounds (Continued):

Compound #:

7575:

Specific Gravity: 1.42
Hardness: 75

8090C:

Specific Gravity: 1.37
Hardness: 80

8804:

Specific Gravity: 1.41
Hardness: 88

9614:

Specific Gravity: 1.40
Hardness: 95

COLORITE PLASTICS CO.: UNICHEM Compounds: General Purpose Extrusion and Molding Compounds(Continued):

Low Gravity General Purpose Compounds:

Compound #:

7033:

Specific Gravity: 1.28
Hardness: 70

8033:

Specific Gravity: 1.32
Hardness: 80

8533:

Specific Gravity: 1.32
Hardness: 85

9033:

Specific Gravity: 1.32
Hardness: 90

COLORITE PLASTICS CO.: UNICHEM Compounds: General Purpose Extrusion and Molding Compounds(Continued):

Solvent Resistant/Non-migrating Compounds:

<u>Compound #:</u>

6742:

Specific Gravity: 1.25
Hardness: 67

6743:

Specific Gravity: 1.23
Hardness: 67

7342:

Specific Gravity: 1.23
Hardness: 73

7580:

Specific Gravity: 1.24
Hardness: 75

8080:

Specific Gravity: 1.16
Hardness: 80

8083:

Specific Gravity: 1.30
Hardness: 80

COLORITE PLASTICS CO.: UNICHEM Compounds: Multi Chem PVC Alloy Compounds:

VINYL-URETHANE

* Greater resiliency for peristaltic action
* Improved elasticity
* Low temperature properties
* Lower extractability
* Molding and extrusion

TYPICAL PHYSICAL PROPERTIES:

Product Name:

6014:

Shore A Hardness	ASTM D-2240	60
Specific Gravity	ASTM D-792	1.15
Tensile Strength, psi	ASTM D-638	1640
Elongation, %	ASTM D-638	540
100% Modulus, psi	ASTM D-638	400

7014:

Shore A Hardness	ASTM D-2240	70
Specific Gravity	ASTM D-792	1.19
Tensile Strength, psi	ASTM D-638	2040
Elongation, %	ASTM D-638	600
100% Modulus, psi	ASTM D-638	625

COLORITE PLASTICS CO.: UNICHEM Compounds: Tubing Compounds:

<u>Compound:</u>

105:

Application: 105C Flame Resistant, Mil C-631D Grade C
Specific Gravity: 1.31
Duro.: 87

5376:

Application: UL 224, Mil-C-631D, Grade C, Clear
Specific Gravity: 1.25
Duro.: 87

7444:

Application: 7444 Class I (extreme low temperature)
Specific Gravity: 1.18
Duro.: 70

7444D:

Application: 7444 Class II
Specific Gravity: 1.20
Duro.: 70

6476:

Application: Mil-I-631D, Grade A & B, Class I, Clear
Specific Gravity: 1.23
Duro.: 73

COLORITE PLASTICS CO.: UNICHEM Compounds: UNICHEM Radiation Sterilizable Compounds:

A series of vinyl compounds identified as the UNICHEM "G" Series has been developed to resist discoloration and degradation when subjected to beta and gamma ray sterilization.

All components used in manufacture of the UNICHEM "G" Series are sanctioned by the US Food & Drug Administration for food packaging applications under Title 21 of the Code of Federal Regulations.

To comply with the exacting requirements of the medical industry, complete traceability and lot identification of all components are maintained.

COMPOUND:

1280G-015:

Specific Gravity: 1.26
Hardness: 78

4011, 4012G-015:

Specific Gravity: 1.16
Hardness: 40

5011, 5012G-015:

Specific Gravity: 1.16
Hardness: 50

6011, 6012G-015:

Specific Gravity: 1.18
Hardness: 60

6511, 6512G-015:

Specific Gravity: 1.19
Hardness: 64

6811, 6812G-015:

Specific Gravity: 1.20
Hardness: 68

COLORITE PLASTICS CO.: UNICHEM Compounds: UNICHEM Radiation Sterilizable Compounds(Continued):

Compound:

7011, 7012G-015:

Specific Gravity: 1.21
Hardness: 70

7511, 7512G-015:

Specific Gravity: 1.22
Hardness: 75

7815, 7812G-015:

Specific Gravity: 1.23
Hardness: 78

8011, 8012G-015:

Specific Gravity: 1.23
Hardness: 80

8311, 8312G-015:

Specific Gravity: 1.24
Hardness: 83

8511, 8512G-015:

Specific Gravity: 1.25
Hardness: 86

9011, 9012G-015:

Specific Gravity: 1.27
Hardness: 90

9512G-015:

Specific Gravity: 1.28
Hardness: 95

COLORITE PLASTICS CO.: UNICHEM Compounds: UNICHEM Radiation Sterilizable Compounds (Continued):

Compound:

9612G-015:

Specific Gravity: 1.29
Hardness: 97

9811G-015:

Specific Gravity: 1.31
Hardness: 66 (D)

9812G-015:

Specific Gravity: 1.31
Hardness: 66 (D)

1116G-015:

Specific Gravity: 1.33
Hardness: 74 (D)

COLORITE PLASTICS CO.: UNICHEM Compounds: UNICHEM RADIOPAQUE Compounds:

UNICHEM RADIOPAQUE Compounds are available in a variety of hardnesses and opacifier types. These compounds are designed for use as radiopaque striping or to be extruded into catheter tubing.

Radiopacity is obtained through the incorporation of either Barium Sulphate or Bismuth Subcarbonate in a non-toxic vinyl base. Complete traceability of all components is maintained to comply with GMP requirements for medical applications.

<u>Compound No.:</u>

7011:

Hardness: 72 (A)
Specific Gravity: 1.48
Opacifier Type: Barium Sulphate

8011:

Hardness: 84 (A)
Specific Gravity: 1.60
Opacifier Type: Barium Sulphate

9011:

Hardness: 91 (A)
Specific Gravity: 1.50
Opacifier Type: Barium Sulphate

9812-10:

Hardness: 60 (D)
Specific Gravity: 1.42
Opacifier Type: Barium Sulphate

9812-15:

Hardness: 45 (D)
Specific Gravity: 1.45
Opacifier Type: Barium Sulphate

9855:

Hardness: 57 (D)
Specific Gravity: 1.84
Opacifier Type: Barium Sulphate

81-403:

Hardness: 38-44 (D) Specific Gravity: 1.54
Opacifier Type: Bismuth Subcarbonate

COLORITE PLASTICS CO.: UNICHEM Compounds: UNICHEM E 11 Non-Toxic Film:

UNIFILM E 11 is an extruded non-toxic PVC film suitable for use in urinary drainage, vaginal irrigation, wound drainage and ostomy devices.

UNIFILM E 11 is produced from raw materials sanctioned by the Food and Drug Administration under Title 21 of the Code of Federal Regulations. For medical use, complete traceability and lot identification of all components is maintained.

UNIFILM E 11 film are available in clears or in desired colors suitably embossed to provide resistance to blocking. In addition, UNIFILM E 11 is provided in any desired slit width.

Specific Gravity (clear)	1.26
Shrinkage (4 hrs at 100F)	less than 5%

COLORITE PLASTICS CO.: UNICHEM Compounds: UNICHEM 66 Series Non-Phthalate Compounds:

The UNICHEM 66 series is a family of non-toxic PVC compounds developed specifically for those medical device, food and infant care items where the use of a non-phthalate plasticizer is desirable.

The UNICHEM 66 series is produced from raw materials that are sanctioned for use under Title XXI of the Code of Federal Regulations. This family of compounds is designed to be molded and extruded under similar processing conditions to compounds plasticized with DEHP. UNICHEM 66 compounds are available in a complete range of hardnesses from very soft to semi-rigid. They are also available in ethylene and radiation sterilizable grades.

As with all UNICHEM non-toxic compounds, complete traceability and lot identification of all raw materials and finished products is maintained.

Typical Physical Properties:

Product Number:

7066:

	ASTM	
Shore Hardness	ASTM D-2240	70A
Specific Gravity	ASTM D-792	1.21
Tensile Strength (psi)	ASTM D-638	2380
Elongation (%)	ASTM D-638	425
100% Modulus (psi)	ASTM D-638	880
Brittle Point (C)	ASTM D-746	-42
Flexible Strength (psi)	ASTM D-790	x
Melt Flow (grams/10 min)	ASTM D-1237	x

7866:

Shore Hardness	ASTM D-2240	78A
Specific Gravity	ASTM D-792	1.23
Tensile Strength (psi)	ASTM D-638	2530
Elongation (%)	ASTM D-638	395
100% Modulus (psi)	ASTM D-638	1170
Brittle Point (C)	ASTM D-746	-26
Flexible Strength (psi)	ASTM D-790	x
Melt Flow (grams/10 min)	ASTM D-1237	x

COLORITE PLASTICS CO.: UNICHEM Compounds: UNICHEM 66 Series Non-Phthalate Compounds (Continued):

Typical Physical Properties:

Product Number:

8566:

	ASTM	
Shore Hardness	ASTM D-2240	85A
Specific Gravity	ASTM D-792	1.25
Tensile Strength (psi)	ASTM D-638	2639
Elongation (%)	ASTM D-638	369
100% Modulus (psi)	ASTM D-638	1453
Brittle Point (C)	ASTM D-638	-25
Flexible Strength (psi)	ASTM D-790	x
Melt Flow (grams/10 min)	ASTM D-1237	x

1066:

Shore Hardness	ASTM D-2240	70D
Specific Gravity	ASTM D-792	1.32
Tensile Strength (psi)	ASTM D-638	4970
Elongation (%)	ASTM D-638	12
100% Modulus (psi)	ASTM D-638	x
Brittle Point (C)	ASTM D-746	x
Flexible Strength (psi)	ASTM D-790	7720
Melt Flow (grams/10 min)	ASTM D-1237	5

COLORITE PLASTICS CO.: UNICHEM Compounds: UNICHEM 77 Series Non-Phthalate Compound:

The UNICHEM 77 Series is a family of compounds designed for those medical device producers that require the use of a non-phthalate plasticizer that has superior resistance to extraction by blood lipids and high fat emulsions.

This family of compounds is designed to be molded and extruded under similar processing conditions to compounds plasticized with DHEP. UNICHEM 77 compounds are available in a range of hardnesses from flexible to semi-rigid. They are also available in ethylene oxide and radiation sterilizable grades.

As with all UNICHEM non-toxic compounds, traceability and lot identification of all raw materials and finished products is maintained.

PHYSICAL PROPERTIES:

<u>Product Name:</u>

7077:	ASTM	
Shore Hardness (+-2)	D-2240	70A
Specific Gravity (+-0.01)	D-792	1.20
100% Modulus (psi)	D-638	985
Tensile Strength (psi)	D-638	2150
Elongation (%)	D-638	425
Brittle Point (C)	D-746	-37

7777:		
Shore Hardness (+-2)	D-2240	78A
Specific Gravity (+-0.01)	D-792	1.22
100% Modulus (psi)	D-638	1375
Tensile Strength (psi)	D-638	2570
Elongation (%)	D-638	400
Brittle Point (C)	D-746	-33

8577:		
Shore Hardness (+-2)	D-2240	85A
Specific Gravity (+-0.01)	D-792	1.24
100% Modulus (psi)	D-638	1600
Tensile Strength (psi)	D-638	2695
Elongation (%)	D-638	355
Brittle Point (C)	D-746	-30

COLORITE PLASTICS CO.: UNICHEM Compounds: UNICHEM 77 Series Non-Phthalate Compounds(Continued):

Product Name:

1077:

	ASTM	
Shore Hardness (+-2)	D-2240	70D
Specific Gravity (+-0.01)	D-792	1.31
100% Modulus (psi)	D-638	----
Tensile Strength (psi)	D-638	4765
Elongation (%)	D-638	14
Brittle Point (C)	D-746	----

CONTINENTAL POLYMERS, INC.: ACRYCAL MP Acrylic Pellets:

ACRYCAL MP Acrylic Pellets offer notable advantages to processor and end user alike:

CLARITY AND BRILLIANCE - Crystal clarity, over 92% light transmission with little haze.

DIMENSIONAL STABILITY - Rugged, durable, unplasticized.

IMPACT RESISTANCE - Resilient, trouble-free service in most applications.

WEATHER RESISTANCE - Time proven, unique in ability to stand environmental rigors.

COLOR CHOICES - Available in most transparent, translucent and opaque colors.

CHEMICAL RESISTANCE - Unaffected by alkalies, non-oxidizing acids and salt water.

ELECTRICAL INSULATION - Excellent insulating properties, good resistance to arcing.

TASTELESS AND ODORLESS - Compatible with human tissue.

GOOD PROCESSABILITY - Wide choice of uniform melt flow grades with excellent stability.

EASE OF DECORATION - Painting, metallizing, hot-stamping easy to perform.

MACHINABILITY - Drilling, threading, tapping, routing, machinining operations easy to perform.

EASE OF ASSEMBLY - Cementing, ultrasonic welding easily provide strong bonds.

SPECIAL UV GRADES - Standard ultraviolet transmitting and absorbing grades as well as special formulations readily available.

IMPACT MODIFIED FORMULATIONS - Impact modified formulations up to 12 times stronger than general purpose grades are commercially available.

CONTINENTAL POLYMERS, INC.: ACRYCAL MP Acrylic Pellets (Continued):

General Purpose ACRYCAL MP Formulations	Characteristics	Melt Flow Rate*
CP-41	Injection Molding Grade Pellets only. Lowest heat resistance and maximum flow.	29.0
CP-51	Injection Molding and Extrusion Grade Pellets. Medium heat resistance, medium flow.	16.5
CP-61	Injection Molding and Extrusion Grade Pellets. High heat resistance, medium flow.	7.0
CP-71	Injection Molding Grade Pellets only. High heat resistance, medium flow	5.0
CP-75	Injection Molding Grade Pellets only. Maximum heat resistance, medium flow.	4.1
CP-80	Extrusion Grade Pellets only. Unlubricated version of CP-81. Highest molecular weight, high heat resistance.	2.5
CP-81	Extrusion Grade Pellets. Highest molecular weight, high heat resistance. Specially designed for extrusion applications.	2.5
CP-82	Injection Molding Grade Pellets. Highest molecular weight, maximum heat resistance. Specially designed for molding applications requiring superior mold release properties.	2.5

* ASTM D-1238 Condition 1g/10 min.

COSMIC PLASTICS, INC.: Diallyl Phthalate Compounds for Military and Commercial Applications:

DIALLYL ORTHO PHTHALATE RESIN (Regular):

Compound Number:

Long Glass Fiber:

D69:
Forms: Flakes
Standard (S) or Flame Retardant (F): S
Specific Gravity: 1.62

D62:
Forms: Flakes
Standard (S) or Flame Retardant (F): F
Specific Gravity: 1.76

D63:
Forms: Flakes
Standard (S) or Flame Retardant (F): S
Specific Gravity: 1.77

Short Glass Fiber:
D33:
Forms: Granules
Standard (S) or Flame Retardant (F): S
Specific Gravity: 1.70

D72:
Forms: Granules
Standard (S) or Flame Retardant(F): F
Specific Gravity: 1.76

Dacron:

224:
Forms: Flakes
Standard (S) or Flame Retardant (F): F
Specific Gravity: 1.42

Orlon:

306:
Forms: Flakes
Standard (S) or Flame Retardant (F): S
Specific Gravity: 1.39

Mineral:

D45(D44):
Forms: Powder
Standard (S) or Flame Retardant (F): S
Specific Gravity: 1.67

COSMIC PLASTICS, INC.: Diallyl Phthalate Compounds for Military and Commercial Applications(Continued):

Diallyl Meta Phthalate Resin:
(High Heat Resistant)

Long Glass Fiber:

K66:
Forms: Flakes
Standard (S) or Flame Retardant (F): F
Specific Gravity: 1.78

K61:
Forms: Flakes
Standard (S) or Flame Retardant (F): S
Specific Gravity: 1.60

Short Glass Fiber:

K77:
Forms: Granules
Standard (S) or Flame Retardant (F): F
Specific Gravity: 1.78

K31:
Forms: Granules
Standard (S) or Flame Retardant (F): S
Specific Gravity: 1.70

Mineral:

K43:
Forms: Powder
Standard (S) or Flame Retardant (F): S
Specific Gravity: 1.67

COSMIC PLASTICS, INC.: Epoxy Molding Materials:

There are two basic types of formulations:

Group I compounds are Electrical Grades which possess:
1. High Heat Resistance
2. Good Mechanical Strength
3. Excellent Electrical Properties

Group II compounds are encapsulation Grades which possess:
1. High Resistance to Thermal Shock
2. Good Dimensional Stability

Other outstanding features of COSMIC Epoxy Compounds include:
1. Low weight loss at elevated temperature
2. Low outgassing under vacuum
3. Retention of good electrical properties at high temperature
4. Unaffected by fungus and most chemicals

Electrical Grade:
Glass-Fiber Filled:

E484:
Specific Gravity: 1.84
Bulk factor: 2.2
Molding temperature: F: 280-350

E486:
Specific Gravity: 1.86
Bulk factor: 2.2
Molding temperature: F: 280-350

E487:
Specific Gravity: 1.85
Bulk factor: 2.2
Molding temperature: F: 280-350

Encapsulation Grade:
Mineral-Glass Filled:

E4930:
Specific Gravity: 1.85
Bulk factor: 2.0
Molding temperature: F: 250-350

E4920:
Specific Gravity: 1.74
Bulk factor: 2.0
Molding temperature: F: 250-350

E4905:
Specific Gravity: 1.90
Bulk factor: 2.0
Molding temperature: F: 250-350

CUSTOM RESINS: High Impact Strength Engineering Plastics:

702 is a plasticized, low viscosity polymer for good flow in molding and extrusion equipment. 702 has exceptional impact strength associated with engineering plastics. 702 has very good flexibility for tubing applications. It is available in black, natural, and heat stabilized versions. Processing conditions should result in melt temperatures in the 450F to 480F range.

711 is a medium viscosity molding and extrusion grade. 711 retains the exceptional notched Izod impact strength of 702 while featuring the increased stiffness often desired in molded articles. 711 is available in natural and heat stabilized versions. Processing conditions may range from 470F to 500F.

721 is a medium viscosity polymer for fast, easy molding. Impact strength is good right out of the mold, without post conditioning. Stiffness is in the range of nylon 6. Molding temperatures should be in the 450F to 500F range.

733 is a high stiffness, high strength molding and extrusion grade with the best impact strength of any of the resins. 733 is available in natural and heat stabilized versions. Processing should be done at 500F to 550F.

PROPERTY:

702:

Specific Gravity (D792): 1.08
Tensile Strength (D638): 9000 psi
Yield Point (D638): 5500 psi

711:

Specific Gravity (D792): 1.08
Tensile Strength (D638): 9500 psi
Yield Point (D638): 6500 psi

721:

Specific Gravity (D792): 1.10
Tensile Strength (D638): 9000 psi
Yield Point (D638): 9000 psi

733:

Specific Gravity (D792): 1.08
Tensile Strength (D638): 10,500 psi
Yield Point (D638): 7000 psi

CUSTOM RESINS: Nylon 401:

Possesses the excellent overall balance of properties generally attributed to polyamides. Recommended for general purpose nylon molding applications in both thin and heavy sections. 401 has high flow characteristics at minimal processing temperatures.

401 is recommended for use in all general purpose applications requiring toughness, heat resistance, hardness, chemical resistance and abrasion properties normally associated with nylon.

401 is produced in continuous polymerization to provide outstanding lot-to-lot uniformity for high qualirty and highly efficient molding. All Custom Resins is vacuum dried to 0.1% maximum moisture content.

Excellent melt thermal stability prevents degradation or "yellowing" and a sharp drop in viscosity during processing. Physicals and appearance are retained when regrind is safely blended.

401, like all Custom Resins Nylon, exhibits certain unique qualities compared with the older 6/6 nylons. Most noticeable, perhaps, is an improved processability due to a broader range of processing temperature, an improved flow at lower temperatures and pressures and a reduced tendency to sink and void. In addition, the lessened tendency toward nozzle drool and stinging is a definite advantage in obtaining consistent cycles and operating efficiency.

PROPERTY:

Specific Gravity (D792-66): 1.14
Apparent Density (D1895-65T): 0.39 oz/cu.in.
Relative Viscosity (D789-66): 30
Water Absorption (D570-63): 1.8%
Tensile Strength (-40F) (D638-64T): 14,000 psi
Tensile Strength (73F): 11,500 psi
Tensile Strength (170F): 7,000 psi
Yield Point (-40F) (D638-64T): 14,000 psi
Yield Point (73F): 11,500 psi
Yield Point (170F): 7,000 psi
Elongation (-40F) (D638-64T): 25%
Elongation (73F): 40%
Elongation (170F): 150%
Impact Strength--Izod (-40F) (D256-56): 0.3 ft.lb./in. notch
Impact Strength--Izod (73F): 0.8
Impact Strength--Izod (170F): 2.0
Shear Strength (D732-46): 8,400 psi
Compression Strength (D695-63T): 9,700 psi
Flexural Strength (D790-66): 15,400 psi
Flexural Stiffness (D747-63): 250,000 psi

CUSTOM RESINS: Nylon 402:

High residual monomer resin. Offers outstanding toughness and flexibility. 402 exhibits excellent flow properties at temperature and pressure settings suitable for a wide variety of general pourpose molding applications.

Custom Resins Nylon 402 is a high monomer resin designed to offer injection molders fast, economical molding cycles on injection machines. Because of having excellent flow and crystalline properties, 402 is recommended for applications of difficult product shapes and multicavity moldings.

The crystallinity of 402 can be controlled to a remarkable extent by processing conditions. A relatively cold mold temperature in the range of 30F to 75F will promote a low degree of crystallinity, thus producing excellent clarity in molded parts having thin cross-sections. Higher mold temperatures will increase the crystallinity and produce parts having increased opaqueness.

The outstanding features of 402 are its excellent toughness and flexibility--possessing a marked superiority to 6/6 nylon in respect to tensile impact strength at equilibium moisture conditions. It exhibits excellent flow at lower temperatures and pressures which makes it suitable for a wide variety of general purpose applications.

PROPERTY:

Specific Gravity (D792-66): 1.13
Apparaent Density (D1895-65T): 0.39 oz./cu.in.
Relative Viscosity (D789-66): 25
Water Absorption (D570-63): 3.5%
Tensile Strength (-40F) (D638-64T): 12,000 psi
Tensile Strength (73F): 7,500 psi
Tensile Strength (170F): 8,700 psi
Yield Point (-40F) (D638-64T): 12,000 psi
Yield Point (73F): 6,700 psi
Yield Point (170F): 4,400 psi
Elongation (-40F) (D638-64T): 35%
Elongation (73F): 300%
Elongation (170F): 350%
Impact Strength--Izod (-40F) (D256-56): 0.2 ft.lb./in. notch
Impact Strength--Izod (73F): 2.4 ft.lb./in. notch
Impact Strenght--Izod (170F): 7.0 ft.lb./in. notch
Shear Strength (D732-46): 8,800 psi
Compression Strength (D695-63T): 7,100 psi
Flexural Strength (D790-66): 8,000 psi
Flexural Stiffness (D747-63): 90,000 psi
Hardness--Rockwell R Scale (D785-63): 101
Heat Distortion 264 psi (D648-56): 133F
Heat Distortion 66 psi: 320F
Melt Point (D789-66): 430F

CUSTOM RESINS: Nylon 404:

Heat stabilized injection molding resin provideing for protection of the finished product against thermal degradation in in high temperature applications. Well suited for both thin and heavy section molding and for multi-cavity production. Has excellent overall balance of properties.

Nylon normally cannot endure constant exposure to temperatures in excess of 175F without suffering some loss of its original physical properties. For this reason, Custom Resins Nylon 404 is offered to designers and processors who need a reliable engineering material to be used in temperatures above this limit. The unique Heats stabilizer system for nylon allows 404 to be used continuously at an elevated temperature of 250F for a period of three months without any significant change in properties.

Since heat aging is cumulative, heat stabilized nylon parts can provide yeare of service when expsoure to elevated temperatures is intermittent. As a result, for intermittent service, temperatures of 300F are not uncommon.

The superior heat resistant and chemical properties of 404 have been proved in several highly critical applications in service over a number of years. 404 is especially recommended for use in automotive fuel filters, coil bobbins, encapsulation, fasteners and other critical engineering applications requiring a uniform and premium quality molding resin. The superior heat stability properties of 404 are achieved at no loss of fine processing characteristics and physical properties.

PROPERTY:

Specific Gravity (D792-66): 1.14
Apparent Density (D1895-65T): 0.39 oz./cu.in.
Relative Viscosity (D789-66): 30
Water Absorption (D570-63): 1.8%
Tensile Strength (-40F) (D638-64T): 14,000 psi
Tensile Strength (73F): 12,400 psi
Tensile Strength (170F): 7,000 psi
Yield Point (-40F) (D638-64T): 14,000 psi
Yield Point (73F): 12,400 psi
Yield Point (170F): 7,000 psi
Elongation (-40F) (D638-64T): 25%
Elongation (73F): 40%
Elongation (170F): 150%
Impact Strength--Izod (-40F) (D256-56): 0.3 ft.lb./in. notch
Impact Strength--Izod (73F): 0.8 ft.lb./in. notch
Impact Strength--Izod (170F): 2.0 ft.lb./in. notch
Shear Strength (D732-46): 8,400 psi
Compression Strength (D695-63T): 9,700 psi
Flexural Strength (D790-66): 15,400 psi
Flexural Stiffness (D747-63): 250,000 psi
Hardness--Rockwell R Scale (D785-63): 118

CUSTOM RESINS: Nylon 406:

New general purpose molding nylon with crystallization initiator to provide fast molding cycles through rapid set-up from the melt. 406 also has excellent balance of properties usually associated with nylon. Unique "X-TAL" nucleation system has brought 406 quick and wide acceptance among fast-cycling molders.

High crystallinity is generally considered desirable as this indicates a regular and compact structure of the molecular chains forming the resin. Moldings made of highly crystalline resin will have a higher yield strength, a higher modulus of elesticity and will be harder at the surface than moldings made of less crystalline (more amorphous) resin.

The crystallization initiator incorporated in 406 promotes faster crystallinity growth in the molding to an improvement of approximately 10% over that found in parts molded from non-nucleated Custom Resins Nylon.

Seconds of cycle time are saved because the 406-melt sets up faster in the mold than conventional nylon melts. This permits earlier ejection of the part form the mold. Moreover, as crystallinity increases, mold shrinkage increases. This often results in outstanding mold release characteristics for parts of varying sections and intricate design.

Shrinkage of 406 moldings is almost comparable to that of nylon 6/6 moldings. For this reason, 406 can generally be used in molds design for nylon 6/6.

406 parts have better dimensional stability in the presence of moisture than parts molded of conventional nylon. At moisture equilibrium (2.5-3.0% water), dimensional shrinkage of 406 parts is approxiamtely 50% less.

Three years of extensive field use have provides the data for the information given above and have underscored the value of the contibution "X-TAL" resins have made to plastics processing.

PROPERTY:

Specific Gravity (D792-66): 1.13
Water Absorption (D570-63): 1.5%
Tensile Strength (D638-64T): 10,900 psi
Yield Point (D638-64T): 10,900 psi
Elongation (D638-64T): 25%
Impact Strength--Izod (D256-56): 0.80 ft./lb./in. Notch
Flexural Stiffness (D747-63): 265,800 psi
Heat Distortion 264 psi (D648-56): 160F
Heat Distortion 66 psi: 360F
Flammability (D635-56T): Self Extinguishing

CUSTOM RESINS: Nylon 409:

New heat stable, nucleated molding resin. Offers rapid cycling through crystallization initiators. Unique heat stabilizer system protects finished parts against high temperature degradation. Well suited for both thin ann heavy section molding.

The crystallization initiators incorporated in 409 promote faster growth in the molding to an improvement of approximately 10% over that found in parts molded from non-nucleated Custom Resins Nylon.

High crystallinity is generally considered desirable as this indicates a regular and compact structure of the molecular chains forming the resin. Moldings made of highly crystalline resin will have a higher yield strength, a higher modulus of elasticity and will be harder at the surface than moldings made of less crystalline (more amorphous) resin. Moldings have high quality white appearance.

Seconds of cycle time are saved because 409-melt sets up faster in hot molds than conventional nylon melts. Faster set up allows earlier ejection and a corresponding reduction in cycle time.

Nylon normally cannot endure constant exposure to temperatures in excess of 175F without suffering some loss of its original physical properties. For this reason, Custom Resins Nylon 409 is offered to designers and processors who need a reliable engineering material to be used in temperatures above this limit. The unique heat stabilizer system for nylon allows 409 to be used continuously at an elevated temperature of 250F for a period of three months without any significant change in properties.

Since heat aging is cumulative, heat stabilized nylon parts can provide years of srevice when expsoure to elevated temperatures is intermittent. As a result, for intermittent service, temperatures of 300F are not uncommon.

409 is listed by Underwriter's Laboratories, Inc.

External pellet lubrication is indicated by suffix letter A.

PROPERTY:

Specific Gravity (D792-66): 1.13
Water Absorption (D570-63): 1.5%
Tensile Strength (D638-64T): 10,900 psi
Yield Point (D638-64T): 10,900 psi
Elongation (D638-64T): 25%
Impact Strength--Izod (D256-56): 0.80 ft./lb./in. Notch
Flexural Stiffness (D747-63): 265,800 psi

CUSTOM RESINS: Nylon 451:

For use in molding applications requiring maximum durability and toughness. High viscosity in combination with outstanding processability. Exceptional fatigue and wear resistance insure long product life to gears, bearings, cams, fasteners and other load bearing parts. Chemically resists oils and grease. Effective in noise control activities.

A balance of excellent properties make 451 particularly adaptable for load-bearing and contact-surface applications.

Tensile strength is outstanding when compared to other thermoplastic materials. When both strength and weight are considered, 451 exceeds even the performance of many metals. For instance, 0.6% carbon cold rolled steel has only four times the yield strength of 451, but approximately seven times the weight.

451 has twice the impact strength of 6/6 nylon at equilibrium moisture condition, yet processes with a notable absence of nozzle drool and stringing. The average Izod value of 451 at 170F is a whopping 15.0.

As 451 has a lower modulus of elasticity when compared to most metals, it will effectively reduce or eliminate the effects of shock or vibration. The lower modulus permits a slight deformation at the points where stresses are concentrated, which has the effect of distributing loads over larger areas. Providing the high compressive strength of 451 is not exceeded, parts will resume their original shape when loads are removed.

These same elastic properties prevent galling of contact surfaces. Parts tend to conform with mating surfaces after a period of time under load. Abrasion is reduced to a minimum.

With the advent of the Walsh-Healey Act, more companies will strive for lower noise levels in their plants. Mechanical parts designed with 451 will give long-wearing properties without the use of lubrication while providing quieter operation than metal parts.

The self-lubricating feature of 451, as with all nylons, is advantageous where lubricant contamination is a problem or where moving parts are normally inaccessible.

Chemical resistance is excellent, Custom Resins Nylon is particularly resistant to alkalies, solvents, petroleums, oils and greases. Like all polyamides, Custom Resins Nylon is degraded by acids.

For parts of highest quality produced under efficient production rates, a hot mold surface is recommended (180F to 200F). Cylinder temperature should permit a melt temperature of 460F to 480F at the nozzle.

CUSTOM RESINS: Nylon 452:

Plasticized injection molding nylon. For use in applications requiring high flexibility, excellent toughness and abrasion resistance. Good balance of viscosity and melt flow in critical moldings.

452 is a high monomer, lubricated resin for use in injection molding. It is especially designed for applications requiring good flexibility and abrasion resistance. Because of its internal lubricant and high monomer content, 452 will perform well in critical molding processes requiring very good melt flow.

The outstanding features of 452 are its excellent toughness, abrasion resistance and high flexibility. It has marked superiority to 6/6 nylon in tensile impact strength. 452 exhibits excellent flow at lower temperatures and pressures, which makes it suitable for a wide range of general applications.

In addition to its excellent melt flow charactreistics, 452 possesses good crystalline properties. The degree of crystalline development can be controlled to a remarkable extent by processing conditions.

PROPERTY:

Specific Gravity (D792-66): 1.12
Apparent Density (D1895-65T): 0.36 oz./cu.in.
Relative Viscosity (D789-66): 45
Water Absorption (D570-63): 3.9%
Tensile Strength (-40F)(D638-64T): 12,000 psi
Tensile Strength (73F): 9,600 psi
Tensile Strength (170F): 9,000 psi
Yield Point (-40F) (D638-64T): 12,000 psi
Yield Point (73F): 6,300 psi
Yield Point (170F): 4,000 psi
Elongation (-40F) (D638-64T): 70%
Elongation (73F): 360%
Elongation (170F): 350%
Impact Strength-Izod (-40F)(D256-56): 0.3 ft.lb./in. notch
Impact Strength-Izod (73F): 3.5 ft.lb./in. notch
Impact Strength-Izod (170F): 7.0 ft.lb./in. notch
Shear Strength (D732-46): 8,600 psi
Compression Strength (D790-66): 6,700 psi
Flexural Strength (D790-66): 6,500 psi
Flexural Stiffness (D747-63): 188,000 psi
Hardness-Rockwell R Scale (D785-65): 96
Heat Distortion (264 psi) (D648-56): 132F
Heat Distortion (66 psi): 305F
Melt Point (D789-66): 430F
Flow Temperature (D569-59): 374F
Coefficient of Linear Thermal Expansion per F (D696-44):
5.4 x 10 -5 in./in.

CUSTOM RESINS: Nylon 453:

Weather resistant injection molding resin providing for protection of the finished product against the embrittling effects of ultra violet light. 453 is a high viscosity nylon possessing excellent strength and toughness.

453 is UV stabilized through the incorporation of approximatelyy 2% carbon black, the msot effective stabilizer for nylon. 453 is available in black only.

The high quality of UV resistance provided by 453 is attributed to highly uniform dispersion of the finely divided carbon black in the resin. The solution light transmission of 453, as measured in percent absorbency, indicates the quality of carbon black dispersion. For a good weather resistant material, an absorbency of 0.18% or above is recommended. 453 possesses an average absorbency of 0.22%.

The incoroporation of carbon black has several beneficial effects other than UV stabilization. Flexural stiffness is increased about 15% over natural nylon. Wear resistance is enhanced as coefficient of friction is lowered.

PROPERTY:

Specific Gravity (D792-66): 1.17
Apparent Density (D1895-65T): 0.40 oz/cu.in.
Relative Viscosity (D789-66): 56
Water Absorption (D570-63): 1.5%
Carbon Content: 2.0%
Tensile Strength (D638-64T): 10,900 psi
Yield Point (D638-64T): 10,900 psi
Elongation (D638-064T): 20%
Impact Strength--Izod (D256-56): 0.7 psi
Flexural Stiffness (D747-63): 300,000 psi
Hardness--Rockwell (D785-65) (R Scale): 121
Heat Distortion--264 psi (D648-56): 121F
Heat Distortion--66 psi: 375F
Deformation Under Load, 122F (2,000 psi): 2.6%
Melt Point (D789-66): 430F
Flow Temperature (D569-66): 420F
Coefficient of Linear Thermal Expansion per F (D696-44):
4.43 x 10 -5 in/in
Solution Light Transmission (Absorbency ICM Cell): 0.22%

CUSTOM RESINS: Nylon 454:

Heat stabilized high viscosity molding resin. Finished products retain superior physical properties at elevated temperatures. Excellent for load-bearing parts.

The unique stabilizer system for 454 provides an additional 50F to 100F to the service temperature of molded parts. Users of gears, bearings, fasteners and other critical parts can safely rely on the physical durability of their parts at extremes of temperatures and environment.

The resistance of 454 to the deteriorating effects of high temperature exposure is significant. Basic physical properties will remain almost constant after sustained exposure at 250F for three months.

Since heat aging is cumulative, heat stabilized nylon parts can provide prolonged periods of service when exposure at elevated temperatures is intermittent. Years of service at intermittent exposure to 300F is not uncommon.

PROPERTY:

Specific Gravity (D792-66): 1.14
Apparent Density (D1895-65T): 0.38 oz./cu.in.
Relative Viscosity (D789-66): 50
Water Absorption (D570-63): 1.9%
Tensile Strength (-40F) (D638-64T): 14,600 psi
Tensile Strength (73F): 12,000 psi
Tensile Strength (170F): 6,600 psi
Yield Point (-40F) (D638-64T): 14,600 psi
Yield Point (73F): 12,000 psi
Yield Point (170F): 6,600 psi
Elongation (-40F) (D638-64T): 40%
Elongation (73F): 70%
Elongation (170F): 150%
Impact Strength--Izod (-40F) (D256-56): 0.4 ft.lb./in. notch
Impact Strength--Izod (73F): 0.7 ft.lb./in. notch
Impact Strength--Izod (170F): 150 ft.lb./in. notch
Shear Strength (D732-46): 8,300 psi
Compression Strength (D695-63T): 8,800 psi
Flexural Strength (D790-66): 14,800 psi
Hardness--Rockwell R Scale (D785-65): 117
Heat Distortion (264 psi) (D648-56): 167F
Heat Distortion (66 psi): 347F
Melt Point (D789-66): 430F
Flow Temperature (D569-59): 420F
Coefficient of Linear Thermal Expansion per F (D696-44): 4.6 x 10 -5 in./in.
Thermal Conductivity: 1.3 BTU/hr./sq.ft./F/in.
Dielectric Strength--short time (D149-64): 440 v/mil
Dielectric Strength--step by step (D149-64): 415 v/mil

CUSTOM RESINS: Nylon 600:

General Purpose Extrusion Resin

Custom Resins Nylon 600 is a general purpose extrusion grade resin having an ASTM relative viscosity of 50. The resin is used in many extrusion applications such as rods, slabs and tubing. Other applications of Custom Resins Nylon 600 are for wire jacketing, centrifugal casting and many uses requiring good dielectric properties, toughness and fatigue resistance.

Because Custom Resins Nylon 600 meets FDA requirements for several food packaging applications, it is particularly recommended for the extrusion of film and the extrusion coating of film substrates in flexible packaging constructions. In addition, it will not pick up odor in direct contact with food.

Custom Resins Nylon 600 has gained wide acceptance in the nylon extrusion field because of its excellent processing ability and superior product properties. It has superior impact strength--being more than twice the impact strength of 6/6 nylon. It has very good hydrolysis stability--withstanding repeated and prolonged sterilization. The resin has excellent overall balance of toughness and processing ease.

For maximum extrusion performance of Custom Resins Nylon 600, extruder barrel and die temperatures should be such as to permit a melt temperature of 480-500F at the die.

PROPERTY:

Specific Gravity (D792-66): 1.14
Apparent Density (D1895-65T): 0.39 oz./cu.in.
Relative Viscosity (D789-66): 50
Water Absorption (D570-63): 1.9%
Tensile Strength (-40F) (D638-64T): 14,600 psi
Tensile Strength (73F): 12,000 psi
Tensile Strength (170F): 6,600 psi
Yield Point (-40F) (D638-64T): 14,600 psi
Yield Point (73F): 12,000 psi
Yield Point (170F): 6,600 psi
Elongation (-40F) (D638-64T): 40%
Elongation (73F): 70%
Elongation (170F): 150%
Impact Strength--Izod (-40F) (D256-56): 0.4 ft.lb./in. notch
Impact Strength--Izod (73F): 0.7 ft.lb./in. notch
Impact Strength--Izod (170F): 4 ft.lb./in. notch
Shear Strength (D732-46): 8,300 psi
Compression Strength (D695-63T): 8,800 psi
Flexural Strength (D790-66): 14,800 psi
Hardness--Rockwell R Scale (D785-65): 117
Heat Distortion (264 psi) (D648-56): 167F
Heat Distortion (66 psi): 347F
Melt Point (D789-66): 430F
Flow Temperature (D569-59): 420F

CUSTOM RESINS: Nylon 601:

Heat Stabilized General Purpose Extrusion Resin

Custom Resins Nylon 601 is a heat stabilized extrusion resin designed and formulated to offer outstanding extrusion characteristics in most applications. 601 exhibits superior toughness; the property of being firm, strong and flexible. 601 offers outstanding toughness at extremes of temperatures and environments. This enables users of rods, tubes, slabs and sheeting, and other critical products such as insulated electrical wire and cable to safely rely on the durability of their extruded materials. The heat stabilizer adds an additional 50F to 100F to the service temperature of the extruded items. This is why 601 is particularly recommended for wire and cable jacketing and for other applications where nylon is subjected to elevated service temperatures. On an intermittent basis, the added temperature protection will be realized for years.

The resistance of Custom Resins Nylon 601 to the deteriorating effects of high temperatures is significant. The basic physical properties will remain almost constant for several months when exposed to a sustained temperature of 250F. A similar period of exposure to 300F shows a slightly drop in these properties; however they do remain serviceable. Custom Resins Nylon 601 will still possess premium strength, durability and other outstanding properties normally associated with Custom Resins Nylon.

601 is similar to Custom Resins Nylon 600 in processing characteristics. However, in order to assure that the heat stabilizing agent is not thermally destroyed during processing, a low melt temperature is recommended. Therefore, extruder barrel temperatures should be adjusted to produce melt temperatures in the range of 475F to 500F.

PROPERTY:

Specific Gravity (D792-66): 1.14
Apparent Density (D1895-65T): 0.39 oz./cu.in.
Relative Viscosity (D789-66): 50
Water Absorption (D570-63): 1.9%
Tensile Strength (-40F) (D638-64T): 14,600 psi
Tensile Strength (73F): 12,000 psi
Tensile Strength (170F): 6,600 psi
Yield Point (-40F) (D638-64T): 14,600 psi
Yield Point (73F): 12,000 psi
Yield Point (170F): 6,600 psi
Elongation (-40F) (D638-64T): 40%
Elongation (73F): 70%
Elongation (170F): 150%
Impact Strength--Izod (-40F) (D256-56): 0.4 ft.lb./in. notch
Impact Strength--Izod (73F): 0.7 ft.lb./in. notch
Impact Strength--Izod (170F): 4 ft.lb./in. notch
Shear Strength (D732-46): 8,300 psi

CUSTOM RESINS: Nylon 603:

For Extrusion of Flexible Products

Custom Resins Nylon 603 is a highly residual monomer resin designed for extrusion applications requiring excellent flexibility, high strength and good abrasion resistance. It is recommended for the extrusion of flexible tubing, cable jacketing and monofilament.

603 has earned a fine reputation in the monofilament field particularly in the extrusion of fish line and netting, sewing thread and weaving thread. With proper machine conditions, Custom Resins Nylon 603 can be extruded into monofilament having high tenacity, good residual elongation and low shrinkage.

603 will process at about 50F lower temperature than 6/6 nylon, and extruder and die temperature should be adjusted to produce a melt temperature of approximately 450F for optimum processing results.

PROPERTY:

Specific Gravity (D792-66): 1.14
Apparent Density (D1895-65T): 0.36 oz./cu.in.
Relative Viscosity (D789-66): 45
Water Absorption (D570-63): 3.2%
Tensile Strength (-40F) (D638-64T): 12,800 psi
Tensile Strength (73F): 12,800 psi
Tensile Strength (170F): 10,100 psi
Yield Point (-40F) (D638-64T): 12,800 psi
Yield Point (73F): 7,300 psi
Yield Point (170F): 6,900 psi
Elongation (-40F) (D638-64T): 70%
Elongation (73F): 250%
Elongation (170F): 270%
Impact Strength--Izod (-40F) (D256-56): 0.4 ft.lb./in. notch
Impact Strength--Izod (73F): 3.5 ft.lb./in. notch
Impact Strength--Izod (170F): 7.0 ft.lb./in. notch
Shear Strength (D732-46): 7,900 psi
Compression Strength (D695-63T): 6,700 psi
Flexural Strength (D790-66): 8,900 psi
Flexural Stiffness (D747-63): 125,000 psi
Hardness--Rockwell R Scale (D785-65): 101
Heat Distortion (264 psi) (D648-56): 160F
Heat Distortion (66 psi): 335F
Melt Point (D789-66): 430F
Flow Temperature (D569-59): 374F
Coefficient of Linear Thermal Expansion per F (D696-44): 7.1 x 10 5 in./in.
Thermal Conductivity: 1.4 BTU/hr./sq.ft./F/in.
Dielectric Strength--short time (D149-64): 1.7 x 10 5 v/mil
Dielectric Strength--step by step (D149-64): 500 v/mil
Volume Resistivity (D257-66): 7.4 x 10 12

DARTCO MANUFACTURING INC.: XYDAR High-Performance Enginerering Resins:

Property Profile:

XYDAR FC filled compounds are intended for general purpose injection molding applications. They offer a combination of high-performance properties and easy processability unmatched by any other thermoplastic and are suggested for evaluation in electrical/electronics, aerospace, automotive, ordnance, lighting and other demanding applications.

XYDAR FC-110:

XYDAR FC-110 is a glass-filled compound providing superior heat resistance in continuous use and a DTUL @ 264 psi of 606F (319C). It also exhibits excellent strength and impact resistance at temperature extremes down to -65F(-54C). XYDAR FC-110 offers outstanding resistance to most chemicals, weathering and radiation in addition to inherent, self-intumescent flame resistance(V-O @ 1/32 in.). It exhibits a low coefficient of thermal expansion and extremely low moisture absorption. XYDAR FC-110 is suggested for electical/electronic, aerospace, safety appliance and other high-performance components.

In injection molding, XYDAR FC-110 allows fast cycles on all types of machinery and excellent flow for thin-wall and difficult-to-fill part configurations. It can be used at conventional regrind ratios with excellent results. Molds designed for other thermoplastics can be used for developmental evaluations. XYDAR FC-110 can be vacuum metallized with or without a base coat.

DARTCO MANUFACTURING INC.: XYDAR High-Performance Engineering Resins(Continued):

Product Name:

XYDAR FC-120:

XYDAR FC-120 is a glass/mineral-filled compound providing superior heat resistance in continuous use and a DTUL @ 264 psi of 602F(316C). It also exhibits excellent stiffness and impact resistance at temperature extremes down to -65F(-54C). XYDAR FC-120 offers good appearance qualities and surface hardness and has the capacity for snap-fit designs. It provides outstanding resistance to most chemicals, weathering and radiation in addition to inherent, self-intumescent flame resistance(V-0 @ 1/16 in.). XYDAR FC-120 is suggested for appliance, aerospace and ordnance applications because of its excellent heat resistance, strength and appearance.

In injection molding, XYDAR FC-120 allows fast cycles on all types of machinery and excellent flow on all types of machinery and excellent flow for thin-wall and difficult-to-fill part configurations. It can be used at conventional regrind ratios with excellent results. Molds designed for other thermoplastics can be used for developmental evaluations. XYDAR FC-120 can be vacuum metallized with or without a base coat.

XYDAR FC-130:

XYDAR FC-130 is a mineral-filled compound providing superior heat resistance in continuous use and a DTUL @ 264 psi of 554F (290C). It also exhibits excellent stiffness and impact resistance at temperature extremes down to -65F(-54C). XYDAR FC-130 offers outstanding resistance to most chemicals, weathering and radiation in addition to inherent, self-intumescent flame resistance(V-0 @ 1/16 in.). It provides excellent flow properties and a superior surface appearance in molded parts for appliance, electrical/electronic, lighting and other high-performance applications.

In injection molding, XYDAR FC-130 allows fast cycles on all types of machinery and excellent flow for long, thin-wall parts and difficult-to-fill configurations. It can be used at conventional regrind ratios with excellent results. Molds designed for other thermoplastics can be used for developmental evaluations. XYDAR FC-130 can be vacuum metallized with or without a base coat, or electroless plated.

DARTCO MANUFACTURING, INC.: XYDAR High Performance Engineering Resins(Continued):

XYDAR RC reinforced composites are intended for general purpose injection molding applications. They offer a combination of high-performance properties at elevated temperatures and easy processability unmatched by any other thermoplastic and are suggested for evaluation in demanding applications across all markets.

Product Name:

XYDAR RC-210:

XYDAR RC-210 is a glass fiber reinforced composite providing superior heat resistance in continuous use and a DTUL @ 264 psi of 655F(346C). It also exhibits excellent strength and impact resistance at temperature extremes down to -65F (-54C). XYDAR RC-210 offers outstanding resistance to most chemicals, weathering and radiation in addition to inherent, self-intumescent flame resistance(V-O @ 1/32 in.). It exhibits a low coefficient of thermal expansion and extremely low moisture absorption. XYDAR RC-210 is suggested for broad based use across all markets where strength at elevated temperatures is important.

In injection molding, XYDAR RC-210 allows fast cycles on all types of machinery and excellent flow for thin-wall and difficult-to-fill part configurations. It can be used for conventional regrind ratios with excellent results. Molds designed for other thermoplastics can be used for developmental evaluations.

DARTCO MANUFACTURING, INC.: XYDAR High-Performance Engineering Resins(Continued):

Product Name:

XYDAR RC-220:

XYDAR RC-220 is a glass and mineral reinforced composite providing superior heat resistance in continuous use and a DTUL @ 264 psi of 626F(330C). It also exhibits excellent strength and impact resistance at temperature extremes down to -65F(-54C). XYDAR RC-220 offers outstanding resistance to most chemicals, weathering and radiation in addition to inherent, self-intumescent flame resistance (V-0 @ 1/32 in.). It exhibits a low coefficient of thermal expansion and extremely low moisture absorption. Originally developed for demanding electrical/electronic applications where absolute flatness is critical, XYDAR RC-220 is suggested for broad based use across all markets where strength at elevated temperatures and dimensional stability is important.

In injection molding, XYDAR RC-220 allows fast cycles on all types of machinery and excellent flow for thin-wall and difficult-to-fill part configurations. It can be used at conventional regrind ratios with excellent results. Molds designed for other thermoplastics can be used for developmental evaluations.

DEXTER PLASTICS: Blends & Alloys:

DEXPRO 101:

RPI 101EP is a flame retardant polypropylene compound designed for applications requiring chemical resistance and a ULV-2 rating. RPI 101EP meets requirements of Type II 37206 as described in ASTM D 2146.

RPI 101EP applications include profile extrusion operations, specifically piping.

Property:

Specific Gravity: 0.93 (D 792)
Water Absorption (24 Hr.): 0.0% (D 570)
Melt Flow Rate: 2 g/10 min (D 1238)
Tensile Strength: 4400 (30) psi (MPa) (D 638)
Elongation @ Break: 300% (D 638)
Flexural Strength: 4500 (31) psi (MPa) (D 790)
Flexural Modulus: 225,000 (1550) psi (MPa) (D 790)
UL Flammability: V-2: UL 94

DEXPRO 106:

RPI 106EP is a flame retardant polypropylene copolymer compound designed for applications requiring good impact strength, chemical resistance, and a ULV-2 rating.

RPI 106EP applications include injection molded parts and profile extrusions.

Property:

Specific Gravity: 0.94 (D 792)
Mold Shrinkage: 0.015-0.020 in/in (mm/mm) (D 955)
Melt Flow Rate: 2-4g/10 min (D1238)
Tensile Strength: 4400 (30) psi (MPa) (D 638)
Elongation @ Break: 400% (D 638)
Flexural Strength: 4800 (33) psi (MPa) (D 790)
Flexural Modulus: 210,000 (1450) psi (MPa) (D790)
UL Flammability: V-2 (UL 94)

DEXPRO 107:

RPI 107EP is a flame retardant polypropylene compound designed for applications requiring good impact strength, dimensional stability, long term heat resistance, chemical resistance, and a ULV-O rating.

RPI 107EP applications include injection molded parts and certain electrical/appliance components.

Property:

Specific Gravity: 1.26 (D 792)
Mold Shrinkage: 0.012 in/in (mm/mm) (D 955)
Melt Flow Rate: 5-7g/10 min (D 1238)
Tensile Strength: 3000 (21) psi(MPa) (D 638)
Elongation @ Break: 110% (D 638)
Flexural Strength: 5100 (35) psi (MPa) (D 790)
Flexural Modulus: 250,000(1720) psi (MPa) (D 790)
UL Flammability: V-0 (UL 94)

DEXTER PLASTICS: Blends & Alloys(Continued):

DEXPRO 109:

RPI 109EP is a thermoplastic olefin elastomer (TPO) designed for exterior gasketing applications requiring a soft but non-sticking surface. Product is UV stabilized for long term outdoor exposure.

RPI 109EP applications include injection molded gasketing parts.

Property:

- Specific Gravity: 0.90 (D 792)
- Water Absorption (24 hr.): <0.10 (D 570)
- Mold Shrinkage: .012 in/in (mm/mm) (D 955)
- Melt Flow Rate: 4.0g/10 min (D1238)
- Tensile Strength: 2500 (17) psi (MPa) (D 638)
- Elongation @ Break: 800% (D 638)
- Flexural Modulus: 30,000 (207) psi (MPa) (D790)
- UL Flammability: HB: UL 94

DEXPRO 110:

RPI 110EP is a 10% glass fiber reinforced propylene compound designed to economically balance tensile, stiffness, impact, and heat deflection properties.

RPI 110EP applications include injection molded parts that maintain cost/performance values coupled with processing ease.

Property:

- Specific Gravity: 0.98 (D 792)
- Mold Shrinkage: 0.007 in/in (mm/mm) (D 955)
- Tensile Strength: 7,000 (48) psi (MPa) (D 638)
- Elongation @ Break: 4% (D 638)
- Flexural Strength: 9,000 (62) psi (MPa) (D790)
- Flexural Modulus: 400,000 (2,760) psi (MPa) (D 790)

DEXPRO 120:

RPI 120EP is a 20% glass fiber reinforced polypropylene compound designed for good tensile, stiffness, and heat deflection properties.

RPI 120EP applications include injection molded parts that provide good strength and heat properties while maintaining processability and chemical resistance.

Property:

- Specific Gravity: 1.04 (D 792)
- Mold Shrinkage: 0.006 in/in (mm/mm) (D 955)
- Tensile Strength: 9,200 (63) psi (MPa) (D 638)
- Elongation @ Break: 3% (D 638)
- Flexural Strength: 10,000 (69) psi (MPa) (D 790)

DEXTER PLASTICS: Blends & Alloys(Continued):

DEXPRO 124:

RPI 124EP is a 20% mineral filled polypropylene copolymer injection molding grade material, designed for applications requiring a good balance of processability, stiffness and impact strength.

RPI 124EP is particularly suited for molded parts used in outdoor environments where good UV stability is required.

Property:

Specific Gravity: 1.05 (D 792)
Water Absorption (24 Hr.): .03% (D 570)
Mold Shrinkage : .012 in/in (mm/mm) (D955)
Melt Flow Rate: 8.0g/10 min (D 1238)
Tensile Strength: 3500 (24) psi (MPa) (D 638)
Elongation @ Break: 125% (D 638)
Flexural Strength: 4500 (31) psi (MPa) (D 790)
Flexural Modulus: 290 (2002) psi (MPa) (D 790)
UL Flammability: HB (UL 94)

DEXPRO 130:

RPI 130EP is a 30% glass fiber reinforced polypropylene compound designed for excellent tensile, stiffness, and heat deflection properties.

RPI 130EP applications include injection molded parts that provide high strength and heat properties while delivering chemical resistance and dimensional stability.

Property:

Specific Gravity: 1.13 (D 792)
Mold Shrinkage: 0.005 in/in(mm/mm) (D 955)
Tensile Strength: 12,000 (83) psi (MPa) (D638)
Elongation @ Break: 3% (D638)
Flexural Strength: 13,000 (90) psi (MPa) (D 790)
Flexural Modulus: 720,000 (4,970) psi (MPa) (D 790)

DEXPRO 132:

RPI 132EP is a high strength olefin compound developed to replace ABS and PVC molding resins. This material exhibits excellent AV stability and chemical resistance.

RPI 132EP applications include injection molded parts requiring good strength, UV stability and chemical resistance. Specific applications include pump housings and pipe fittings.

Property:

Specific Gravity: 1.12 (D 792)
Mold Shrinkage: 0.006-0.008 in/in (mm/mm) (D 955)
Tensile Strength: 6500 (45) psi (MPa) (D 638)
Elongation @ Break: 3-4% (D 638)
Flexural Strength: 7100 (50) psi (MPa) (D790)
Flexural Modulus: 340,000 (2340) psi (MPa) (D 790)

DEXTER PLASTICS: Blends & Alloys(Continued):

DEXPRO 135:

RPI 135EP is a reinforced polypropylene compound designed for high strength, stiffness and heat properties.

RPI 135EP applications include injection molded parts requiring high strength, good dimensional stability, and excellent chemical resistance.

Property:

Specific Gravity: 1.13 (D 792)
Mold Shrinkage: 0.005-0.007 in/in(mm/mm) (D 955)
Tensile Strength: 5300 (40) psi (MPa) (D 638)
Elongation @ Break: 4% (D 638)
Flexural Strength: 7500 (52) psi (MPa) (D 790)
Flexural Modulus: 450,000 (3100) psi (MPa) (D 790)

DEXPRO 136:

RPI 136EP is a 30% glass fiber reinforced polyolefin copolymer alloy characterized by high heat distortion, good heat aging, low specific gravity, and uniform processability.

RPI 136EP applications include injection molded parts that require high heat distortion and dimensional stability coupled with cost performance value. Examples include small tool housings and automobile under the hood parts.

Property:

Specific Gravity: 1.14 (D 792)
Water Absorption (24 hr.): 0.03% (D 570)
Mold Shrinkage: 0.005 in/in (mm/mm) (D 955)
Tensile Strength: 13,500 (93) psi (MPa) (D 638)
Elongation @ Break: 3% (D 638)
Flexural Strength: 11,000 (76) psi (MPa) (D 790)
Flexural Modulus: 720,000 (4,970) psi (MPa) (D 790)

DEXPRO-142:

RPI 142EP is a polyolefin sheet grade resin developed to compete with ABS for outdoor applications. RPI 142EP has an excellent combination of: high stiffness, impact strength, and UV stability. RPI 142EP maintains a matte finish upon extrusion into sheet and exhibits good melt strength during vacuum forming.

Property:

Flexural Modulus: 1795 MPa (D-790)
(260,000) (psi)
Tensile Strength: 28 MPa (D-638)
(4,000) (psi)
Elongation @ Break: 400% (D-638)
Hardness: 105 Rockwell R (D-2240)
Specific Gravity: 1.13 gms/cc (D-792)

DEXTER PLASTICS: Blends & Alloys(Continued):

DEXPRO 144:

RPI 144EP is a calcium carbonate filled polypropylene compound designed for sheet extrusion applications. It shows good impact strength, stiffness and heat resistance.

RPI 144EP is recommended for good contact applications, possessing excellent chemical resistance and low odor.

Property:

Specific Gravity: 1.23 (D 792)
Mold Shrinkage: .006-0.010 in/in(mm/mm) (D 955)
Melt Flow Rate: 5g/10 min (D 1238)
Tensile Strength: 3700 (26) psi (MPa) (D 638)
Elongation @ Break: 200% (D 638)
Flexural Strength: 6000 (41) psi (MPa) (D 790)
Flexural Modulus: 380,000 (2620) psi (MPa) (D790)

DEXLOY 201:

RPI 210EP is an ABS alloy designed to optimize impact strength and heat distortion while enhancing performance in stress cracking environments.

RPI 201EP applications include extrusion operations, specifically sheet extrusion where vacuum formed parts are subject to chemical cleaning agants.

Physical:

Specific Gravity: 1.12 (D 792)
Water Absorption (24 hr.): 0.15% (D570)
Mold Shrinkage: 0.005-0.006 in/in (mm/mm) (D 955)
Tensile Strength: 9300 (64) psi (MPa) (D 638)
Elongation @ Break: 110% (D 638)
Flexural Strength: 13,600 (94) psi (MPa) (D790)
Flexural Modulus: 375,000 (2,590) psi (MPa) (D 790)

DEXLOY 202:

RPI 202EP is an ABS alloy designed to optimize impact strength and heat distortion while enhancing performance in stress cracking environments.

RPI 202EP applications include injection molded parts and is particularly suited for parts subject to chemical cleaning agents where ABS compounds fail.

Property:

Specific Gravity: 1.12 (D 792)
Water Absorption (24 hr.): 0.15% (D 570)
Mold Shrinkage: 0.005-0.006 in/in (mm/mm) (D 955)
Tensile Strength: 9,000 (62) psi (MPa) (D 638)
Elongation @ Break: 100% (D 638)
Flexural Strength: 13,400 (92) psi (MPa) (D 790)
Flexural Modulus: 370,000 (2,550) psi (MPa) (D 790)

DEXTER PLASTICS: Blends & Alloys(Continued):

DEXLOY 203:

RPI 203EP is an ABS/PC alloy designed to provide a good balance of toughness, exceptional heat distortion and processability.

RPI 203EP applications include injection molded parts where a balance of mechanical properties and processability insures uniformity in all dimensions.

Physical:

Water Absorption (24 hr.): 0.16% (D 570)
Mold Shrinkage: 0.005-0.006 in/in(mm/mm) (D 955)
Tensile Strength: 8,800 (61) psi (MPa) (D 638)
Elongation @ Break: 40% (D 638)
Flexural Strength: 13,500 (93) psi (MPa) (D790)
Flexural Modulus: 410,000 (2,830) psi (MPa) (D790)

DEXLOY 204:

RPI 204EP is an ABS alloy designed for excellent impact and heat distortion while maintaining uniform processability.

RPI 204EP applications include injection molded parts where high impact and heat distortion requirements must be coupled with processability and gloss.

Property:

Specific Gravity: 1.12 (D 792)
Water Absorption (24 hr.): 0.15% (D 570)
Mold Shrinkage: 0.005-0.006 in/in(mm/mm) (D 955)
Tensile Strength: 9,100 (63) psi (MPa) (D 638)
Elongation @ Break: 80% (D 638)
Flexural Strength: 13,400 (92) psi (MPa) (D 790)
Flexural Modulus: 380,000 (2,620) psi (MPa) (D 790)

DEXLOY 204-FR:

RPI 204EP FR is a flame retardant ABS alloy designed for high impact, excellent heat resistance, and a UL 94 V-O rating.

RPI 204EP FR applications include injection molded parts where mechanical strength, heat resistance, and aesthetics are combined to service a UL 94 V-O rating.

Property:

Specific Gravity: 1.22 (D 792)
Water Absorption (24 hr.): 0.16% (D 570)
Mold Shrinkage: 0.005 in/in (mm/mm) (D 955)
Tensile Strength: 9,800 (68) psi (MPa) (D 638)
Elongation @ Break: 50% (D 638)
Flexural Strength: 13,700 (95) psi (MPa) (D 790)
Flexural Modulus: 420,000 (2,900) psi (MPa) (D 790)
UL Flammability: V-O (UL 94)

DEXTER PLASTICS: Blends & Alloys(Continued):

DEXLOY 207:

RPI 207EP is an ABS/PC alloy designed to improve impact and heat distortion while maintaining processability.

RPI 207EP applications include injection molded parts, specifically automotive, where the balance of properties insures paint and plateability.

Property:

Specific Gravity: 1.08 (D 792)
Water Absorption (24 hr.): 0.15 (D 570)
Mold Shrinkage: 0.006-0.008 in/in(mm/mm) (D 955)
Tensile Strength: 7800 (54) psi (MPa) (D 638)
Elongation @ Break: 40% (D 638)
Flexural Strength: 12,500 (86) psi (MPa) (D 790)
Flexural Modulus: 380,000 (2620) psi (MPa) (D 790)

DEXLOY 300 Series:
Flame Retardant PBT

This series of products are polybutylene terephthalate compounds tailored for a combination of excellent processability, physical properties and V-O flame retardance.

RPI 310:

Glass Fiber, %: 7.5
Specific Gravity: 1.43
Tensile Strength (psi): 9,800
Elongation (%): 3
Flexural Modulus (psi): 550,000
Heat Deflection (F) @ 264 psi: 344

RPI 320:

Glass Fiber, %: 20
Specific Gravity: 1.60
Tensile Strength (psi): 16,000
Elongation (%): 3
Flexural Modulus (psi) 1,000,000
Heat Deflection (F) @ 264 psi: 370

RPI 330:

Glass Fiber, %: 30
Specific Gravity: 1.65
Tensile Strength (psi) 17,500
Elongation (%): 3
Flexural Modulus (psi): 1,180,00
Heat Deflection (F) @ 264 psi: 400

DEXTER PLASTICS: Blends & Alloys(Continued):

DEXLOY 345

RPI 345EP is a high modulus polyester compound designed for high heat resistance, excellent dimensional stability, and improved noise abatement.

RPI 345EP applications include injection molded parts, specifically automotive underhood components where noise reduction and heat resistance are design criteria.

Property:

Specific Gravity: 1.68 (D 792)
Water Absorption (24 hr.): 0.06% (D 570)
Mold Shrinkage: 0.004-0.005 in/in (mm/mm) (D 955)
Tensile Strength: 18,000 (124) psi (MPa) (D 638)
Elongation @ Break: 3% (D 638)
Flexural Strength: 27,500 (190) psi (MPa) (D 790)
Flexural Modulus 1,250,000 (8,620) psi (MPa) (D 790)

DEXLOY 347:

DEXLOY 347 is a high modulus polyester compound designed for high heat resistance, excellent dimensional stability, and minimal warpage.

DEXLOY 347 applications include injection molded parts in high load and high heat environments that require tight dimensional tolerances. Automotive use includes panels and underhood applications.

Property:

Specific Gravity: 1.65 (D 792)
Water Absorption (24 hr): 0.06% (D 570)
Mold Shrinkage: 0.004-0.006 in/in (mm/mm) (D 955)
Tensile Strength: 17,500 (121) psi (MPa) (D 638)
Elongation @ Break: 3% (D 638)
Flexural Strength: 24,000 (165) psi (MPa) (D 790)
Flexural Modulus: 1,100,000 (7,590) psi (MPa) (D 790)
U L Flammability: HB (UL 94)

DEXLOY BEE-15:

RPI BEE-15 is a high impact polyester elastomer alloy designed for applications requiring good flexibility and easy processing. This material exhibits excellent chemical resistance, paintability, and superior low temperature impact strength.

RPI BEE-15 applications include injection molded parts, specifically automotive components.

Property:

Specific Gravity: 1.15 (D 792)
Water Absorption (24 hr.): 0.12% (D 570)
Tensile Strength: 3,200 (22) psi (MPa) (D 638)
Elongation @ Break: 210% (D 638)
Flexural Strength: 2,500 (17) psi (MPa) (D 790)
Flexural Modulus: 48,000 (330) psi (MPa) (D 790)

DEXTER PLASTICS: Blends & Alloys(Continued):

DEXLOY BEE-23:

RPI BEE-23 is a high impact polyester elastomer alloy designed for applications requiring good flexibility and easy processing. This material exhibits excellent chemical resistance, paintability, and superior low temperature impact strength.

RPI BEE-23 applications include injection molded parts, specifically automotive components.

Property:
- Specific Gravity: 1.18 (D 792)
- Water Absorption (24 hr.): 0.14% (D 570)
- Mold Shrinkage: 0.012 in/in (mm/mm) (D955)
- Tensile Strength: 3,900 (27) psi (MPa) (D 638)
- Elongation @ Break: 300% (D638)
- Flexural Strength: 4,200 (29) psi (MPa) (D 790)
- Flexural Modulus: 105,000 (720) psi (MPa) (D 790)

DEXCARB 500 Series:
Flame Retardant Polycarbonate:

A series of flame retardant polycarbonate compounds developed with a combination of excellent processability, physical characteristics, stress crack resistance and V-O flammability.

RPI 510:
- Glass Fiber, %: 10
- Specific Gravity: 1.30
- Tensile Strength (psi): 9,000
- Elongation (%): 2.0
- Flexural Modulus (psi): 500,000
- Heat Deflection (F) @ 264 psi: 280

RPI 520:
- Glass Fiber, %: 20
- Specific Gravity: 1.36
- Tensile Strength (psi): 16,000
- Elongation (%): 2.0
- Flexural Modulus (psi): 750,000
- Heat Deflection (F) @ 264 psi: 290

RPI 530:
- Glass Fiber, %: 30
- Specific Gravity: 1.45
- Tensile Strength (psi): 18,800
- Elongation (%): 2.0
- Flexural Modulus (psi): 1,100,000
- Heat Deflection (F) @ 264 psi: 300

DEXTER PLASTICS: Blends & Alloys(Continued):

DEXCARB 504:

RPI 504EP is a polycarbonate alloy designed for high heat resistance, good impact strength, and chemical resistance. RPI 504EP is superior to polycarbonate type resins in resistance to cleaning compounds and solvents.

RPI 504EP applications include injection molded parts, specifically where autoclaving in the presence of cleaning compounds causes failure in polycarbonate.

Property:

Specific Gravity: 1.18 (D 792)
Water Absorption (24 hr.): 0.4% (D 570)
Mold Shrinkage : 0.005-0.006 in/in (mm/mm) (D 955)
Tensile Strength: 9,200 (64) psi (MPa) (D 638)
Elongation @ Break: 100% (D638)
Flexural Strength: 12,500 (86) psi (MPa) (D 790)
Flexural Modulus: 360,000 (2,480) psi (MPa) (D 790)

DEXLON 609:

RPI 609EP is a low modulus non-plasticized nylon developed to balance heat and chemical resistance with processing ease.

RPI 609EP applications include extrusion operations and specifically tubing components where plasticized nylon II systems fail due to extraction at elevated temperatures.

Property:

Specific Gravity: 1.00 (D 792)
Water Absorption (24 Hr.): 0.3% (D 570)
Mold Shrinkage: 0.007 in/in (mm/mm) (D 955)
Tensile Strength: 2200 (15) psi (MPa) (D638)
Elongation @ Break: 350% (D 638)
Flexural Modulus: 42,000 (290) psi (MPa) (D 790)

DEXLON 603:

RPI 603EP is a high impact 6,6 nylon alloy developed to reduce moisture absorption. This reduction in moisture absorption allows this product to maintain a good combination of tensile and stiffness properties over a wide range of environmental conditions.

RPI 603EP applications include injection molded parts and, due to excellent melt strength, extrusion or blow molding operations.

Property:

Specific Gravity: 1.04 (D 792)
Water Absorption (24 hr.): 0.5% (D 570)
Mold Shrinkage: 0.010-0.012 in/in(mm/mm) (D 955)
Tensile Strength: 6,500 (45) psi (MPa) (D 638)
Elongation @ Break: 160% (D 638)
Flexural Strength: 10,500 (72) psi (MPa) (D 790)
Flexural Modulus: 250,000 (1,720) psi (Mpa) (D 790)

DEXTER PLASTICS: Blends & Alloys(Continued):

DEXLON 602:

RPI 602EP is a high impact 6,6 nylon alloy designed to provide good tensile and stiffness properties while delivering the excellent chemical resistance associated with nylons.

RPI 602EP applications include injection molded parts where environmental conditions require excellent toughness and chemical resistance.

Property:
- Specific Gravity: 1.13 (D 792)
- Water Absorption (24 hr.): 1.1% (D 570)
- Mold Shrinkage: 0.015 in/in (mm/mm) (D 955)
- Tensile Strength: 7,800 (54) psi (MPa) (D 638)
- Elongation @ Break: 110% (D 638)
- Flexural Strength: 12,000 (83) psi (MPa) (D 790)
- Flexural Modulus: 300,000 (2,070) psi (MPa) (D 790)

Nylon Compounds:

Nylon 6:

RPI 611 13% Glass Filled:
- Tensile Strength (psi): 14,000
- Flex Modulus (psi): 650K
- Notched Izod (Ft. lbs/in): 1.0
- HDT (F @ 264 psi): 375

RPI 631 33% Glass Filled:
- Tensile Strength (psi): 22,000
- Flex Modulus (psi): 1,100K
- Notched Izod (Ft. lbs/in): 2.0
- HDT (F @ 264 psi): 410

RPI 633 Low Moisture, Glass Filled:
- Tensile Strength (psi): 22,000
- Flex Modulus (psi): 990K
- Notched Izod (Ft. lbs./in): 2.0
- HDT (F @ 264 psi): 400

RPI 635 Mineral Filled:
- Tensile Strength (psi): 11,500
- Flex Modulus (psi): 1,000K
- Notched Izod (Ft. lbs/in): 0.8
- HDT (F @ 264 psi): 300

RPI 637 Low Moisture, Mineral Filled:
- Tensile Strength (psi): 10,900
- Flex Modulus (psi): 980K
- Notched Izod (Ft. lbs/in): 0.8
- HDT (F @ 264 psi): 290

DEXTER PLASTICS: Blends & Alloys(Continued):

Nylon Compounds(Continued):

Nylon 6/6:

RPI 600 Medium Impact:
Tensile Strength (psi): 9,000
Flex Modulus (psi): 320K
Notched Izod (Ft. lbs/in): 3.2
HDT (F @ 264 psi): 167

RPI 602 High Impact:
Tensile Strength (psi): 8,800
Flex Modulus (psi): 253K
Notched Izod (Ft. lbs/in): 18.0
HDT (Ft. lbs/in): 162

RPI 610 13% Glass Filled:
Tensile Strength (psi): 15,000
Flex Modulus: (psi) 700K
Notched Izod (Ft. lbs/in): 1.0
HDT (F @ 264 psi): 470

RPI 630 33% Glass Filled:
Tensile Strength (psi) 22,500
Flex Modulus (psi): 1,100K
Notched Izod (Ft. lbs/in): 2.0
HDT (F @ 264 psi): 482

RPI 632 Low Moisture Glass Filled:
Tensile Strength (psi): 22,000
Flex Modulus (psi): 990K
Notched Izod (Ft. lbs/in): 1.9
HDT (F @ 264 psi): 475

RPI 636 High Impact, Glass Filled:
Tensile Strength (psi): 13,300
Flex Modulus (psi): 500K
Notched Izod (Ft. lbs/in): 2.3
HDT (F @ 264 psi): 459

RPI 634 Mineral Filled:
Tensile Strength (psi): 14,000
Flex Modulus (psi): 1,000K
Notched Izod (Ft. lbs/in): 0.5
HDT (F @ 264 psi): 479

RPI 639 Low Moisture, Mineral Filled:
Tensile Strength (psi): 13,000
Flex Modulus (psi): 900K
Notched Izod (Ft. lbs/in): 0.7
HDT (F @ 264 psi): 468

DEXTER PLASTICS: Blends & Alloys(Continued):

Nylon Compounds(Continued):

Nylon 6/6(Continued):

RPI 638 High Impact Mineral Filled:
Tensile Strength (psi): 12,700
Flex Modulus (psi): 750K
Notched Izod (Ft. lbs/in): 1.5
HDT (F @ 264 psi): 450

Low Modulus Alloy:

609:
Tensile Strength (psi): 2,200
Flex Modulus (psi): 42
Notched Izod (Ft. lbs/in): N.B.

BEE-18:
Tensile Strength (psi): 6,000
Flex Modulus (psi): 110K
Notched Izod (Ft. lbs/in): N.B.
HDT (F @ 264 psi): 150

DEXCARB 531:

RPI 531EP is a 30% glass reinforced polycarbonate compound developed to balance glass dimensional stability, heat resistance, and impact strength while maintaining processability.

RPI 531EP applications include injection molded parts where dimensional and thermal expansion constraints are critical.

Property:

Specific Gravity: 143 (D 792)
Water Absorption (24 hr.): 0.14 (D 570)
Mold Shrinkage: 0.001 in/in (mm/mm) (D955)
Tensile Strength: 19,000 (130) psi (MPa) (D638)
Elongation @ Break: 3-5% (D 638)
Flexural Strength: 24,000 (165) psi (Mpa) (D 790)
Flexural Modulus: 1,100,000 (7590) psi (MPa) (D 790)

DEXCARB 507:

Description: RPI 507EP is a polycarbonate alloy developed for injection molding applications requiring excellent chemical resistance, high impact strength and easy processing.

Molding Conditions: The nature of this particular alloy allows it to be injection molded at temperatures lower than standard polycarbonates. A target melt temperature of 480 is recommended for best results.

Mold temperature settings of 120-140 are recommended.

Injection speeds and pressures should follow normal polycarbonate conditions.

DEXTER PLASTICS: Blends & Alloys(Continued):

DEXLON 632:

RPI 632EP is a glass reinforced 6,6 nylon designed to reduce inherent moisture absorption. The reduction in moisture absorption imparts dimensional stability over a wide range of environmental conditions.

RPI 632EP applications include injection molded parts where environmental conditions require stringent dimensional tolerance.

Property:

Specific Gravity: 1.35 (D 792)
Water Absorption (24 hr.): 0.7% (D570)
Mold Shrinkage: 0.002 in/in (mm/mm) (D 955)
Tensile Strength: 24,000 (166) psi (MPa) (D 638)
Elongation @ Break: 3% (D 638)
Flexural Strength: 38,000 (262) psi (MPa) (D 790)
Flexural Modulus: 1,000,000 (6,900) psi (Mpa) (D 790)

DEXLON 633HI:

RPI 633HI is a 33% glass filled nylon 6 alloy with improved impact resistance and reduced moisture absorption.

RPI 633HI is recommended for injection molding applications requiring high heat resistance, good dimensional stability and good resistance to impact.

Property:

Specific Gravity: 1.40 (D 792)
Water Absorption (24 hr.): 0.9% (D 570)
Mold Shrinkage: .003 in/in (mm/mm) (D 955)
Tensile Strength: 23,500 (162) psi (MPa) (D 638)
Elongation @ Break: 3% (D 638)
Flexural Strength: 34,000 (234) psi (MPa) (D 638)
Flexural Modulus: 1.1 x 10 6 (7590) psi (MPa) (D 790)
UL Flammability: HB (UL 94)

726:

RPI 726EP is a polyolefin compound which balances stiffness and impact while delivering excellent UV stability and chemical resistance.

RPI 726EP applications include injection molded parts that require a tough, stiff, and dimensionally stable material that retains properties despite UV and chemical exposure.

Property:

Specific Gravity: 1.12 (D 792)
Mold Shrinkage: 0.0007-0.009 in/in (mm/mm) (D 955)
Tensile Strength: 4200 (29) psi (MPa) (D638)
Flexural Strength: 4700 (33) psi (MPa) (D790)
Flexural Modulus: 250,000 (1730) psi (MPa) (D 790)

DEXTER PLASTICS: Blends & Alloys(Continued):

726V-O:

RPI 726V-0 is a flame retardant olefinic compound designed to balance impact, stiffness, and dimensional properties while delivering a UL V-O rating.

RPI 726V-O is designed for injection molding applications, specifically appliances, where appearance and performance demands are met while maintaining processability.

Property:

- Specific Gravity: 1.27 (D 792)
- Mold Shrinkage: 0.008-0.010 in/in (mm/mm) (D 955)
- Melt Flow Rate: 6-8g/10 min (D 1238)
- Tensile Strength: 3000 (21) psi (MPa) (D 638)
- Elongation @ Break: 100% (D 638)
- Flexural Strength: 4200 (29) psi (MPa) (D 790)
- Flexural Modulus: 220,000 (1520) psi (MPa) (D 790)
- UL Flammability: V-O (UL 94)

RPI 645EP*:

- Tensile Strength (psi): 11,000
- Elongation (%): 14
- Flexural Modulus (psi): 800,000
- Notched Izod Impact (ft.lbs./in): 2.0
- Deflection Temperature (F)
  - @ 66 psi: 410
  - 264 psi: 350
- Rockwell Hardness R Scale: 110
- Specific Gravity: 1.40
- Water Absorption - 24 Hours (%): 0.5
- Mold Shrinkage: 12

* Properties for Dam Test Parts

DOW CHEMICAL U.S.A.: CALIBRE Polycarbonate Resins:

CALIBRE polycarbonate resins, from the world's largest manufacturer of thermoplastics, are made with proprietary, Dow-developed technology. CALIBRE polycarbonate resins offer exceptional impact resistance, high heat resistance, outstanding clarity, and excellent dimensional stability.

The combination of properties, and easy processability, make CALIBRE polycarbonate resins the material of choice for a growing variety of applications in demanding service environments. Typical applications are in the appliance, automotive, business machine, electronics and communications, housewares, lighting, medical and health care, recreation, and safety industries. And more applications are in development.

200 Series:
FDA Compliance

Resin Designations:

200-22:

Melt flow rate (D-1238): 22
Deflection temp., F unannealed (D-648) @ 264 psi: 258
Vicat softening temp, F: (D-1525): 306
Tensile strength, psi (D-638):
  Yield: 9000
  Ultimate: 9500
Tensile modulus, 10 5 psi (D-638): 3.4
Izod impact strength, ft-lb/in (D-256):
  Notched, 1/8" thick: 14

200-15:

Melt flow rate (D-1238): 15
Deflection temp., F unannealed (D-648) @ 264 psi: 260
Vicat Softening temp, F: (D-525): 309
Tensile Strength, psi (D-638):
  Yield: 9000
  Modulus: 10,300
Tensile modulus, 10 5 psi (D-638): 3.4
Izod impact strength, ft-lb/in (D-256):
  Notched, 1/8" thick: 16

DOW CHEMICAL U.S.A.: CALIBRE Polycarbonate Resins(Continued):

200 Series(Continued):
FDA Compliance

Resin Designation:

200-10:

Melt flow rate (D-1238): 10
Deflection temp., F unannealed (D-648) @ 264 psi: 263
Vicat softening temp, F (D-1525): 312
Tensile strength, psi (D-638):
  Yield: 9000
  Ultimate: 10,300
Tensile Modulus, 10 5 psi (D-638): 3.5
Izod impact strength, ft-lb/in (D-256):
  Notched, 1/8" thick: 17

200-6:

Melt flow rate (D-1238): 6
Deflection temp., F unannealed (D-648) @ 264 psi: 265
Vicat softening temp, F (D-1525): 314
Tensile Strength, psi (D-638):
  Yield: 9000
  Ultimate: 10,500
Tensile Modulus, 10 5 psi (D-638): 3.5
Izod impact strength, ft-lb/in (D-256):
  Notched, 1/8" thick: 17

200-4:

Melt flow rate (D-1238): 4
Deflection temp., F unannealed (D648) @ 264 psi: 270
Vicat softening temp., F (D-1525): 316
Tensile Strength, psi (D-638):
  Yield: 9000
  Ultimate: 10,500
Tensile Modulus, 10 5 psi (D-638): 3.6
Izod impact strength, ft-lb/in (D-256):
  Notched, 1/8" thick: 18

DOW CHEMICAL U.S.A.: CALIBRE Polycarbonate Resins(Continued):

300 Series:
General Purpose

Resin Designation:

300-22:

Melt flow rate (D-1238): 22
Deflection temp., F unannealed (D-648) @ 264 psi: 285
Vicat softening temp. F (D-1525): 306
Tensile strength, psi (D-638):
Yield: 9000
Ultimate: 9500
Tensile Modulus, 10 5 psi (D-638): 3.4
Izod impact strength, ft-lb/in (D-256):
Notched, 1/8" thick: 14

300-15:

Melt flow rate (D-1238): 15
Deflection temp., F unannealed (D-648) @ 264 psi: 260
Vicat softening temp. F (D-1525): 309
Tensile strength, psi (D-638):
Yield: 9000
Ultimate: 10,300
Tensile modulus, 10 5 psi (D-638): 3.4
Izod impact strength, ft-lb/in (D-256):
Notched, 1/8" thick: 16

300-10:

Melt flow rate (D-1238): 10
Deflection temp., F unannealed (D-648) @ 264 psi: 263
Vicat softening temp., F (D-1525): 312
Tensile strength, psi (D-638):
Yield: 9000
Ultimate: 10,300
Tensile modulus, 10 5 psi (D-638): 3.5
Izod impact strength, ft-lb/in (D-256):
Notched, 1/8" thick: 17

DOW CHEMICAL U.S.A.: CALIBRE Polycarbonate Resins(Continued):

300 Series(Continued):
General Purpose

Resin Designation:

300-6:

Melt flow rate (D-1238): 6
Deflection temp., F unannealed (D-648) @ 264 psi: 265
Vicat softening temp., F (D-1525): 314
Tensile strength, psi (D-638):
Yield: 9000
Ultimate: 10,500
Tensile modulus, 10^5 psi (D-638): 3.5
Izod impact strength, ft-lb/in (D-256):
Notched, 1/8" thick: 17

300-4:

Melt flow rate (D-1238): 4
Deflection temp., F unannealed (D-648) @ 264 psi: 270
Vicat softening temp., F (1525): 316
Tensile strength, psi (D-638):
Yield: 9000
Ultimate: 10,500
Tensile modulus, 10^5 psi (D-638): 3.6
Izod impact strength, ft-lb/in (D-256):
Notched, 1/8" thick: 18

DOW CHEMICAL U.S.A.: CALIBRE Polycarbonate Resins(Continued):

500 Series
Glass Reinforced

Product Designation:

510:
550:

Deflection temp., F unannealed (D-648) @ 264 psi: 285
Vicat softening temp., F (D-1525): 320
Tensile Strength, psi (D-638):
  Yield: 8500-9500
  Ultimate: 7000-9100
Tensile modulus, 10 5 psi (D-638): 4.5-5.3
Izod impact strength, ft-lb/in (D-256):
  Notched, 1/8" thick: 2-4

700 Series:
Ignition Resistance/Clarity

700-15:

Melt flow rate (D-1238): 15
Deflection temp., F unannealed (D-648) @ 264 psi: 260
Vicat softening temp., F (D-1525): 309
Tensile strength, psi (D-638):
  Yield: 8800
  Ultimate: 9500
Tensile modulus, 10 5 psi (D-638): 3.25
Izod Impact strength, ft-lb/in (D-256):
  Notched, 1/8" thick: 16

700-10:

Melt flow rate (D-1238): 10
Deflection temp., F unannealed (D-648) @ 264 psi: 263
Vicat softening temp. F (D-1525): 312
Tensile strength, psi (D-638):
  Yield: 8800
  Ultimate: 9500
Tensile modulus, 10 5 psi (D-638): 3.25
Izod impact strength, ft-lb/in (D-256):
  Notched, 1/8" thick: 17

DOW CHEMICAL U.S.A.: CALIBRE Polycarbonate Resins(Continued):

700 Series(Continued):
Ignition Resistance/Clarity

<u>Product</u> <u>Designation:</u>

700-6:

Melt flow rate (D-1238): 6
Deflection temp., F unannealed (D-648) @ 264 psi: 265
Vicat softening temp. F (D-1525): 314
Tensile strength, psi (D-638):
  Yield: 8800
  Ultimate: 9500
Tensile modulus, 10 5 psi (D-638): 3.25
Izod Impact strength, ft lb/in (D-256):
  Notched, 1/8" thick: 17

700-4:

Melt flow rate (D-1238): 4
Deflection temp., F unannealed (D-648) @ 264 psi: 270
Vicat softening temp. F (D-1525): 316
Tensile strength, psi (D-638):
  Yield: 8800
  Ultimate: 9500
Tensile modulus, 10 5 psi (D-638): 3.25
Izod impact strength, ft-lb/in (D-256):
  Notched, 1/8" thick: 18

DOW CHEMICAL U.S.A.: CALIBRE Polycarbonate Resins(Continued):

800 Series:
Superior Ignition Resistance

Resin Designation:

800-15:

Melt flow rate (D-1238): 15
Deflection temp., unannealed (D-648) @ 264 psi: 266
Vicat softening temp., F (D-1525): 300
Tensile strength, psi (D-638):
- Yield: 8700
- Ultimate: 8500

Tensile modulus, 10 5 psi (D-638): 3.3
Izod impact strength, ft-lb/in (D-256):
- Notched, 1/8" thick: 10

800-10:

Melt flow rate (D-1238): 10
Deflection temp., F unannealed (D-648) @ 264 psi: 266
Vicat softening temp., F (D-1525): 305
Tensile strength, psi (D-638):
- Yield: 8700
- Ultimate: 8500

Tensile modulus, 10 5 psi (D-638): 3.3
Izod impact strength, ft-lb/in (D-256):
- Notched, 1/8" thick: 12

800-6:

Melt flow rate (D-1238): 6
Deflection temp., F unannealed (D-648) @ 264 psi: 266
Vicat softening temp., F (D-1525): 305
Tensile strength, psi (D-638):
- Yield: 8700
- Ultimate: 8500

Tensile modulus, 10 5 psi (D-638): 3.3
Izod impact strength, ft-lb/in (D-256):
- Notched, 1/8" thick: 12

DOW CHEMICAL U.S.A.: CALIBRE Polycarbonate Resins(Continued):

800 Series(Continued):
Superior Ignition Resistance

Resin Designation:

800-4:

- Melt flow rate (D-1238): 4
- Deflection temp., F unannealed (D-648) @ 264 psi: 266
- Vicat softening temp., F (D-1525): 310
- Tensile strength, psi (D-638):
  - Yield: 8700
  - Ultimate: 8500
- Tensile modulus, 10 5 psi (D-638): 3.3
- Izod impact strength, ft-lb/in (D-256):
  - Notched, 1/8" thick: 12

DOW CHEMICAL U.S.A.: CALIBRE Polycarbonate Resins(Continued):

2000 Series:
For the Health Care Industry

2060-6:

- Melt flow rate (D-1238): 6
- Deflection temp., F unannealed (D-648) @ 264 psi: 265
- Vicat softening temp., F (D-1525): 314
- Tensile strength, psi (D-638):
  - Yield: 9000
  - Ultimate: 10,500
- Tensile modulus, 10 5 psi (D-638): 3.5
- Izod impact strength, ft-lb/in (D-256):
  - Notched, 1/8" thick: 17

2060-4:

- Melt flow rate (D-1238): 4
- Deflection temp., unannealed (D-648) @ 264 psi: 270
- Vicat softening temp., F (D-1525): 316
- Tensile strength, psi (D-638):
  - Yield: 9000
  - Ultimate: 10,500
- Tensile modulus, 10 5 psi (D-638): 3.6
- Izod impact strength, ft-lb/in (D-256):
  - Notched, 1/8" thick: 18

2080:
2080-15:

- Melt flow rate (D-1238): 15
- Deflection temp., F unannealed (D-648) @ 264 psi: 252
- Vicat softening temp., F (D-638):
  - Yield: 9000
  - Ultimate: 9800
- Tensile modulus, 10 5 psi (D-638): 3.2
- Izod impact strength, ft-lb/in (D-256):
  - Notched, 1/8" thick: 14

DOW CHEMICAL U.S.A.: CALIBRE Polycarbonate Resins(Continued):

2000 Series(Continued):
For the Health Care Industry

Resin Designation:

2080:
2080-4:

Melt flow rate (D-1238): 4
Deflection temp., F unannealed (D-648) @ 264 psi: 260
Vicat softening temp., F (D-1525): 305
Tensile strength, psi (D-638):
  Yield: 9000
  Ultimate: 10,000
Tensile modulus, 10 5 psi (D-638): 3.4
Izod impact strength, ft-lb/in (D-256):
  Notched, 1/8" thick: 18

7000 Series:
Structural Foam:

7070:
.250":

Deflection temp., F unannealed (D-648) @ 264 psi: 273
Tensile strength, psi (D-638):
  Yield: 6500
Tensile modulus, 10 5 psi (D-638): 3.3

7070:
.150":

Deflection temp., F unannealed (D-648) @ 264 psi: 274
Tensile strength, psi (D-638):
  Yield: 7600
Tensile modulus, 10 5 psi (D-638): 3.6

DOW CHEMICAL U.S.A.: MAGNUS ABS Resins:

The new family of MAGNUS ABS resins from Dow can meet all your ABS resin needs.

Whatever your injection molding applications - appliances, automotive, power tools, telecommunications, electronics, recreation, housewares or medical--there is a MAGNUM ABS resin that can provide the specific performance properties and the outstannding advantages you're looking for.

Outstanding High Gloss:

97+ Gardner gloss
90+ Dorigon gloss

MAGNUM high gloss resins are available in medium impact, high impact and superior impact to satisfy your specific toughness requirements. And, what's more, they provide you with the highest high gloss available today over a range of practical molding conditions.

Outstanding Low Gloss:

Proven technology for surface aesthetics that demand a low gloss, matte finish.

MAGNUM low gloss resins, manufactured using unique Dow ABS technology, are the proven, industry leading material of choice for producing molded-in low gloss surfaces on those parts where a matte finish is desired. Consistent 60 Gardner gloss readings of 10 to 20 are readily achievable using efficient, everyday molding techniques.

Special ABS Resins for Special Needs:

The line of MAGNUM resins also includes ABS products specifically designed for transparency, ignition resistance, high heat resistance and plating applications.

DOW CHEMICAL U.S.A.: PELLETHANE Polyurethane Elastomers:

PELLETHANE brand thermoplastic elastomers are fully reacted polycaprolactone, polyester, and polyether-type polyurethane polymers that are available in a broad range of hardnesses. They can be fabricated into various shapes and forms using conventional injection molding, extrusion, blow molding, compression and transfer molding, and vacuum forming techniques. PELLETHANE polymers are proven to be commercially suitable for hundreds of applications in dozens of industries.

Customers will find that--compared to competitive polyurethane elastomers--PELLETHANE polymers commonly offer lighter color, improved transparency, and improved or at least equivalent physicals and processability. The breadth of the PELLETHANE product line is unmatched. The injection molding grades often provide shorter molding cycles; the extrusion grades commonly offer higher service temperatures.

PELLETHANE Products:

Series 2102: Polyester (polycaprolactone)
oil resistance, good hydrolytic stability

Series 2202: Polyester (polycaprolactone)
Special Compound

Series 2354: Polyester (polycaprolactone)
paintable, exterior automotive uses

Series 2103: Polyether (polytetramethylene glycol ether)
excellent hydrolytic stability, lowest extractables

Series 2363: Polyether (polytetramethylene glycol ether)
biomedical, drug admin. uses

Series 2355: Polyester (polyadipate)
superior transparency, color, good oil and chemical resistance

Series 2351: Polyether hybrid
Special Compound

Series 2352: Polyether-polyester hybrid
Special Compound

DOW CHEMICAL U.S.A.: PELLETHANE Polyurethane Elastomers (Continued):

Series 2102:

Polycaprolactone Ester Elastomers:

Hardness, Shore: 75A:

Specific Gravity(ASTM D792): 1.17
Tensile Modulus(ASTM D412):
PSI @ 50% Elongation: 430
PSI @ 100% Elongation: 550
PSI @ 300% Elongation: 1250
Ultimate Tensile Strength, PSI(ASTM D412): 5000
Ultimate Elongation, %(ASTM D412): 620

Hardness, Shore: 80A:

Specific Gravity(ASTM D792): 1.18
Tensile Modulus(ASTM D412):
PSI @ 50% Elongation: 560
PSI @ 100% Elongation: 800
PSI @ 300% Elongation: 1800
Ultimate Tensile Strength, PSI(ASTM D412): 5500
Ultimate Elongation, %(ASTM D412): 550

Hardness, Shore: 80AE:

Specific Gravity(ASTM D792): 1.18
Tensile Modulus(ASTM D412):
PSI @ 50% Elongation: 560
PSI @ 100% Elongation: 800
PSI @ 300% Elongation: 1800
Ultimate Tensile Strength, PSI(ASTM D412): 5500
Ultimate Elongation, %(ASTM D412): 550

Hardness, Shore: 80PF:

Specific Gravity(ASTM D792): 1.18
Tensile Modulus(ASTM D412):
PSI @ 50% Elongation: 560
PSI @ 100% Elongation: 800
PSI @ 300% Elongation: 1800
Ultimate Tensile Strength, PSI(ASTM D412): 4250
Ultimate Elongation, %(ASTM D412): 550

DOW CHEMICAL U.S.A.: PELLETHANE Polyurethane Elastomers (Continued):

Series 2102(Continued):

Polycaprolactone Ester Elastomers(Continued):

Hardness, Shore: 85A:

Specific Gravity(ASTM D792): 1.18
Tensile Modulus(ASTM D412):
PSI @ 50% Elongation: 850
PSI @ 100% Elongation: 1100
PSI @ 300% Elongation: 2100
Ultimate Tensile Strength, PSI(ASTM D412): 5300
Ultimate Elongation, %(ASTM D412): 600

Hardness, Shore: 90PF:

Specific Gravity(ASTM D792): 1.20
Tensile Modulus(ASTM D412):
PSI @ 50% Elongation: 1150
PSI @ 100% Elongation: 1280
PSI @ 300% Elongation: 2100
Ultimate Tensile Strength, PSI(ASTM D412): 5850
Ultimate Elongation, %(ASTM D412): 580

Hardness, Shore: 90A:

Specific Gravity(ASTM D792): 1.20
Tensile Modulus(ASTM D412):
PSI @ 50% Elongation: 1220
PSI @ 100% Elongation: 1550
PSI @ 300% Elongation: 3600
Ultimate Tensile Strength, PSI(ASTM D412): 6000
Ultimate Elongation, %(ASTM D412): 500

Hardness, Shore: 90AR:

Specific Gravity(ASTM D792): 1.20
Tensile Modulus(ASTM D412):
PSI @ 50% Elongation: 1220
PSI @ 100% Elongation: 1550
PSI @ 300% Elongation: 3600
Ultimate Tensile Strength, PSI(ASTM D412): 6000
Ultimate Elongation, %(ASTM D412): 500

DOW CHEMICAL U.S.A.: PELLETHANE Polyurethane Elastomers (Continued):

Series 2102(Continued):

Polycaprolactone Ester Elastomers(Continued):

Hardness, Shore: 90AE:

Specific Gravity(ASTM D792): 1.20
Tensile Modulus(ASTM D412):
PSI @ 50% Elongation: 1220
PSI @ 100% Elongation: 1550
PSI @ 300% Elongation: 3600
Ultimate Tensile Strength, PSI(ASTM D412): 5900
Ultimate Elongation, %(ASTM D412): 500

Hardness, Shore: 55D:

Specific Gravity(ASTM D792): 1.21
Tensile Modulus(ASTM D412):
PSI @ 50% Elongation: 2220
PSI @ 100% Elongation: 2750
PSI @ 300% Elongation: 5000
Ultimate Tensile Strength, PSI(ASTM D412): 6600
Ultimate Elongation, %(ASTM D412): 440

Hardness, Shore: 65D:

Specific Gravity(ASTM D792): 1.22
Tensile Modulus(ASTM D412):
PSI @ 50% Elongation: 3000
PSI @ 100% Elongation: 3300
PSI @ 300% Elongation: 5000
Ultimate Tensile Strength, PSI(ASTM D412): 6000
Ultimate Elongation, %(ASTM D412): 430

DOW CHEMICAL U.S.A.: PELLETHANE Polyurethane Elastomers (Continued):

Series 2354:

Polycaprolactone Ester Elastomers:

Hardness, Shore: 45D/45DG:

Specific Gravity(ASTM D792): 1.21
Tensile Modulus(ASTM D412):
PSI @ 50% Elongation: 1200
PSI @ 100% Elongation: 1550
PSI @ 300% Elongation: 3100
Ultimate Tensile Strength, PSI(ASTM D412): 5600
Ultimate Elongation, %(ASTM D412): 500

Hardness, Shore: 45DGA:

Specific Gravity(ASTM D792): 1.19
Tensile Modulus(ASTM D412):
PSI @ 50% Elongation: 1150
PSI @ 100% Elongation: 1400
PSI @ 300% Elongation: 2600
Ultimate Tensile Strength, PSI(ASTM D412): 5600
Ultimate Elongation, %(ASTM D412): 580

Hardness, Shore: 55D/55DG:

Specific Gravity(ASTM D792): 1.21
Tensile Modulus(ASTM D412):
PSI @ 50% Elongation: 1800
PSI @ 100% Elongation: 2000
PSI @ 300% Elongation: 3600
Ultimate Tensile Strength, PSI(ASTM D412): 5500
Ultimate Elongation, %(ASTM D412): 500

DOW CHEMICAL U.S.A.: PELLETHANE Polyurethane Elastomers (Continued):

Series 2355:

Polyadipate Ester Elastomers:

Hardness, Shore: 75A:

Specific Gravity(ASTM D792): 1.19
Tensile Modulus(ASTM D412):
PSI @ 50% Elongation: 430
PSI @ 100% Elongation: 610
PSI @ 300% Elongation: 1530
Ultimate Tensile Strength, PSI(ASTM D412): 4630
Ultimate Elongation, %(ASTM D412): 620

Hardness, Shore: 80A:

Specific Gravity(ASTM D792): 1.18
Tensile Modulus(ASTM D412):
PSI @ 50% Elongation: 560
PSI @ 100% Elongation: 800
PSI @ 300% Elongation: 1800
Ultimate Tensile Strength, PSI(ASTM D412): 5500
Ultimate Elongation, %(ASTM D412): 550

Hardness, Shore: 80AE:

Specific Gravity(ASTM D792): 1.18
Tensile Modulus(ASTM D412):
PSI @ 50% Elongation: 560
PSI @ 100% Elongation: 800
PSI @ 300% Elongation: 1800
Ultimate Tensile Strength, PSI(ASTM D412): 5500
Ultimate Elongation, %(ASTM D412): 550

Hardness, Shore: 80AEF:

Specific Gravity(ASTM D792): 1.18
Tensile Modulus(ASTM D412):
PSI @ 50% Elongation: 560
PSI @ 100% Elongation: 800
PSI @ 300% Elongation: 1800
Ultimate Tensile Strength, PSI(ASTM D412): 5500
Ultimate Elongation, %(ASTM D412): 550

DOW CHEMICAL U.S.A.: PELLETHANE Polyurethane Elastomers (Continued):

Series 2355 (Continued):

Polyadipate Ester Elastomers (Continued):

Hardness, Shore: 85ABR:

Specific Gravity(ASTM D792): 1.18
Tensile Modulus(ASTM D412):
PSI @ 50% Elongation: 800
PSI @ 100% Elongation: 900
PSI @ 300% Elongation: 1400
Ultimate Tensile Strength, PSI(ASTM D412): 5000
Ultimate Elongation, %(ASTM D412): 600

Hardness, Shore: 90A:

Specific Gravity(ASTM D792): 1.20
Tensile Modulus(ASTM D412):
PSI @ 50% Elongation: 1220
PSI @ 100% Elongation: 1550
PSI @ 300% Elongation: 3600
Ultimate Tensile Strength, PSI(ASTM D412): 6000
Ultimate Elongation, %(ASTM D412): 500

Hardness, Shore: 90AE:

Specific Gravity(ASTM D792): 1.19
Tensile Modulus(ASTM D412):
PSI @ 50% Elongation: 1100
PSI @ 100% Elongation: 1400
PSI @ 300% Elongation: 2800
Ultimate Tensile Strength, PSI(ASTM D412): 6000
Ultimate Elongation, %(ASTM D412): 550

Hardness, Shore: 95AE:

Specific Gravity(ASTM D792): 1.22
Tensile Modulus(ASTM D412):
PSI @ 50% Elongation: 1500
PSI @ 100% Elongation: 1700
PSI @ 300% Elongation: 3500
Ultimate Tensile Strength, PSI(ASTM D412): 5000
Ultimate Elongation, %(ASTM D412): 480

DOW CHEMICAL U.S.A.: PELLETHANE Polyurethane Elastomers (Continued):

Series 2355 (Continued):

Polyadipate Ester Elastomers (Continued):

Hardness, Shore: 95AEF:

Specific Gravity(ASTM D792): 1.22
Tensile Modulus(ASTM D412):
PSI @ 50% Elongation: 1500
PSI @ 100% Elongation: 1700
PSI @ 300% Elongation: 3500
Ultimate Tensile Strength, PSI(ASTM D412): 5000
Ultimate Elongation, % (ASTM D412): 480

Hardness, Shore: 95AESP:

Specific Gravity(ASTM D792): 1.22
Tensile Modulus(ASTM D412):
PSI @ 50% Elongation: 1500
PSI @ 100% Elongation: 1700
PSI @ 300% Elongation: 3500
Ultimate Tensile Strength, PSI(ASTM D412): 4800
Ultimate Elongation, %(ASTM D412): 490

Hardness, Shore: 95ACR:

Specific Gravity(ASTM D792): 1.41
Tensile Modulus(ASTM D412):
PSI @ 50% Elongation: 1200
PSI @ 100% Elongation: 1350
PSI @ 300% Elongation: 2300
Ultimate Tensile Strength, PSI(ASTM D412): 4700
Ultimate Elongation, %(ASTM D412): 450

Hardness, Shore: 55D:

Specific Gravity(ASTM D792): 1.21
Tensile Modulus(ASTM D412):
PSI @ 50% Elongation: 2250
PSI @ 100% Elongation: 2750
PSI @ 300% Elongation: 5000
Ultimate Tensile Strength, PSI(ASTM D412): 6500
Ultimate Elongation, %(ASTM D412): 440

DOW CHEMICAL U.S.A.: PELLETHANE Polyurethane Elastomers (Continued):

Series 2355 (Continued)

Polyadipate Ester Elastomer (Continued):

Hardness, Shore: 55DE:

Specific Gravity(ASTM D792): 1.19
Tensile Modulus(ASTM D412):
PSI @ 50% Elongation: 1350
PSI @ 100% Elongation: 1750
PSI @ 300% Elongation: 3600
Ultimate Tensile Strength, PSI(ASTM D412): 6000
Ultimate Elongation, %(ASTM D412): 500

Hardness, Shore: 65D:

Specific Gravity(ASTM D792): 1.22
Tensile Modulus(ASTM D412):
PSI @ 50% Elongation: 2800
PSI @ 100% Elongation: 3200
PSI @ 300% Elongation: 4500
Ultimate Tensile Strength, PSI(ASTM D412): 5900
Ultimate Elongation, %(ASTM D412): 450

DOW CHEMICAL U.S.A.: PELLETHANE Polyurethane Elastomers (Continued):

Series 2103:

Polyether Elastomers:

Hardness, Shore: 70A:

Specific Gravity(ASTM D792): 1.06
Tensile Modulus(ASTM D412):
PSI @ 50% Elongation: 400
PSI @ 100% Elongation: 500
PSI @ 300% Elongation: 800
Ultimate Tensile Strength, PSI(ASTM D412): 3500
Ultimate Elongation, %(ASTM D412): 700

Hardness, Shore: 80WC:

Specific Gravity(ASTM D792): 1.10
Tensile Modulus(ASTM D412):
PSI @ 50% Elongation: 600
PSI @ 100% Elongation: 730
PSI @ 300% Elongation: 1050
Ultimate Tensile Strength, PSI(ASTM D412): 4000
Ultimate Elongation, %(ASTM D412): 700

Hardness, Shore: 80A:

Specific Gravity(ASTM D792): 1.13
Tensile Modulus(ASTM D412):
PSI @ 50% Elongation: 550
PSI @ 100% Elongation: 700
PSI @ 300% Elongation: 1600
Ultimate Tensile Strength, PSI(ASTM D412): 5500
Ultimate Elongation, %(ASTM D412): 550

Hardness, Shore: 80AE:

Specific Gravity(ASTM D792): 1.13
Tensile Modulus(ASTM D412):
PSI @ 50% Elongation: 600
PSI @ 100% Elongation: 800
PSI @ 300% Elongation: 1700
Ultimate Tensile Strength, PSI(ASTM D412): 5200
Ultimate Elongation, %(ASTM D412): 550

DOW CHEMICAL U.S.A.: PELLETHANE Polyurethane Elastomers (Continued):

Series 2103 (Continued):

Polyether Elastomers (Continued):

Hardness, Shore: 80AEF:

Specific Gravity(ASTM D792): 1.13
Tensile Modulus(ASTM D412):
PSI @ 50% Elongation: 700
PSI @ 100% Elongation: 900
PSI @ 300% Elongation: 1400
Ultimate Tensile Strength, PSI(ASTM D412): 5500
Ultimate Elongation, %(ASTM D412): 540

Hardness, Shore: 80AES:

Specific Gravity(ASTM D792): 1.14
Tensile Modulus(ASTM D412):
PSI @ 50% Elongation: 600
PSI @ 100% Elongation: 800
PSI @ 300% Elongation: 1400
Ultimate Tensile Strength, PSI(ASTM D412): 5500
Ultimate Elongation, %(ASTM D412): 500

Hardness, Shore: 85AE:

Specific Gravity(ASTM D792): 1.14
Tensile Modulus(ASTM D412):
PSI @ 50% Elongation: 950
PSI @ 100% Elongation: 1150
PSI @ 300% Elongation: 1800
Ultimate Tensile Strength, PSI(ASTM D412): 5000
Ultimate Elongation, %(ASTM D412): 600

Hardness, Shore: 80PF:

Specific Gravity(ASTM D792): 1.10
Tensile Modulus(ASTM D412):
PSI @ 50% Elongation: 600
PSI @ 100% Elongation: 750
PSI @ 300% Elongation: 1100
Ultimate Tensile Strength, PSI(ASTM D412): 5000
Ultimate Elongation, %(ASTM D412): 700

DOW CHEMICAL CO.: PELLETHANE Polyurethane Elastomers (Continued):

Series 2103 (Continued):

Polyether Elastomers (Continued):

Hardness, Shore: 85ACR:

Specific Gravity(ASTM D792): 1.30
Tensile Modulus(ASTM D412):
PSI @ 50% Elongation: 500
PSI @ 100% Elongation: 600
PSI @ 300% Elongation: 900
Ultimate Tensile Strength, PSI(ASTM D412): 4000
Ultimate Elongation, %(ASTM D412): 600

Hardness, Shore: 90A:

Specific Gravity(ASTM D792): 1.14
Tensile Modulus(ASTM D412):
PSI @ 50% Elongation: 1100
PSI @ 100% Elongation: 1530
PSI @ 300% Elongation: 3400
Ultimate Tensile Strength, PSI(ASTM D412): 6200
Ultimate Elongation, %(ASTM D412): 430

Hardness, Shore: 90AE:

Specific Gravity(ASTM D792): 1.14
Tensile Modulus(ASTM D412):
PSI @ 50% Elongation: 920
PSI @ 100% Elongation: 1250
PSI @ 300% Elongation: 2300
Ultimate Tensile Strength, PSI(ASTM D412): 5400
Ultimate Elongation, %(ASTM D412): 560

Hardness, Shore: 90AEF:

Specific Gravity(ASTM D792): 1.14
Tensile Modulus(ASTM D412):
PSI @ 50% Elongation: 920
PSI @ 100% Elongation: 1250
PSI @ 300% Elongation: 2300
Ultimate Tensile Strength, PSI(ASTM D412): 5400
Ultimate Elongation, %(ASTM D412): 560

DOW CHEMICAL U.S.A.: PELLETHANE Polyurethane Elastomers (Continued):

Series 2103 (Continued):

Polyether Elastomers (Continued):

Hardness, Shore: 90ACR:

Specific Gravity(ASTM D792): 1.30
Tensile Modulus(ASTM D412):
PSI @ 100% Elongation: 1100
PSI @ 300% Elongation: 1700
Ultimate Tensile Strength, PSI (ASTM D412): 4500
Ultimate Elongation, %(ASTM D412): 450

Hardness, Shore: 55D:

Specific Gravity(ASTM D792): 1.15
Tensile Modulus(ASTM D412):
PSI @ 50% Elongation: 1950
PSI @ 100% Elongation: 2600
PSI @ 300% Elongation: 4900
Ultimate Tensile Strength, PSI(ASTM D412): 6500
Ultimate Elongation, %(ASTM D412): 400

Hardness, Shore: 55DE:

Specific Gravity(ASTM D792): 1.15
Tensile Modulus(ASTM D412):
PSI @ 50% Elongation: 1540
PSI @ 100% Elongation: 2100
PSI @ 300% Elongation: 4340
Ultimate Tensile Strength, PSI(ASTM D412): 6000
Ultimate Elongation, %(ASTM D412): 450

Hardness, Shore: 65D:

Specific Gravity(ASTM D792): 1.17
Tensile Modulus(ASTM D412):
PSI @ 50% Elongation: 2500
PSI @ 100% Elongation: 3000
PSI @ 300% Elongation: 5000
Ultimate Tensile Strength, PSI(ASTM D412): 6500
Ultimate Elongation, %(ASTM D412): 420

DOW CHEMICAL U.S.A.: PELLETHANE Polyurethane Elastomers (Continued):

Series 2363:

Polyether Elastomers:

Hardness, Shore: 75A:

Specific Gravity(ASTM D792): 1.09
Tensile Modulus(ASTM D412):
PSI @ 50% Elongation: 320
PSI @ 100% Elongation: 450
PSI @ 300% Elongation: 910
Ultimate Tensile Strength, PSI(ASTM D412): 4650
Ultimate Elongation, %(ASTM D412): 650

Hardness, Shore: 80A:

Specific Gravity(ASTM D792): 1.13
Tensile Modulus(ASTM D412):
PSI @ 50% Elongation: 550
PSI @ 100% Elongation: 770
PSI @ 300% Elongation: 1600
Ultimate Tensile Strength, PSI(ASTM D412): 5500
Ultimate Elongation, %(ASTM D412): 550

Hardness, Shore: 80AE:

Specific Gravity(ASTM D792): 1.12
Tensile Modulus(ASTM D412):
PSI @ 50% Elongation: 700
PSI @ 100% Elongation: 850
PSI @ 300% Elongation: 1300
Ultimate Tensile Strength, PSI(ASTM D412): 4500
Ultimate Elongation, %(ASTM D412): 650

Hardness, Shore: 90A:

Specific Gravity(ASTM D792): 1.14
Tensile Modulus(ASTM D412):
PSI @ 50% Elongation: 1120
PSI @ 100% Elongation: 1540
PSI @ 300% Elongation: 3400
Ultimate Tensile Strength, PSI(ASTM D412): 6500
Ultimate Elongation, %(ASTM D412): 460

DOW CHEMICAL U.S.A.: PELLETHANE Polyurethane Elastomers (Continued):

Series 2363 (Continued):

Polyether Elastomers (Continued):

Hardness, Shore: 90AE:

Specific Gravity(ASTM D792): 1.14
Tensile Modulus(ASTM D412):
PSI @ 50% Elongation: 920
PSI @ 100% Elongation: 1250
PSI @ 300% Elongation: 2300
Ultimate Tensile Strength, PSI(ASTM D412): 5400
Ultimate Elongation, %(ASTM D412): 560

Hardness, Shore: 55D:

Specific Gravity(ASTM D792): 1.15
Tensile Modulus(ASTM D412):
PSI @ 50% Elongation: 1900
PSI @ 100% Elongation: 2500
PSI @ 300% Elongation: 4930
Ultimate Tensile Strength, PSI(ASTM D412): 6640
Ultimate Elongation, %(ASTM D412): 390

Hardness, Shore: 55DE:

Specific Gravity(ASTM D792): 1.15
Tensile Modulus(ASTM D412):
PSI @ 50% Elongation: 1540
PSI @ 100% Elongation: 2100
PSI @ 300% Elongation: 4340
Ultimate Tensile Strength, PSI(ASTM D412): 6000
Ultimate Elongation, %(ASTM D412): 450

Hardness, Shore: 65D:

Specific Gravity(ASTM D792): 1.17
Tensile Modulus(ASTM D412):
PSI @ 50% Elongation: 2370
PSI @ 100% Elongation: 2840
PSI @ 300% Elongation: 4550
Ultimate Tensile Strength, PSI(ASTM D412): 6270
Ultimate Elongation, %(ASTM D412): 450

Hardness, Shore: 75D:

Specific Gravity(ASTM D792): 1.21
Ultimate Tensile Strength, PSI(ASTM D412): 6980
Ultimate Elongation, %(ASTM D412): 250

DOW CHEMICAL U.S.A.: PELLETHANE Polyurethane Elastomers (Continued):

Specialty Compounds:

2202:

Polycaprolactone-Ester Elastomer:

Hardness, Shore: 93A:

Specific Gravity(ASTM D792): 1.20
Tensile Modulus(ASTM D412):
PSI @ 50% Elongation: 1100
PSI @ 100% Elongation: 1500
PSI @ 300% Elongation: 2500
Ultimate Tensile Strength: PSI(ASTM D412): 5000
Ultimate Elongation, %(ASTM D412): 400

2351:

Polyether Hybrid Elastomer

Hardness, Shore: 85AE:

Specific Gravity(ASTM D792): 1.13
Tensile Modulus(ASTM D412):
PSI @ 50% Elongation: 820
PSI @ 100% Elongation: 1050
PSI @ 300% Elongation: 1410
Ultimate Tensile Strength, PSI(ASTM D412): 4000
Ultimate Elongation, %(ASTM D412): 650

Series 2352:

Poly(ester-ether) Elastomer:

Hardness, Shore: 70A:

Specific Gravity(ASTM D792): 1.12
Tensile Modulus(ASTM D412):
PSI @ 50% Elongation: 270
PSI @ 100% Elongation: 380
PSI @ 300% Elongation: 600
Ultimate Tensile Strength, PSI (ASTM D412): 1500
Ultimate Elongation, %(ASTM D412): 900

DOW CHEMICAL U.S.A.: PULSE Engineering Resins:

Unique Engineering Thermoplastic Resins
Designed for Automotive Applications

PULSE 830 and PULSE 869 Engineering Resins were recently evaluated in an automotive instrument panel (IP) retainer application. The purpose of this evaluation was to compare the balance of properties and processability of these two PULSE products with competitive resins that are either being used in IP production or are targeted for use in the manufacture of future IP retainers.

Test samples were cut from the IP retainers. The data obtained from these samples were compared with data obtained from a previous molding trial in which the same mold was used.

The recent evaluation of PULSE 830 and PULSE 869 Engineering Resins in automotive IP retainers has shown that these PULSE products have an excellent combination of properties and processability. High quality IP retainers can be molded with PULSE Engineering Resins in molding cycle times 10-15% shorter than with competitive resins.

DOW CHEMICAL U.S.A.: ROVEL Weatherable Polymers:

For a Wide Range of Outdoor Applications
Spas--RV Parts--Automotive--Construction Profiles

ROVEL high impact weatherable polymers are engineering thermoplastics that have outstanding resistance to weather aging, an excellent balance of physical and mechanical properties, and are processable by conventional extrusion and injection molding techniques.

These polymers feature:

* Retention of physical properties outdoors, without need for painting or capping with an acrylic sheet.

* Excellent high heat distortion.

* Excellent impact strength.

* Retention of color and appearance outdoors.

* Resistance to acids, alkalies, and salts.

* Molding and extrusion grades.

* Ability to be foamed, filled, painted or plated.

ROVEL polymers have excellent resistance to outdoor UV exposure. This property of ROVEL is superior to that of most other engineering thermoplastics. Extensive outdoor testing in Florida has confirmed the exceptional outdoor stability of ROVEL.

Coex Sheet of ROVEL and ABS:

ROVEL polymers can be applied to ABS sheet during fabrication using conventional coextrusion equipment.

Color Retention:

Color changes often associated with outdoor exposure of plastics are minimized with ROVEL.

Pigmentation:

ROVEL polymers are readily colored, and pigmented end products are very resistant to fading and chalking.

Balanced Physical, Mechanical and Electrical Properties:

ROVEL polymers exhibit an exceptional balance of impact, tensile, and hardness properties.

DOW CHEMICAL U.S.A.: ROVEL Weatherable Polymers(Continued):

Product Name:

ROVEL 401 Polymer:
For Sheet Extrusion

This polymer provides excellent weatherability, impact, and appearance retention. It is readily extruded into monolayer or coextruded sheet.

* Izod Impact: 15.0
* Vicat Softening Point: 221
* Melt Flow Rate: 0.9
* Weatherability: Excellent

ROVEL 501 Polymer:
For Extruding Profiles for the Building Industry

This polymer provides extraordinary high heat distortion, impact retention, and weatherability properties in easily extruded or coextruded profiles.

* Izod Impact: 14.0
* Vicat Softening Pt: 222.0
* Melt Flow Rate: 1.1
* Weatherability: Excellent

ROVEL 701 Polymer:
For Injection Molding

This high impact, easy flow grade is designed for most molding applications. It offers excellent weatherability and chemical resistance.

* Izod Impact: 13.0
* Vicat Softening Point: 225
* Melt Flow Rate: 3.0
* Weatherability: Excellent

ROVEL 705 Polymer:
For Injection Molding

This polymer provides superior chemical resistance, melt flow rate, and tensile properties, in addition to impact retention and weatherability.

* Izod Impact: 12.0
* Vicat Softening Point: 226
* Melt Flow Rate: 3.2
* Weatherability: Excellent

DOW CHEMICAL U.S.A.: SPECTRIM MM:

RIM Systems for high-performance structural composites.

Proven RIM technology advances reinforced composite performance.

SPECTRIM MM products are fast-reacting, low viscosity systems designed specifically for mat molding of structural composites. This mat molding RIM process uses conventional RIM equipment, long fiber mat reinforcement, and easy-to-manage RIM process technology. SPECTRIM MM products provide specialized properties, such as enhanced impact resistance or heat stability.

Design flexibility simplifies complex parts.
SPECTRIM MM expands both design and production capabilities.

Uniform reinforcement means high performance.

Sperior toughness and high impact strength.
Conventional reinforced RIM (RRIM) uses milled or flaked glass mixed with system components.

Excellent thermal stability.
Composites made with SPECTRIM MM products can tolerate temperatures of 400F for 30 minutes, without experiencing heat sag.

Familiar RIM Processing increases productivity, cuts costs.
Parts can be made using SPECTRIM MM on existing RIM molding equipment.

Wide processing latitude.
SPECTRIM MM products rum well on existing RIM equipment, with predictable performance, both in the mold and in the finished composite.

SPECTRIM MM 300 Reaction Moldable Product:

SPECTRIM MM300 reaction moldable product is a two-component, fast-reacting, low viscosity polycarbamate system ideally suited for mat molding of structural composites. The mat molding RIM (MM/RIM) process uses conventional RIM equipment to produce structural composites by injection of a reaction moldable system onto a preplaced fiber reinforcement.

The two components of the polycarbonate system are SPECTRIM MM 300-A Isocyanate, a specialty isocyanate, and SPECTRIM MM 300-B polyol, a fully formulated B-side.

Applications:
SPECTRIM MM products were developed for load-bearing applications requiring high impact strength. A broad range of fiberglass reinforcements may be used.
Fiber reinforced composites made with SPECTRIM MM 300 display outstanding toughness and high modulus.

DOW CHEMICAL U.S.A.: SPECTRIM MM(Continued):

Mechanical Properties of Composites Made With SPECTRIM MM 300
1/8" Thickness:

% Glass: 25.5
- Spec. Grav.: 1.37
- Flex. Mod. x 10 3 psi:
  - at 77F: 734
  - at 158D: 627
- Flex. Str. x 10 3 psi:
  - at 77F: 26.0
  - at 158F: 20.0
- Tensile Str. x 10 3 psi: 12.1
- Izod Impact Notched
  - ft. lbs./in.: 9.6
- HDT, 264 psi, F: 286

SPECTRIM MM 353 Reaction Moldable Product:

SPECTRIM MM 353 reaction moldable product is a two component, fast-reacting, low viscosity polyisocyanurate system ideally suited for mat molding of structural composites. The mat molding RIM (MM/RIM) process uses conventional RIM equipment to produce structural composites by injection of a reaction moldable system onto a preplaced fiber reinforcement.

Applications:

SPECTRIM MM products were developed for load-bearing applications which require excellent stability at elevated temperatures. A broad range of fiberglass reinforcements may be used.

Fiber reinforced composites made with SPECTRIM MM 353 display high modulus and excellent thermal stability.

Mechanical Properties of Composites Made With SPECTRIM MM 353
1/8" Thickness:

% Glass: 25
- Spec. Grav.: 1.42%
- Flex. Mod. X 10 3 psi: 900
- Flex. Str. X 10 3 psi: 29.6
- Tensile Str. X 10 3 psi: 15.1
- Elongation, %: 5.9
- Izod Impact Notched ft. lbs./in.: 6.4
- HDT, 264 psi F: 351
- HDT, 264 psi C: 177

% Glass: 57
- Spec. Grav.: 1.74
- Flex. Mod. X 10 3 psi: 1820
- Flex. Str. X 10 3 psi: 55.2
- Tensile Str. X 10 3 psi: 28.8
- Elongation, %: 7.4
- Izod Impact Notched ft.lbs./in.: 13.7
- HDT, 264 psi F: 484
- HDT, 264 psi F: 251

DOW CHEMICAL U.S.A.: SPECTRIM RIM Systems for Automotive Exteriors:

RIM makes the plastic exterior concept a commercial reality.

Exterior plastic body parts have become an accepted design solution because they provide key marketing benefits: better fit/finish, elimination of minor collision damage, and the ability to produce short-run specialty vehicles. They also provide manufacturing economies through parts consolidation and reduced labor requirements.

RIM: Everything you've always liked about plastic, but couldn't achieve in production.

Only RIM has been able to meet the combined requirements for surface quality, complexity, engineering properties, and productivity on a broad enough scale to make the all-plastic exterior a manufacturable reality.

RIM cost-effectiveness means the lowest total cost per part.

RIM has evolved as the preferred process to produce high quality, large vertical body panel parts at the lowest total per-part cost.

That's because RIM provides:

* Surface quality
* Productivity
* Paintability
* Part performance

SPECTRIM...a family of RIM systems advances the technology of automotive exterior design and production.

The family of SPECTRIM reaction moldable polymers provides a wide range of engineering material solutions for automotive exteriors. These include products specifically designed for fascia and body panel applications, where RIM can provide improved performance as well as reduced costs.

Fascia: Improving the pioneer RIM application with IMR.

Polyurea products for the future of RIM.

HT polyurea reaction moldable products under development by Dow represent state-of-the-art capabilities for producing large body panels.

They provide:

* Full range of required mechanical properties.
* High thermal and impact performance.
* High productivity and capability for automated RIM molding.
* Internal Mold Release (IMR).
* Elimination of post-cure.
* Robotic part removal potential.
* Thermal stability for on-line painting.

HF Polyurea: High productivity RIM with current equipment.

DOW CHEMICAL U.S.A.: STYRON Polystyrene Advanced Performance Resins:

STYRON Polystyrene Advanced Performance Resins for the Injection Molding Market

Strong, Consistent, Clear Resins Having Excellent Processing, Property, and Economic Advantages

STYRON polystyrene Advanced Performance Resins offer plastics molders a superior combination of properties that meet - and exceed - end-product requirements. Improved, state-of-the-art process innovation has given these clear, strong, versatile resins a combination of properties that can add up to significant cost savings for the molder, the manufacturer, and the user.

The benefits of STYRON Advanced Performance Resins present an attractive price-value balance that most molders will use to their advantage - and in several cases, already have. These benefits include:

* Superior processability...high melt flow rates, coupled with superior heat distortion temperatures and Vicat softening points, provide excellent fabrication, promote faster mold filling, and reduce cycle times. Further, injection molding and more efficient mold filling of thin-walled parts can lower materials costs through use of less resin.

* Improved product toughness...outstanding dimensional stability and high tensile and flexural strength provide excellent end-product durability and performance, with less liklehood of breakage.

* Economic advantages...easier flow and reduced molding pressures can decrease machine wear and increase energy efficiency. Increased yields of quality molded parts can dramatically improve overall productivity.

* Optimum lot-to-lot consistency...a continuous process proprietary to Dow ensures superior consistency in clarity, processability, and reliable end-product performance and uniformity.

* Excellent general properties...exceptional clarity, attractive light-transmission capabilities, biological inertness, chemical resistance, and important radiation stability all pose advantages in a variety of applications.

DOW CHEMICAL U.S.A.: STYRON Polystyrene Advanced Performance Resins(Continued):

Resin Types and Characteristics:

STYRON polystyrene Advanced Performance Resins are available in easy flow and general molding grade resins designed for injection molding, injection blow molding, and extrusion.

Product Types:

HIGH HEAT:
* For Injection Molding
* For Injection Blow Molding

Key Characteristics
* Clear, Good Hot Melt Strength
* Clear, Rigid, Good Hot Melt Strength

Markets/End Uses:
* Tape Reels
* Packaging Containers

EASY FLOW:
* For Injection Molding
* For Coextrusion

Key Characteristics:
* Good Colorability, Good Melt Flow
* Biologically Inert
* Clear, Rigid
* Good Melt Flow

Markets/End Uses:
* Disposable Dinnerware
* Medical Ware
* Toys
* Packaging, Thin-Walled Containers
* Packaging

GENERAL:
* For Injection Molding
* For Injection Blow Molding
* For Extrusion

Key Characteristics:
* Clear, Good Flow
* Biologically Inert
* Good Balance of Flow and Heat Resistance

Markets/End Uses:
* Tumblers
* Toys, Packaging
* Food Packaging
* Medical Ware
* Bottles

DOW CHEMICAL U.S.A.: STYRON Polystyrene Advanced Performance Resins(Continued):

<u>Product Number:</u>

615APR:

STYRON 615APR Advanced Performance Resin is an easy-flow polystyrene resin designed to reduce cycle times and to provide high heat resistance in finished parts. It is especially suited for medical, pharmaceutical and food uses.

* Melt Flow Rate: 15.0
* Vicat Softening Point: 212F

Properties

* Easy flowing, fast filling thin sections
* Fast cycle
* Excellent clarity
* FDA Compliance

Process:

* Injection molding
* Coextrusion
* Extrusion blending

666APR:

STYRON 666APR Advanced Performance Resin is an intermediate-flow polystyrene resin. It is designed for injection molding applications, particularly for medical, pharmaceutical and food uses.

* Melt Flow Rate: 9.0
* Vicat Softening Point: 224F

Properties:

* High heat, intermediate flow
* Excellent clarity
* Fast cycles
* FDA Compliance

Process:

* Injection molding
* Injection blow molding
* Extrusion

DOW CHEMICAL U.S.A.: STYRON Polystyrene Advanced Performance Resins(Continued):

Product Number:

675APR:

STYRON 675APR Advanced Performance Resin is an intermediate-flow resin designed for injection molding applications, particularly for electronic, consumer products, and food service uses.

* Melt Flow Rate: 9.0
* Vicat Softening Point: 224F

Properties:

* High heat, intermediate flow
* Excellent clarity
* Fast cycles
* FDA Compliance

Process:

* Injection molding
* Injection blow molding
* Extrusion

DOW CHEMICAL U.S.A.: TYRIL Styrene-Acrylonitrile Resins:

TYRIL Styrene Acrylonitrile (SAN) is a clear, hard, thermoplastic resin formulated to produce an excellent balance of strength, chemical resistance, heat resistance, clarity and processability. The overall combination of physical properties exhibited by TYRIL resins is unique in polymers that are also transparent.

How do you know when you should use TYRIL resins? When your product requires any three of the six characteristics explained below, you should consider TYRIL.

When you need clarity

Products made with TYRIL exhibit clarity and sparkle.

When high strength and physical toughness are required

The high flexural and tensile strengths of TYRIL resins make them prime candidates for applications where load bearing capabilities are a must.

When you need chemical resistance

In general, TYRIL resins offer better chemical resistance than other clear thermoplastics--many of which cost more.

When your product has to take the heat and still maintain dimensional stability

TYRIL resins offer excellent dimensional stability, even at continuous working temperatures up to 200F.

When you want excellent processability

TYRIL resins are one of the easiest to process engineering thermoplastics available today. They have high Vicat softening points and an excellent melt flow rate.

When you want to reduce costs or suspect over-design

Designers often recommend a plastic having properties that exceed those required.

DOW CHEMICAL U.S.A.: TYRIL Styrene-Acrylonitrile Resins (Continued):

Product Name:

TYRIL 880 Resin:

This SAN copolymer offers excellent physical properties, incorporating the highest levels of chemical resistance, heat resistance and overall strength of any TYRIL resin.

* High Heat Resistance
* High Strength
* Superior Chemical Resistance
* Superior Properties

TYRIL 880B Resin:

This SAN copolymer combines the high levels of chemical resistance, heat resistance and overall strength found in TYRIL 880 resin, with improved mold flow and mold release characteristics.

* High Heat Resistance
* Superior Chemical Resistance
* High Strength
* Improved Flow
* Improved Mold Release

TYRIL 1000 Resin:

This SAN copolymer provides an excellent balance of physical properties, processability and chemical resistance. It can be injection molded or extruded and has excellent optical properties, superior heat resistance and good practical toughness.

* High Heat Resistance
* Excellent Processability
* Superior Clarity
* High Strength

DOW CHEMICAL U.S.A.: TYRIL Styrene-Acrylonitrile Resins (Continued):

Product Name:

TYRIL 1000B Resin:

This SAN copolymer offers the best balance of physical properties of all TYRIL resins. It also provides excellent processability and chemical resistance. It can be injection molded or extruded and has excellent optical properties, superior heat resistance and good practical toughness.

* High Heat Resistance
* Excellent Chemical Resistance
* Excellent Processability
* Minimal Contaminates
* Improved Mold Release
* Superior Clarity
* High Strength

TYRIL 1020 Resin:

This copolymer has been stabilized for outdoor applications. Available in crystone color only.

* Zero Transmittance of UV
* Excellent Solar Heat Transmittance
* Low Re-radiation Transmittance
* Good Chemical Resistance
* Excellent Weathering Characteristics
* Non-yellowing

TYRIL 1021 Resin:

This copolymer has been designed specifically for injection molding and provides an excellent balance of properties.

* Injection Molding Grade
* Excellent Weathering Characteristics
* Non-yellowing
* High Heat Resistance
* Excellent Processability
* High Strength

DOW CHEMICAL U.S.A.: TYRIL Styrene-Acrylonitrile Resins (Continued):

Product Name:

TYRIL 1031 Resin:

This light stabilized general purpose injection molding resin is available in crystone color only.

* Excellent Weathering Characteristics
* Non-yellowing
* Good Chemical Resistance
* Excellent Processability

DOW SAN 125 Blending Resin:

This SAN resin is specifically formulated for blending and is an ideal carrier material for use in making color concentrates.

* Excellent Chemical Resistance
* High Melt Flow
* High Performance Blend Base

TYRIL 1030 Resin:

This light stabilized SAN resin contains an antistatic additive and is available in transparent crystone color only. It is especially suited for uses where dust attraction or static build-up may be a problem.

* Non-yellowing
* Excellent Weathering Characteristics
* Good Chemical Resistance
* Superior Antistatic Properties
* Crystone Tint

E.I. DUPONT DE NEMOURS & CO., INC.: ARYLON Polyarylate Resin:

With end-use benefits and processing advantages superior to many other amorphous resins and available at a lower cost, you'll find ARYLON polyarylate resin may be your best choice for future and current designs. ARYLON offers high-performance properties at significantly lower costs than competitive resins!

ARYLON offers an exiting array of high performance properties, such as dimensional stability, heat resistance, impact strength and good electricals. It also has inherent UV stability for excellent weatherability.

ARYLON brings you closer to the future for:

* Electrical/electronic housings, components and lighting
* Appliance housings and trim
* Consumer goods
* Automotive hardware, interior components, and accessories

ARYLON LP 101 NC10 Polyarylate Resin
- Moisture Absorption (D570) (%)
  - 2 hr, 50% RH: 0.02
  - 24 hr, 50% RH: 0.10
- Specific Gravity (72/72) (D792): 1.21
- Flamability (UL94):
  - UL Class: 1/16": V-2
  - 1/8": V-O
- LOI: %: 37
- Melt Viscosity: Poise: 6000

ARYLON 401 NC10 Polyarylate Resin: Unreinforced
- Melt Viscosity
  - 626F (330C), 1000 1/sec
  - L = 0.6 in (0.15 mm): Poise: 4250
  - D = 0.03 in (0.00762 mm): Pa-S: 425
- Specific Gravity (D-792): 1.19
- Rockwell Hardness (D-785-65):
  - M = 66
  - R = 122
- Moisture Absorption (D-955) (%)
  - 2 hr. 50% RH: 0.02
  - 24 hr. 50% RH: 0.09
- Flammability (UL-94)
  - UL Class: 1/16 inch (1.6 mm): HB
  - UL Class: 1/8 inch (3.18 mm): HB

ARYLON 501 BK103 Polyarylate Resin:
- Melt Viscosity: 626F (330C), 1000 1/sec
  - L = 0.6 in (0.15 mm): Poise: 5500
  - D = 0.03 in (0.00762 mm): Pa-S: 550
- Specific Gravity (D-792): 1.44
- Rockwell Hardness (D785-65)
  - M: 85
  - R: 122
- Flammability (UL94):
  - UL Class: 1/16 in (1.6 mm): V-O
  - UL Class: 1/8 in (3.18 mm): V-O

E.I. DUPONT DE NEMOURS & CO., INC.: DELRIN Acetal Resin:

DELRIN is DuPont's registered trademark for its acetal resins.

DELRIN acetal resin is a crystalline plastic made by the polymerization of formaldehyde. It offers an excellent balance of desirable properties that bridge the gap between metals and plastics; in fact; DELRIN has replaced metals in many important mechanical and structural applications.

The outstanding characteristics of DELRIN resins are: stiffness, tensile strength and creep resistance under a wide range of temperature and humidity conditions, high fatigue endurance, corrosion resistance and resilience for applications requiring springiness and toughness.

Applications

Chemical
* Garden chemical sprayers
* Carburetor venturi
* Lipstick containers
* Farm machinery
* Household water softeners
* Soap dispensers
* Paint mixing paddles, canisters.
* Scrubbing disc-power rug cleaner
* Irrigation sprinkler components

Mechanical
* Wear surfaces - many types of bushings, bearings, cams.
* Gears - spur, cluster, helical, bevel, worm.
* Conveyors - flat top conveyor chain, wear strips, conveyor links.
* Pump impellers - submersible, gear, centrifugal, proportioning, diaphraghm, jet well.
* Fan and blower blades.
* Handles and knobs - automotive window and marine hardware, auto brake release handles, knife handles, movie and slide projector knobs.
* Springs - tape embossers, camera, print wheels, staple removers, drip emitters, tubing clamps.

Electrical:
* Automotive switches
* Telephone terminal strips
* Coil forms
* Relay components
* Electrical switch components, buttons and knobs
* Audio and video tape cartridge platforms and wear parts
* Electromechanical counter frames

Environmental Conditions
* Hot air - heater blowers, clothes dryers
* Water - showerheads, faucet aerators, lawn sprinklers
* Salt water - marine fittings, fishing reels

E.I. DUPONT DE NEMOURS & CO., INC.: DELRIN Acetal Resin (Continued):

Standard Compositions:

DELRIN 100(1,2)
DELRIN 107(1)
DELRIN 100F(1,2)
DELRIN 100D(1)
Usual Methods of Working: Injection Molding
DELRIN 150SA(1,2)
Usual Methods of Working: Extrusion
Characteristics: Special purpose molding or extrusion; high viscosity, hence suitable for easy-to-fill molds. Maximum toughness in an unmodified resin. Surface lubricated. DELRIN 107 is a UV stabilized product like DELRIN 100. Delrin 100F, and 100D are designed for improved processing.
Uses: Molded or extruded parts, such as highly loaded gears, requiring greater toughness or elongation than available in DELRIN 500 without a sacrifice in other mechanical properties. Mill shapes for machining production parts, blown bottles, sheeting, tube and rod.

DELRIN 100AF(3)
Usual Methods of Working: Injection Molding or Extrusion
Characteristics: DELRIN 100 base resin containing TEFLON TFE fibers, for reduction in coefficient in friction.
Uses: Low wear/low friction applications where an improvement in toughness and resistance to fatigue are important.

DELRIN 500(1,2)
DELRIN 507(1)
Usual Methods of Working: Injection Molding or Extrusion
DELRIN 500F(1,2)
Usual Methods of Working: Injection Molding
DELRIN 500D(1)
Usual Methods of Working: Extrusion
DELRIN 500HP(1)
Usual Methods of Working: Injection Molding
DELRIN 550SA(1,2)
Usual Methods of Working: Extrusion
Characteristics: General purpose molding or extrusion; surface lubricated. DELRIN 507 is similar to DELRIN 500 but contains a UV stabilizer. DELRIN 500F, 500D and 500HP are designed for improved processing.
Uses: General mechanical parts, mill shapes for production machining, rod stock for prototype fabrication. DELRIN 507 is used in parts which require improved weathering. DELRIN 500F, 500D, and 500HP are used in parts for which surface appearance is critical.

E.I. DUPONT DE NEMOURS & CO., INC.: DELRIN Acetal Resin (Continued):

Standard Compositions:

DELRIN 500AF(3)
Usual Methods Of Working: Injection Molding or Extrusion
Characteristics: DELRIN 500 base resin containing TEFLON TFE fibers of fillers.
Uses: Parts where a low coefficient of friction, high resistance to abrasion and wear are required.
DELRIN 500CL(3)
Usual Methods of Working: Injection Molding or Extrusion
Characteristics: DELRIN 500 base resin containing an internal lubricant.
Uses: This is the medium viscosity analog of DELRIN 100CL.
DELRIN 570(1,3)
DELRIN 577(1,3)
Usual Methods of Working: Injection Molding or Extrusion
Characteristics: Special glass-filled composition offering very high stiffness, low warpage and low creep, particularly in use above room temperature. DELRIN 577 is a weatherable black version.
Uses: General mechanical parts, mill shapes, rod and slabs, where improved stiffness, etc. are needed.
DELRIN 900(1,2)
DELRIN 900F(1,2)
DELRIN 900D(1)
DELRIN 907F(1)
DELRIN 907HP(1)
Usual Methods of Working: Injection Molding
Characteristics: Low viscosity, hence better melt flow than DELRIN 500. Surface lubricated. DELRIN 907F is UV stabilized. DELRIN 900D, 900HP, 900F are designed for improved processing.
Uses: Multicavity molds and thin sections which are difficult to fill. Mechanical properties similar to DELRIN 500, except DELRIN 900 grades have somewhat lower tensile elongation and toughness
DELRIN 200PL(1,2)
Usual Methods of Working: Injection Moldin
Characteristics: Super tough acetal-elastomer alloy.
Uses: Mechanical parts where the ultimate in impact resistance is needed. There is some reduction in other mechanical properties.
DELRIN 500T
Usual Method of Working: Injection Molding
Characteristics: Toughened general purpose acetal-elastomer alloy. DELRIN 500T also shows improved wear resistance.
Uses: Applications where an improvement in resistance to impact is required but with the other mechanical properties of an acetal.
Footnotes to Composition:
(1) Meets FDA food contact requirements up to 250F(121C).
(2) Meets NSF requirements: contact with potable water to 180F.
(3) Meets NSF requirements: contact with potable water to 99F.

E.I. DUPONT DE NEMOURS & CO., INC.: HYTREL Polyester Elastomer:

HYTREL is DuPont's registered trademark for its family of polyester thermoplastic elastomers.

Properties and Characteristics

HYTREL is an engineering thermoplastic elastomer which combines many of the desirable characteristics of high-performance elastomers and flexible plastics. It features: exceptional toughness and resilience; high resistance to creep, impact and flex fatigue; flexibility at low temperatures; and good retention of properties at elevated temperatures. In addition, it resists deterioration from many industrial chemicals, oils and solvents.

Processing

HYTREL can be readily formed into high-performance products by a variety of thermoplastic processing techniques, including injection molding, extrusion, blow moldings, rotational molding and melt castings. HYTREL is processed at temperatures between 177 and 260C (350 and 500F), depending on the process and polymer type. All grades have a sharp melting point and very good melt stability.

Composition

HYTREL polyester elastomers are all block copolymers, consisting of a hard (crystalline) segment of polybutylene terephthalate and a soft (amorphous) segment based on long-chain polyether glycols. Properties are determined by the ratio of hard to soft segments and by the make-up of the segments.

The available grades of HYTREL are grouped into four categories, by performance:

1) High-Productivity grades exhibit more versatile processing characteristics, are lowest in cost and are suitable for most applications;
2) High-Performance grades provide an extra margin of strength or serviceability for the most demanding applications;
3) Specialty grades provide special properties or processing characteristics for particular applications;
4) Concentrates contain relatively high concentrations of specific property-enhancing additives for blending with other grades of HYTREL.

E.I. DUPONT DE NEMOURS & CO., INC.: HYTREL Polyester Elastomer (Continued):

High-Productivity Grades:

These grades offer the best balance of properties and cost.

HYTREL G-3548:

Description: Lowest modulus molding and extrusion grade. Contains color-stable antioxidants and some UV stabilizer.

Characteristics: Most flexible grade of HYTREL. Excellent flex-fatigue resistance, especially at low temperatures. Moldable even in thin sections. Can be used in light colored products.

Typical Uses: Applications requiring flex life coupled with good flexibility at low temperatures. Thin, flexible membranes.

HYTREL G-4074:

Description: Low modulus molding and extrusion grade. Contains a discoloring antioxidant.

Characteristics: Excellent heat-aging resistance and resistance to oils at high temperatures. Best low modulus molding and extrusion grade.

Typical Uses: Tubing-Hose jackets-Wire and cable jackets-Film sheeting-Molded products

HYTREL G-4078:

Description: Low modulus molding and extrusion grade. Contains color-stable antioxidants and some UV stabilizer.

Characteristics: Like HYTREL G-4074, except that heat-aging resistance is reduced (similar to HYTREL 4056). Can be used in light-colored products.

Typical Uses: Same as HYTREL G-4074.

HYTREL G-4774:

Description: Medium-low modulus molding and extrusion grade. Contains a discoloring antioxidant

HYTREL G-5544:

Description: Medium modulus molding and extrusion grade. Contains a discoloring antioxidant.

Characteristics: Excellent heat-aging resistance and resistance to oils at high temperatures. Good resistance to oils, fuels and solvents.

Typical Uses: Tubing-Hose jackets-Wire and cable jackets-Profiles-Molded products.

E.I. DUPONT DE NEMOURS & CO., INC.: HYTREL Polyester Elastomer (Continued):

High-Performance Grades:

These grades provide an extra measure of strength or serviceability in the most demanding applications; can all be used in light-colored products(except HYTREL 5555HS).

HYTREL 4056:
Description: Low modulus extrusion grade. Contains color-stable antioxidants.
Characteristics: Excellent low-temperature properties. Excellent flex-fatigue resistance. Excellent creep resistance.
Typical Uses: Hose jackets-Wire and cable jackets-Film and sheeting-Belting-Seals

HYTREL 4059 FG:
Description: Low modulus molding and extrusion grade. Contains a color-stable antioxidant which is suitable for applications covered by FDA regulations.
Characteristics: Low modulus food grade similar to HYTREL 4056, except a higher melting point and reduced heat aging resistance.
Typical Uses: Same as HYTREL 4056. Can be used in food-handling applications, such as conveyor belts and tubing.

HYTREL 5526:
Description: Medium modulus molding grade. Contains color-stable antioxidants.
Characteristics: Combine the best balance or properties of the product line.
Typical Uses: Seals, packing and gaskets. Gears and bearings.

HYTREL 5556:
Description: Medium modulus extrusion grade. Contains color-stable antioxidants.
Characteristics: Combine the best balance or properties of the product line.
Typical Uses: Tubing and hose-Wire and cable jackets-Film and sheeting-Belting

HYTREL 5555 HS:
Description: Heat-stabilized grade of HYTREL 5556. Contains a discoloring antioxidant.
Characteristics: Combine the best balance or properties of the product line.
Typical Uses: Same as above. Used where increased heat-aging stability is required.

E.I. DUPONT DE NEMOURS & CO., INC.: HYTREL Polyester Elastomer (Continued):

High-Performance Grades:

HYTREL 6356:
Description: Medium-high modulus molding and extrusion grade. Contains color-stable antioxidants.

HYTREL 6359 FG:
Description: Same as HYTREL 6356, except contains a color-stable antioxidant which is suitable for applications covered by FDA regulations.

Characteristics: Very good resistance to oils, hydraulic fluids and fuels. Very good resistance to permeation by gases and liquids. HYTREL 6359FG has reduced heat aging resistance compared to HYTREL 6356.
Typical Uses: Tubing and hose-Film-Profiles-Seals-Gears and sprockets-Fuel tanks

HYTREL 7246:
Description: Highest modulus molding and extrusion grade. Contains color-stable antioxidants.
Characteristics: Highest service temperature. Retains good low-temperature flexibity. Best resistance to oils, fuels and solvents. Lowest fuel permeability.
Typical Uses: Tubing-Wire and cable jackets-Gears and sprockets-Oil field parts

Specialty Grades:

HTR-5612:
Description: Medium-low modulus, high viscosity grades. Contains a discoloring antioxidant

HTR-5612BK:
Description: HTR-5612B is pigmented black. Particularly suitable for extrusion blow molding and extrusion. May be injection blow-molded.

HTR-4275:
HTR-4275BK:
Description: Medium modulus, high viscosity grades. Contains a discoloring antioxidant. HTR-4275B is pigmented black. Particularly suitable for extrusion blow molding and extrusion.

Characteristics: Good balance of properties combined with high viscosity for extrusion and blow molding applications.
Typical Uses: Hollow thin-walled parts-Blown film and sheeting-Large diameter tubing-Hose mandrels-Profiles

E.I. DUPONT DE NEMOURS & CO., INC.: HYTREL Polyester Elastomer (Continued):

Specialty Grades(Continued):

HTR-5861BK:

Description: Medium-low modulus, high-viscosity grade. Contains a discoloring antioxidant and is pigmented black. Particularly suitable for extrusion blow molding and extrusion.

Characteristics: Good balance of properties combined with high viscosity for extrusion and blow molding applications.

Typical Uess: Hollow thin-walled parts-Blown film and sheeting-Large diameter tubing-Hose mandrels-Profiles

HTR-5735:

Description: Medium modulus grade. Contains a low level of color-stable antioxidants.

Characteristics: Provides stain resistance and low dirt pickup. Low permeability to oils, fuels and plasticizers. High clarity in thin films.

Typical Uses: Thin barrier coating over various substrates. Co-extruded coatings for liners and sheet goods.

HTR-6108:

Description: Medium modulus grade. Contains color-stable antioxidants.

Characteristics: Low permeability to oils, fuels and plasticizers--approx. one-third that of other grades of similar stiffness. High clarity in thin films.

Typical Uses: Applications requiring good flexibility, coupled with low permeability to fuels, oils and plasticizers. Coextrudable barrier membrane over more permeable substrates.

E.I. DUPONT DE NEMOURS & CO., INC.: HYTREL Polyester Elastomer (Continued):

Concentrates:

These concentrates are in a base of HYTREL 4056.

HYTREL 40 CB:
Description: Concentrate of a fine particle size carbon black.

HYTREL 20 UV:
Description: Ultraviolet (UV) light stabilizer concentrate.

Characteristics and Typical Uses:
HYTREL must be protected against degradation from exposure to UV light when used outdoors or when exposed to sunlight.

HYTREL 40 CB provides the most effective protection. Recommended let-down ratio for direct outdoor exposure is 7:1.

HYTREL 20 UV is used for protection of light-colored parts against UV degradation. Recommmended let-down ratio is 25:1 or less.

HYTREL 30 HS:
Description: Heat stabilizer concentrate.

Characteristics and Typical Uses: For blending with other grades of HYTREL to retard thermal oxidative degradation and extend useful life at elevated temperatures. Recommmended let-down ratio is between 16:1 and 40:1, usually about 20:1.

HYTREL 10 MS:
Description: Hydrolytic stabilizer concentrate.

Characteristics and Typical Uses: For blending with other grades of HYTREL to improve serviceability in hot, moist environments. Recommended let-down ratio is 9:1.

E.I. DUPONT DE NEMOURS & CO., INC.: MINLON Engineering Thermoplastic Resin:

MINLON is DuPont's trademark for engineering thermoplastic resins that combine DuPont nylon with a mineral or mineral/glass combination. It is important to point out that these compositions are reinforced, not filled resins, meaning that there is a chemical bond between the nylon and the mineral or mineral/glass particles, thus greatly enhancing the tensile and stiffness properties.

Various grades of MINLON are designed to give a range of enhanced stiffness, strength, heat resistance and impact strength. All grades can be painted and machined, and one can be chrome-plated.

MINLON engineering plastics are noted for their good processability. All are fast cycling materials. The addition of mineral (or mineral/glass) decreases the set-up time of the nylon, and this, combined with easy flow, makes MINLON a definite choice for high productivity.

The properties of MINLON resins summarized should be useful in choosing the most suitable resin. Choose MINLON 10B for high stiffness, MINLON 11C for stiffness with impact strength, and MINLON 12T for optimum impact strength.

MINLON 12T has all the recognized advantages of mineral-reinforced nylons, including low warpage and good stiffness, while offering superior toughness in terms of all recognized impact strength measurements.

MINLON 20B and 22C are mineral/glass-reinforced nylons with increased stiffness and heat resistance.

The broad range of physical properties offered by these product variations give MINLON the ability to replace metals, as well as other reinforced and unreinforced engineering thermoplastics in engine components, housings, and exterior automotive body parts.

E.I. DUPONT DE NEMOURS & CO., INC.: RYNITE Thermoplastic Polyester Resin:

RYNITE is DuPont's registered trademark for its thermoplastic polyester resins.

Current grades of RYNITE resins contain uniformly dispersed glass or mica/glass combinations in polyethylene terephthalate, specifically formulated for rapid crystallization during the injection molding process. It is important to point out that these compositions are reinforced, not filled. There is a chemical bond established between the polyethylene terephthalate and the glass or mica/glass particles, thus greatly enhancing strength and stiffness properties. Additionally, RYNITE resins offer excellent dimensional stability, outstanding chemical and heat resistance, and inherently good electrical properties. This, coupled with a competitive price, leads to high value-in-use.

Four different grades of RYNITE thermoplastic polyester resins are available: General Purpose; Low Warp; Toughened; and Flame Retarded.

RYNITE thermoplastic polyester resins are noted for their easy fill and excellent melt flow characteristics in the injection molding process. This leads to close molding tolerance and high productivity from multicavity tools, and therefore lower part costs.

Variety of Compositions

Choose a General Purpose grade when you need a combination of stiffness, strength and toughness; a Low Warp grade for stiffness, strength and dimensional control; a Toughened Grade for outstanding impact resistance combined with stiffness; and a Flame Retarded grade when V-O characteristics and high temperature resistance are important

Within these grades, a total of eleven product compositions are available. The following are of particular interest.

RYNITE SST 35, Stiffened Super Tough thermoplastic polyester has the highest combined stiffness/toughness of any thermoplastic. It also offers outstanding surface appearance, high and low temperature performance, electrical properties and dimensional stability.

RYNITE FR-543, a Flame Retarded resin, has a UL 94 Temperature Index of 155C and is equivalent to many thermosets.

E.I. DUPONT DE NEMOURS & CO, INC.: RYNITE Thermoplastic Polyester Resin(Continued):

GENERAL PURPOSE GRADES:

Standard Compositions:

RYNITE 530:

Characteristics: 30% glass-reinforced modified polyethylene terephthalate-outstanding balance of strength, stiffness and toughness, excellent electrical properties, surface appearance and chemical resistance.

Candidate Uses: Electrical/electronic parts such as ignition components, relay bases, lamp sockets, bobbins; housings and other parts for pumps; mechanical components including gears, sprockets, vacuum cleaner parts, motor end bells; chair arms, casters and other furniture parts.

RYNITE 545:

Characteristics: 45% glass-reinforced modified polyethylene terephthalate--greater strength and stiffness, excellent dimensional stability and creep resistance.

Candidate Uses: Lamp housings, compressor housings, fuel, air and temperature sensor housings, sun roof frames, spools, bobbins, transmission components, medical devices.

RYNITE 555:

Characteristics: 55% glass-reinforced, modified polyethylene terephthalate--superior stiffness, dimensional stability, heat resistance, and outstanding resistance to creep.

Candidate Uses: Structural support brackets, housings and covers, auto parts, bicycle components, propellors.

LOW WARP GRADES:

RYNITE 935:

Characteristics: 35% mica/glass reinforced, modified polyethylene terephthalate-exceptionally low warpage, excellent electrical properties, high stiffness, high heat resistance.

Candidate Uses: Exterior body parts, structural housings and frames, irrigation components including transformer and ignition coil housings.

RYNITE 940:

Characteristics: 40% mica/glass-reinforced modified polyethylene terephthalate--greater strength, stiffness and low warpage.

Candidate Uses: Frames, exterior body parts. Structural supports.

E.I. DUPONT DE NEMOURS & CO., INC.: RYNITE Thermoplastic Polyester Resin(Continued):

LOW WARP GRADES:
Standard Compositions:

RYNITE 940 FB:
Characteristics: Foam injection molding version of RYNITE 940 for minimum sinks opposite ribs and bosses.
Candidate Uses: Frames, exterior body parts.

TOUGHENED GRADES:

RYNITE 430:
Characteristics: 30% glass-reinforced modified polyethylene terephthlate--improved impact resistance--Excellent balance of strength, stiffness, and temperature resistance.
Candidate Uses: Water pump housings, structural housings and brackets, windshield wiper blade supports.

RYNITE SST 35:
Characteristics: 35% Stiffened Super Tough glass-reinforced modified polyethylene terephthalate--superior combination of toughness and stiffness. Excellent surface appearance, moldability and temperature resistance.
Candidate Uses: Automotive parts, wheels, yard and shop tools, sporting goods, luggage components, appliance housings, structural furniture components.

FLAME-RETARDED GRADES:

RYNITE FR-530:
Characteristics: Flame-retarded, 30% glass reinforced modified polyethylene terephthalate. Recognized by UL as 94 V-O with temperature indexing of 150C. Outstanding balance of properties and excellent flow characteristics.
Candidate Uses: Electrical and electronic connectors and other components requiring flame retarded characteristics. Used in applications employing vapor phase and wave soldering techniques.

RYNITE FR-543:
Characteristics: Flame-retarded 43% glass-reinforced polyethylene terephthalate. 155C temperature index--equivalent to many thermosets.
Candidate Uses: Electrical/electronic applications such as relays switches, lighting ballasts, and terminal blocks.

RYNITE FR-945:
Characteristics: Flame-retarded, 45% mineral/glass reinforced modified polyethylene terephthalate. Recognized by UL as 94 V-O at 1/32". Low warpage, high stiffness and economical price.
Candidate Uses: Electrical & electronic components. Economical for large parts requiring flame-retarded characteristics, such as motor housings, bobbins, terminal blocks, fans.

E. I. DUPONT DE NEMOURS & CO., INC.: ZYTEL Nylon Resins:

ZYTEL is the DuPont Trademark for the many different Nylon resins which the company makes. ZYTEL nylon resins are thermoplastic polyamides.

The principal ZYTEL nylon resins may be divided by chemical composition into four basic groups--66 nylon, 612 nylon, 6 nylon and copolymers, all of which may be modified to give special properties.

ZYTEL Nylon Resins for Moldings and Extrusions

Designation

66 Nylons-Melt at 491F(255C)-Stiff and strong over a wide range of temperatures. Excellent toughness and chemical resistance.

ZYTEL 101:
Description: General Purpose--Unlubricated
Characteristics and Major Uses: Basic 66 nylon. Unmodified 66 nylon of molding viscosity. The industry standard.

ZYTEL 101 L:
Description: General Purpose--Lubricated
Characteristics and Major Uses: A 66 nylon lubricated for improved machine feed and mold release characteristics. Widely used in injection molding for mechanical parts, consumer products, etc.

ZYTEL 101 F:
Description: General Purpose--Fast Cycle
Characteristics and Major Uses: A non-nucleated 66 nylon for optimum molding performance.

ZYTEL 103 HS-L:
Description: Heat Stabilized--Lubricated
Characteristics and Major Uses: New, improved heat stabilized 66 nylon designed to retard embrittlement at high service temperatures. Has a 130C UL rating for electrical use. Optimum stabilization for heat life and good electrical properties. Lubricated for improved machine feed and mold release.

ZYTEL 105 BK-10A:
Description: Weather Resistant
Characteristics and Major Uses: Contains well-dispersed carbon black for maximum resistance to weathering.

E.I. DUPONT DE NEMOURS & CO., INC.: ZYTEL Nylon Resins (Continued):

66 Nylons(Continued):

Designation:

ZYTEL 122 L:
Description: Hydrolysis Resistant
Characteristics and Major Uses: Stabilized to resist hydrolysis and oxidation in long-term exposure to hot water. Lubricated for improved machine feed and mold release.

ZYTEL 42:
Description: High Viscosity for Extrusion
Characteristics and Major Uses: For extrusion into rod, tubing and complex shapes. Can be molded into parts requiring high impact resistance.

Modified 66 nylons--Melt at 491F(255C)--Like 66 nylon with added impact resistance and flexibility.

ZYTEL 408:
Description: General Purpose
Characteristics and Major Uses: Modified resin with superior toughness and moldability.

ZYTEL 408 HS:
Description: Heat Stabilized
Characteristics and Major Uses: A new, improved heat stabilized modified 66 nylon.

ZYTEL 408 L:
Description: General Purpose--Lubricated
Characteristics and Major Uses: A lubricated version of ZYTEL 408, for improved mold release.

ZYTEL 3189:
Description: General Purpose
Characteristics and Major Uses: Impact strength between ZYTEL 408 and ZYTEL ST 801.

Super Tough Nylons--Melt at 491F(255C)--Highest impact resistance of any engineering thermoplastic.

ZYTEL ST 801:
Description: General Purpose
Characteristics and Major Uses: Outstanding impact resistance. Good moldability.

E.I. DUPONT DE NEMOURS & CO., INC.: ZYTEL Nylon Resins (Continued):

Super Tough Nylons (Continued):

Designation:

ZYTEL ST 801 BK-10:
Description: Weather Resistant
Characteristics and Major Uses: Contains well dispersed carbon black for maximum resistance to weathering; outstanding impact resistance.

ZYTEL ST 801 HS:
Description: Heat Stabilized
Characteristics and Major Uses: Heat stabilized version of ZYTEL ST 801.

612 nylons--Melt at 414F (212C)--Low moisture absorption and excellent dimensional stability:

ZYTEL 151 L:
Description: General Purpose--Lubricated
Characteristics and Major Uses: A 612 nylon lubricated for improved machine feed and mold release.

ZYTEL 158 L:
Description: General Purpose-Lubricated
Characteristics and Major Uses: Higher melt viscosity and greater toughness than ZYTEL 151 L. Lubricated for improved machine feed and mold release.

ZYTEL 153 HS-L:
Description: Heat Stabilized-Lubricated
Characteristics and Major Uses: Heat stabilized ZYTEL 158 L to retard embrittlement at high service temperatures. Primarily for wire jacketing.

ZYTEL 157 HS-L BK-10:
Description: Weather and Heat Resistant-Lubricated
Characteristics and Major Uses: Contains well-dispersed carbon black for maximum resistance to weathering. Heat stabilized. Lubricated for improved machine feed and mold release.

E.I. DUPONT DE NEMOURS & CO., INC.: ZYTEL Nylon Resins (Continued):

Designation:

Glass-reinforced Nylons--Very high strength, stiffness and toughness. Excellent creep resistance and dimensional stability.

ZYTEL 70G L:
Description: General Purpose-Lubricated
Characteristics and Major Uses: 66 nylon reinforced with short glass fibers. Available in 13, 33 and 43% nominal glass content by weight.

ZYTEL 70G HS1-L:
Description: Heat Stabilized-Lubricated
Characteristics and Major Uses: Heat stabilized ZYTEL 70 G L. Available in 13 and 33% nominal glass content by weight.

ZYTEL 70G HR-L:
Description: Hydrolysis Resistant-Lubricated
Characteristics and Major Uses: ZYTEL 70G L with added resistance to hydrolysis and oxidation. Available in 33% nomimal glass content by weight.

ZYTEL 71GL:
Description: General Purpose-Lubricated
Characteristics and Major Uses: Modified 66 nylon with short glass fibers. Provides additional toughness and outstanding dimensional stability. Available in 13 and 33% nominal glass content by weight.

ZYTEL 77G L:
Description: General Purpose-Lubricated
Characteristics and Major Uses: 612 nylon reinforced with short glass fibers. Excellent toughness and outstanding dimensional stability. Available in 33 and 43% nominal glass content by weight.

E.I. DUPONT DE NEMOURS & CO., INC.: ZYTEL Nylon Resins (Continued):

Designation:

Miscellaneous Products-including copolymers, modified polymers, blends and unextracted 6 nylon. Properties tailored for specific uses.

ZYTEL 91 HS1-L:

Description: Heat Stabilized

Characteristics and Major Uses: A plasticized resin for flexible tubing and cable jacketing. Has low permeability to "FREON" for use in air conditioner hose.

ZYTEL 109 L:

Description: General Purpose-Color Stabilized, Nucleated, Lubricated.

Characteristics and Major Uses: Easy processing at the expense of stiffness and high temperature properties. Excellent for heavy section moldings. Color stabilized and nucleated. Lubricated for machine feed and mold release.

ZYTEL 211:

Description: General Purpose-Unextracted 6 Nylon

Characteristics and Major Uses: Relatively flexible for high impact uses. Melts at 410F(210C). For moldings and extrusion requiring flexibility and toughness.

Flame Retarded Nylons:

Description: General Purpose-Flame Retarded

Characteristics and Major Uses: Several compositions are available, rated 94V-O and 94V-1 by Underwriters' Laboratories, Inc.

E.I. DUPONT DE NEMOURS & CO., INC.: GRZ Glass-Reinforced ZYTEL Nylon Resins:

There are three basic types of GRZ resins:

ZYTEL 70G for high strength.
This series is based on 66 nylon. There are three glass loadings--13,33 and 43%, all lubricated for improved feed and mold release, as are all GRZ nylons. As might be expected, stiffness and tensile strength increase with higher glass contents in the ZYTEL 70G series.

ZYTEL 71G for high impact strength.
This series is based on a modified 66 nylon designed for increased impact strength. It has outstanding resistance to impact with only moderate reduction of the mechanical and load carrying properties found in the 70G series. There are two glass loadings-13 and 33%.

ZYTEL 77G for dimensional stability.
This series is based on a 612 nylon, and consequently absorbs less water than the other two groups. This family of GRZ resins offers its own significant property advantages compared with those using other base resins. Of major importance is the outstanding retention of key physical and electrical properties over a wide humidity range. Maximum dimensional stability is achieved by combining the excellent low moisture absorption characteristics of the base resin with glass fiber reinforcement.

In general, the GRZ nylon resins offer these advantages over unreinforced nylons:

* superior tensile strength (over 2 times greater)
* greater stiffness (3 times greater)
* higher impact strength
* excellent fatigue endurance
* enhanced creep resistance
* excellent retention of tensile strength and stiffness at high temperatures
* dimensional stability better than unrinforced 66 nylon
* superior retention of physical properties when exposed to high temperatures, hot oils, greases and lubricants
* low thermal expansion (similar to metals)

E.I. DUPONT DE NEMOURS & CO., INC.: GRZ Glass-Reinforced ZYTEL Nylon Resins(Continued):

<u>Designation:</u>

High Strength 70G Series Based on 66 Nylons:

ZYTEL 70G 13 L:
General purpose with 13% glass reinforcement. Lubricated for improved machine feed and mold release.

ZYTEL 70G 13 HS1-L:
Contains 13% glass reinforcement. Heat stabilized for long term service at high temperatures. Lubricated for improved machine feed and mold release.

ZYTEL 70G 33 L:
Contains 33% glass reinforcement. High strength. Lubricated.

ZYTEL 70G 33 HS1-L:
Contains 33% glass reinforcement. High strength. Lubricated. Heat stabilized.

ZYTEL 70G 43 L:
Contains 43% glass reinforcement. Maximum strength. Lubricated.

ZYTEL 70G 43 HR-L:
Contains 33% glass reinforcement. High strength. Has an unusually high resistance to hot water and oxidation. Lubricated.

High Impact 71G Series Based on Modified 66 Nylons:

ZYTEL 71G 13L:
Contains 13% glass reinforcement. Lubricated. Used where shock loading is encountered and maximum toughness is required.

ZYTEL 71G 13 HS1-L:
Contains 13% glass reinforcement. Lubricated. Is heat stabized and used where shock loading is encountered, especially at elevated service temperatures.

ZYTEL 71G 33L:
Contains 33% glass reinforcement. Lubricated. Has an outstanding resistance to impact with only moderate reduction of the mechanical and load carrying properties found in the 70G Series.

E.I. DUPONT DE NEMOURS & CO., INC.: GRZ Glass-Reinforced ZYTEL Nylon Resins(Continued):

Designation:

Low Moisture Absorption 77G Series Based on 612 Nylons:

ZYTEL 77G 33 L:
Contains 33% glass reinforcement. Lubricated. Lower moisture absorption than in comparable members of the 70G and 71G series results in higher dimensional stability. Excellent electrical properties.

ZYTEL 77G 43 L:
Contains 43% glass reinforcement. Lubricated. Highest dimensional stability in the glass-reinforced nylons. Excellent electrical properties over a wide range of humidities.

EAGLE PICHER PLASTICS DIVISION: Products:

EP-BGC-184:

Acid & Heat Resistant B.M.C.

This bulk molding compound has an excellent history of performance in hostile environments at continuous elevated temperatures. Its retention of mechanical strengths at these temperatures is outstanding.

Physical Properties:
- Flexural Strength: 18,070 P.S.I. (D-790)
- Flexural Strength @ 130C: 16,900 P.S.I. (D-790)
- Flexural Strength after 1 wk. @ 150C: 18,290 P.S.I. (D-790)
- Flexural Modulus: 1.6 x 10 6 P.S.I. (D-790)
- Tensile Strength: 4,660 P.S.I. (D-638)
- Impact Strength, Notched: 10.24 (D-256)
- Compressive Strength: 24,950 (D-695)

EP-BLE-70-12:

Electrical Grade, Self-Extinguishing B.M.C.

A self-extinguishing fiberglass reinforced polyester, low-profile BMC that combines excellent electrical properties with high strength and extremely fine dimensional stability and surface appearance. It may be compression, transfer or injection molded.

This material is designed for parts requiring good appearance, close dimensional tolerance and retains 80% of mechanical properties after 2 week immersion in 90F water bath. This compound is pigmentable to meet most color requirements.

Physical Properties:

- Flexural Strength: 12,000 P.S.I. (D-790-63)
- Flexural Modulus: 1.5 x 10 6 P.S.I.
- Tensile Strength: 5,000 P.S.I. (D-651)
- Notched Izod Impact: 4.0 ft. lbs. (D-256-56)
- Heat Distortion Temperature: 300F D-648 (266 PSI)
- Water Absorption (24 hrs.): .38% (D-570)
- Specific Gravity: 1.95
- Flammability: Self-Extinguishing
- Compressive Strength: 20,100 P.S.I.
- Color - Mottled Off-White

EAGLE-PICHER PLASTICS DIVISION: Products(Continued):

EP-SG-154:

High Strength H.M.C.

This compound has been formulated to offer designers an alternative to hot rolled steel for the manufacture of structural components. Weight per unit volume is 25% of that of steel, and 65% of aluminum, making possible significant reduction of product weight. The proven corrosion resistance and processability of FRP is retained in this compound.

Physical Properties:
- Flexural Strength: 43-46,000 P.S.I. (D-790)
- Flexural Modulus: 1.7 x 10 6 P.S.I. (D-790)
- Tensile Strength: 28-30,000 P.S.I. (D-638)
- Notched Izod Impact: 12-14 ft.lbs. (D-256)
- Heat Distortion Temperature: 500F+ (D-648)
- Water Absorption (24 hrs.): .38% (D-570)
- Specific Gravity: 1.75 (D-695)
- Compressive Strength: 35-40,000 P.S.I.
- Flex Fatigue (after 10 6 cycles): 11,200 P.S.I.

EP-SL-142:

Acid & Fire Resistant S.M.C.

This compound is a low shrink, low profile S.M.C. that is designed to work around acid batteries and other types of harsh environments. EP-SL-142 is acid and fire resistant and maintains its ability to maintain close dimensional tolerances.

Typical uses for this material include battery support stands and bases, battery trays and covers, etc.

Physical Properties:
- Flexural Strength: 19-22,000 P.S.I. (ASTM D-790)
- Flexural Tangent Modulus: 1.1 x 10 6 P.S.I. (ASTM D-790)
- Flexural Strength after 100 hours @ 122F SP. GR. H2SO4: 19-20,000 P.S.I. (ASTM D-790)
- Tensile Strength: 10,000 P.S.I. (ASTM D-638)
- Notched Izod Impact: 7-8 ft. lbs. (ASTM D-256)
- Water Absorption: .058% (ASTM D-570)
- Specific Gravity: 1.70
- Fiberglass Content: 32-34% (Burnout)

EAGLE-PICHER PLASTICS DIVISION: Products(Continued):

EP-SL-186:
High Strength Low Shrink S.M.C.

This compound is a high strength, low shrink material. It yields excellent appearance and close dimensional tolerances.

Typical uses for this material would be load floor, floor pan, or any other application where high strength and close tolerances are a factor.

Physical Properties:
- Flexural Strength: 40-44,000 P.S.I. (ASTM D-790)
- Flexural Tangent Modulus: 1.75-1.9 x 10 6 P.S.I. (ASTM D-790)
- Tensile Strength: 21-25,000 P.S.I. (ASTM D-638)
- Unnotched Izod Impact: 29-31 ft. lbs. (ASTM D-256)
- Fiberglass Content: 49-51% (Burnout)

EP-SL-208:
A low shrink, low profile fiberglass reinforced vinyl-ester S.M.C. This material exhibits extremely low moisture absorption, has excellent chemical resistance combined with high strength.

Typical uses for this material are in underground applications, high moisture areas, etc.

Physical Properties:
- Flexural Strength: 21-23,000 P.S.I. (ASTM D-790)
- Flexural Modulus: 1.61 x 10 6 (ASTM D-790)
- Tensile Strength: 10-11,000 P.S.I. (ASTM D-638)
- Unnotched Izod Impact: 16-17.0 ft. lbs. (ASTM D-256)
- Fiberglass Content: 26-28% (Burnout)
- % L.O.I.: 31-33% (Burnout)
- Filler Content: 41-43% (Burnout)

EP-SL-218-1:
High Strength Low Shrink S.M.C.

Low shrink, low profile fiberglass reinforced polyester S.M.C. that combines high strength, good surface appearance and good dimensional stability. This material is also paintable and bondable.

Physical Properties:
- Flexural Strength: 28,582 P.S.I. (D-790)
- Flexural Modulus: 1.62 x 10 6 P.S.I. (D-790)
- Tensile Strength: 15,317 P.S.I. (D-638)
- Izod Impact, Unnotched: 21.0 ft.lb./in. (D-256)
- Water Absorption: .34% (D-570)
- Specific Gravity: 1.855 (D-792)
- Hardness, Barcol: 45-55 (D-2583)
- Fiberglass Content: 35% (Burnout)

EAGLE-PICHER PLASTICS DIVISION: Products(Continued):

EP-SLE-114-1:

Low Shrink, High Glass, Self-Extinguishing S.M.C.

An electrical grade SMC material. It is a low shrink, self-extinguishing, high glass content material.

This material is designed for electrical enclosures, bus supports and other applications requiring high strength and good electrical properties.

This material will give excellent appearance and close dimensional tolerances.

Physical Properties:
- Flexural Strength: 22-24,000 (D-790-63)
- Tensile Strength: 10-12,000 (D-651)
- Unnotched Izod Impact: 12-15 ft.lbs. (D-256-56)
- Heat Distortion Temperature: 400F+ (D-648 (266 PSI))
- Water Absorption (24 hrs): Less 0.2% (D-570)
- Specific Gravity: 1.75
- Flammability: Self-Extinguishing
- Compressive Strength: 30-32,000 (D-695)
- Color - Off-White

EP-SI-221:

Structural Grade Automotive SMC

A structural grade automotive SMC formulated with vinyl ester resin. The physical properties meet or surpass structural specifications such as GM's CMP 203XA.

Physical Properties:
- Glass Content: 65+-3 (Burnout)
- Specific Gravity: 1.74
- Barcol Hardness: 50
- Izod Impact Strength--Unnotched: 1600 J/M (30 ft.-lbs.) (ASTM 256)
- Tensile Strength: 210 MPa (30,500 psi) (ASTM 638)
- Flexural Strength: 373 Mpa (54,000 psi) (ASTM 790)
- Flexural Modulus: 13,000 Mpa (1.9 x 10 psi) (ASTM 790)
- Water Absorption: 0.105%

EAGLE-PICHER PLASTICS DIVISION: Products(Continued):

EP-SLI-224:

This compound is a low shrink, low profile SMC material that yields excellent surface appearance and close dimensional tolerances.

Typical uses for this material would be exterior body panels such as hoods, roofs, deck lids and doors.

Physical Properties:
- Flexural Strength: 20-22,000 P.S.I. (D-790)
- Flexural Modulus: 1.0-1.2 x 10^6 P.S.I (D-790)
- Tensile Strength: 10-12,000 P.S.I. (D-638)
- Unnotched Izod Impact: 18-20 ft. lbs. (D-256)
- Fiberglass Content: 28% - 30% (Burnout)
- Filler Content: 35% - 40% (Burnout)
- L.O.I. %: 27% - 40% (Burnout)
- Specific Gravity: 1.75 (D-792, Method A-1)
- Budd Surface Number: 40-100

EP-SLE-174:

Low Shrink Low Glass Paintable SMC

An electrical grade S.M.C. It is a low shrink, self-extinguishing, low glass content material.

This material is designed for electrical enclosures, cabinets and housings and other applications where U.L. approval and good appearance are necessary.

Physical Properties:
- Flexural Strength: 16-18,000 P.S.I. (D-790)
- Flexural Tangent Modulus: 1.7 x 10^6 P.S.I. (D-790)
- Tensile Strength: 6-8,000 P.S.I. (D-638)
- Unnotched Izod Impact: 9-11 ft. lbs. (D-256)
- Water Absorption: .12% (D-570)
- Specific Gravity: 1.725
- Fiberglass Content: 21-23% (Burnout)
- L.O.I. %: 43-45%
- Filler %: 33-35%

Color: Pigmentable and paintable to meet a variety of colors!

EASTMAN CHEMICALS DIVISION: EKTAR MB Polymers:

EKTAR MB polymers alloy polyester materials with an array of other materials to provide a wide range of properties. EKTAR MB polymers can be customized to meet special requirements.

A wide assortment of EKTAR MB polymers are engineered to meet needs for chemical resistance, clarity, gloss, toughness, and processability.

Applications: Business machine and appliance parts
Features: Clarity, good impact

Applications: Vertical body panels for automobiles
Features: Chemical resistance, toughness

Applications: Trays and tote bins
Features: Gloss, toughness, excellent forming characteristics

Applications: Lawn equipment housings
Features: Toughness, chemical resistance

Thermal Properties:
Deflection temperature
at 66 psi (0.455 MPa) fiber stress
F: 180-243
C: 82-117
at 264 psi (1.82 MPa) fiber stress
F: 160-192
C: 71-89

Mechanical Properties:
Tensile strength at break
psi: 4,300-8,600 D 638
MPa: 30-59
Tensile strength at yield
psi: 5,800-8,100 D 638
MPa: 40-56

Miscellaneous Properties:
Specific gravity at 73F (23C): 1.13-1.26 D 792
Water absorption, %, 24 hr, 73F (23C): <0.20 D 570
Mold shrinkage, %, 1/8 in. (3.2 mm) thickness: 0.5-0.7 D 955

EASTMAN CHEMICALS DIVISION: EKTAR Thermoplastic Polyesters:

Performance Plastics EXTAR thermoplastic polyester family features polyesters and copolyesters exhibiting excellent properties. Formulations offer a combination of toughness and chemical resistance, attractive gloss properties and/or high clarity.

EKTAR PET thermoplastic polyesters are available in amorphous or crystallized formulations for blends and alloys.

EKTAR PCTG copolyester 5445 and EKTAR PETG copolyester 6763 are clear amorphous polymers. These materials offer excellent clarity, gloss, toughness and good melt strength at processing temperatures. These properties permit use in a variety of processing techniques and applications.

Injection Molding
- Applications - Protective cases and covers, lenses, pump housings, telephone housings and other electronic and appliance parts
- Features - Toughness, clarity, gloss, and chemical resistance

Sheet and Formed Parts
- Applications - Sheet for protective cases and covers, laminants and other electronic and appliance parts
- Features - Gloss, transparency, toughness, fast forming cycles, ease of forming and reusable scrap

Shape and Tubing Extrusion
- Applications - Protective sleeves and various profiles
- Features - Clarity and toughness

Property, Units:
- Density, g/cm 3: 1.30-1.40 D 1505
- Molecular weight
  - Number average, Mn: 19,000-24,000
  - Weight average, Mw: 38,000-48,000
- Intrinsic viscosity: 0.63-0.76

EASTMAN CHEMICALS DIVISION: EKTAR TO Polymers:

EKTAR TO polymers, Performance Plastics' thermoplastic polyolefin, combine a variety of additives, modulus/modifiers, reinforcements and fillers. EKTAR TO polymers are engineered to meet needs for toughness, stiffness, ease of processing, and paintability.

APPLICATIONS: Bumper fascia
Features: Low temperature flexibility, paintability

APPLICATIONS: Speaker covers
Features: Stiffness

APPLICATIONS: Truck step covers
Features: Dimensional stability

THERMAL PROPERTIES:

Property, Units:		
Flow rate (condition 230/2.16), g/10 min:	2-10	D 1238
Brittleness temperature, C	-40 to 0	D 746
Coefficient of linear thermal expansion,		
(-22 to 73F), 10 -5 in./in./F:	2.8-8.3	D 696
(-30 to 23C), 10 -5 m/m/C:	5-15	

Mechanical Properties:
The most outstanding mechanical properties are flexibility at low temperatures and excellent impact strength.

EKTAR FB Melt Blends:
For service in a high-temperature environment, EKTAR FB polymers combine dimensional stability and high-strength properties with resistance to warpage and chemical attack. They may be formulated to be flame retardant, low warp, and they may be customized to meet a variety of specifications.

EKTAR FB polymers are available in an assortment of base materials including polypropylene, polymer melt blends, polyesters and copolyesters. These base polymers, combined with filler, additives, and modifiers, offer a wide variety of physical properties including excellent dimensional stability and high strength at elevated temperatures.

Applications	Features
Electrical connections	Stiffness, dimensional stability
Molded wiring devices	Low warpage, wave solderable
Terminal blocks	High strength at elevated temperatures
Automotive parts	Chemical resistance, dimensional stability

EKTAR FB PCT polymers possess excellent thermal properties which allow use in applications requiring wave soldering.

EL PASO PRODUCTS CO.: Polypropylenes:

Type:

PP 9140:
Description: Impact copolymer offering good low temperature and drop impact properties.
Applications: Extrusion; thermoforming
Typical Properties:
Melt Flow Rate: D 1238: 1.5 Dg/Min
Density: D 1505: 0.900 g/cm 3

PP 9170:
Description: Impact copolymer offering good low temperature and drop impact properties.
Applications: Extrusion; thermoforming
Typical Properties:
Melt Flow Rate: D 1238: 1.5 Dg/Min
Density: D 1505: 0.900 g/cm 3

PP 9230:
Description: Impact copolymer offering good low temperature and drop impact properties.
Applications: Extrusion; thermoforming
Typical Properties:
Melt Flow Rate: D 1238: 2.0 Dg/Min
Density: D 1505: 0.900 g/cm 3

PP 9400:
Description: Impact copolymer offering good low temperature and drop impact properties.
Applications: Extrusion; thermoforming
Typical Properties:
Melt Flow Rate: D 1238: 1.0 Dg/Min
Density: D1505: 0.901 g/cm 3

PP 9401:
Description: Impact copolymer offering good low temperature and drop impact properties.
Applications: Extrusion; thermoforming
Typical Properties:
Melt Flow Rate: D 1238: 1.2 Dg/Min
Density: D 1505: 0.902 g/cm 3

PP 9402:
Description: Impact copolymer offering good low temperature and drop impact properties.
Applications: Extrusion; thermoforming
Typical Properties:
Melt Flow Rate: D 1238: 1.5 Dg/Min
Density: D 1505: 0.903 g/cm 3

PP 9500:
Description: Impact copolymer offering good low temperature and drop impact properties
Applications: Extrusion; thermoforming
Typical Properties:
Melt Flow Rate: D 1238: 1.2 Dg/Min
Density: D 1505: 0.902 g/cm 3

ENRON CHEMICAL CO.: NORTUFF Engineering Polypropylene:

What makes NORTUFF engineering polypropylene different from standard reinforced polypropylene is the combination of:

1) High flexural modulus
2) Excellent impact strength (Izod and Gardner)
3) Dimensional stability
4) High heat stability
5) Good wear/abrasion resistance
6) Ease of processing

These performance characteristics make the NORTUFF family of engineering resins new candidates for applications that have been the province of "traditional" engineering resins, but at a much-reduced cost. While these cost reductions are dependent on the competitive resin, they could reach 65%, and commonly have been in the 35-50% range.

EXTRUSION GRADES:

NFA Prefix:

1700 MO:
Melt Flow Rate, gms/10 min. Condition L (D-1238): 1.5
Specific Gravity (D-792): 1.17
Tensile Strength, psi (MPa) (D-638): 2700 (18.6)
Elongation, % (D-638): 3.0
Outstanding Material Characteristics: Low Warpage, Good Dimensional Stability. FDA
Typical Applications: Sheet Extrusion Thermoforming, Profile Extrusion

1800 TO:
Melt Flow Rate, gms/10 min. Condition L (D-1238): 1.0
Specific Gravity (D-792): 1.23
Tensile Strength, psi (MPa) (D-638): 3100 (21.4)
Elongation, % (D-638): 9.0
Outstanding Material Characteristics: Low Warpage, Good Toughness, Good Dimensional Stability. FDA.
Typical Appllications: Sheet Extrusion Thermoforming, Profile Extrusion

2400 CO:
Melt Flow Rate, gms/10 min. Condition L (D-1238): 2.5
Specific Gravity (D-792): 1.04
Tensile Strength, psi (MPa) (D-638): 7000 (48.3)
Elongation, % (D-638): 7.0
Outstanding Material Characteristics: High Modulus, High Impact Heat Resistance, Dimensional Stability. Can be used for inj. Molding when impact is key factor.
Typical Applications: Profile Extrusion Rods, Tubes, Etc.

ENRON CHEMICAL CO.: NORTUFF Engineering Polypropylene (Continued):

EXTRUSION GRADES:

<u>NFA</u> <u>Prefix:</u>

2600 CO:
Melt Flow Rate, gms/10 min. Condition L (D-1238): 1.5
Specific Gravity (D-792): 1.12
Tensile Strength, psi (MPa) (D-638): 9000 (62.0)
Elongation, % (D-638): 7.0
Outstanding Material Characteristics: High Modulus, High Impact Heat Resistance, Dimensional Stability. Can be used for Inj. Molding when impact is key factor.
Typical Applications: Profile Extrusion Rods, Tubes, Etc.

INJECTION MOLDING GRADES:

<u>NFC</u> <u>Prefix:</u>

1700 MO:
Melt Flow Rate, gms./10 min. Condition L (D-1238): 7.0
Specific Gravity (D-792): 1.17
Tensile Strength, psi (MPa) (D-638): 2500 (17.2)
Elongation, % (D-638): 1.8
Outstanding Material Characteristics: Low Warpage, Good Heat Resistance FDA
Typical Applications: Small Appliances Housewares

1800 TO:
Melt Flow Rate, gms./10 min. Condition L (D-1238): 4.0
Specific Gravity (D-792): 1.23
Tensile Strength, psi (MPa) (D-638): 3100 (21.4)
Elongation, % (D-638): 4.0
Outstanding Material Characteristics: Toughness, High Modulus, Good Dimensional Stability FDA
Typical Applications: Lawn and Garden Equipment, Power Tool Accessories, Covers.

2400 CO:
Melt Flow Rate, gms./10 min. Condition L (D-1238): 7.0
Specific Gravity (D-792): 1.04
Tensile Strength, psi (MPa) (D-638): 7500 (51.7)
Elongation, % (D-638): 7.0
Outstanding Material Characteristics: Light Weight, Good Dimensional Stability, High Heat and Wear Resistance.
Typical Applications: Gears, Cams, Electrical Enclosures, Power Tool Components.

ENRON CHEMICAL CO.: NORTUFF Engineering Polypropylene (Continued):

INJECTION MOLDING GRADES:

NFC Prefix:

2600 CO:
Melt Flow Rate, gms./10 min. Condition L (D-1238): 6.0
Specific Gravity (D-792): 1.12
Tensile Strength, psi (MPa): 9000 (62.0)
Elongation, % (D-638): 7.0
Outstanding Material Characteristics: Light Weight, Good Dimensional Stability, High Heat and Wear Resistance
Typical Applications: Gears, Cams, Electrical Enclosures, Power Tool Components.

NFH 2800 CO:
Melt Flow Rate, gms./10 min. Condition L (D-1238): 5.0
Specific Gravity (D-792): 1.26
Tensile Strength, psi (MPa): 13,000 (89.6)
Elongation, % (D-638): 9.0
Outstanding Material Characteristics: Light Weight, Good Dimensional Stability, High Heat and Wear Resistance
Typical Applications: Gears, Cams, Electrical Enclosures, Power Tool Components.

FLAME RETARDANT GRADES:

EXTRUSION INJECTION MOLDING:

NFA Prefix:

4400 TO:
Melt Flow Rate, gms./10 min. Condition L (D-1238): 1.7
Specific Gravity (D-792): 1.26
Tensile Strength, psi (MPa) (D-638): 2600 (18.0)
Elongation, % (D-638): 12
Outstanding Material Characteristics: V-O Flammability, Toughness, Moderate Rigidity, Chemical Resistance.
Typical Applications: Grilles, Small Appliance Components/ Housings, Smoke Detectors.

4000 FR:
Melt Flow Rate, gms./10 min. Condition L (D-1238): 12
Specific Gravity (D-792): 0.98
Tensile Strength, psi (MPa) (D-638): 2400 (16.5)
Elongation, %: 14
Outstanding Material Characteristics: V-2 Flammability, Light Weight, Excellent Processibility, Good Toughness.
Typical Applications: Applications Requiring Flame Retardance, High Impact Resistance and Good Flexibility.

ENRON CHEMICAL CO.: NORTUFF Engineering Polypropylene (Continued):

FLAME RETARDANT GRADES:

INJECTION MOLDING:

NFC Prefix:

4400 CO:

Melt Flow Rate, gms/10 min. Condition L (D-1238): 5.0
Specific Gravity (D-792): 1.34
Tensile Strength, psi (MPa) (D-638): 8100 (55.8)
Elongation, % (D-638): 4.0
Outstanding Material Characteristics: V-O Flammability, Excellent Dimensional Stability, Impact, Wear and Heat/Humidity Tolerance, Processing Ease
Typical Applications: Gears, Cams, Electrical/Electronic Housings, Power Tool Components.

4600 CO:

Melt Flow Rate, gms/10 min. Condition L (D-1238): 5.0
Specific Gravity (D-792): 1.48
Tensile Strength, psi (MPa) (D-638): 9500 (65.5)
Elongation, % (D-638): 5.0
Outstanding Material Characteristics: V-O Flammability, Excellent Dimensional Stability, Impact, Wear and Heat/Humidity Tolerance, Processing Ease.
Typical Applications: Gears, Cams, Electrical/Electronic Housings, Power Tool Components.

EVAL COMPANY OF AMERICA: EVAL Resin Ethylene Vinyl Alcohol Copolymer: EP-E Series: 44 mol % Ethylene:

High Barrier Resins for Multilayer Barrier Structures

PROPERTIES:

* High gas barrier
* Excellent formability
* Excellent odor and aroma barrier
* Excellent processability
* FDA compliance

APPLICATIONS:

EVAL EP-E Series resins are excellent materials for cast co-extrusion applications. They also provide deep draw forming capability for both melt and solid phase thermoforming applications. EVAL EP-E resins provide very good oxygen barrier properties at high relative humidities as well as processing ease and recycleability.

CERTIFICATIONS:

EVAL Resins are approved for use in applications involving direct food contact as outlined by the Food and Drug Administration regulations, 21 CFR, Section 177.1360. Approval of EVAL resins for indirect food contact (adhesive resins) is outlined in 21 CFR, Section 175.105.

EVAL resins are also approved by the U.S. Department of Agriculture and the Canadian Ministery of Agriculture for use in contact with meat and poultry products.

FABRICATION TECHNIQUES:

Suggested extrusion conditions are as follows:

Extruder Barrel Temperature, C:		Adaptor Temperature, 195C
	C1:170	Die Temperature, 195C
	C2:190	
	C3:195	
	C4:205	
	C5:210	

PROPERTY	TEST METHOD	VALUE
RESIN		
Density, g/cc	ASTM D1505	1.14
Melting Point, C	----	164
Crystallization Temperature, C	----	142
Glass Transition Point, C	----	55
Extruder Melt Temperature Range, C	----	185-250

EVAL COMPANY OF AMERICA: EVAL Resin Ethylene Vinyl Alcohol Copolymer: EP-F Series: 32 mol % Ethylene:

High Barrier Resins for Multilayer Barrier Structures

PROPERTIES:

* High gas barrier
* Excellent solvent barrier
* Excellent odor and aroma barrier
* FDA compliance

APPLICATIONS:

EVAL EP-F Series resins are designed for use in coextrusion blow molding, film profile, sheet and melt phase thermoforming applications. This series offers superior gas barrier properties over a wide range of environmental conditions. Ease of processing, the ability to recycle and excellent physical properties make the EP-F Series an excellent choice for barrier packaging applications.

CERTIFICATIONS:

EVAL resins are approved for use in applications involving direct food contact as outlined by the Food and Drug Administration regulations, 21 CFR, Section 177.1360. Approval of EVAL resins for indirect food contact (adhesive resins) is outlined in 21 CFR, Section 175.105.

EVAL resins are also approved by the U.S. Department of Agriculture and the Canadian Ministery of Agriculture for use in contact with meat and poultry products.

FABRICATION TECHNIQUES:

Suggested extrusion conditions are as follows:

Extruder Barrel Temperature, C		Adaptor temperature, 215C
	C1: 180	Die temperature, 215C
	C2: 200	
	C3: 205	
	C4: 215	
	C5: 220	

PROPERTY	TEST METHOD	VALUE
RESIN		
Density, g/cc	ASTM D1505	1.19
Melting Point, C	----	181
Crystallization Temperature, C	----	161
Glass Transition Point, C	----	69
Extruder Melt Temperature Range, C	----	200-240

EVAL COMPANY OF AMERICA: EVAL Resin Ethylene Vinyl Alcohol Copolymer: EP-G Series: 48 mol % Ethylene:

High Barrier Resins for Multilayer Barrier Structures

PROPERTIES

* High gas barrier
* High yield
* Excellent odor and aroma barrier
* Biaxial orientation
* FDA compliance

APPLICATIONS:

EVAL EP-G Series are designed for use in biaxial oriented coextruded structures or applications where a low viscosity material is needed. These resins offer an excellent yield, good gas and odor barrier properties, recycleability and high output rates. As with other EVAL resins, the EP-G Series offer an excellent balance of optical and physical properties.

CERTIFICATIONS:

EVAL resins are approved for use in applications involving direct food contact as outlined by the Food and Drug Administation regulations, 21 CFR, Section 177.1360. Approval of EVAL resins for indirect food contact (adhesive resins) is outlined in 21 CFR, Section 175.105.

EVAL resins are also approved by the U.S. Department of Agriculture and the Canadian Ministery of Agriculture for use in contact with meat and poultry products.

FABRICATION TECHNIQUES:

Suggested extrusion conditions are as follows:

Extruder Barrel Temperature, C		Adaptor Temperature, 190C
	C1:165	Die Temperature, 190C
	C2:185	
	C3:190	
	C4:200	
	C5:205	

PROPERTY	TEST METHOD	VALUE
RESIN		
Density, g/cc	ASTM D1505	1.12
Melting Point, C	----	156
Crystallization Temperature, C	----	134
Glass Transition Point, C	----	48
Extruder Melt Temperature Range, C	----	180-250

EVAL COMPANY OF AMERICA: EVAL Resin Ethylene Vinyl Alcohol Copolymer: EP-H Series: 38 mol % Ethylene:

High Barrier Resins for Multilayer Barrier Structures

PROPERTIES:

* High gas barrier
* Ease of processing
* Excellent odor and aroma barrier
* Excellent oil and grease barrier
* FDA compliance

APPLICATIONS:

EVAL EP-H Series resins are designed for use in blown and cast coextrusion applications. These resins offer high barrier properties to gases, solvents, odors and aromas. EP-H Series resins exhibit very good processability, recycleability and excellent optical and physical properties.

CERTIFICATIONS:

EVAL resins are approved for use in applications involving direct food contact as outlined by the Food and Drug Administration regulations, 21 CFR, Section 177.1360. Approval of EVAL resins for indirect food contact (adhesive resins) is outlined in 21 CFR, Section 175.105.

EVAL resins are also approved by the U.S. Department of Agriculture and the Canadian Ministery of Agriculture for use in contact with meat and poultry products.

FABRICATION TECHNIQUES:

Suggested extrusion conditions are as follows:

Extruder Barrel Temperature, C		Adaptor Temperature, 210C
	C1: 175	Die Temperature, 210C
	C2: 195	
	C3: 205	
	C4: 215	
	C5: 220	

PROPERTY	TEST METHOD	VALUE
RESIN		
Density, g/cc	ASTM D1505	1.17
Melting Point, C	----	175
Crystallization Temperature, C	----	151
Glass Transition Point, C	----	62
Extruder Melt Temperature Range, C	----	190-245

EVAL COMPANY OF AMERICA: EVAL Resin Ethylene Vinyl Alcohol Copolymer: EP-K Series: 38 mol % Ethylene:

High Barrier Resins for Multilayer Barrier Structures

PROPERTIES:

* High gas barrier
* SPPF forming
* Excellent odor and aroma barrier
* Recycleability
* FDA compliance

APPLICATIONS:

EVAL EP-K resin is specifically designed for Solid Phase Pressure Forming (SPPF) applications. The broad molecular weight distribution of this resin provides deep draw capabilities when used with polypropylene coextruded structures. As with other EVAL resins, EVAL EP-K resin offers excellent processability, recycleability and superior optical and physical properties.

CERTIFICATIONS:

EVAL resins are approved for use in applications involving direct food contact as outlined by the Food and Drug Administration regulations, 21 CFR, Section 177.1360. Approval of EVAL resins for indirect food contact (adhesive resins) is outlined in 21 CFR, Section 175.105.

FABRICATION TECHNIQUES:

Suggested extrusion conditions are as follows:

Extruder Barrel Temperature, C		
C1:	175	Adaptor Temperature, 210C
C2:	195	Die Temperature, 210C
C3:	205	
C4:	215	
C5:	220	

PROPERTY	TEST METHOD	VALUE
RESIN		
Density, g/cc	ASTM D1505	1.17
Melting Point, C	----	175
Crystallization Temperature, C	----	151
Glass Transition Point, C	----	62
Extruder Melt Temperature Range, C	----	190-245

EVAL COMPANY OF AMERICA: EVAL Resin Ethylene Vinyl Alcohol Copolymer: SC-E Series: 44 mol % Ethylene:

High Barrier Resins for Multilayer Barrier Structures

PROPERTIES:

* High gas barrier
* Excellent formability
* Excellent odor and aroma barrier
* Excellent processability
* FDA compliance

APPLICATIONS:

EVAL SC-E Series resins are excellent materials for cast coextrusion applications. They also provide deep draw forming capability for both melt and solid phase thermoforming applications. EVAL SC-E resins provide very good oxygen barrier properties at high relative humidities as well as processing ease and recyleability.

CERTIFICATIONS:

EVAL resins are approved for use in applications involving direct food contact as outlined by the Food and Drug Administration regulations, 21 CFR, Section 177.1360. Approval of EVAL resins for indirect food contact (adhesive resins) is outlined in 21 CFR, Section 175.105.

EVAL Resins comply with food additive regulations covering material used for the packaging of foods sterilized under retort conditions.

EVAL resins are also approved by the U.S. Department of Agriculture and the Canadian Ministery of Agriculture for use in contact with meat and poultry products.

PROPERTY	TEST METHOD	VALUE
RESIN		
Density, g/cc	ASTM D1505	1.14
Melting Point, C	----	164
Crystallization Temperature, C	----	142
Glass Transition Point, C	----	55
Extruder Melt Temperature Range, C	----	185-250

EVAL COMPANY OF AMERICA: EVAL Resin Ethylene Vinyl Alcohol Copolymer: SC-F Series: 32 mol % Ethylene:

High Barrier Resins for Multilayer Barrier Structures

PROPERTIES:

* High gas barrier
* Excellent solvent barrier
* Excellent odor and aroma barrier
* FDA compliance

APPLICATIONS:

EVAL SC-F Series resins are designed for use in coextrusion blow molding, film profile, sheet and melt phase thermoforming applications. This series offers superior gas barrier properties over a wide range of environmental conditions. Ease of processing, the ability to recycle and excellent physical properties make the SC-F Series an excellent choice for barrier packaging applications.

CERTIFICATIONS:

EVAL resins are approved for use in applications involving direct food contact as outlined by the Food and Drug Administration regulations, 21 CFR, Section 177.1360. Approval of EVAL resins for indirect food contact (adhesive resins) is outlined in 21 CFR, Section 175.105.

EVAL Resins comply with food additive regulations covering material used for the packaging of foods sterilized under retort conditions.

EVAL resins are also approved by the U.S. Department of Agriculture and the Canadian Ministery of Agriculture for use in contact with meat and poultry products.

FABRICATION TECHNIQUES:

Suggested extrusion conditions are as follows:

Extruder Barrel Temperature, C		Adaptor Temperature, 215C
	C1: 180	Die Temperature, 215C
	C2: 200	
	C3: 205	
	C4: 215	
	C5: 220	

PROPERTY	TEST METHOD	VALUE
RESIN		
Density, g/cc	ASTM D1505	1.19
Melting Point, C	----	181
Crystallization Temperature, C	----	161
Glass Transition Point, C	----	69
Extruder Melt Temperature Range, C	----	200-240

EVAL COMPANY OF AMERICA: EVAL Resin Ethylene Vinyl Alcohol Copolymer: SC-H Series: 38% mol % Ethylene:

High Barrier Resins for Multilayer Barrier Structures

PROPERTIES:

* High gas barrier
* Ease of processing
* Excellent odor and aroma barrier
* Excellent oil and grease barrier
* FDA compliance

APPLICATIONS

EVAL SC-H Series resins are designed for use in blown and cast coextrusion applications. These resins offer high barrier properties to gases, solvents, odors and aromas. SC-H Series resins exhibit very good processability, recycleability and excellent optical and physical properties

CERTIFICATIONS:

EVAL resins are approved for use in applications involving direct food contact as outlined by the Food and Drug Administration regulations, 21 CFR, Section 177.1360. Approval of EVAL resins for indirect food contact (adhesive resins) is outlined in 21 CFR, Section 175.105.

EVAL Resins comply with food additive regulations covering material used for the packaging of foods steriilized under retort conditions.

EVAL resins are also approved by the U.S. Department of Agriculture and the Canadian Minístery of Agriculture for use in contact with meat and poultry products.

FABRICATION TECHNIQUES:

Suggested extrusion conditions are as follows:

Extruder Barrel Temperature, C		
	C1: 175	Adaptor Temperature, 210C
	C2: 195	Die Temperature, 210C
	C3: 205	
	C4: 215	
	C5: 220	

PROPERTY	TEST METHOD	VALUE
RESIN		
Density, g/cc	ASTM D1505	1.17
Melting Point, C	----	175
Crystallization Temperature, C	----	151
Glass Transition Point, C	----	62
Extruder Melt Temperature Range, C	----	190-245

FIBERITE: Epoxy Molding Compounds:

Encapsulation Grades:

E 8382:
Reinforcement: Glass/Mineral
Resin System: Epoxy

E 8379:
Reinforcement: Mineral
Resin System: Epoxy

E 8411:
Reinforcement: Mineral
Resin System: Epoxy

E 8414:
Reinforcement: Mineral
Resin System: Epoxy

Electro-Mechanical:

E 8625:
Reinforcement: Mineral
Resin System: Epoxy

E 8639:
Reinforcement: Glass/Mineral
Resin System: Epoxy

E 8353:
Reinforcement: Glass/Mineral
Resin System: Epoxy

E 8354:
Reinforcement: Glass/Mineral
Resin System: Epoxy

E 8354-19189:
Reinforcement: Glass/Mineral
Resin System: Epoxy

E 2748:
Reinforcement: Short Glass
Resin System: Epoxy

E2748-9641CB:
Reinforcement: Short Glass
Resin System: Epoxy

FIBERITE CORP.: Epoxy Molding Compounds(Continued):

Electro-Mech. Long Fiber:

E 7300:
Reinforcement: 1/2" Chop. Glass
Resin System: Epoxy

E 264H:
Reinforcement: 1/4" Chop. Glass
Resin System: Epoxy

Structural Long Fiber:

E 260H:
Reinforcement: 1/2" Chop. Glass
Resin System: Epoxy

E 261H:
Reinforcment: 1" Chop. Glass
Resin System: Epoxy

E 20328:
Reinforcement: 1/2" Chop. Kevlar
Resin System: Epoxy

E 21718x60:
Reinforcement: 1/2" Chop. ST Carbon
Resin System: Epoxy

E 21718x61:
Reinforcement: 1/2" Chop. HS Carbon
Resin System: Epoxy

E 21718x65:
Reinforcement: 1/2" Chop. HM Carbon
Resin System: Epoxy

Specialty Grade:

X-28057:
Reinforcement: Mineral
Resin System: Epoxy

FIBERITE: Phenolic Molding Compounds:

Electro-Mechanical Fabric and Fiber Reinforcements:

Compound Number:

FM 510:
Reinforcement: Cellulose
Resin System: Two Stage Phenolic

FM 1132:
Reinforcement: Cellulose
Resin System: Two Stage Phenolic

FM 3510:
Reinforcement: Cellulose
Resin System: Two Stage Phenolic

FM 10365:
Reinforcement: Cellulose
Resin System: Two Stage Phenolic

FM 9294:
Reinforcement: Cellulose
Resin System: Two Stage Phenolic

FM 5367:
Reinforcement: Cellulose & Glass
Resin System: Two Stage Phenolic

FM 32168:
Reinforcement: Cellulose
Resin System: Single Stage Phenolic

FM 1303-8880:
Reinforcement: Nylon Flock
Resin System: Two Stage Phenolic

Electro-Mechanical Short Fiber:

FM 4070:
Compound Number: Short Glass
Resin System: Two Stage Phenolic

FM 4029F:
Compound Number: Short Glass
Resin System: Single Stage Phenolic

FM 6030:
Compound Number: Short Glass
Resin System: Two Stage Phenolic

FM 4004:
Compound Number: Short Glass
Resin System: Two Stage Phenolic

FIBERITE: Phenolic Molding Compounds(Continued):

Electro-Mechanical Short Fiber(Continued):

FM 4055:
Reinforcement: Short Glass
Resin System: Two Stage Phenolic

FM 4024:
Reinforcement: Short Glass
Resin System: Two Stage Phenolic

FM 4007:
Reinforcement: Short Glass
Resin System: Two Stage Phenolic

FM 4009:
Reinforcement: Short Glass
Resin System: Two Stage Phenolic

FM 4008:
Reinforcement: Short Glass
Resin System: Single Stage Phenolic

FM 4005:
Reinforcement: Short Glass
Resin System: Two Stage Phenolic

4005F:
Reinforcement: Short Glass
Resin System: Two Stage Phenolic

FM 4036:
Reinforcement: Short Glass
Resin System: Single Stage Phenolic

FM 4002:
Reinforcenment: Short Glass
Resin System: Two Stage Phenolic

FM 4011:
Reinforcement: Short Glass
Resin System: Two Stage Phenolic

FM 4056:
Reinforcement: Short Glass
Resin System: Two Stage Phenolic

FM 4060:
Reinforcement: Short Glass
Resin System: Two Stage Phenolic

FIBERITE: Phenolic Molding Compounds (Continued):

Electro-Mechanical Long Fiber:

FM 5064:
Reinforcement: 1/2" Chop Glass
Resin System: Two Stage Phenolic

FM 8130:
Reinforcement: 1/2" Chop Glass
Resin System: One Stage Phenolic

FM 16771:
Reinforcement: 1/2" Chop Glass
Resin System: One Stage Phenolic

FM 21288:
Reinforcement: 1/2" Chop Glass
Resin System: One Stage Phenolic

Structural Long Fiber:

FM 5085:
Compound Number: 1/2" Chop Glass
Resin System: Two Stage Phenolic

FM 8130E:
Compound Number: 1/2" Chop Glass
Resin System: Two Stage Phenolic

FM 32780:
Compound Number: 1/2" Chop St Carbon
Resin System: Two Stage Phenolic

FM 50501:
Compound Number: 1/2" Chop St Carbon
Resin System: Two Stage Phenolic

Special Properties:
FM 4005 x 4648:
Reinforcement: Short Glass
Resin System: Two Stage Phenolic

FM 32175:
Reinforcement: Short Glass Aluminum Flake
Resin System: Two Stage Phenolic

FM 4555:
Reinforcement: Glass
Resin Stystem: Two Stage Phenolic

FM 8720:
Reinforcement: Metal Powder
Resin System: Two Stage Phenolic

FIBERITE: Phenolic Molding Compounds (Continued):

Wear Resistance Compounds:

Compound Number:

FM 3000:
Reinforcement: Cellulose-Teflon
Resin System: Two Stage Phenolic

FM 3001:
Reinforcement: Cellulose-Teflon
Resin System: Two Stage Phenolic

FM 3002:
Reinforcement: Cellulose-Teflon
Resin System: Two Stage Phenolic

FIBERITE: Specialty Molding Compounds:

Melamines:

Compound Number:

Electro-Mech. Mineral:
M 6204:
Reinforcement: Mineral
Resin System: Melamine

Electo-Mech. Short Glass:
M 2840:
Reinforcement: Glass
Resin System: Melamine

Electro-Mech. Long Glass:
M 2037:
Reinforcement: Glass
Resin System: Melamine

Polyester-Vinylester:

Electro-Mechanical Long Fiber:

S 660:
Reinforcement: 1/2" Chop Glass
Resin System: Polyester

S 706:
Reinforcement: 1/2" Chop Glass
Resin System: Polyester

S 706M:
Reinforcement: 1/2" Chop Glass
Resin System: Polyester

S 680:
Reinforcement: 1/2" Chop Glass
Resin System: Polyester

S 695:
Reinforcement: 1/2" Chop Glass
Resin System: Polyester

FIBERITE: Specialty Molding Compounds(Continued):

Polyvinyl-Vinylester:

Electro-Mechanical Long Fiber(Continued):

S 690:
Reinforcement: 1/2" Chop Glass
Resin System: Polyester

Structural Long-Fiber:
VE49595:
Reinforcement: 1" Chop Glass
Resin System: Vinyl Ester

Silicone:
Electro-Mechanical:

SI-9002:
Reinforcement: Glass-Mineral
Resin: Silicone

SI-9041:
Reinforcement: Long Glass
Resin: Silicone

Polyimide:
High Temperature Resistance:

PI-730:
Reinforcement: Glass-Mineral
Resin System: Polyimide

PI-740:
Reinforcement: Short Glass
Resin System: Polyimide

PI-745:
Reinforcement: Med. Glass
Resin System: Polyimide

PI-750:
Reinforcement: Long Glass
Resin System: Polyimide

FINA OIL AND CHEMICAL CO.: DYPRO Polypropylenes:

Product Name:

B255:

Excellent hot fill properties of blow molded bottles produced with this copolymer resin are combined with good product compatibility. An excellent choice for packaging maple syrups, other favored syrups and food items. Highly recommended for coextrusion. Also available with nucleation (K222Z).

Resin Properties:	Value	ASTM Method
Melt Flow, g/10 min.	1.9	D-1238 Condition "L"
Density, g/cc	0.900	D-1505
Melting Point, F	297	DSC
C	147	

B-943 Homopolymer:

Excellent color stability and processability are featured in this homopolymer resin for injection blow molding. Applications include food, cosmetic and pharmaceutical packaging.

This resin is also available with antistat (B-943Z).

Resin Properties:	Value	ASTM Method
Melt Flow, g/10 min.	2.0	D-1238 Condition "L"
Density, g/cc	0.905	D-1505
Melting Point, F	330	DSC
C	165	

K-122 Copolymer:

Excellent clarity, high gloss, good impact and low odor copolymer for conventional blow molding of containers for packaging cosmetics, toiletries and food. Also acceptable in medical applications.

Resin Properties:	Value	ASTM Method
Melt Flow, g/10 min.	1.5	D-1238 Condition "L"
Density, g/cc	0.900	D-1505
Melting Point, F	297	DSC
C	147	

FINA OIL AND CHEMICAL CO.: DYPRO Polypropylenes(Continued):

Product Name:

K-193 Copolymer:

Excellent color, clarity and chemical inertness of articles for medical applications produced via the extrusion blow molding process. Also suitable for oriented containers exhibiting enhanced stiffness, impact, barrier properties and clarity for medical applications.

Resin Properties:	Value	ASTM Method
Melt Flow, g/10 min.	2.5	D-1238 Condition "L"
Density, g/cc	0.900	D-1505
Melting Point, F	297	DSC
C	147	

K-222Z Copolymer:

Excellent hot fill properties of blow molded bottles produced with this copolymer resin are combined with good product compatibility. An excellent choice for packaging maple syrups, other flavored syrups and food items.

Resin Properties:	Value	ASTM Method
Melt Flow, g/10 min.	1.5	D-1238 Condition "L"
Density, g/cc	0.900	D-1505
Melting Point, F	297	DSC
	147	

Q532 Homopolymer:

DYPRO Q532 polypropylene features excellent processability and the good physical properties necessary for tapes, carpet backing, monofilaments and multifilaments. It is formulated to resist gas fading while maintaining excellent processing stability.

Resin Properties:	Value	ASTM Method
Melt Flow, g/10 min.	12	D-1238 Condition "L"
Density, g/cc	0.905	D-1505
Melting Point, F	330	DSC
C	165	

FINA OIL & CHEMICAL CO.: DYPRO Polypropylenes(Continued):

Product Name:

S040 Injection Molding Utility Grade Homopolymer:

Good Melt Flow Uniformity - DYPRO S040 has good melt flow uniformity for more consistent molding of parts for utility applications.

Recommended Applications:

DYPRO S040 is recommended for use in molding parts calling for a low melt flow and where FDA compliance is not required.

Resin Properties:	Value	ASTM Method
Melt Flow, g/10 min.	3-5	D-1238 Condition "L"
Density, g/cc	0.905	D-1505
Melting Point, F	330	DSC
C	165	
Form	Pellets	

S055 Injection Molding Utility Grade Copolymer:

Good Melt Flow Uniformity - DYPRO S055 has good melt flow uniformity for more consistent molding of copolymer parts for utility applications.

Recommended Applications:

DYPRO S055 is recommended for use in molding parts calling for a medium melt flow copolymer and where FDA compliance is not required.

Resin Properties:	Value	ASTM Method
Melt Flow, g/10 min.	4-6	D-1238 Condition "L"
Density, g/cc	0.900	D-1505
Form	Pellets	

FINA OIL & CHEMICAL CO.: DYPRO Polypropylenes(Continued):

Product Name:

S120 Injection Molding Utility Grade Homopolymer:

Easy Molding - DYPRO S120 is easy processing and will mold well in utility applications.

Recommended Applications:

DYPRO S120 is recommended for use in utility molding where FDA compliance is not required.

Resin Properties:	Value	ASTM Method
Melt Flow, g/10 min.	8-14	D-1238 Condition "L"
Density, g/cc	0.905	D-1505
Melting Point, F	330	DSC
C	165	
Form	Pellets	

S350 High Melt Flow Concentrate Base Homopolymer:

Dispersibility - DYPRO S350 is designed for better dispersibility and higher pigment loadings than are obtainable with typical polypropylenes.

Compatibility - DYPRO S350 has excellent compatibility with low, medium and high melt flow polypropylene homo- and copolymers.

Surface Gloss - For fabricators, DYPRO S350 has the advantage of excellent retention of surface gloss of extrusions when used as a concentrate base.

Recommended Applications:

DYPRO S350 is recommended for use as a vehicle in color concentrate compounds.

Resin Properties:	Value	ASTM Method
Melt Flow, g/10 min.	27-50	D-1238 Condition "L"
Density, g/cc	0.905	D-1505
Melting Point, F	330	DSC
C	165	
Form	Pellets	

FINA OIL & CHEMICAL CO.: DYPRO Polypropylenes(Continued):

Product Name:

W-431 Copolymer:

Low melting, high ethylene random copolymer film resin with a low, broad heat seal range and improved optical and impact properties. Excellent for packaging applications requiring clarity, sparkle and improved heat sealability. Available with custom slip and anti-block to meet end-use requirements.

Resin Properties:	Value	ASTM Test Method
Melt Flow, g/10 min	3.0	D-1238 Condition "L"
Density, g/cc	0.900	D-1505
Melting Point, F	275	DSC
C	135	

W-756 Copolymer:

Low melting, high ethylene random copolymer film resin with a low, broad heat seal range and improved optical and impact properties. Excellent for packaging applications requiring clarity, sparkle and improved heat sealability. Available with custom slip and anti-block to meet end-use requirements.

Resin Properties	Value	ASTM Test Method
Melt Flow, g/10 min.	7.0	D-1238 Condition "L"
Density, g/cc	0.900	D-1505
Melting Point, F	275	DSC
C	135	

Z7450 Random Copolymer Heat Sealable Film Grade for Medical Use:

Color/Clarity/Impact - DYPRO Z7450 is a fourth generation low melting, high ethylene random copolymer with improved color, optics and impact properties.

Heat Sealable - The very low melting point of DYPRO Z7450 makes it an excellent heat sealable layer for oriented films.

Slip/Antiblock - DYPRO Z7450 is available with custom slip and antiblock packages.

FDA - DYPRO Z7450 complies with FDA regulations Title 21CFR Section 177.1520(Z) 3.1 and may be used under these provisions for food contact and packaging.

FINA OIL & CHEMICAL CO.: DYPRO Polypropylenes(Continued):

Product Name:

Z7450 Random Copolymer Heat Sealable Film Grade For Medical Use (Continued):

Recommended Applications:

DYPRO Z7450 is recommended for use in either non oriented cast film processes for manufacture of packaging films that require improved optical and impact properties or as a heat sealable layer for oriented films.

Resin Properties:	Value	ASTM Method
Melt Flow, g/10 min.	3.0	D-1238 230C/2160 g
Density, g/cc	0.89	D-1505
Melting Point, F	264	DSC
C	129	

Z7650 Random Copolymer Heat Sealable Film Grade for Medical Use:

Color/Clarity/Impact - DYPRO Z7650 is a fourth generation low melting, high ethylene random copolymer with improved color, optics and impact properties.

Heat Sealable - The very low melting point of DYPRO Z7650 makes it an excellent heat sealable layer for oriented films.

Slip/Antiblock - DYPRO Z7650 is available with custom slip and antiblock packages.

FDA - DYPRO Z7650 complies with FDA regulations Title 21CFR Section 177.1520(c) 3.1 and may be used under these provisions for food contact and packaging.

Recommended Applications:

DYPRO Z7650 is recommended for use in either non oriented cast film processes for manufacture of packaging films that require improved optical and impact properties or as a heat sealable layer for oriented films.

Resin Properties:	Value	ASTM Method
Melt Flow, g/10 min.	5.0	D-1238 230C/2160g
Density, g/cc	0.89	D-1505
Melting Point, F	264	DSC
C	129	

FINA OIL & CHEMICAL CO.: DYPRO Polypropylenes(Continued):

Product Name:

Z7670 Random Copolymer Heat Sealable Film Grade:

Color/Clarity/Impact - DYPRO Z7670 is a fourth generation low melting, high ethylene random coploymer with improved color, optics and impact properties.

Heat Sealable - The very low melting point of DYPRO Z7670 makes it an excellent heat sealable layer for oriented films.

Slip/Antiblock - DYPRO Z7670 is available with custom slip and antiblock packages.

FDA - DYPRO Z7670 complies with FDA regulation Title 21CFR Section 177.1520(c) 3.1 and may be used under these provisions for food contact and packaging.

Recommended Applications:

DYPRO Z7670 is recommended for use in either non oriented cast film processes for manufacture of packaging films that require improved optical and impact properties or as a heat sealable layer for oriented films.

Resin Properties	Value	ASTM Method
Melt Flow, g/10 min	5.0	D-1238 230C/ 2160g
Density, g/cc	0.89	D-1505
Melting Point, F	264	DSC
C	129	

860T2 Homopolymer, Talc Filled:

Good stiffness and heat distortion resistance, 20% talc filled homopolymer resin for injection molding automotive and appliance parts with a UL 110C (230F) yellow card rating, excellent heat aging resistance and low shrinkage and warpage.

Resin Properties:	Value	ASTM Method
Melt Flow, g/10 min.	5.0	D-1238 Condition "L"
Density, g/cc	1.06	D-1505
Melting Point, F	330	DSC
C	165	

FINA OIL & CHEMICAL CO.: DYPRO Polypropylenes(Continued):

Product Name:

3230/3230Z Homopolymer Injection Blow Molding Grade:

Lubricated - DYPRO 3230 and 3230Z offer superior core rod release, eliminating the need for additional lubricant. DYPRO 3230Z is the same as 3230, but with antistat to minimize dust accumulation for shelf cleanliness as well as aid in mold release.

High Purity - Low catalyst residues in DYPRO 3230 and 3230Z minimize odor and optimize melt stability for superior color and processability.

FDA - DYPRO 3230 and 3230Z comply with all applicable FDA regulations and may be used under these provisions for food contact and packaging.

Recommended Applications:

DYPRO 3230 and 3230Z are ideal for food, cosmetic and pharmaceutical injection blow molding containers. DYPRO 3230Z is particularly well suited when shelf dust is a problem.

Resin Properties:	Value	ASTM Method
Melt Flow, g/10 min.	2.0	D-1238 Condition "L"
Density, g/cc	0.905	D-1505
Melting Point, F	330	DSC
C	165	

3275 Homopolymer Film Grade:

Low Residuals - DYPRO 3275 utilizes high-yield catalyst technology plus a wash/de-ash process to provide extremely low residuals for improved color stability and clarity.

Slip/Antiblock - DYPRO 3275 is available with custom slip and antiblock packages.

FDA - DYPRO 3275 complies with all applicable FDA regulations for food contact applications.

Recommended Applications:

DYPRO 3275 is recommended in orientation processes for manufacture of packaging tapes and films.

Resin Properties:	Value	ASTM Method
Melt Flow, g/10 min.	1.7	D-1238 230C/2160 g
Density, g/cc	0.905	D-1505
Melting Point, F, C	330, 165	DSC

FINA OIL & CHEMICAL CO.: DYPRO Polypropylenes (Continued):

Product Name:

3366 Minimal UV Homopolymer Fiber Grade For Industrial Bags:

Low Water Carry-over - Excellent low water carry-over properties allow DYPRO 3366 to be processed at high rates into water quenched slit tape.

U.V. Stability - DYPRO 3366 will resist a minimum of 200 hours carbon arc weatherometer exposure until 70% strength retention when extruded into 2 x 100 mil slit tape.

FDA - DYPRO 3366 complies with all applicable FDA regulations for food contact applications.

Recommended Applications:

DYPRO 3366 is recommended for slit tape applications, such as grain sacks, requiring FDA acceptance and a moderate degree of U.V. stability.

Resin Properties	Value	ASTM Method
Melt Flow, g/10 min.	3.0	D-1238 Condition "L"
Density, g/cc	0.905	D-1505
Melting Point, F	330	DSC
C	165	

3374 Homopolymer Film Grade:

Low Residuals - DYPRO 3374 utilizes high-yield catalyst technology plus a wash/de-ash process to provide extremely low residuals for improved color stability and clarity.

Slip/Antiblock - DYPRO 3374 is available with custom slip and antiblock packages.

FDA - DYPRO 3374 complies with all applicable FDA requlations for food contact applications.

Recommended Applications:

DYPRO 3374 is recommended in orientation processes for manufacture of packaging tapes and films.

Resin Properties:	Value	ASTM Method
Melt Flow, g/10 min.	2.5	D-1238 230C/2160 g
Density, g/cc	0.905	D-1505
Melting Point, F	330	DSC
C	165	

FINA OIL & CHEMICAL CO.: DYPRO Polypropylenes(Continued):

Product Name:

3462 Homopolymer Fiber Grade with Increased Process Stability for Tapes and Monofilaments:

Process Stability - The uniform processability of DAPRO 3462 yields good, consistent physical properties for extruded tapes and monofilaments.

Low Filterables - Very low catalyst residues and excellent polymer stability of DYPRO 3462 allow extended processing runs without screen pack blinding.

FDA - DYPRO 3462 complies with all applicable FDA regulations and may be used under these provisions for food contact applications.

Recommended Applications:

DYPRO 3462 is recommended for extruded tapes and monofilaments for carpet backing, produce bags, rope and cordage products.

Resin Properties:	Value	ASTM Method
Melt Flow, g/10 min.	4.0	D-1238 Condition "L"
Density, g/cc	0.905	D-1505
Melting Point, F	330	DSC
C	165	

3471 Homopolymer Film Grade:

Low Residuals - DYPRO 3471 utilizes high-yield catalyst technology plus a wash/de-ash process to provide extremely low residuals for improved color stability and clarity.

Slip/Antiblock - DYPRO 3471 is available with custom slip and antiblock packages.

FDA - DYPRO 3471 complies with all applicable FDA regulations for food contact applications.

Recommended Applications:

DYPRO 3471 is recommended in orientation processes for manufacture of packaging tapes and films.

Resin Properties:	Value	ASTM Method
Melt Flow, g/10 min.	4.5	D-1238 230C/2160C
Density, g/cc	0.905	D-1505
Melting Point, F	330	DSC
C	165	

FINA OIL & CHEMICAL CO.: DYPRO Polypropylenes(Continued):

Product Name:

3571 Homopolymer Film Grade:

Low Residuals - DYPRO 3571 utilizes high-yield catalyst technology plus a wash/de-ash process to provide extremely low residuals for improved color stability and clarity.

Slip/Antiblock - DYPRO 3571 is available with custom slip and antiblock packages.

FDA - DYPRO 3571 complies with all applicable FDA regulations for food contact applications.

Recommended Applications:

DYPRO 3571 is for use in non oriented cast film processes for manufacture of packaging films.

Resin Properties:	Value	ASTM Method
Melt Flow, g/10 min.	6.5	D-1238 230C/2160g
Density, g/cc	0.905	D-1505
Melting Point, F	330	DSC
C	165	

3576BC Homopolymer Film Grade:

Low Residuals - DYPRO 3576BC utilizes high-yield catalyst technology plus a wash/de-ash process to provide extremely low residuals for improved color stability and clarity.

Slip/Antiblock - DYPRO 3576 is available with custom slip and antiblock packages.

FDA - DYPRO 3576BC complies with all applicable FDA regulations for food contact applications.

Recommended Applications:

DYPRO 3576BC is recommended for use in non oriented cast film processes for manufacture of packaging films.

Resin Properties:	Value	ASTM Method
Melt Flow, g/10 min.	9.0	D-1238 230C/2160 g
Density, g/cc	0.905	D-1505
Melting Point, F	330	DSC
C	165	

FINA OIL & CHEMICAL CO.: DYPRO Polypropylenes(Continued):

Product Name:

3661 Homopolymer Fiber Grade for Staple Fiber Multifilament Yarns:

Fading Resistance - DYPRO 3661 is formulated to resist gas fading while maintaining excellent processing stability.

Process Stability - DYPRO 3661 features excellent processability and the good physical properties necessary for tapes, monofilaments and multifilaments.

FDA - DYPRO 3661 complies with all applicable FDA regulations for food contact applications.

Recommended Applications:

DYPRO 3661 is recommended for staple fibers and bulk continuous filament (BCF) yarns.

Resin Properties:	Value	ASTM Method
Melt Flow, g/10 min.	12	D-1238 Condition "L"
Density, g/cc	0.905	D-1505
Melting Point, F	330	DSC
C	165	

3861 High Melt Flow Homopolymer Fiber Grade for Melt Spinning:

Narrow Molecular Weight Distribution - The uniform rheology and low swell characteristics of DYPRO 3861 provide for easy drawing and denier control.

Fade Resistant - DYPRO 3861 is formulated for minimal gas fading.

FDA - DYPRO 3861 complies with all applicable FDA regulations for food contact applications.

Recommended Applications:

DYPRO 3861 is recommended for multifilament and non-woven fiber extrusion applications requiring a low viscosity polymer.

Resin Properties:	Value	ASTM Method
Melt Flow, g/10 min.	35.0	D-1238 Condition "L"
Density, g/cc	0.905	D-1505
Melting Point, F	330	DSC
C	165	

FINA OIL & CHEMICAL CO.: DYPRO Polypropylenes(Continued):

Product Name:

6234Z/6234WZ Random Copolymer Extrusion Blow Molding Grade:

Impact Strength - DYPRO 6234Z and DYPRO 6234WZ offer good clarity and room temperature impact strength, combined with top load strength for hot fill applications.

Antistat - DYPRO 6234Z and 6234WZ contain antistat to minimize dust accumulation for shelf cleanliness, as well as improved mold release.

Nucleation - DYPRO 6234Z is non-nucleated for excellent barrier resin compatibility in co-extrusion applications. DYPRO 6234WZ is nucleated to provide a fast cycle and superior top load strength.

High Purity - Low catalyst residues in DYPRO 6234Z and 6234WZ minimize odor and optimize melt stability for superior color and processability.

FDA - DYPRO 6234Z and 6234WZ comply with all applicable FDA regulations and may be used under these provisions for food contact and packaging.

Recommended Applications:

DYPRO 6234Z and 6234WZ are ideal for hot fill, single layer extrusion of blow molded containers for syrups and other food items.

DYPRO 6234Z is also an excellent choice for multi-layer extrusion blow molded containers.

Resin Properties:	Value	ASTM Method
Melt Flow, g/10 min.	1.5	D-1238 Condition "L"
Density, g/cc	0.900	D-1505
Melting Point, F	297	DSC
C	147	

FINA OIL & CHEMICAL CO.: DYPRO Polypropylenes (Continued):

Product Name:

6571 Random Copolymer Film Grade:

Color - Low catalyst residuals in DYPRO 6571 result in improved color stability.

Clarity/Impact - The random copolymer structure of DYPRO 6571 improves its clarity and impact resistance over conventional homopolymer films.

Slip/Antiblock - DYPRO 6571 is available with custom slip and antiblock packages.

FDA - DYPRO 6571 complies with all applicable FDA regulations for food contact applications.

Recommended Applications:

DYPRO 6571 is recommended for use in non oriented cast film processes for manufacture of packaging films that require improved optical and impact properties.

Resin Properties:	Value	ASTM Method
Melt Flow, g/10 min.	8.0	D-1238 230C/2160 g
Density, g/cc	0.90	D-1505
Melting Point, F	305	DSC
C	152	

FINA OIL & CHEMICAL CO.: DYPRO Polypropylenes(Continued):

Product Name:

7231 Random Copolymer Injection Blow Molding Grade with Improved Impact and Clarity:

Impact Strength - DYPRO 7231 offers excellent impact strength, clarity and gloss.

Lubricated - DYPRO 7231 offers superior core rod release, eliminating the need for additional lubricant.

High Purity - Low catalyst residues in DYPRO 7231 minimize odor and optimize melt stability and superior color and processability.

FDA - DYPRO 7231 complies with all applicable FDA regulations and may be used under these provisions for food contact and packaging.

Recommended Applications:

DYPRO 7231 is recommended for injection blow molded containers for food, drug, cosmetic and toiletry applications requiring superior impact, strength and clarity.

Resin Properties:	Value	ASTM Method
Melt Flow, g/10 min.	2.5	D-1238 Condition "L"
Density, g/cc	0.900	D-1505
Melting Point, F	289	DSC
C	143	

FINA OIL & CHEMICAL CO.: DYPRO Polypropylenes(Continued):

Product Name:

7232W Random Copolymer Extrusion Blow Molding Grade for Medical Applications:

Impact Strength - DYPRO 7232W offers excellent impact strength, clarity and gloss.

Nucleation - DYPRO 7232W is nucleated to allow shorter cooling time for high production rates.

Chemical Inertness - DYPRO 7232W has been formulated specifically for medical applications requiring low level extractables.

High Purity - Low catalyst residues in DYPRO 7232W minimize odor and optimize melt stability for superior color and processability.

FDA - DYPRO 7232W complies with all applicable FDA regulations and may be used under these provisions for food contact and packaging.

Recommended Applications:

DYPRO 7232W is ideal for extrusion blow molding of medical, food and other containers requiring low extractables.

Resin Properties:	Value	ASTM Method
Melt Flow, g/10 min.	1.5	D-1238 Condition "L"
Density, g/cc	0.900	D-1505
Melting Point, F	289	DSC
C	143	

FINA OIL & CHEMICAL CO.: DYPRO Polypropylenes (Continued):

Product Name:

7235 Random Copolymer Extrusion Blow Molding Grade with Improved Impact and Clarity:

Impact Strength - DYPRO 7235 offers excellent impact strength, clarity and gloss. The outstanding parison strength of DYPRO 7235 allows for large container sizes.

High Purity - Low catalyst residues in DYPRO 7235 minimize odor and optimize melt stability for superior color and processability.

FDA - DYPRO 7235 complies with all applicable FDA regulations and may be used under these provisions for food contact and packaging.

Recommended Applications:

DYPRO 7235 is ideal for extrusion blow molding containers for food, drug, cosmetic and toiletry applications requiring superior impact, strength and clarity.

Resin Properties:	Value	ASTM Method
Melt Flow, g/10 min.	1.5	D-1238 Condition "L"
Density, g/cc	0.900	D-1505
Melting Point, F	289	DSC
C	143	

FINA OIL & CHEMICAL CO.: DYPRO Polypropylenes(Continued):

Product Name:

7371 Random Copolymer Heat Sealable Film Grade:

Clarity/Impact - DYPRO 7371 is a second generation random copolymer with improved clarity and impact properties over homopolymer and first generation EP copolymers.

Heat Sealablity - The reduced melting point of DYPRO 7371 over first generation EP copolymers makes it suitable as a heat sealable layer for oriented films.

Slip/Antiblock - DYPRO 7371 is available with custom slip and antiblock packages.

FDA - DYPRO 7371 complies with all applicable FDA regulations for food contact applications.

Recommended Applications:

DYPRO 7371 is recommended for use in either non oriented cast film processes for manufacture of packaging films that require improved optical and impact properties or as a heat sealable layer for oriented films.

Resin Properties:	Value	ASTM Method
Melt Flow, g/10 min.	3.0	D-1238 230C/2160 g
Density, g/cc	0.90	D-1505
Melting Point, F	293	DSC
C	145	

FINA OIL & CHEMICAL CO.: DYPRO Polypropylenes(Continued):

Product Name:

7450 Random Copolymer Heat Sealable Film Grade for Medical Use:

Clarity/Impact - DYPRO 7450 is a second generation random copolymer with improved clarity and impact properties over homopolymer and first generation EP copolymers.

Heat Sealability - The reduced melting point of DYPRO 7450 over first generation EP copolymers makes it suitable as a heat sealable layer for oriented films.

Slip/Antiblock - DYPRO 7450 is available with custom slip and antiblock packages.

FDA - DYPRO 7450 complies with all applicable FDA regulations for food contact applications.

Recommended Applications:

DYPRO 7450 is recommended for use in either non oriented cast film processes for manufacture of packaging films that require improved optical and impact properties or as a heat sealable layer for oriented films.

Resin Properties:	Value	ASTM Method
Melt Flow, g/10 min.	5.0	D-1238 230C/2160 g
Density, g/cc	0.90	D-1505
Melting Point, F	293	DSC
C	145	

FINA OIL & CHEMICAL CO.: DYPRO Polypropylenes(Continued):

Product Name:

8473 Random Copolymer Heat Sealable Film Grade:

Color/Clarity/Impact - DYPRO 8473 is a third generation low melting, high ethylene random copolymer with improved color, optics and impact properties.

Heat Sealable - The low melting point of DYPRO 8473 makes it a very good heat sealable layer for oriented films.

Slip/Antiblock - DYPRO 8473 is available with custom slip and antiblock packages.

FDA - DYPRO 8473 complies with FDA regulations Title 21 CFR Section 177.1520(c) 3.1 and may be used under these provisions for food contact and packaging.

Recommended Applications:

DYPRO 8473 is recommended for use in either non oriented cast film processes for manufacture of packaging films that require improved optical and impact properties or as a heat sealable layer for oriented films.

Resin Properties:	Value	ASTM Method
Melt Flow, g/10 min.	4.5	D-1238 230C/2160 g
Density, g/cc	0.895	D-1505
Melting Point, F	275	DSC
C	135	

8485 Homopolymer:

High molecular weight homopolymer with faster extrusion processing for sheet and strapping without sacrifice of strength properties, melt strength or process stability.

Resin Properties:	Value	ASTM Method
Melt Flow, g/10 min.	1.2	D-1238 Condition "L"
Density, g/cc	0.905	D-1505
Melting Point, F	330	DSC
C	165	

FINA OIL & CHEMICAL CO.: DYPRO Polypropylenes(Continued):

Product Name:

8486 Low Melt Flow Homopolymer for Sheet and Strapping:

Melt Strength - High melt strength DYPRO 8486 allows uniform drawdown during processing, resulting in maximum line speeds and sheet and strapping with a good balance of physical properties.

Process Stability - Excellent polymer stability of DYPRO 8486 produces consistent product properties during extrusion, even with the use of regrind.

FDA - DYPRO 8486 complies with all applicable FDA regulations and may be used under these provisions for food contact applications.

Recommended Applications:

DYPRO 8486 is recommended for sheet and strapping applications where high melt strength and high extrusion speeds are required.

Resin Properties:	Value	ASTM Method
Melt Flow, g/10 min	1.1	D-1238 Condition "L"
Density, g/cc	0.905	D-1505
Melting Point, F	330	DSC
C	165	

FINA OIL & CHEMICAL CO.: DYPRO Polypropylenes(Continued):

Product Name:

8573 Random Copolymer Heat Sealable Film Grade:

Color/Clarity/Impact - DYPRO 8573 is a third generation low melting, high ethylene random copolymer with improved color, optics and impact properties.

Heat Sealable - The low melting point of DYPRO 8573 makes it a very good heat sealable layer for oriented films.

Slip/Antiblock - DYPRO 8573 is available with custom slip and antiblock packages.

FDA - DYPRO 8573 complies with FDA regulations Title 21 CFR Section 177.1520(c) 3.1 and may be used under these provisions for food contact and packaging.

Recommended Applications:

DYPRO 8573 is recommended for use in either non oriented cast film processes for manufacture of packaging films that require improved optical and impact properties or as a heat sealable layer for oriented films.

Resin Properties:	Value	ASTM Method
Melt Flow, g/10 min	7.0	D-1238 230C/2160 g
Density, g/cc	0.895	D-1505
Melting Point, F	275	DSC
C	135	

FINA OIL & CHEMICAL CO.: DYPRO Polypropylenes(Continued):

Product Name:

8667 Film Fiber Homopolymer with Low Water Carry-Over:

Low Water Carry-Over - DYPRO 8667 exhibits excellent low water carry-over performance on water quench slit film extrusion lines, which allows operation at maximum line speeds.

Low Filterables - Low catalyst residuals in DYPRO 8667 allow extended processing runs without screen pack pluggage.

FDA - DYPRO 8667 complies with all applicable FDA regulations and may be used under these provisions for food contact applications.

Recommended Applications:

DYPRO 8667 is recommended for water quench slit films and monofilaments for carpet backing, rope and cordage products.

Resin Properties:	Value	ASTM Method
Melt Flow, g/10 min.	4.0	D-1238 Condition "L"
Density, g/cc	0.905	D-1505
Melting Point, F	330	DSC
C	165	

8670 Homopolymer:

High modulus homopolymer resin for production of cast or crisp oriented films with excellent machineability and superior clarity and sparkle for packaging a wide variety of consumer products. Available with custom slip and anti-block to meet processing and end-use requirements.

Resin Properties:	Value	ASTM Method
Melt Flow, g/10 min.	4.0	D-1238 Condition "L"
Density, g/cc	0.905	D-1505
Melting Point, F	330	DSC
C	165	

FINA OIL & CHEMICAL CO.: DYPRO Polypropylenes(Continued):

Product Name:

8771 Homopolymer:

Superior clarity and sparkle cast film produced via high speed extrusion. Excellent for use in packaging and processing foods. Available with custom slip and anti-block additives.

Resin Properties:	Value	ASTM Method
Melt Flow, g/10 min.	9.0	D-1238 Condition "L"
Density, g/cc	0.905	D-1505
Melting Point, F	330	DSC
C	165	

8921 Easy Flow LTHA Injection Molding Homopolymer:

LTHA - DYPRO 8921 is designed with an LTHA additive system that gives excellent leach resistance in hot-wet environments that contain bleach and laundry detergents.

Easy Flow - DYPRO 8921 is easy flowing and yet has physical properties similar to 4.0 melt flow polypropylene homopolymers.

UL Rating - DYPRO 8921 has a 115C (240F) UL Yellow Card rating and can be used for appliance parts.

FDA - DYPRO 8921 complies with all applicable FDA regulations for food contact applications.

Recommended Applications:

DYPRO 8921 has excellent resistance to hot-wet environments containing bleach and laundry detergents and is particularly suited for washing machine and dishwasher parts.

Resin Properties:	Value	ASTM Method
Melt Flow, g/10 min.	30.0	D-1238 Condition "L"
Density, g/cc	0.905	D-1505
Melting Point, F	330	DSC
C	165	

FINA OIL & CHEMICAL CO.: DYPRO Polypropylenes(Continued):

Product Name:

9670 Copolymer:

Cast and oriented high ethylene random copolymer film resin with excellent machineability balanced with good low temperature impact performance. Excellent for packaging applications requiring clarity, sparkle and excellent heat sealability. Available with custom slip and anti-block to meet end-use requirements.

Resin Properties:	Value	ASTM Method
Melt Flow, g/10 min.	4.0	D-1238 Condition "L"
Density, g/cc	0.900	D-1505
Melting Point, F	293	DSC
C	145	

FINA OIL & CHEMICAL CO.: FINA Polypropylenes:

Product Name:

3422/3422Z General Purpose Injection Molding and Extrusion Homopolymer:

Process Stability - FINA 3422 and 3422Z offer good process stability for injection molding and extrusion. FINA 3422Z is the same as FINA 3422 but with antistat to reduce process static and minimize dust accumulation on finished parts.

FDA - FINA 3422 and 3422Z comply with all applicable FDA regulations and may be used under those provisions for food contact applications.

Recommended Applications:

FINA 3422 is recommended for the injection molding of food containers. FINA 3422Z is particularly well suited for the extrusion of drinking straws.

Resin Properties:	Value	ASTM Method
Melt Flow, g/10 min.	4.0	D-1238 Condition "L"
Density, g/cc	0.905	D-1505
Melting Point, F	330	DSC
C	165	

3423 LTHA Injection Molding Homopolymer:

LTHA - FINA 3423 is engineered with an LTHA additive system that gives excellent leach resistance in hot/wet environments that contain bleach and laundry detergents.

UL Rating - FINA 3423 has a 115C (240F) UL Yellow Card Rating making it suitable for appliance parts.

FDA - FINA 3423 complies with all applicable FDA regulations for food contact applications.

Recommended Applications:

FINA 3423 is recommended for appliance parts and is particularly suited for parts that are exposed to a hot-wet environment such as: washing machine agitation, dishwasher silverware baskets and coffee machine parts.

Resin Properties:	Value	ASTM Method
Melt Flow, g/10 min.	5.0	D-1238 Condition "L"
Density, g/cc	0.905	D-1505
Melting Point, F	330	DSC
C	165	

FINA OIL & CHEMICAL CO.: FINA Polypropylenes (Continued):

Product Name:

3620WZ/3620Z Injection Molding Homopolymer:

Nucleation - FINA 3620WZ is engineered with nucleation to provide fast cycle time and improve contact clarity in thin wall multi-cavity molds.

Antistat - FINA 3620WZ and 3620Z contain an anti-stat that will help protect molded parts from dust accumulation.

FDA - FINA 3620WZ and 3620Z comply with all applicable FDA regulations for food contact applications.

Recommended Applications:

FINA 3620WZ and 3620Z are particularly well suited for caps, closures, medical syringes and thin wall containers.

Resin Properties:	Value	ASTM Method
Melt Flow, g/10 min.	12.0	D-1238 Condition "L"
Density, g/cc	0.905	D-1505
Melting Point, F	330	DSC
C	165	

3621WZ Fast Cycle Injection Molding Homopolymer with Easy Part Release:

Fast Cycling - FINA 3621WZ is engineered with nucleation to provide fast cycle times and improved clarity in thin wall parts.

Easy Part Release - FINA 3621WZ is formulated with a high level of antistat for shelf cleanliness and easy part release in complex multi-cavity molds.

FDA - FINA 3621WZ complies with all applicable FDA regulations for food contact applications.

Recommended Applications:

FINA 3621WZ is recommended for cutlery and allied applications. FINA 3621WZ polypropylene should not be used in parts that utilize dark colors or post mold decoration techniques.

Resin Properties:	Value	ASTM Method
Melt Flow, g/10 min.	12.0	D-1238 Condition "L"
Density, g/cc	0.905	D-1505
Melting Point, F	330	DSC
C	165	

FINA OIL & CHEMICAL CO.: FINA Polypropylenes(Continued):

Product Name:

3622 General Purpose Injection Molding Homopolymer:

Good Processability - FINA 3622 is engineered for good processability in thin wall multi-cavity molds.

Good Clarity - FINA 3622 has good contact/see-through clarity and is easily printed with silk screen or hot stamping techniques.

FDA - FINA 3622 complies with all applicable FDA regulations and may be used under those provisions for food contact applications.

Recommended Applications:

FINA 3622 is recommended for the injection of housewares, utility boxes, and containers.

Resin Properties:	Value	ASTM Method
Melt Flow, g/10 min.	12.0	D-1238 Condition "L"
Density, g/cc	0.905	D-1505
Melting Point, F	330	DSC
C	165	

3720W/3720WZ High Flow Reactor Injection Molding Homopolymer: (Contains antistat)

High Flow - FINA 3720W and 3720WZ have high flow characteristics that result in ease of filling thin wall parts.

Nucleation - FINA 3720W and 3720WZ are nucleated to provide fast cycle time and improved contact clarity in thin wall, multi-cavity molds.

Antistat - FINA 3720WZ contains an antistat that will help protect molded parts from dust accumulation.

FDA - FINA 3720WZ and 3720W comply with all applicable FDA regulations for food contact applications.

Recommended Applications:

FINA 3720WZ is recommended for cups, closures and thin wall containers.

FINA 3720W is recommended for caps, closures and thin wall containers.

Resin Properties:	Value	ASTM Method
Melt Flow, g/10 min.	20	D-1238 Condition "L"
Density, g/cc	0.905	D-1505
Melting Point, F	330	DSC
C	165	

FINA OIL & CHEMICAL CO.: FINA Polypropylenes(Continued):

Product Name:

3721WZ/3721Z High Flow Reactor Homopolymer with a High Level of Antistat:

High Flow - FINA 3721WZ and 3721Z are designed with high flow characteristics for ease of filling thin wall parts.

High Antistat - FINA 3721WZ and 3721Z are engineered with a high level of antistat for shelf cleanliness and mold release.

FDA - FINA 3721WZ and 3721Z comply with all applicable FDA regulations for food contact applications.

Nucleation - FINA 3721WZ is nucleated to provide fast cycle time and improve contact clarity in thin wall, multi-cavity molds.

RECOMMENDED APPLICATIONS:

FINA 3721WZ is ideal for cutlery and other thin wall multi-cavity, fast cycling applications.

FINA 3721Z is well suited for caps, closures and thin wall parts, particularly when high antistat is required.

Resin Properties:	Value	ASTM Method
Melt Flow, g/10 min.	20	D-1238 Condition "L"
Density, g/cc	0.905	D-1505
Melting Point, F	330	DSC
C	165	

FINA OIL & CHEMICAL CO.: FINA Polypropylenes(Continued):

Product Name:

3824/3824W/3824WZ Easy Flow Injection Molding Homopolymer:

Easy Flow - FINA 3824, 3824W and 3824WZ exhibit exceptionally easy flow combined with physical properties similar to 4.0 melt polypropylene homopolymer.

FINA 3824W is nucleated and FINA 3824WZ is nucleated and antistated. Nucleation will reduce cycle time and improve contact clarity in thin wall parts. Antistat helps prevent accumulation of dust on the the finished parts.

FDA - FINA 3824, 3824W and 3824WZ comply with all applicable FDA regulations for food contact applications.

Recommended Applications:

FINA 3824, 3824W and 3824WZ are recommended for large thin wall parts, caps and closures.

Resin Properties:	Value	ASTM Method
Melt Flow, g/10 min.	30.0	D-1238 Condition "L"
Density, g/cc	0.905	D-1505
Melting Point, F	330	DSC
C	165	

3840 Improved Radiation High Flow Homopolymer For Injection Molding:

Radiation Resistant - FINA 3840 is a second-generation radiation grade designed to maintain good physical properties when exposed to radiation used for sterilization.

Very High Flow - FINA 3840 has very high flow characteristics for ease of filling thin wall parts.

FDA - FINA 3840 complies with all applicable FDA regulations for food contact applications.

Recommended Applications:

FINA 3840 is recommended for injection molded medical parts where radiation is used for sterilization.

Resin Properties:	Value	ASTM Method
Melt Flow, g/10 min.	30.0	D-1238 Condition "L"
Density, g/cc	0.905	D-1505
Melting Point, F	330	DSC
C	165	

FINA OIL & CHEMICAL CO.: FINA Polypropylenes(Continued):

Product Name:

3841 Radiation Resistant High Flow Homopolymer for Injection Molding:

Radiation Resistant - FINA 3841 is designed to maintain good physical properties when radiation below 2.0 MRADS is used for sterilization.

High Flow - FINA 3841 has high flow characteristics for ease of filling thin wall parts.

FDA - FINA 3841 complies with all applicable FDA regulations for food contact applications.

Recommended Applications:

FINA 3841 is recommended for injection molded medical parts where radiation doses below 2.0 MRADS are used for sterilization.

Resin Properties:	Value	ASTM Method
Melt Flow, g/10 min.	20.0	D-1238 Condition "L"
Density, g/cc	0.905	D-1505
Melting Point, F	330	DSC
C	165	

7425 Random Copolymer for Injection Molding:

Clarity - FINA 7425, a random copolymer, exhibits excellent clarity when molded in polished molds.

Impact - FINA 7425 produces molded parts with measurably improved impact compared with parts molded with parts from polypropylene homopolymers.

FDA - FINA 7425 complies with all applicable FDA regulations for food contact applications.

Recommended Applications:

FINA 7425 is recommended for injection molding medical devices and disposables, clear food containers and specialty housewares.

Resin Properties:	Value	ASTM Method
Melt Flow, g/10 min.	4.0	D-1238 Condition "L"
Density, g/cc	0.900	D-1505
Melting Point, F	293	DSC
C	145	

FINA OIL & CHEMICAL CO.: FINA Polypropylenes(Continued):

Product Name:

7622MZ Clarified Random Copolymer for Injection Molding:

Clarity - FINA 7622MZ, exhibits exceptionally high clarity in molded parts.

Impact - FINA 7622MZ produces molded parts with improved impact compared with parts molded from homopolymer polypropylene.

Antistat - FINA 7622MZ contains an antistat to help protect molded parts from dust accumulation.

FDA - FINA 7622MZ complies with all applicable FDA regulations for food contact applications.

Recommended Applications:

FINA 7622MZ is recommended for injection molding houseware, cassette cases and other items calling for good impact and high clarity.

Resin Properties:	Value	ASTM Method
Melt Flow, g/10 min.	4.0	D-1238 Condition "L"
Density, g/cc	0.900	D-1505
Melting Point, F	293	DSC
C	145	

FINA OIL & CHEMICAL CO.: FINA & DYPRO Polypropylenes:

Product Name:

7771 Copolymer:

Superior sparkle and clarity, cast film grade medium ethylene random copolymer for packaging a wide variety of foods, dry goods, confections, pastries, etc. Melt rheology is designed for high speed cast film extrusion. Available with custom slip and anti-block to meet processing and end-use requirements.

Resin Properties:	Value	ASTM Method
Melt Flow, g/10 min.	8.0	D-1238 Condition "L"
Density, g/cc	0.900	D-1505
Melting Point, F	306	DSC
C	152	

8068 Homopolymer:

High melt flow and low viscosity for high shear spinning operations. Multi-filament and non-woven fiber via extrusion processing. Uniform rheological and low swell characteristics provide for easy drawing and denier control.

Resin Properties:	Value	ASTM Method
Melt Flow, g/10 min.	36.0	D-1238 Condition "L"
Density, g/cc	0.905	D-1505
Melting Point, F	330	DSC
C	165	

FREEMAN CHEMICAL CO.: ACPOL Polymer Engineering Resin System:

A series of thermosetting alloy resins...very cost effective in molding processes that require extremely fast cure cycles even in parts with thick wall cross sections. They are frequently used to replace thermosetting polyester resins in applications which need higher heat performance or extended chemical or corrosion resistance.

TYPICAL END USE APPLICATIONS:

* Automotive: Class A body panels
* Electrical: circuit breakers - switch-gear - insulating varnish
* Cultured marble - sanitary ware
* Construction: structural I-beams - channels
* Recreation: boat hulls - water skis - laminated bows-fishing rods
* Architectural: building facia - translucent panels - insulating & security glass
* Consumer products: microwave cookware - food trays - satellite dish antennas

TYPICAL PROCESS APPLICATIONS:

1 SMC/HMC compression molding
2 BMC injection molding
3 Pultrusion/pulforming
4 Filament winding
5 Spray-up/hand lay-up
6 RTM
7 Polymer concrete
8 Casting - filled & clear
9 Bag molding
10 Glass laminating

FREEMAN CHEMICAL CO.: ACPOL Polymer Engineering Resin System (Continued):

TYPE: ACPOL 42-2640 is a modified thermosetting acrylic resin.

USES: To prepare SMC or BMC when used in conjunction with STYPOL 40-7263.

DISTINGUISHING CHARACTERISTICS:

Good high strength properties at 77F (25C)
Good retention of physical properties at elevated temperature
Excellent hydrolytic properties
Low water absorption
Excellent fatigue properties
Good surface appearance

LIQUID PROPERTIES:

Test	Value	Method
NVM	69-72%	
77F (25C) Viscosity LVF #2 @ 30 rpm	200-400 cps	Brookfield
Acid Value	7-13	
Color	4 Maximum	Gardner
Uncatalyzed Life @ 77F (25C)	3 months	

FORMULATION I (HIGH GLASS)

Components	Parts by Weight
ACPOL 42-2640	16.49
STYPOL 40-0087 (Low Profile Additive)	7.07
t-Butylperbenzoate	0.24
Zinc Stearate	0.82
Calcium Carbonate	24.58
STYPOL 40-7263	6.80
% Glass	44.00

FREEMAN CHEMICAL CO.: INTERPOL Polymer Engineering Resin System:

INTERPOL (IPN) resins are at the leading edge of engineering polymer technology. This series is specifically designed to maintain superior fatigue resistance in humid environments, even at high temperatures. Advanced composites fabricated with INTERPOL systems reinforced with aramid and carbon fiber result in very high strength laminates. This unique combination of polymers has exceptional wetting characteristics. For this reason, advanced composites can be fabricated with very high fiber loadings with no loss of interfacial bonding.

TYPICAL END USE APPLICATIONS:

* Aircraft/aerospace: Structural members
* Electrical/electronic: Housings
* Transportation: Structural members

TYPICAL PROCESS APPLICATIONS:

1 SMC/HMC compression molding
2 BMC injection molding
3 Pultrusion/pulforming
4 Filament winding
6 RTM
8 Casting - filled & clear
9 Bag molding

FREEMAN CHEMICAL CO.: INTERPOL Polymer Engineering Resin System (Continued):

Product Name:

INTERPOL 047-1000:

TYPE: INTERPOL 047-1000 is a modified thermosetting acrylic resin.

USES: For pultrusion parts when used in conjunction with STYPOL 040-7263 and glass, aramid or graphite fiber reinforcement.

DISTINGUISHING CHARACTERISTICS:

Low viscosity
Good retention of physical properties at elevated temperature
Excellent wet-out of KEVLAR and graphite
Good surface appearance

LIQUID PROPERTIES:

Test	Value	Method
NVM	62-64	
77F (25C) Viscosity LVF #2 @ 30 rpm	90-150 cps	Brookfield
Acid Value	7-10	
Color	4 Maximum	Gardner
Uncatalyzed Life @ 77F (25C)	3 months	
Hydroxyl No.	95	

PULTRUSION FORMULATION:

Components	Parts by Weight
INTERPOL 047-1000	82.30
t-Butylperbenzoate	0.50
t-Butylperoctoate	1.50
Zinc stearate or ZELEC UN	1.00
STYPOL 040-7263	14.70

FREEMAN CHEMICAL CO.: INTERPOL Polymer Engineering Resin System (Continued):

Product Name:

INTERPOL 047-1050:

TYPE: Polyester; Isophthalic Type, Styrene Monomer, a Clear Light Amber Liquid.

USES: INTERPOL 047-1050 is a medium reactive, resilient resin, specifically designed for pultrusion when used in conjunction with STYPOL 040-7263 and high strength composite materials.

DISTINGUISHING CHARACTERISTICS:

Excellent Resistance to Cracking in Resin Rich Areas
Resistance to Warping
Excellent Finish and Gloss
Fast Cure Time

TYPICAL PROPERTIES:

Uncured Resin

Test	Value	Method
Non-volatile Material (NVM)	60-62	
Viscosity, 25C	90-150 cps	Brookfield
Acid Value	1-5	
Hydroxyl No.	70-80	

FREEMAN CHEMICAL CO.: NUPOL Polymer Engineering Resin System:

A series of engineering thermosetting acrylic vinyl ester polymers with excellent high temperature performance and exceptional chemical and corrosion resistance. They are frequently used for applications which specify excellent alkali resistance at high temperatures. Inherent low viscosities allow very fast wet-out and superior interfacial bonding to fillers and fiber reinforcements.

TYPICAL END USE APPLICATIONS:

* Automotive: Class A body panels
* Electrical: circuit breakers - switch-gear - insulating varnish
* Cultured marble - sanitary ware
* Construction: structural I-beams - channels
* Recreation: boat hulls - water skis - laminated bows - fishing rods
* Architectural: building facia - translucent panels - insulating & security glass
* Consumer products: microwave cookware - food trays - satellite dish antennas

PROCESS APPLICATIONS:

1 SMC/HMC compression molding
2 BMC injection molding
3 Pultrusion/pulforming
4 Filament winding
5 Spray-up/hand lay-up
6 RTM
7 Polymer concrete
8 Casting - filled & clear
9 Bag molding
10 Glass laminating

FREEMAN CHEMICAL CORP.: STYPOL Polymer Engineering Resin System:

A family of engineering thermoset polyester resins with high temperature performance, chemical & corrosion resistance, inherent flame resistance and outstanding electric insulating properties.

TYPICAL END USE APPLICATIONS:

* Automotive: Class A body panels
* Electrical: circuit breakers - switch-gear - insulating varnish
* Cultured marble - sanitary ware
* Construction: structural I-beams - channels
* Recreation: boat hulls - water skis - laminated bows - fishing rods
* Architectural: building facia - translucent panels - insulating & security glass
* Consumer products: microwave cookware - food trays - satellite dish antennas

TYPICAL PROCESS APPLICATIONS:

1 SMC/HMC compression molding
2 BMC injection molding
3 Pultrusion/pulforming
4 Filament winding
5 Spray-up/hand lay-up
6 RTM
7 Polymer concrete
8 Casting - filled & clear
9 Bag molding
10 Glass laminating

FREEMAN CHEMICAL CORP.: STYPOL Polymer Engineering Resin System (Continued):

Product Name:

STYPOL 40-0087:

STYPOL 40-0087 is used as a low-shrink additive in fiber-reinforced unsaturated polyester molding compounds. This additive imparts improved dimensional stability to polyester moldings. SMC formulations containing low-profile additive can be pigmented with a variety of colors.

LIQUID PROPERTIES

Test	Value
Appearance	White Viscous Liquid
Total Solids (% by Weight)	39 - 41
77C (25C) Brookfield Viscosity	8,000-12,000 cps*
Uncatalyzed Life @ 77F (25C)	3 months

* LVF #3 spindle at 6 RPM

STYPOL 40-7263:

TYPE: Modified Polyisocyanate

DESCRIPTION: Liquid, modified polyisocyanate

TYPICAL PROPERTIES:

TEST	VALUE
NCO Content, %	17.7-20.9
Viscosity at 25C(77F)	110-170 cps
Flash Point	87F(30.6C)
Crystallization Point	40F(4.4C)

GENERAL ELECTRIC CO.: GELOY Resin Engineered Blow Molding Grades:

Easy procesability and superior color and property retention combine to make GELOY resin the choice for applications involving prolonged outdoor exposure. GELOY blow moldable resin also offers an excellent balance of physical properties, and its color selection range, in dark as well as light tones, allows for a broad spectrum of design opportunities.

TYPICAL PROPERTY VALUES:

GELOY BG10 resin:

Specific Gravity (D792): 1.06
HDT @ 264 psi, F (D648): 180
Flexural Modulus, psi (D790): 240,000
Flexural Strength, psi (D790): 8,200
Tensile Strength, psi (D638): 5,500
Notched Izod Impact Strength, ft-lbs/in (D256):
@ 73F: 6
Elongation, % (D638): 25
Falling Dart Impact Strength (1/8") in-lbs:
@ 73F: 210
Flammability (UL94): HB

GENERAL ELECTRIC CO.: GELOY Resin Weatherable Resins:

GELOY resins are a family of outdoor-weatherable polymers based on an advanced terpolymer of acrylic-styrene-acrylonitrile (ASA). These resins offer a superior combination of color and property retention as well as ease of processing on conventional equipment.

Available in sheet extrusion, profile extrusion, blow molding and injection molding grades, GELOY resins have demonstrated excellent property and color retention in both light and dark colors after prolonged outdoor exposure.

EXTRUSION GRADES:

GELOY 1020 resin is an ASA granular powder perfect for blending with PVC. GELOY 1020 resin is easily mixed with PVC powder, providing a blend with improved heat distortion temperature, impact, and dark-color weatherability while maintaining excellent processability using equipment and tooling designed for PVC. Products and colors can thus be formulated to fit the manufacturer's exact material needs for extrusion applications such as window profiles, mobile home skirting and siding. The blends can also be easily coextruded or laminated over PVC, providing a cost-effective weatherable capstock.

GELOY 1120 resin is a pelletized form of GELOY 1020 grade. It is used for applications such as truck tops, RV parts, spas, and pool steps where high heat, gloss, and mechanical properties are needed in addition to excellent outdoor-weatherability. GELOY 1120 resin can be used in profile extrusion, sheet extrusion (monolithic or coextruded over ABS), and in extrusion blow molding (door panels, shutters) where color and property retention is desired, eliminating the need for painting.

GELOY 1220 resin is an alloy of ASA and PVC provided in pellet form. It is perfect for profile extruded parts that need higher heat and/or outdoor-weatherability in either dark or light colors. Typical applications include window and door profiles, gutter systems, fencing, vending machine trim, and automotive exterior trim.

GENERAL ELECTRIC CO.: GELOY Resin Weatherable Resins (Continued):

INJECTION MOLDING GRADES:

GELOY XP1001 resin is an injection molding ASA resin offering excellent weatherability. Typical applications are interior or exterior automotive/truck trim and parts, providing color stability without painting.

GELOY XP2001 resin is an ASA/PVC alloy offering flame retardance along with good low-temperature impact. It is used for outdoor electrical boxes, NIDs, and appliances.

GELOY XP2003 resin is an ASA/PVC alloy with excellent weatherability and chemical resistance. RV parts and accessories, marine parts, and automotive injection trim are typical applications.

GELOY XP4001 resin is an ASA alloy with polycarbonate. It offers higher heat resistance and impact strength than GELOY XP1001 resin, along with good weatherability. Unpainted interior and exterior automotive parts, either injection molded or blow molded, are good applications for GELOY XP4001 resin. Both low gloss and high gloss versions are available.

PRELIMINARY PROPERTY PROFILE:

GELOY XP1001 resin:
- Deflection Temperature, F(C) (D648):
  - @ 66 psi, 1/4" (0.46 MPa, 6.4 mm): 205(96)
  - @264 psi, 1/4" (1.82 MPa, 6.4 mm): 185(85)

GELOY XP2001 resin:
- Deflection Temperature, F(C) (D648):
  - @ 66 psi, 1/4" (0.46 MPa, 6.4 mm): 175(79)
  - @264 psi, 1/4" (1.82 MPa, 6.4 mm): 160(70)

GELOY XP2003 resin:
- Deflection Temperature, F(C) (D648):
  - @ 66 psi, 1/4" (0.46 MPa, 6.4 mm): 180(82)
  - @264 psi, 1/4" (1.82 MPa, 6.4 mm): 165(73)

GELOY XP4001HG resin:
- Deflection Temperature, F(C) (D648):
  - @ 66 psi, 1/4" (0.46 MPa, 6.4 mm): 230(110)
  - @264 psi, 1/4" (1.82 MPa, 6.4 mm): 210(99)

GELOY XP4001LG resin:
- Deflection Temperature, F(C) (D648):
  - @ 66 psi, 1/4" (0.46 MPa, 6.4 mm): 225(107)
  - @264 psi, 1/4" (1.82 MPa, 6.4 mm): 205(96)

GENERAL ELECTRIC CO.: LEXAN Business Equipment (BE) Resins:

LEXAN Business Equipment resins are flame retardant polycarbonate products offering improved processing features compared to the LEXAN 900 resin series of materials.

While providing slightly reduced property performance, LEXAN BE resin grades meet the thermal and overall toughness requirements of most business equipment application designs.

Having an outstanding cost-to-performance balance, LEXAN BE resins are ideal materials for today's computer cabinetry and enclosures.

Each LEXAN BE resin grade includes a unique package of features, including:

LEXAN BE1230:

* High flow; Melt flow: 25
* High flame retardancy: UL94 V-O @ .063"
* Reduced gloss: 70 untextured 60

LEXAN BE2130 resin:

* Same as BE1230 except with standard gloss: 90 untextured 60

LEXAN BE3030:

* Super high flow: Melt flow: 45
* Medium flame retardance: UL94 V-O @ .090"
* Reduced gloss: 70 untextured 60

Typical Property Values:

LEXAN BE1230 resin:

Viscosity Midpoint (Melt Flow Rate) g/10 min (D1238): 25

LEXAN BE2130 resin:

Viscosity Midpoint (Melt Flow Rate) g/10 min (D1238): 25

LEXAN BE3030 resin:

Viscosity Midpoint (Melt Flow Rate) g/10 min (D1238): 45

GENERAL ELECTRIC CO.: LEXAN Engineering Structural Foam Resins:

Foamable LEXAN resin is an ideal choice for structural components where load bearing capability at elevated temperatures is a key requirement. It exhibits outstanding impact strength, high heat resistance, flexural characteristics, creep resistance and processability. In addition, it is an excellent alternative to metal for large components with broad application potential in appliance, automotive, telecommunications, material handling and business machine industries. The material combines rigitity, impact strength and toughness with UL 94 V-O and 5V listings.

LEXAN Foam Typical Property Values:

LEXAN FL400 resin:

Solid Specific Gravity/Density (oz/in 3) (D792): 1.20 (0.69)
Foam Molded Specific Gravity (D792): 0.95
Weight Reduction, % (GE Test Method): 20

LEXAN FL410 resin:

Solid Specific Gravity/Density (oz/in 3) (D792): 1.25 (0.72)
Foam Molded Specific Gravity (D792): 1.0
Weight Reduction, % (GE Test Method): 20

LEXAN FL478 resin:

Solid Specific Gravity/Density (oz/in 3) (D792): 1.26 (0.73)
Foam Molded Specific Gravity (D792): 0.90
Weight Reduction, % (GE Test Method): 28

LEXAN FL900 resin:

Solid Specific Gravity/Density (oz/in 3) (D792): 1.21 (0.70)
Foam Molded Specific Gravity (D792): 0.90
Weight Reduction, % (GE Test Method): 26

LEXAN FL910 resin:

Solid Specific Gravity/Density (oz/in 3) (D792): 1.25 (0.72)
Foam Molded Specific Gravity (D792): 0.95
Weight Reduction, % (GE Test Method): 24

GENERAL ELECTRIC CO.: LEXAN Engineering Structural Foam Resins (Continued):

LEXAN Foam Typical Property Values(Continued):

LEXAN FL930 resin:

Solid Specific Gravity (oz/in 3) (D792) : 1.43 (0.83)
Foam Molded Specific Gravity: 1.05
Weight Reduction, %: 27

LEXAN FL1600 resin:

Solid Specific Gravity (oz/in 3) (D792): 1.21 (0.70)
Foam Molded Specific Gravity: 1.10
Weight Reduction, %: 10

LEXAN FL2000 resin:

Solid Specific Gravity (oz/in 3) (D792): 1.21 (0.70)
Foam Molded Specific Gravity: 1.10
Weight Reduction, %: 10

GENERAL ELECTRIC CO.: LEXAN Polycarbonate Film:

LEXAN polycarbonate films offer a unique combination of features including clarity, dimensional stability, toughness, flexibility, heat resistance and excellent dielectric performance. Virtually haze-free, LEXAN films are available in clear grades as well as a broad choice of textures and colors. With the introduction of thin-gauge, hard-coated and improved combustion-resistant grades, LEXAN films can meet virtually every requirement for high-performance membrane switches, graphics, packaging and bright trim applications.

LEXAN CF Film is a one-side flexible hard-coated polycarbonate film with outstanding abrasion and chemical resistance on the coated side. It is designed for reverse printed applications.

LEXAN 8800-112 is a clear, thin gauge (.001"-.005") (0.026 mm - 0.13 mm) polycarbonate film for the industrial and electrical markets. It offers excellent physical, thermal, and electrical properties as well as tight dimensional and flatness tolerances.

UL rated 94 V-O and 94 VTM-O. Lexan FR Films are flame retardant polycarbonate films. They are available in textures and clear and in thicknesses from .001" to .030".

LEXAN film also is available in graphic quality, standard, UV-stabilized, non UV-stabilized and FDA-approved grades, with a wide choice of mattes, colors, textures and gauges.

GENERAL ELECTRIC CO.: LEXAN Resin Blow Molding 150/1500 Series:

LEXAN 150 series resins for extrusion blow molding exhibit exceptional melt strength and are designed to provide high programming responsiveness and optimum wall distribution because of their non-Newtonian flow characteristics. LEXAN 150 series resins are well suited for water bottle applications requiring high impact strength, glass-like transparency, high gloss, and high heat deflection temperature. LEXAN resin has the highest combination of rigidity and toughness available in any thermoplastic. Grades are 151 (General Purpose), 153 (UV Stabilized) and 154 (Covered by Federal Food Additive Regulations, principally 21 CFR 177.1580).

LEXAN 1500 resin is a grade specially developed for the injection blow molding process. An FDA material exhibiting the traditional balance of LEXAN resin properties, LEXAN 1500 resin can also be used in extrusion blow molding when the high melt strength of LEXAN 150 series resins is not required.

TYPICAL PROPERTY VALUES:

Specific Gravity (D792): 1.20
Specific Volume, in 3/lb (cm 3/g): 23.1(0.83)
Weight/Volume, lbs/in 3 (g/cm 3): 0.043(1.20)
Water Absorption, % (D570):
- 24 hrs @ 73F (23C): 0.15
- Equilibrium, 73F (23C): 0.35
- Equilibrium, 212F (100C): 0.58

Mold Shrinkage, in/in (mm/mm) (D955): 0.005-0.007
Transmittance, % (D1003): 88-89
Haze, % (D1003): 1-2
Refractive Index: 1.586

GENERAL ELECTRIC CO.: LEXAN Resin Flame Retardant 920/940/950 Series:

LEXAN 920, 940 and 950 resins are low, medium and high viscosity, flame retardant grades developed to meet stringent industry regulations.

All three grades provide properties similar to standard grades of LEXAN resin. In addition, LEXAN 920 and 940 resins offer:

* V-O flame retardance rating per UL Standard 94 down to 0.062" in opaque colors only. LEXAN 940 grade resin is also rated V-O down to 0.044".
* Decreased smoke emission and toxic gas evolution relative to traditional FR grades.

LEXAN 950 resin offers:

* 5V flame retardance listing per UL Standard 94 down to 0.125" in all opaque colors.

The properties and processing characteristics of LEXAN 920, 940 and 950 resins make them ideal candidates for electrical, electronic, appliance, aircraft, lighting, computer, communications, business equipment and hardware applications.

LEXAN 920A, 940A AND 950A grades provide high flame retardance at a lower cost, combined with improved surface appearance.

The LEXAN "A" products offer:

* V-O flame retardance rating per UL Standard 94 down to 0.125" in opaque colors as well as in transparent and translucent tint colors.

The exceptional surface appearance provided by the LEXAN "A" flame retardant products make them ideally suited for high-visibility parts, including electrical, electronic, appliance, computer and business equipment housings, and aircraft interiors.

UV stabilization option:

LEXAN 920, 940 and 950 series resins are available with enhanced UV stabilization for outdoor applications. These grades are designated as 923, 923A; 943, 943A; 953, 953A.

Mold release option:

All 900 series resins are also available with additional mold release, designated by the addition of "R" following the basic grade number (i.e., 920R, 920AR, 923R, etc.).

GENERAL ELECTRIC CO.: LEXAN Resin General Purpose Grades:

LEXAN polycarbonate resin spans a wide range of physical properties that combine to make it one of the toughest, most versatile of all engineering thermoplastics. This combination of properties provides the design engineer with exceptional freedom in creating functional, attractive, cost effective products. The processing, performance and economic advantages of LEXAN resin over thermosets, metals, glass and other materials are reasons for its wide use in industries such as appliance, automotive, packaging, electrical and electronic, and many more.

LEXAN general purpose grade resins are now available in a continuum of tightly controlled viscosities that aid in optimizing resin selection. By expanding the standard LEXAN resin offering from three grades(121, 141, 101) to six(121, 141L, 141, 161, 101, 181), mold and design versatility is greatly enhanced. Each grade is tightly controlled to minimize lot-to-lot variations and thereby facilitate easier and more consistent molding from run to run. These refinements allow even greater efficiency and productivity gains with the use of automatic molding equipment.

Each controlled viscosity, general purpose LEXAN resin grade is available with a variety of additive options, including UV stabilization, FDA compliance and mold release.

LEXAN 121 grade resin is a low viscosity material available in natural and transparent colors only, for use in molding intricate, hard-to-fill parts.

LEXAN 141L grade resin is a new lower viscosity resin for enhanced flow versus 141 grade.

LEXAN 141 grade resin is a medium viscosity material most commonly used for injection molding.

LEXAN 161 grade resin is a new viscosity grade offering a balance between LEXAN 141 and 101 grade resins.

LEXAN 101 grade resin is a high viscosity material for molding thicker sections without sinks.

GENERAL ELECTRIC CO.: LEXAN Resin General Purpose Grades (Continued):

LEXAN 181 grade resin is a very high viscosity material offering superior toughness.

LEXAN 1X4 resin designations are manufactured in compliance with Food Additive Regulations governing polycarbonate resins for food contact applications: 21 CFR 177.1580.

LEXAN 1X3 resin designations are UV-stabilized grades for outdoor and lighting applications.

LEXAN 303 grade resin is a high viscosity material designed to service severe environments of HID lighting.

Typical Physical Values:

Product Number:

LEXAN 121 resin:

Viscosity Midpoint(Melt Flow Rate) g/10 min.: 16.5
Specific Gravity: 1.20

LEXAN 141L resin:

Viscosity Midpoint(Melt Flow Rate) g/10 min.: 11.5
Specific Gravity: 1.20

LEXAN 141 resin:

Viscosity Midpoint(Melt Flow Rate) g/10 min.: 9.5
Specific Gravity: 1.20

LEXAN 161 resin:

Viscosity Midpoint(Melt Flow Rate) g/10 min.: 8.0
Specific Gravity: 1.20

LEXAN 101 resin:

Viscosity Midpoint(Melt Flow Rate) g/10 min.: 6.5
Specific Gravity: 1.20

LEXAN 181 resin:

Viscosity Midpoint(Melt Flow rate) g/10 min.: 5.3
Specific Gravity: 1.20

GENERAL ELECTRIC CO.: LEXAN Resin Glass Reinforced Grades:

LEXAN resin reinforced by glass fiber is available in three standard grades. The 3412 products contain 20% glass, the 3413 products contain 30% glass, and the 3414 products 40% glass. Unreinforced LEXAN resin is outstanding among engineering thermoplastics in the areas of impact strength, heat resistance, dimensional stability, and creep resistance. Reinforcement with glass fiber raises performance to levels approaching those of metals while retaining the basic plastic attributes of low cost processing, dielectric character, resistance to corrosion and inherent color.

The outstanding improvements in reinforced (40% glass) over standard LEXAN resin are:

Mechanical:

* Tensile modulus is as much as five times greater
* Flexural and tensile strengths are as much as double
* Compressive strength increases as much as 65%
* Fatigue endurance is dramatically changed, increasing more than 7 times to permit tolerance of a 6500 psi (45MPa) load at 1800 cycles per minute
* Deformation under load, a more than satisfactory 0.2% for standard resin, drops to 0.1% under 4000 psi (27MPa) at 73F (23C)

Thermal:

* Coefficient of thermal expansion, reduced nearly 75%, equals that of many metals

Molding:

* Mold shrinkage is reduced to about 1/3 that of standard LEXAN resin, improving the ability to produce very precise parts.

TYPICAL PROPERTY VALUES:

<u>Product Name:</u>

LEXAN 3412 resin:

Specific Gravity (D792): 1.35
Specific Volume, in 3/lb (cm 3/g): 20.5(0.74)
Weight Volume, lbs/in 3 (g/cm 3): 0.049 (1.35)
Water Absorption, %:
  24 hrs @ 73F (23C): 0.16
  Equilibrium, 73F (23C): 0.29
Mold Shrinkage, in/in (mm/mm) (D955): 0.001-0.003

GENERAL ELECTRIC CO.: LEXAN Resin Glass Reinforced Grades (Continued):

TYPICAL PROPERTY VALUES:

Product Name:

LEXAN 3413 resin:

Specific Gravity (D792): 1.43
Specific Volume, in 3/lb (cm 3/g): 19.3(0.70)
Weight/Volume, lbs/in 3 (g/cm 3): 0.052 (1.43)
Water Absorption, % (D570):
24 hrs @ 73F (23C): 0.14
Equilibrium, 73F (23C): 0.26
Mold Shrinkage, in/in (mm/mm) (D955): 0.0015-0.0025

LEXAN 3414 resin:

Specific Gravity (D792): 1.52
Specific Volume, in 3/lb (cm 3/g): 18.2 (0.66)
Weight/Volume, lbs/in 3 (g/cm 3): 0.055 (1.52)
Water Absorption, % (D570):
24 hrs @ 73F (23C): 0.12
Equilibrium, 73F (23C): 0.23
Mold Shrinkage, in/in (mm/mm): 0.001/0.002

GENERAL ELECTRIC CO.: LEXAN Resin Healthcare Product (HP) Resins:

All LEXAN Healthcare Product (HP and HPS) resins are formulated and manufactured to meet the performance requirements of medical and health-related applications. Specifically, LEXAN HP and HPS resins are controlled in viscosity, color and contaminants, and are FDA and USP Class VI compliant. Additionally, mold release is formulated into each grade to facilitate part ejection.

Gamma Radiation Stabilized Resins

A specially formulated LEXAN HP resin family has been developed to control resin color shift after gamma radiation sterilization. This new LEXAN resin family, LEXAN HPS resins (Healthcare Products with enhanced Stability) combine the tight manufacturing tolerances common to the LEXAN HP resin family with proprietary color enhancement technology. The effect of this advancement is to reduce the resin color shift caused by gamma radiation by over 50% compared with unstabilized LEXAN HP resin grades. As a result of this significant improvement, initial color levels can be reduced to a light blue tint in order to maintain a neutral post-irradiated color. A new LEXAN resin color, number 1136, has been created to optimize the beenfits of this technology.

Colors:

LEXAN HP resin: All LEXAN HP resin viscosities are avilable in natural, transparent tints, and custom opaque colors. Caution is recommended in specifying heavily pigmented colors in low viscosity grades (LEXAN HP1 and HP2 resins) in notch-sensitive applications. When the level of pigment additives increases as the viscosity is reduced, the material becomes increasingly notch-sensitive.

LEXAN HPS resin: LEXAN HPS resin viscosities are offered in three standard tints - 1125, 1136, and 1124. Because gamma radiation affects different pigment additives in varying degrees, additional tints and colors are available, but only on a special request basis.

GENERAL ELECTRIC CO.: LEXAN Resin Healthcare Product (HP) Resins (Continued):

Typical Property Values:

Product Name:

LEXAN HP1/HPS1 resins:

Viscosity Midpoint(Melt Flow Rate) g/10 min. (D1238): 22.0
Specific Gravity (D792): 1.20
Mold Shrinkage, in/in (D955): 0.005-0.007

LEXAN HP2/HPS2 resins:

Viscosity Midpoint (Melt Flow Rate) g/10 min. (D1238): 16.5
Specific Gravity (D792): 1.20
Mold Shrinkage, in/in (D955): 0.005-0.007

LEXAN HP3/HPS3 resins:

Viscosity Midpoint (Melt Flow Rate) g/10 min. (D1238): 11.5
Specific Gravity (D792): 1.20
Mold Shrinkage, in/in (D955): 0.005-0.007

LEXAN HP4 resin:

Viscosity Midpoint (Melt Flow Rate) g/10 min. (D1238): 9.5
Specific Gravity (D792): 1.20
Mold Shrinkage, in/in (D955): 0.005-0.007

LEXAN HP5 resin:

Viscosity Midpoint (Melt Flow Rate) g/10 min. (D1238): 8.0
Specific Gravity (D792): 1.20
Mold Shrinkage, in/in (D955): 0.005-0.007

GENERAL ELECTRIC CO.: LEXAN Resin High Flow (HF) Resins:

The LEXAN family of High Flow resins: 1) is low in viscosity (melt flow rate midpoint of 22) to provide enhanced processability and longer flow lengths, allowing thinner wall designs than with any of the other standard, general purpose LEXAN grades; 2) has a polymer modification to provide superior long-term property retention compared to unmodified resins at the same viscosity; and 3) includes inherent mold release to promote part ejection. These benefits can be utilized to achieve greater cost effectiveness in molded parts.

LEXAN High Flow resins offer superior flow compared to higher viscosity LEXAN resin grades. However, while LEXAN HF resins do maintain excellent heat resistance, rigidity, and dimensional stability characteristics, they do not display the overall toughness over time of the higher viscosity LEXAN resin grades. Care should be taken in the specification and use of LEXAN HF resins. Consideration should be given to the application's end-use environment and a performance evaluation based on end-use testing is encouraged.

LEXAN HF1 resin family has been developed to provide ease of processing and good property retention capability. Its application is ideal for disposables or where product life cycles are not demanding, and in applications where clarity, heat resistance and outstanding dimensional stability are the key criteria required.

HF1 is available in a standard, general purpose grade (HF1110), a UV stabilized grade (HF1130), and an FDA compliant grade (HF1140).

Typical Property Values:

Viscosity Midpoint(Melt Flow Rate) g/10 min: 22.0
Mold Shrinkage, in/in: 0.005-0.007

GENERAL ELECTRIC CO.: LEXAN Resin High Heat PPC Series:

LEXAN PPC (Polyphthalate carbonate) resins are a new copolymer resin series engineered for high heat and moisture resistance.

LEXAN PPC resins provide properties similar to standard grades of LEXAN polycarbonate resin:

* 85% transmittance and water-clear transparency
* Falling dart impact strength greater than 110 ft-lbs
* Flexural modulus of 30,000 psi (2070 MPa)

In addition:

* PPC 4501 is engineered for good thermal capability at 305F (152C) along with improved processing characteristics and resistance to repeated autoclaving.

* PPC 4701, with a heat deflection temperature of 325F (163C) @ 264 psi (1.82 MPa), provides exceptional thermal performance.

For special end-use requirements, impact-modified LEXAN PPC resins are available, as are grades which comply with FDA and USP Class VI standards.

TYPICAL PROPERTY VALUES:

LEXAN PPC 4501 resin:

Specific Gravity (D792): 1.2
Water Absorption, % (24 hrs @ 73F (23C)) (D570): 0.16
Mold Shrinkage, in/in (mm/mm) (D955): 0.007-0.008
Transmittance, % (D1003): 85
Haze, % (D1003): 1-2
Refractive Index: 1.6

LEXAN PPC 4701 resin:

Specific Gravity (D792): 1.2
Water Absorption, % (24 hrs @ 73F (23C)) (D570): 0.19
Mold Shrinkage, in/in (mm/mm): 0.008-0.010
Transmittance, % (D1003): 85
Haze, % (D1003): 1-2
Refractive Index: 1.6

GENERAL ELECTRIC CO.: LEXAN Resin High Modulus 500 Series:

LEXAN 500 resin is a member of the LEXAN family of polycarbonate resins particularly suited to applications requiring high rigidity combined with toughness and impact strength.

LEXAN 500 resin has the highest combination of rigidity and impact strength available in any thermoplastic, plus excellent dimensional stability, low mold shrinkage, and colorability.

LEXAN 500 resin also has these additional advantages:

Mechanical:

* Flexural modulus increased nearly 50% over standard LEXAN resins

Flammability:

* Listed 94 V-O at 1/16" (1.6 mm) per UL Standard 94
* Listed 94-5V at 1/8" (3.2 mm) per UL Standard 94

Molding:

* Reduced mold shrinkage allows for the production of more precise parts.

LEXAN 503 resin is a UV stabilized grade for outdoor applications.

Typical Property Values:

Specific Gravity (D792): 1.25
Specific Volume, $in^3/lb (cm^3/g)$: 22.2(0.80)
Weight/Volume $lbs/in^3 (g/cm^3)$: 0.045 (1.25)
Water Absorption, % (D570):
 24 hrs @ 73F (23C): 0.12
 Equilibrium, 73F (23C): 0.31
Mold Shrinkage, in/in (mm/mm) (D955): 0.002-0.004

GENERAL ELECTRIC CO.: LEXAN Resin Optical Quality (OQ) Resins:

LEXAN Optical Quality (OQ) resins represent an emerging family of resins being specially formulated and manufactured to provide improved optical clarity and resin purity beyond standard, general purpose grades for the most demanding optical applications.

LEXAN OQ1 resin, the initial family offering, has been developed to meet the stringent requirements of the compact audio disc market. An ultra-low viscosity resin, LEXAN OQ1 ensures microscopic pit reproduction and low birefringence to optimize sound reproduction. While the LEXAN OQ1 product also provides far superior mechanical properties compared to acrylic and glass, it is not designed for use in traditional LEXAN polycarbonate applications.

LEXAN OQ1 resin marks the first step in what will be an evolving line addressing selected optics-sensitive market segments.

TYPICAL PROPERTY VALUES:

Specific Gravity (D792): 1.20
Mold Shrinkage, in/in (mm/mm) (D995): 0.005-0.007
Light Transmittance, % (D1003): 90
Tensile Strength, psi (MPa) (D638):
  Yield: 9,100 (63)
Elongation, % (D638):
  Rupture: 40-80
Flexural Strength, psi (MPa) (D790): 14,200 (98)
Flexural Modulus, 10 5 psi (D790): 3.19 (2,200)
Izod Impact Strength, ft-lbs/in (J/m) (D256):
  Notched, 1/8" thick: 2 (100)
  Unnotched: No failure
Falling Dart Impact Strength, ft-lbs (J)
  1/8" thick (3.2 mm): >125 (>169)

GENERAL ELECTRIC CO: LEXAN Resin Wear Resistant (WR) Resins:

LEXAN Wear-Resistant (WR) resins are specifically formulated to provide greater lubricity and reduced friction and wear between moving parts for extended life. Particularly suited for the business equipment and electrical/electronics markets, typical LEXAN WR resin applications include keyboard frames, swivel bases, paper drives, and printers.

The unique capabilities of LEXAN WR resins also make them applicable for use in other mechanical devices, such as camera and instrument internal parts, where long-term performance is essential.

GRADES:

The LEXAN WR1 resin family is based on polytetrafluoroethylene (PTFE) filler technology.

Two LEXAN WR1 resins are available:

* A general-purpose grade - LEXAN WR1210 resin
* An FDA-compliant grade - LEXAN WR1240 resin

The LEXAN WR2 resin family is based on alternative polymeric wear additive technology. LEXAN WR2 resin products do not contain silicones and offer superior processability and part appearance compared with WR1.

The LEXAN WR2 resin family is available in a variety of grades including:

* Two general-purpose viscosities
  - WR2210 with a flow similar to LEXAN 121 resin
  - WR2310 with a flow similar to LEXAN 141 resin

* A flame-retardant grade
  - WR2310A - recognized UL94 V-O @ 0.125"

* Two glass-filled grades
  - WR2214, 20% glass filled
  - WR2216, 30% glass filled

GENERAL ELECTRIC CO.: LEXAN Resin Wear Resistant (WR) Resins (Continued):

Typical Property Values:

Product Name:

LEXAN WR2210, WR2310 resins:

Specific Gravity (D792): 1.19
Mold Shrinkage, in/in (D955): 0.005-0.007
Heat Deflection Temperature, F @ 264 psi (D648): 270
Flammability Rating:
@ 0.125" (UL94):
@ 0.060" (UL94): HB

LEXAN WR2310A resin:

Specific Gravity (D792): 1.20
Mold Shrinkage, in/in (D955): 0.005-0.007
Heat Deflection Temperature, F @ 264 psi (D648): 270
Flammability Rating:
@ 0.125" (UL94): V-O
@ 0.060" (UL94): V-1

LEXAN WR2214 resin:

Specific Gravity (D792): 1.34
Mold Shrinkage, in/in (D955): 0.001-0.004
Heat Deflection Temperature, F @ 264 psi (D648): 270
Flammability Rating:
@ 0.125" (UL94):
@ 0.060" (UL94): HB

LEXAN WR2216 resin:

Specific Gravity (D792): 1.42
Mold Shrinkage, in/in (D955): 0.001-0.003
Heat Deflection Temperature, F @ 264 psi (D648): 270
Flammability Rating:
@ 0.125" (UL94):
@ 0.060" (UL94): HB

LEXAN WR1210, 1240 resins:

Specific Gravity (D792): 1.22
Mold Shrinkage, in/in (D955): 0.005-0.007
Heat Deflection Temperature, F @ 264 psi (D648): 270
Flammability Rating:
@ 0.125" (UL 94):
@ 0.060" (UL 94): HB

GENERAL ELECTRIC CO.: LOMOD Engineering Elastomers:

Combining the impact resistance of thermoset rubber and processability of thermoplastics, LOMOD engineering elastomers provide a wide range of stiffness, strength and thermal properties in many colors, with UL94 V-0 and HB flammability listings.

With flexural moduli from 12,000-95,000 psi, LOMOD engineering elastomers also exhibit excellent resistance to abrasion, cut growth and chemicals, retaining property performance at both high and low temperatures. They meet a broad spectrum of flexible design requirements in automotive, electrical/electronic, sporting goods, industrial, wire coating and similar applications.

LOMOD grades are generally selected by their flexural modulus (stiffness) and flame retardant characteristics. The numbering system for LOMOD helps you easily identify each grade through its proper designation.

High Heat Grades--These grades offer the best retention of mechanical properties such as compression set resistance and retention of stiffness at elevated temperatures. They also exhibit lower hysteresis and fast cycling.

Flame Retarded Grades--These grades meet UL94 V-O testing requirements. BO150 and BO852 are recognized as V-O by the Underwriters Laboratories in both black and natural. BO250 is currently under evaluation at the time of this printing.

General Purpose Grades, BO100, BO200 and BO800--Offer the outstanding performance of a copolyester elastomer. They can offer attractive cost savings where other other copolyester elastomers are over specified. Like all the other LOMOD grades, they should not be exposed to temperatures above 250F for extended periods of time.

GENERAL ELECTRIC CO.: LOMOD Engineering Elastomers(Continued):

PRELIMINARY DATA:

LOMOD BO100:

Specific Gravity (D792): 1.16
Water Absorption, % (D570): 0.37

LOMOD BO150:

Specific Gravity (D792): 1.32
Water Absorption, % (D570): 0.28

LOMOD BO200:

Specific Gravity (D792): 1.21
Water Absorption, % (D570): 0.28

LOMOD BO220:

Specific Gravity (D792): 1.19
Water Absorption, % (D570): 0.81

LOMOD BO250:

Specific Gravity (D792): 1.34
Water Absorption, % (D570): 0.21

LOMOD BO320:

Specific Gravity (D792): 1.22
Water Absorption, % (D570): 0.72

LOMOD BO520:

Specific Gravity (D792): 1.25
Water Absorption, % (D570): 0.41

LOMOD BO800:

Specific Gravity (D792): 1.23
Water Absorption, % (D570): 0.11

LOMOD BO852:

Specific Gravity (D792): 1.39
Water Absorption, % (D570): 0.15

GENERAL ELECTRIC CO.: LR 3320 Resin:

A block copolymer of LEXAN polycarbonate and silicone rubber, LR resin is a transparent thermoplastic elastomer. Chemical linking of two premium-performance resins yields a unique material combining the most useful properties of each polymer in an otherwise unobtainable combination. The silicone component contributes elasticity and excellent low-temperature properties, while the polycarbonate contributes higher-temperature capabilities and transparency plus the key property of thermoplasticity.

LR resin can be processed by all conventional thermoplastic processes: injection molding, extrusion, blow molding. In addition, its amorphous nature allows casting into very thin films.

LR resin is also blendable with a variety of polymers to contribute impact resistance and improve low-temperature and flammability performance. Intrinsicaly flame retardant, LR resin is listed V-1 by UL Standard 94 at 1/8" (3.2 mm), and can readily be made V-0 by the addition of special non-halogenated additives, which do not affect its transparency.

TYPICAL PROPERTY VALUES:

Specific Gravity (D792): 1.115
Water Absorption, %:
  Equilibrium, 73F (D570): 0.09
Brittleness Temeperature, F (D1882): <-76
Vicat Softening Point, F (D1525): 133
Flammability Rating (UL94): V-1
Glass Transition Temperature, C: -120/+70

GENERAL ELECTRIC CO.: NORYL C&BE, Extrudable Grades:

New NORYL resins for computers and business equipment offer varying levels of flame and ultraviolet light resistance. Whether you're designing home video games, personal computers, business computers or stationary or portable business equipment, there's a NORYL resin grade to meet your specific agency listing requirements and other needs -- most economically.

EXTRUDABLE GRADES:

NORYL EN185 resin combines a heat distortion temperature resistance of 185F at 264 psi with excellent impact resistance even in thin walls. NORYL EN185 resin is typically used in electronics housings, beverage cases and for transportation component applications.

NORYL EN212 extends the range of extruded thermoplastic applications to high-temperature environments that requires both toughness and UL approval for use in communications and electronic housings.

NORYL EN265 resin is ideal for higher temperature electrical housing applications where flame retardance is needed. Listed UL94 V-1 with a continuous use temperature of 105C, the material provides high mechanical strength, toughness and low water absorption.

EXTRUSION GRADES:

TYPICAL PROPERTY VALUES:

NORYL EN185 resin:

Heat Deflection Temperature, F(C) (D648):
@ 264 psi (1.82 MPa): 185(85)
@ 66 psi (0.46 MPa): 200(93)

NORYL EN212 resin:

Heat Deflection Temperature, F(C) (D648):
@ 264 psi (1.82 MPa): 212(100)
@ 66 psi (0.46 MPa): 230(110)

GENERAL ELECTRIC CO.: NORYL C&BE, Extrudable Grades (Continued):

EXTRUSION GRADES:

TYPICAL PROPERTY VALUES:

NORYL EN265 resin:

Heat Deflection Temperature, F(C) (D648):
@ 264 psi (1.82 MPa): 265(129)
@ 66 psi (0.46 MPa): 279(137)

NORYL PC180 resin:

Heat Deflection Temperature, F(C) (D648):
@ 264 psi (1.82 MPa): 180(82)
@ 66 psi (0.46 MPa): 200(96)

NORYL UV180 resin:

Heat Deflection Temperature, F(C) (D648):
@ 264 psi (1.82 MPa): 180(82)
@ 66 psi (0.46 MPa): 195(94)

GENERAL ELECTRIC CO.: NORYL Engineering Structural Foam Resins:

Foamable NORYL resin exhibits outstanding versatility in the combination of heat resistance, dimensional stability and impact resistance. As adjuncts to these superior properties it also demonstrates low creep, high mechanical strength, and low water absorption. And foamable NORYL resin is offered with UL 94 V-O and 5V listings. The combination of these properties makes the materials ideal for business machine housings and their structural bases, weather resistant electrical enclosures, and lightweight structural components for the transportation industry which resist the "greenhouse" effect.

NORYL Foam Typical Property Values:

NORYL FN150 resin:

Solid Specific Gravity/Density (oz/in 3) (D792): 1.13 (0.66)
Foam Molded Specific Gravity (D792): 0.90
Weight Reduction, % (GE Test Method): 20

NORYL FN170 resin:

Solid Specific Gravity/Density (oz/in 3) (D792): 1.10 (0.64)
Foam Molded Specific Gravity (D792): 0.85
Weight Reduction, % (GE Test Method): 23

NORYL FN210 resin:

Solid Specific Gravity/Density (oz/in 3) (D792): 1.10 (0.64)
Foam Molded Specific Gravity (D792): 0.85
Weight Reduction, % (GE Test Method) 23

NORYL FN215 resin:

Solid Specific Gravity/Density (oz/in 3) (D792): 1.10 (0.64)
Foam Molded Specific Gravity (D792): 0.85
Weight Reduction, % (GE Test Method): 23

NORYL FN700 resin:

Solid Specific Gravity/Density (oz/in 3) (D792): 1.12 (0.65)
Foam Molded Specific Gravity (D792): 0.85
Weight Reduction, % (GE Test Method): 23

GENERAL ELECTRIC CO.: NORYL GTX Resin:

NORYL GTX resin is a revolutionary new class of materials specifically designed to offer excellent chemical resistance to many hostile environments, excellent mechanical performance, high heat resistance and dimensional stability. These materials are particularly suited for applications such as automotive body panels and exterior hardware where high heat capability for on-line paintability and chemical resistance to auto fluids is required. Reinforced grades meet the performance needs of fluid and material handling systems, while excellent mechanical properties position these products in furniture and other consumer applications.

TYPICAL PROPERTY VALUES:

NORYL GTX810 resin:

Specific Gravity (black color) (D792): 1.16
Weight/Volume, lb/in 3 (g/cm 3): 0.042(1.16)
Rockwell Hardness (D785): R119

NORYL GTX820 resin:

Specific Gravity (black color) (D792): 1.24
Weight/Volume, lb/in 3 (g/cm 3): 0.045(1.24)
Rockwell Hardness (D785): R119

NORYL GTX830 resin:

Specific Gravity (black color) (D792): 1.33
Weight/Volume, lb/in 3 (g/cm 3): 0.048(1.33)
Rockwell Hardness (D785): R120
UL Properties: Flammability: HB

NORYL GTX901 resin:

Specific Gravity (natural color) (D792): 1.10
Weight/Volume, lb/in 3 (g/cm 3): 0.040(1.10)
Rockwell Hardness (D785): R116
UL Properties: Flammability: HB

NORYL GTX910 resin:

Specific Gravity (natural color) (D792): 1.10
Weight/Volume, lb/in 3 (g/cm 3): 0.040(1.10)
Rockwell Hardness (D785): R116
UL Properties: Flammability: HB

GENERAL ELECTRIC CO.: NORYL Resin Automotive Grades:

A family of NORYL resin grades has been developed specifically for the automotive industry offering a far wider range of superior properties combinations than ABS and SMA:

These include such property combinations as heat resistance and impact strength for instrument panels, speaker grilles and glove box doors; electrical insulating capabilities, dimensional stability and resistance to moisture for connectors, fuse blocks and instrument panal components; and low-temperature impact strength and electroplatability for mirrors, exterior trim, grilles and wheel covers.

TYPICAL PROPERTY VALUES:

NORYL 722 resin:
Heat Deflection Temperature, F(C) (D648):
@ 264 psi (1.82 MPa): 210(99)

NORYL 844 resin:
Heat Deflection Temperature, F(C) (D648):
@ 264 psi (1.82 MPa): 235(113)

NORYL PN235 resin:
Heat Deflection Temperature, F(C) (D648):
@ 264 psi (1.82 MPa): 235(113)

NORYL 1222 resin:
Heat Deflection Temperature, F(C) (D648):
@ 264 psi (1.82 MPa): 235(113)

NORYL 888 resin:
Heat Deflection Temperature, F(C) (D648):
@ 264 psi (1.82 MPa): 245(118)

NORYL PX1265 resin:
Heat Deflection Temperature, F(C) (D648):
@ 264 psi (1.82 MPa): 265(129)

NORYL 1390 resin:
Heat Deflection Temperature, F(C) (D648):
@ 264 psi (1.82 MPa): 290(143)

NORYL 1391 resin:
Heat Deflection Temperature, F(C) (D648):
@ 264 psi (1.82 MPa): 300(149)

GENERAL ELECTRIC CO.: NORYL Resin Energy Management Series:

NORYL resin energy management grades for automotive interior applications utilize new technology to improve flow characteristics while maintaining impact and toughness over a broad range of temperatures. They offer all the properties of PX-designated NORYL resins, which have long been industry standards, together with superior processability featuring consistent melt behavior and longer flow lengths.

Energy management grades of NORYL resin and their equivalent standard grades (in parentheses) are as follows: NORYL EM5100 resin (PX0844); NORYL EM5101 (PX1222); NORYL EM5102 resin (PX0888); NORYL EM5103 resin (PX1265); NORYL EM5104 resin (PX1391).

NORYL EM7101 resin for automotive interior applications offers 210F heat distortion and was specifically designed for consoles, interior trim, door speaker grilles and glove box doors. The product incorporates a new material technology that provides excellent processability for outstanding value. EM7101 resin offers high impact and mechanical properties that are ideally suited to a variety of interior applications.

TYPICAL PROPERTY VALUES:

NORYL EM5100 resin:
Heat Deflection Temperature, F(C) (D648):
@ 66 psi (0.46 N/mm 2): 225(124)
@ 264 psi (1.82 N/mm 2): 235(112)

NONYL EM5101 resin:
Heat Deflection Temperature, F(C) (D648):
@ 66 psi (0.46 N/mm 2): 255(124)
@ 264 psi (1.82 N/mm 2): 235(112)

NONYL EM5102 resin:
Heat Deflection Temperature, F(C) (D648):
@ 66 psi (0.46 N/mm 2): 265(130)
@ 264 psi (1.82 N/mm 2): 245(118)

NONYL EM5103 resin:
Heat Deflection Temperature, F(C) (D648):
@ 66 psi (0.46 N/mm 2): 285(141)
@ 264 psi (1.82 N/mm 2): 265(130)

NONYL EM5104 resin:
Heat Deflection Temperature, F(C) (D648):
@ 66 psi (0.46 N/mm 2): 320(160)
@ 264 psi (1.82 N/mm 2): 300(150)

NONYL EM7101 resin:
Heat Deflection Temperature, F(C) (D648):
@ 264 psi (1.82 N/mm 2): 210(100)

GENERAL ELECTRIC CO.: NORYL Resin Energy Management Series (Continued):

NORYL EM1600 resin incorporates a new material technology for demanding automotive interior applications such as lower instrument panels, trim and glove box doors. The improved flow characteristics lead to overall processing economies based on shorter cycle times. The product offers 230F heat distortion to meet various application requirements. Additionally, EM6100 resin provides high impact properties and better property retention after thermal aging to meet the changing requirements of the automotive industry.

NORYL EMX6101 resin incorporates a new material technology for demanding automotive interior applications such as instrument panels, retainers and trim. The improved flow characteristics lead to overall processing economies based on shorter cycle times. EMX6101 resin provides 240F heat distortion to meet increasing temperature requirements and better property retention to meet the changing requirements of the automotive industry.

NORYL EMX7100 resin for automotive interior applications offers 200F heat distortion and was specifically designed for interior trim and instrument panel components. The product incorporates a new material technology that provides excellent processability for outstanding value, and offers good impact and mechanical properties that will perform favorably versus styrenic materials in interior components.

NORYL EMX7302 resin incorporates a new material technology for automotive interior applications and was specifically designed for glass-reinforced instrument panels and consoles. The product offers a 250F heat distortion to meet increasing temperature requirements for instrument panels. Its outstanding flow characteristics lead to processing economies based on shorter cycle times and improved productivity. The good ductility present in this 15% glass-reinforced grade is designed to reduce in-process breakage during handling and secondary operations. These combinations of properties make EMX7302 the value choice for glass reinforced applications.

PRELIMINARY PROPERTY PROFILE:

NORYL EM6100 resin:
  Heat Deflection Temperature, F(C) (D648):
    @ 66 psi (0.46 N/mm 2): 255(124)
NORYL EMX6101 resin:
  Heat Deflection Temperature, F(C) (D648):
    @ 264 psi (1.82 N/mm 2): 240(116)
NORYL EMX7100 resin:
  Heat Deflection Temperature, F(C) (D648):
    @ 264 psi (1.82 N/mm 2): 200(93)
NORYL EMX7302 resin:
  Heat Deflection Temperature, F(C) (D648):
    @ 264 psi (1.82 N/mm 2): 250(121)

GENERAL ELECTRIC CO.: NORYL Resin Engineered Blow Molding Grades:

The NORYL family of blow moldable thermoplastic resins offers a wide range of performance and processing benefits. High heat resistance and impact strength make it the choice for computers, appliances and business equipment. Its dimensional stability and low temperature impact have many uses in the automotive industry, and fluid handling and telecommunications equipment benefit from its resistance to moisture and electrical insulating capabilities. And all grades offer a virtually infinite range of custom colors which can be expertly matched to your needs.

There is a NORYL resin grade to meet your specific agency listing and other requirements with performance and economy.

TYPICAL PROPERTY VALUES:

NORYL BN11 resin:

Specific Gravity (D792): 1.06
HDT @ 264 psi, F (D648): 210
Flexural Modulus, psi (D790): 250,000
Flexural Strength, psi (D790): 8,000
Tensile Strength, psi (D638): 5,700

NORYL BN13 resin:

Specific Gravity (D792): 1.06
HDT @ 264 psi, F (D648): 235
Flexural Modulus, psi (D790): 285,000
Flexural Strength, psi (D790): 8,500
Tensile Strength, psi (D638): 6,500

NORYL BN15 resin:

Specific Gravity (D792): 1.06
HDT @ 264 psi, F (D648): 265
Flexural Modulus, psi (D790): 325,000
Flexural Strength, psi (D790): 12,000
Tensile Strength, psi (D638): 9,000

NORYL BN25 resin:

Specific Gravity (D792): 1.24
HDT @ 264 psi, F (D648): 230
Flexural Modulus, psi (D790): 500,000
Flexural Strength, psi (D790): 16,000
Tensile Strength, psi (D638): 10,300

GENERAL ELECTRIC CO.: NORYL Resin Engineered Blow Molding Grades(Continued):

TYPICAL PROPERTY VALUES(Continued):

NORYL BN30 resin:

Specific Gravity (D792): 1.09
HDT @ 264 psi, F (D648): 180
Flexural Modulus, psi (D790): 325,000
Flexural Strength, psi (D790): 8,600
Tensile Strength, psi (D638): 6,200

NORYL BN31 resin:

Specific Gravity (D792): 1.10
HDT @ 264 psi, F (D648): 180
Flexural Modulus, psi (D790): 350,000
Flexural Strength, psi (D790): 10,500
Tensile Strength, psi (D638): 7,000

NORYL BN41 resin:

Specific Gravity (D792): 1.09
HDT @ 264 psi, F (D648): 185
Flexural Modulus, psi (D790): 250,000
Flexural Strength, psi (D790): 9,000
Tensile Strength, psi (D638): 6,500

NORYL BN43 resin:

Specific Gravity (D792): 1.08
HDT @ 264 psi, F (D648): 212
Flexural Modulus, psi (D790): 275,000
Flexural Strength, psi (D790): 9,000
Tensile Strength, psi (D638): 7,000

GENERAL ELECTRIC CO.: NORYL Resin Foamable and Extrudable Grades:

NORYL FN215 resin offers a 205F heat distortion temperature at 66 psi, with UL94 V-O and 5V ratings.

It also combines excellent dimensional stability with exceptional impact strength and low creep and water absorption rates.

TYPICAL PROPERTY VALUES:

Solid Specific Gravity, oz/in 3 (gm/cc)(D792): 0.64(1.10)
Foam Molded Specific Gravity, oz/in 3 (gm/cc) (D792): 0.49(0.85)
Weight Reduction, % (GE Test Method): 23
Water Absorption, 24 hrs. 73F, % (23C) (D570): 0.06
@ Equilibrium, 73F, % (23C): 0.07
Mold Shrinkage, 10 -3 in/in(mm/mm) (D1299):
Flow: 6-8

GENERAL ELECTRIC CO.: NORYL Resin Injection Molding Grades:

Easily processed on conventional tooling with no plate-out or mold corrosion, NORYL injection molding resins offer continuous use temperatures of up to 100C and notched Izod impact strengths of up to 10 ft.-lbs./in. @ 73F. Mold shrinkage is consistent, as little as 1 -3 x 10 -3 in./in. The NORYL resin family of injection moldable resins includes: general purpose grades, glass reinforced grades, high strength grades and grades developed specifically for computer and business equipment applications.

Injection moldable, extrudable, foamable NORYL resins provide far more performance and processing benefits than ABS, FRABS, nylon, polysulfone, SMA, metal, thermoset and others - yet demand a cubic inch price extremely competitive with these materials. A virtually infinite range of custom colors is available expertly matched to your environment.

UNFILLED GRADES:

TYPICAL PROPERTY VALUES:

NORYL N190 resin:
Heat Deflection Temperature, F(C) (D648):
@ 264 psi(1.82 MPa): 190(88)
@ 66 psi(0.46 MPa): 205(96)

NORYL SE100 resin:
Heat Deflection Temperature, F(C) (D648):
@ 264 psi(1.82 MPa): 212(100)
@ 66 psi(0.46 MPa): 230(110)

NORYL N225 resin:
Heat Deflection Temperature, F(C) (D648):
@ 264 psi(1.82 MPa): 225(107)
@ 66 psi(0.46 MPa): 245(118)

NORYL 731 resin:
Heat Deflection Temperature, F(C) (D648):
@ 264 psi(1.82 MPa): 265(129)
@ 66 psi(0.46 MPa): 279(137)

NORYL SE1 resin:
Heat Deflection Temperature, F(C) (D648):
@ 264 psi(1.82 MPa): 265(129)
@ 66 psi(0.46 MPa): 279(137)

NORYL N300 resin:
Heat Deflection Temperature, F(C) (D648):
@ 264 psi(1.82 MPa): 300(149)
@ 66 psi(0.46 MPa): 315(157)

GENERAL ELECTRIC CO.: NORYL Resin Injection Molding Grades (Continued):

REINFORCED GRADES:

TYPICAL PROPERTY VALUES:

NORYL GFN2 resin:

Heat Deflection Temperature, F(C) (D648):
@ 264 psi (1.82 MPa): 290(143)
@ 66 psi (0.46 MPa): 293(145)

NORYL GFN3 resin:

Heat Deflection Temperature, F(C) (D648):
@ 264 psi (1.82 MPa): 300(149)
@ 66 psi (0.46 MPa): 317(158)

NORYL SE1-GFN2 resin:

Heat Deflection Temperature, F(C) (D648):
@ 264 psi (1.82 MPa): 270(132)
@ 66 psi (0.46 MPa): 280(138)

NORYL SE1-GFN3 resin:

Heat Deflection Temperature, F(C) (D648):
@ 264 psi (1.82 MPa): 275(135)
@ 66 psi (0.46 MPa): 285(140)

NORYL HS1000 resin:

Heat Deflection Temperature, F(C) (D648):
@ 264 psi (1.82 MPa): 190(88)

NORYL HS2000 resin:

Heat Deflection Temperature, F(C) (D648):
@ 264 psi (1.82 MPa): 230(110)
@ 66 psi (0.46 MPa): 250(121)

GENERAL ELECTRIC CO.: NORYL Resin Superior Productivity:

NORYL SPN resin - Superior productivity in an unreinforced resin that maintains its property profile through a broad processing window. Improved flow even at reduced temperatures increases design flexibility while reducing clamp pressure requirements. Available in general purpose and flame retardant grades.

NORYL RFN resin - Reinforcement technology that opens new dimensions in metal replacement, providing higher strength and modulus with the wide processing parameters and thin-wall performance to meet the most innovative product design needs. Available in general purpose and flame retardant grades.

TYPICAL PROPERTY VALUES:

NORYL SPN410 resin:

Specific Gravity (D792): 1.06
Water Absorption After 24 Hours, % (D570): .07

NORYL SPN410H resin:

Specific Gravity (D792): 1.06
Water Absorption After 24 Hours, % (D570): .07

NORYL SPN420 resin:

Specific Gravity (D792): 1.07
Water Absorption After 24 Hours, % (D570): .07

NORYL SPN422L resin:

Specific Gravity (D792): 1.08
Water Absorption After 24 Hours, % (D570): .06

NORYL RFN420 resin:

Specific Gravity (D792): 1.21
Water Absorption After 24 Hours, % (D570): .06

NORYL RFN430 resin:

Specific Gravity (D792): 1.27
Water Absorption After 24 Hours, % (D570): .06

GENERAL ELECTRIC CO.: ULTEM Resins Engineering Structural Foam Resins:

Available in unreinforced as well as 10% and 30% glass reinforced grades, foamable ULTEM resins exhibit excellent strength and modulus. These high-performance resins also offer high heat deflection temperatures, broad chemical resistance and UL 94 listings of V-0 and 5V. With such an outstanding combination of properties, ULTEM resins are ideally suited for applications in the aircraft and automotive industries, as well as other areas that demand structural integrity and resistance to chemical attack.

ULTEM Foam Typical Property Values:

ULTEM FXU100 resin:

Solid Specific Gravity/Density (oz/in 3) (D792): 1.27 (0.73)
Foam Molded Specific Gravity (D792): 0.90
Weight Reduction, % (GE Test Method): 29

ULTEM FXU210 resin:

Solid Specific Gravity/Density (oz/in 3) (D792): 1.34 (0.77)
Foam Molded Specific Gravity (D792): 0.95
Weight Reduction, % (GE Test Method): 29

ULTEM FXU230 resin:

Solid Specific Gravity/Density (oz/in 3) (D792): 1.51 (0.87)
Foam Molded Specific Gravity (D792): 1.10
Weight Reduction, % (GE Test Method): 27

GENERAL ELECTRIC CO.: ULTEM Resin 1000 Series Unreinforced Grades:

Unreinforced grades offering a heat deflection temperature of up to 392F (200C) at 264 psi (1.82 MPa) and excellent flame resistance. Available in transparent amber as well as opaque colors.

ULTEM 1000 resin:

An unreinforced grade offering a heat deflection temperature of 392F (200C) at 264 psi (1.82 MPa), and a UL94 flammability rating of V-O of 0.016 inch (0.41 mm) thickness in all colors.

ULTEM 1010 resin:

A lower viscosity grade for use in molding intricate, hard-to-fill parts. Provides approximately 25% higher flow with minimal reduction in impact, thermal, chemical resistance.

Both products are available in a transparent amber color (natural) as well as a variety of opaque colors.

TYPICAL PROPERTY VALUES:

ULTEM 1000 resin:

Specific Gravity (D792): 1.27
Mold Shrinkage (D955) (in/in)(m/m)): 0.007
Water Absorption (D570) %:
- 24 hours, 73F (23C): 0.25
- Equilibrium, 73F (23C): 1.25

ULTEM 1010 resin:

Mold Shrinkage (D955) (in/in)(m/m)): 0.007
Water Absorption (D570) %:
- 24 hours, 73F (23C): 0.25
- Equilibrium, 73F (23C): 1.25

GENERAL ELECTRIC CO: ULTEM Resin 2000 Series Glass Reinforced Grades:

ULTEM 2000 Series products are glass reinforced resins which complement the excellent properties of ULTEM 1000 series by providing even greater rigidity and improved dimensional stability while retaining excellent processability.

ULTEM 2100: 10% glass reinforced resin
ULTEM 2200: 20% glass reinforced resin
ULTEM 2300: 30% glass reinforced resin
ULTEM 2400: 40% glass reinforced resin

TYPICAL PROPERTY VALUES:

ULTEM 2100 Resin:

Specific Gravity (D792): 1.34
Mold Shrinkage (in/in (mm/mm)) (D955): 0.005-0.006
Water Absoprtion (D570) %:
24 hours, 73F (23C): 0.21
Equilibrium, 73F (23C): 1.20

ULTEM 2200 Resin:

Specific Gravity (D792): 1.42
Mold Shrinkage (in/in (mm/mm)) (D955): 0.003-0.005
Water Absorption (D570) %:
24 hours, 73F (23C): 0.19
Equilibrium, 73F (23C): 1.10

ULTEM 2300 Resin:

Specific Gravity (D792): 1.51
Mold Shrinkage (in/in (mm/mm)) (D955): 0.002-0.004
Water Absorption (D570) %:
24 hours, 73F (23C): 0.16
Equilibrium, 73F (23C): 0.90

ULTEM 2400 Resin:

Specific Gravity (D792): 1.61
Mold Shrinkage (in/in (mm/mm)) (D955): 0.001-0.003
Water Absorption (D570) %:
24 hours, 73F (23C): 0.13
Equilibrium, 73F (23C): 0.90

GENERAL ELECTRIC CO.: ULTEM Resin 4000 Series Wear Resistant Grades:

ULTEM 4000 Series resins are designed for applications requiring enhanced wear resistance and reduced coefficient of friction. This resin provides:

* Low coefficient of friction and wear rate
* PV limit of 60,000 and 80,000 psi-fpm at 100 fpm
* Broad lubricant compatibility
* Service temperature in excess of 350F
* Excellent dimensional stability and creep resistance

ULTEM 4000: Reinforced, with an internal lubricant
ULTEM 4001: Unreinforced, with an internal lubricant

TYPICAL PROPERTY VALUES:

ULTEM 4000 resin:

Specific Gravity (D792): 1.70
Mold Shrinkage (%): 0.002-0.003
Water Absorption (D570) (%):
24 hours, 73F (23C): 0.11

ULTEM 4001 resin:

Specific Gravity (D792): 1.33
Mold Shrinkage (%): 0.005-0.007

GENERAL ELECTRIC CO.: ULTEM Resin 6000 Series High Temperature Copolymer Grades:

ULTEM 6000 Series resins offer the highest temperature resistance of any ULTEM product. Based on a polyetherimide copolymer, the resins have heat deflection temperatures of up to 437F (225C). They also offer enhanced chemical resistance to partially halogenated solvents.

ULTEM 6000 resins can be processed on standard injection molding equipment, with the same tool designs used for other amorphous thermoplastics. Unlike some high temperature polymers, ULTEM 6000 products do not crosslink during processing and do not require post curing.

ULTEM 6000: Unreinforced high temperature resin
ULTEM 6100: 10% glass reinforced high temperature resin
ULTEM 6200: 20% glass reinforced high temperature resin
ULTEM 6202: 20% mineral filled high temperature resin

TYPICAL PROPERTY VALUES:

ULTEM 6000 resin:

Specific Gravity (D792): 1.29
Mold Shrinkage (in/in (m/m)) (D955): 0.005-0.007
Water Absorption 24 hours, 73F (23C) (D570) (%): 0.28

ULTEM 6100 resin:

Specific Gravity (D792): 1.35
Mold Shrinkage (in/in (m/m)) (D955): 0.005-0.006
Water Absorption 24 hours, 73F (23C) (D570) (%): 0.24

ULTEM 6200 resin:

Specific Gravity (D792): 1.43
Mold Shrinkage (in/in (m/m)) (D955): 0.003-0.005
Water Absorption 24 hours, 73F (23C) (D570) (%): 0.22

ULTEM 6202 resin:

Specific Gravity (D792): 1.42
Mold Shrinkage (in/in (m/m)) (D955): 0.005-0.007
Water Absorption 24 hours, 73F (23C) (D570) (%): 0.22

GENERAL ELECTRIC CO.: VALOX Engineering Structural Foam Resins:

Foamable VALOX resins offer highly desirable thermoplastic polyester properties including excellent heat distortion temperature under load, chemical and solvent resistance, good fatigue endurance and high flexural strength and modulus. In combination with good moldability, these materials exhibit an ideal fit where high heat endurance and good chemical resistance are requirements, such as in electronic enclosures, processing control instruments and exterior transportation applications. Foamable VALOX resins are also available in UL94 V-O and 5V listed grades.

VALOX Foam Typical Property Values:

VALOX FV608 resin:

Solid Specific Gravity/Density (oz/in 3) (D792): 1.50 (0.87)
Foam Molded Specific Gravity (D792): 1.10
Weight Reduction, %: 27

VALOX FV609 resin:

Solid Specific Gravity/Density (oz/in 3) (D792): 1.30 (0.75)
Foam Molded Specific Gravity (D792): 1.00
Weight Reduction, % : 23

VALOX FV620 resin:

Solid Specific Gravity/Density (oz/in 3) (D792): 1.62 (0.94)
Foam Molded Specific Gravity (D792): 1.20
Weight Reduction, %: 26

VALOX FV649 resin:

Solid Specific Gravity/Density (oz/in 3) (D792): 1.26 (0.73)
Foam Molded Specific Gravity (D792): 1.00
Weight Reduction, %: 21

VALOX FV699 resin:

Solid Specific Gravity/Density (oz/in 3) (D792): 1.27 (0.73)
Foam Molded Specific Gravity (D792): 1.00
Weight Reduction, %: 21

GENERAL ELECTRIC CO.: VALOX Resin 300 Family Engineering Thermoplastics:

300 Family. The most complete line of unreinforced engineering thermoplastic resins in the industry offering:

Property Advantages:
* Dimensional/Physical stability in moist environments
* Chemical resistance to most oils and greases
* High surface gloss
* Excellent lubricity
* UL 94 V-O capability
* Notched Izod impact strength to 16 ft.-lb./in.(850 J/m)

Processing Advantages
* Fast cycles * Excellent flow * Easy mold release
* Similar mold shrinkage in flow and cross flow direction
* Low tool/machine wear

TYPICAL PROPERTY VALUES:

VALOX 325 resin Unreinforced PBT:

Specific Gravity (D792): 1.31
Specific Volume (in 3/lb(cm 3/kg)): 21.3 (766)

VALOX 310-SEO resin Unreinforced And UL 94 V-O Recognized:

Specific Gravity (D792): 1.39
Specific Volume (in 3/lb (cm 3/kg)): 20.0(720)

VALOX 357 resin Unreinforced Impact Modified And UL 94 V-O Recognized:

Specific Gravity (D792): 1.34
Specific Volume (in 3/lb (cm 3/kg)): 20.8(748)

VALOX 365 resin Unreinforced Impact Modified and UL 94 V-O Recognized Low Shrink:

Specific Gravity (D792): 1.31
Specific Volume (in 3/lb (cm 3/kg)): 20.8(751)

GENERAL ELECTRIC CO.: VALOX Resin 400 Family Engineering Thermoplastics:

400 Family. A full series of glass reinforced products offering:

Property Advantages:
* HDT's from 360-405F(182-207C) at 264 psi(182 MPa)
* Low moisture absorption
* UL 94 V-O capability
* Dimensional stability in harsh environments
* Excellent mechanical strength and stiffness

Processing Advantages
* Fast Cycles * Good surface * Good color stability
* Low surface marking from ejector pins

TYPICAL PROPERTY VALUES:

VALOX DR-51 resin 15% Glass Reinforced:

Specific Gravity (D792): 1.41
Specific Volume (in 3/lb (cm 3/kg)): 19.7(709)

VALOX DR-48 resin 15% Glass Reinforced and UL 94 V-O Recognized:

Specific Gravity (D792): 1.53
Specific Volume (in 3/lb (cm 3/kg)): 18.2(655)

VALOX 420 resin 30% Glass Reinforced:

Specific Gravity (D792): 1.53
Specific Volume (in 3/lb (cm 3/kg)): 18.2(655)

VALOX 420-SEO 30% Glass Reinforced and UL 94 V-O Recognized:

Specific Gravity (D792): 1.62
Specific Volume (in 3/lb (cm 3/kg)): 17.2(620)

VALOX 430 resin 30% Glass Reinforced and Impact Modified:

Specific Gravity (D792): 1.52
Specific Volume (in 3/lb (cm 3/kg)): 18.2(655)

VALOX 414 resin 40% Glass Reinforced:

Specific Gravity (D792): 1.63
Specific Volume (in 3/lb (cm 3/kg)): 17.0(612)

GENERAL ELECTRIC CO.: VALOX Resin 500 Family Engineering Thermoplastics:

500 Family. A 30% glass reinforced thermoplastic polyester alloy offering:

Property Advantages:
* Improved flatness
* Appearance
* Minimum sink marks
* Low coefficient of thermal expansion
* UL 94 V-O capabilities

Processing Advantages:
* Good flow

TYPICAL PROPERTY VALUES:

VALOX 508 resin 30% Glass Reinforced PBT Alloy:

Specific Gravity (D792): 1.50
Specific Volume (in 3/lb(cm 3/kg)): 18.5(668)
Water Absorption 24 hours (%) (D570): .06

VALOX 553 resin 30% Glass Reinforced PBT Alloy UL 94 V-O Recognized:

Specific Gravity (D792): 1.58
Specific Volume (in 3/lb(cm 3/kg)): 17.6(635)
Water Absorption 24 hours (%) (D570): .07

GENERAL ELECTRIC CO.: VALOX Resin 700 Family Engineering Thermoplastics:

700 Family. A broad line of mineral and mineral glass reinforced materials offering:

Property Advantages:
* Superior electric performance
* Improved flatness
* Improved stiffness versus unreinforced grades
* UL 94 V-O and V-2 capability
* Zero arc track performance

Processing Advantages:
* Reduced shrinkage versus unreinforced grades
* Good flow

TYPICAL PROPERTY VALUES:

VALOX 701 resin 35% Glass/Mineral Reinforced:

Specific Gravity (D792): 1.59
Specific Volume (in 3/lb (cm 3/kg)): 17.4(626)

VALOX 730 resin 35% Glass/Mineral Reinforced:

Specific Gravity (D792): 1.55
Specific Volume (in 3/lb (cm 3/kg)): 18.0(649)

VALOX 735 resin 40% Glass/Mineral Reinforced:

Specific Gravity (D792): 1.62
Specific Volume (in 3/lb (cm 3/kg)): 17.2(618)

VALOX 736 resin 45% Glass/Mineral Reinforced:

Specific Gravity (D792): 1.69
Specific Volume (in 3/lb (cm 3/kg)): 16.4(590)

VALOX 752 resin 45% Glass/Mineral Reinforced UL 94 V-O Recognized:

Specific Gravity (D792): 1.83
Specific Volume (in 3/lb (cm 3/kg)): 15.1(547)

VALOX 780 resin 40% Glass/Mineral Reinforced UL 94 V-O Recognized:

Specific Gravity (D792): 1.77
Specific Volume (in 3/lb (cm 3/kg)): 15.7(566)

GENERAL ELECTRIC CO.: VALOX Resin 800 Family Engineering Thermoplastics:

800 Family. A unique series of glass reinforced polyester alloys offering:

Property Advantages:
* HDT's from 320--380F(160--193C) at 264 psi(1.82 MPa)
* Improved surface gloss
* Mechanical strength and stiffness
* Good colorability
* UL 94 V-O and V-2 capability

Processing Advantages
* Superior flow(thin walls)
* Good mold release

TYPICAL PROPERTY VALUES:

VALOX 815 resin 15% Glass Reinforced PBT Alloy:

Specific Gravity (D792): 1.43
Specific Volume (in 3/lb (cm 3/kg)): 19.5(701)

VALOX 830 resin 30% Glass Reinforced PBT Alloy:

Specific Gravity (D792): 1.54
Specific Volume (in 3/lb (cm 3/kg)): 18.6(669)

VALOX 850 resin 15% Glass Reinforced PBT Alloy UL 94 V-2:

Specific Gravity (D792): 1.51
Specific Volume (in 3/lb (cm 3/kg)): 18.3(659)

VALOX 855 resin 15% Glass Reinforced PBT Alloy UL 94 V-O:

Specific Gravity (D792): 1.54
Specific Volume (in 3/lb (cm 3/kg)): 18.0(649)

VALOX 865 resin 30% Glass Reinforced PBT Alloy UL 94 V-O Recognized:

Specific Gravity (D792): 1.66
Specific Volume (in 3/lb (cm 3/kg)): 16.8(607)

GENERAL ELECTRIC CO.: XENOY Resin Engineered Blow Molding Grades:

Originally developed to meet the automotive industry's stringent requirements for exterior body parts, XENOY resin delivers outstanding cost and performance advantages. Dramatically improved chemical resistance and physical strength, along with excellent dimensional stability allow unprecedented opportunities for innovative design. And with the quality surface finish and options of molded-in color or transparency, new uses for the XENOY family of resins are being developed constantly in applications ranging from fluid handling equipment and medical products to lightweight lawnmower casings.

TYPICAL PROPERTY VALUES:

XENOY 1402B resin:

Specific Gravity (D792): 1.19
HDT @ 264 psi, F (D648): 205
Flexural Modulus, psi (D790): 290,000
Flexural Strength, psi (D790): 10,000
Tensile Strength, psi (D638): 7,000
Notched Izod Impact Strength, ft-lbs/in (D256):
- @ 73F: 13
- @ -20F: 10

Gardner Impact Strenth, in-lbs (D3029):
- @ 73F: >400
- @ -20F: >400

Blow Molded Shrinkage, mils/in: 8-9

XENOY 6625B resin:

Specific Gravity (D792): 1.21
HDT @ 264 psi, F (D648): 168
Flexural Modulus, psi (D790): 275,000
Flexural Strength, psi (D790): 10,000
Tensile Strength, psi (D638): 6,200
Notched Izod Impact Strength, ft-lbs/in (D256):
- @ 73F: 20
- @ -20F: 14

Elongation, % (D638): 110

GENERAL ELECTRIC CO.: XENOY Resin 2000 & 5000 Series Thermoplastic Alloy:

XENOY 2000 resins combine good chemical resistance with excellent impact strength and heat resistance.

* XENOY 2230 resin is an unreinforced blend which offers excellent impact at moderate temperatures.
* XENOY DX2735 resin is an unreinforced blend which offers excellent low-temperature impact performance.

XENOY 5000 resins combine outstanding impact strength with excellent chemical resistance superior to that of the 2000 series resins.

* XENOY 5220 resin is an unreinforced blend offering excellent mechanical and impact properties.
* XENOY DX5230 resin is an unreinforced blend offering excellent mechanical properties and good impact.
* XENOY DX5720 resin is an unreinforced blend offering excellent mechanical properties combined with superior low-temperature impact toughness.

TYPICAL PROPERTY VALUES:

XENOY 2230 Impact Modified resin:

Specific Gravity (D792): 1.22
Specific Volume (in 3/lb (cm 3/kg)): 22.8 (820)

XENOY DX2735 Impact Modified resin:

Specific Gravity (D792): 1.21
Specific Volume (in 3/lb (cm 3/kg)): 23 (827)

XENOY 5220 Impact Modified resin:

Specific Gravity (D792): 1.21
Specific Volume (in 3/lb (cm 3/kg)): 23 (827)

XENOY DX5230 Impact Modified resin:

Specific Gravity (D792): 1.22
Specific Volume (in 3/lb (cm 3/kg)): 22.8 (820)

XENOY DX5720 Impact Modified resin:

Specific Gravity (D792): 1.17
Specific Volume (in 3/lb (cm 3/kg)): 23.7 (853)

GENERAL ELECTRIC CO.: XENOY Resin 6000 Series Thermoplastic Alloy:

XENOY 6000 resins offer the highest level of chemical resistance in the XENOY product line. This superior chemical resistance is combined with excellent mechanical, impact and electrical properties. Higher stiffness is offered in the glass-reinforced grades.

TYPICAL PROPERTY VALUES:

XENOY 6120 Impact Modified resin:

Specific Gravity (D792): 1.25
Specific Volume (in 3/lb (cm 3/kg)): 22.3 (801)
Water Absorption 24 hours % (D570): .08

XENOY 6123 Impact Modified resin:

Specific Gravity (D792): 1.24
Specific Volume (in 3/lb (cm 3/kg)): 22.4 (806)
Water Absorption 24 hours % (D570): .11

XENOY 6620 Impact Modified resin:

Specific Gravity (D792): 1.20
Specific Volume (in 3/lb (cm 3/kg)): 23.2 (835)
Water Absorption 24 hours % (D570): .08

XENOY 6240 10% Glass Reinforced Impact Modified resin:

Specific Gravity (D792): 1.30
Specific Volume (in 3/lb (cm 3/kg)): 21.3 (770)
Water Absorption 24 hours % (D570): .11

XENOY 6370 30% Glass Reinforced Impact Modified resin:

Specific Gravity (D792): 1.44
Specific Volume (in 3/lb (cm 3/kg)): 19.3 (695)
Water Absorption 24 hours % (D570): .09

XENOY 6380 30% Glass Reinforced Impact Modified resin:

Specific Gravity (D792): 1.51
Specific Volume (in 3/lb (cm 3/kg)): 18.3 (658)
Water Absorption 24 hours % (D570): .10

GEORGIA GULF CORP.: Special Purpose Molding Materials:

Special Purpose/High Flow:

SP-7107:

A special purpose high flow material designed for applications requiring excellent thermal stability, and very good mechanical properties. UL Thermal Index - 95C.

Specific Gravity ASTM D-792: 1.35
Mold Shrinkage 10 -3 in/in ASTM D-955: 3-4
Hardness Rockwell R: 100
Heat Deflection C ASTM D648 264 psi: 70
Tensile Strength psi ASTM D-638: 6,000
Flammability UL94: V-O

HM-7054:

A high modulus special purpose molding material with very good surface hardness characteristics. Should be used where load bearing is an important consideration.

Specific Gravity ASTM D-792: 1.38
Mold Shrinkage 10 -3 ASTM D-955: 2-3
Hardness Rockwell R: 104
Heat Deflection Temperature C ASTM D-648 264 psi: 71
Tensile Strength psi ASTM D-638: 7,000
Flammability UL 94: V-O

MP-7085:

A maximum property special purpose product to be used when maximum physical properties are required. UL Thermal Index - 95C

Specific Gravity ASTM D-792: 1.33
Mold Shrinkage 10 -3 in/in ASTM D-955: 3-4
Hardness Rockwell R: 102
Heat Deflection Temperature C ASTM D-648 264 psi: 70
Tensile Strength psi ASTM D-638: 6,200
Flammability UL 94: V-O

FG-7108:

A food grade, FDA opaque molding compound formulated for applications requiring FDA with a good balance of physical properties.

Specific Gravity ASTM D-792: 1.35
Mold Shrinkage 10 -3 in/in ASTM D-955: 3-4
Hardness Rockwell R: 100
Heat Deflection C ASTM D-648 264 psi: 70
Tensile Strength psi ASTM D-638: 6,000
Flammability UL 94: V-O

GEORGIA GULF CORP.: Special Purpose Molding Materials (Continued):

High Heat Resistant Alloys:

HH-7160:

A high heat resistant, very high flow, U.V. stable material exhibiting better heat resistance than UV-7160.
Specific Gravity ASTM D-792: 1.31
Mold Shrinkage 10 -3 in/in ASTM D-955: 3-5
Hardness Rockwell R: 100
Heat Deflection Temperature C ASTM D-648 264 psi: 75
Tensile Strength psi ASTM D-638: 6,000
Flammabilty UL94: V-0/5-V

HF-2230:

A high flow, heat resistant alloy, offering excellent processing characteristics along with a UL 94 5V rating.
Specific Gravity ASTM D-792: 1.25
Mold Shrinkage 10 -3 in/in ASTM D-955: 80
Hardness Rockwell R: 80
Heat Deflection C ASTM D-648 264 psi: 80
Tensile Strength psi ASTM D-638: 6,000
Flammability UL94: V-O/5-V

HM-2230:

A high modulus, heat resistant alloy to be used when stiffness and load bearing requirements are important.
Specific Gravity ASTM D-792: 1.26
Mold Shrinkage 10 -3 in/in ASTM D-955: 4-6
Hardness Rockwell R: 100
Heat Deflection Temperature C ASTM D-648 264 psi: 82
Tensile Strength psi ASTM D-638: 6,200
Flammability UL94: V-0/5-V

U.V Resistant/Weatherable:

UV-7160:
An Ultra-Violet resistant high flow super tough blend with a UL 94 5V rating. Specially formulated for enclosures requiring a good balance of excellent physical properties, and U.V. resistance.
Specific Gravity ASTM D-792: 1.35
Mold Shrinkage 10 -3 in/in ASTM D-955: 3-4
Hardness Rockwell R: 100
Heat Deflection Temperature C ASTM D-648 264 psi: 70
Tensile Strength psi ASTM D-638: 6,000
Flammability UL 94: V-O/5-V

GEORGIA GULF CORP.: Special Purpose Molding Materials (Continued):

U.V. Resistant/Weatherable(Continued):

MW-7164:

A maximum weather resistant material offering maximum U.V. resistance with medium flow and very good physical properties

Specific Gravity ASTM D-792: 1.39
Mold Shrinkage 10 -3 in/in ASTM D-955: 3-4
Hardness Rockwell R: 95
Heat Deflection Temperature C ASTM D-648 264 psi: 67
Tensile Strength psi ASTM D-638: 6,500

Transparent:

CL-7053:

A clear, high flow special purpose injection molding grade with good impact and excellent tensile and flexural values.

Specific Gravity ASTM D-792: 1.33
Mold Shrinkage 10 -3 in/in ASTM D-955: 97
Hardness Rockwell R: 97
Heat Deflection C ASTM D-648 264 psi: 65
Tensile Strength psi ASTM D-638: 6,200
Flammability UL 94: V-O

CL-7049:

A clear medium flow special purpose injection molding grade for FDA applications. Offers excellent impact strength. Optica % transmission is 91.4; haze is 5.0; yellowness index is - 1.3

Specific Gravity ASTM D-792: 1.29
Mold Shrinkage 10 -3 in/in: 3-4
Hardness Rockwell R: 97
Heat Deflection Temperature C ASTM D-648 264 psi: 65
Tensile Strength psi ASTM D-638: 5,300
Flammability UL 94: V-O

GR-7025:

A gamma-resistant transparent compound formulated to resist 2.5 mega rads. A very good medical grade compound with excelle stability and high flow characteristics.

Specific Gravity ASTM D-792: 1.29
Mold Shrinkage 10 -3 in/in ASTM D-955: 3-4
Hardness Rockwell R: 97
Heat Deflection Temperature C ASTM D-648 264 psi: 58
Tensile Strength psi ASTM D-638: 5,400

GEORGIA GULF CORP.: Special Purpose Molding Materials (Continued):

Transparent(Continued):

GR-7050:

A gamma-resistant transparent compound formulated to resist 5.0 mega rads. A very good medical grade compound exhibiting excellent stability and high flow characteristics.
Specific Gravity ASTM D-792: 1.29
Mold Shrinkage 10 -3 in/in ASTM D-955: 3-4
Hardness Rockwell R: 97
Heat Deflection C ASTM D-648 264 psi: 56
Tensile Strength psi ASTM D-638: 5,400

Conduit/Pipe Fittings:

5009:

A pipe-fitting material rated T-1, G-1. NSF approved for potable water - drain, waste and vent systems. Recommended for other industrial applications where maximum chemical resistance and/or tensile strength are product requirements.
Specific Gravity ASTM D-792: 1.39
Mold Shrinkage 10 -3 in/in ASTM D-955: 2-3
Hardness Rockwell R: 115
Heat Deflection Temperature C ASTM D-648 264 psi: 71
Tensile Strength psi ASTM D-638: 7,300
Flammability UL94: V-O

7008:

A high impact compound rated T-II, G-1. NSF approved conduit and pipe fitting material. Offers maximum physical properties with good processing characteristics.
Specific Gravity ASTM D-792: 1.33
Mold Shrinkage 10 -3 in/in ASTM D-955: 4-5
Hardness Rockwell R: 105
Heat Deflection Temperature C ASTM D-648 264 psi: 70
Tensile Strength psi ASTM D-638: 6,100
Flammability UL94: V-O

THE BF GOODRICH CO.: GEON 8700A:

GEON 8700A is a general purpose, high impact, rigid PVC profile extrusion cube compound designed for interior applications. It possesses excellent dimensional stability which facilitates sizing of complex profiles.

TYPICAL PROPERTIES
All tests run at 23C (73F) except where indicated.

GENERAL PROPERTIES	TEST METHOD	ENGLISH VALUES
Cell Classification	ASTM D 1784	16344
Specific gravity, g/cc,+-.02	ASTM D 792	1.38
Hardness-Durometer D,+-3	ASTM D 2240	78
Rockwell R	ASTM D 758	111
Heat Deflection temperature at 1.82 MPa (264 psi)	ASTM D 648	163F
Compressive strength	ASTM D695	8,100 psi
Tensile Properties		
Strength	ASTM D 638	6,200 psi
Modulus	ASTM D 638	390,000 psi
Flexural Properties		
Strength	ASTM D 790	11,000 psi
Modulus	ASTM D 790	350,000 psi
Impact Strength		
Notched Izod at 23C (73F)	ASTM D 256	17.0 ft-lb/inch
0C (32F)		1.6 ft-lb/inch
-18C (0F)		1.1 ft-lb/inch
Modified Izod, unnotched at	ASTM D 256	
23C(73F)		64.0+ ft-lb/inch
0C(32F)		64.0+ ft-lb/inch
-18C(0F)		42.0+ ft-lb/inch

THE BF GOODRICH CO.: GEON 85885-385:

GEON 85885-385 is a weatherable, dark brown, rigid PVC cube compound for extrusions.

TYPICAL PROPERTIES

All tests run at 23C(73F).

GENERAL PROPERTIES	TEST METHOD	ENGLISH VALUES
Cell classification	ASTM D 1784	13444
Specific gravity, +-.02	ASTM D 792	1.35
Hardness-Durometer D, +-3	ASTM D 2240	86
Heat deflection temperature at 1.82 MPa (264 psi)	ASTM D 648	158F
Coefficient of linear expansion	ASTM D 696	3.6 10-5 in/in,F
TENSILE PROPERTIES		
Strength	ASTM D 638	7,570 psi
Modulus	ASTM D 638	390,000 psi
FLEXURAL PROPERTIES		
Strength	ASTM D 790	13,600 psi
Modulus	ASTM D 790	460,000 psi
IMPACT STRENGTH		
Notched Izod with 1/8" thick samples	ASTM D 256	2.4 ft-lb/inch
DROP IMPACT RESISTANCE		
0.125 inch conical dart Minimum failure energy (Procedure A)	ASTM D 4226	1.2 in-lb/mil
Brittle-ductile failure energy		3.8+ in-lb/mil

THE BF GOODRICH CO.: GEON 87256:

GEON 87256 is a general purpose, high impact, rigid PVC, interior-grade profile extrusion cube compound designed for running thin profiles at high linear rates.

TYPICAL PROPERTIES
All tests run at 23C (73F) except where indicated.

GENERAL PROPERTIES	TEST METHOD	ENGLISH VALUES
Cell classification	ASTM D 1784	16354
Specific gravity,+-.02	ASTM D 792	1.39
Hardness - Durometer D,+-3	ASTM D 2240	74
Rockwell R	ASTM D 785	113
Heat Deflection temperature at 1.82 MPa (264 psi)	ASTM D 648	164F
Coefficient of linear expansion	ASTM D 638	3.3 10-5 in/in,F
TENSILE PROPERTIES		
Strength	ASTM D 638	6,600 psi
Modulus	ASTM D 638	400,000 psi
FLEXURAL PROPERTIES		
Strength	ASTM D 790	12,700 psi
Modulus	ASTM D 790	469,000 psi
IMPACT STRENGTH		
Notched Izod	ASTM D 256	
23C (73F)		18.6 ft-lb/inch
0C(32F)		2.6 ft-lb/inch
-18C(0F)		1.4 ft-lb/inch
-28C(-20F)		1.3 ft-lb/inch
-40C(-40F)		1.1 ft-lb/inch
Drop Impact Resistance		
23C(73F)	ASTM D3029	3.0 in-lb/mil
-28C(-20F)	(Modified)	0.3 in-lb/mil

THE BF GOODRICH Co.: GEON 87416:

GEON 87416 is an exterior grade, impact modified, rigid PVC profile extrusion cube compound with good sizing, outstanding toughness and resistance to pinking.

TYPICAL PROPERTIES

All tests run at 23C (73F) except where indicated.

GENERAL PROPERTIES	TEST METHOD	ENGLISH VALUES
Cell classification	ASTM D 1784	13343
Specific gravity, +-.02	ASTM D 792	1.45
Hardness-Durometer D, +- 3	ASTM D 2240	82
Heat deflection temperature at 1.82 MPa (264 psi)	ASTM D 648	163F
Coefficient of linear expansion	ASTM D 696	3.4 10-5 in/in,F
TENSILE PROPERTIES		
Strength	ASTM D 638	6,400 psi
Modulus	ASTM D 638	360,000 psi
FLEXURAL PROPERTIES		
Strength	ASTM D 790	12,300 psi
Modulus	ASTM D 790	430,000 psi
IMPACT STRENGTH		
Notched Izod at 23C (73F)	ASTM D 256	
with 1/4" thick samples		3.5 ft-lb/inch
with 1/8" thick samples		4.0-20.0 ft-lb/inch

THE BF GOODRICH CO.: GEON 87446:

GEON 87446 is a general purpose, interior grade, impact modified, transparent rigid PVC profile extrusion cube compound offering excellent clarity and toughness.

TYPICAL PROPERTIES
All tests run at 23C(73F).

GENERAL PROPERTIES	TEST METHOD	ENGLISH VALUES
Cell classification	ASTM D 1784	13443
Specific gravity, +-.02	ASTM D 792	1.33
Hardness-Durometer D, +-3	ASTM D 2240	86
Heat deflection temperature at 1.82 MPa (264 psi)	ASTM D 648	145F
Coefficient of linear expansion	ASTM D 696	3.6 10-5 in/in,F
TENSILE PROPERTIES		
Strength	ASTM D 638	7,600 psi
Modulus	ASTM D 638	360,000 psi
FLEXURAL PROPERTIES		
Strength	ASTM D 790	13,900 psi
Modulus	ASTM D 790	448,000 psi
IMPACT STRENGTH		
Notched Izod with 1/8" thick samples	ASTM D 256	1.9 ft-lb/inch
DROP IMPACT RESISTANCE*	ASTM D 4226	
0.25 inch hemispherical dart Minimum failure energy (Procedure A)		3.4 in-lb/mil
Brittle-ductile failure energy (Procedure B)		3.7+ in-lb/mil
0.125 inch Conical Dart Minimum failure energy (Procedure A)		1.0 in-lb/mil
Brittle-ductile failure energy (Procedure B)		4.0+ in-lb/mil

* At 370F extrusion melt temperature.

HAYSITE REINFORCED PLASTICS: Bulk Molding Compounds:

Product Name:

Corrosion Resistant: High Strength 16100-(   )-(   ):

Property	ASTM Test Method	Typical Data
Water Absorption, %	D570	.1
Specific Gravity	----	1.80
Tensile Strength, psi	D638	5,000
Flexural Strength, psi	D790	18,000
Compressive Strength, flatwise, psi	D695	15,000
Impact Strength, Izod, edgewise ft-lbs/in. notch	D256	8.0
Arc Resistance, seconds	D495	150
Dielectric Strength, perpendicular, short time, VPM in oil	D149	350
Dielectric Constant @ 60 Hz	D150	5.0
Dissipation Factor @ 60 Hz	D150	2.5
Glass Content, percent for data shown	----	30
Range of Glass Content available %	----	30
Mold Shrinkage, in/in.	----	.002

HAYSITE REINFORCED PLASTICS: Bulk Molding Compounds (Continued):

Product Name:

Corrosion Resistant: U.L. Recognized 16600-(   )-(   ):

Property	ASTM Test Method	Typical Data
Water Absorption, %	D570	.07
Specific Gravity	----	1.84
UL 94 Flame Classification - UL File E-27875	----	94V-O
Tensile Strength, psi	D638	3,500
Flexural Strength, psi	D790	10,500
Compressive Strength, flatwise, psi	D695	10,500
Impact Strength, Izod, edgewise ft-lbs/in. notch	D256	3.5
Arc Resistance, seconds	D495	186
Track Resistance, minutes	D2303	>600
Dielectric Strength, perpendicular, short time, VPM in Oil	D149	400
Dielectric Constant @ 60 Hz	D150	4.8
Dissipation Factor @ 60 Hz	D150	2.6
Glass Content, percent for data shown	----	15
Range of Glass Content available %	----	5-35
Mold Shrinkage, in/in.	----	.002

HAYSITE REINFORCED PLASTICS: Bulk Molding Compounds (Continued):

Product Name:

General Purpose 12000-(   )-(   ):

Property	ASTM Test Method	Typical Data
Water Absorption, %	D570	.4
Specific Gravity	----	2.1
Tensile Strength, psi	D638	5,000
Flexural Strength, psi	D790	15,000
Compressive Strength, flatwise, psi	D695	20,000
Impact Strength, Izod, edgewise ft-lbs/in. notch	D256	3.0
Arc Resistance, seconds	D495	185
Track Resistance, minutes	D2303	150
Dielectric Strength, perpendicular, short time, VPM in Oil	D149	400
Dielectric Constant @ 60 Hz	D150	6.5
Dissipation Factor @ 60 Hz	D150	.5
Glass Content, percent for data shown	----	15
Range of Glass Content available %	----	5-35
Mold Shrinkage, in/in.	----	.002

HAYSITE REINFORCED PLASTICS: Bulk Molding Compounds (Continued):

Product Name:

Special Purpose: High Temperature 18500-(   )-(   ):

Property	ASTM Test Method	Typical Data
Water Absorption, %	D570	.11
Specific Gravity	----	1.86
Tensile Strength, psi	D638	6,300
Flexural Strength, psi	D790	14,000
Compressive Strength, flatwise, psi	D695	15,000
Impact Strength, Izod, edgewise ft-lbs/in. notch	D256	4.0
Arc Resistance, seconds	D495	188
Dielectric Strength, perpendicular, short time, VPM in Oil	D149	400
Dielectric Constant @ 60 Hz	D150	8.5
Dissipation Factor @ 60 Hz	D150	6.7
Glass Content, percent for data shown	----	15
Range of Glass Content available %	----	5-35
Mold Shrinkage, in/in.	----	.0022

HAYSITE REINFORCED PLASTICS: Bulk Molding Compounds (Continued):

Product Name:

Special Purpose: MIL-M-14:MAI-60/MAT-30 - 270 FR II**:

Property	ASTM Test Method	Typical Data
Water Absoprtion, %	D570	.23
Specific Gravity	----	1.75
UL 94 Flame Classification - UL File E-27875	----	94V-O
Tensile Strength, psi	D638	8,500
Flexural Strength, psi	D790	17,000
Compressive Strength, flatwise, psi	D695	19,000
Impact Strength, Izod, edgewise ft-lbs/in. notch	D256	12
Arc Resistance, seconds	D495	185
Track Resistance, minutes	D2303	300+
Dielectric Strength, perpendicular, short time, VPM in Oil	D149	375
Dielectric Constant @ 60 Hz	D150	4.62
Dissipation Factor @ 60 Hz	D150	1.00
Mold Shrinkage, in/in.	----	.0012

**270FRII is approved per MIL-M-14 Type MAI 60/MAT 30.

HAYSITE REINFORCED PLASTICS: Bulk Molding Compounds (Continued):

Product Name:

Special Purpose: Phenolic Substitute 30000-(   )-(   ):

Property	ASTM Test Method	Typical Data
Water Absorption, %	D570	.15
Specific Gravity	----	2.15
Tensile Strength, psi	D638	3,000
Flexural Strength, psi	D790	6,500
Compressive Strength, flatwise, psi	D695	18,000
Impact Strength, Izod, edgewise ft-lbs/in. notch	D256	2.5
Arc Resistance, seconds	D495	180
Track Resistance, minutes	D2303	>600
Dielectric Strength, perpendicular, short time, VPM in Oil	D149	325
Glass Content, percent for data shown	----	5
Range of Glass Content available %	----	5
Mold Shrinkage, in/in.	----	.002

HAYSITE REINFORCED PLASTICS: Bulk Molding Compounds (Continued):

Product Name:

U.L. Recognized Electrical 13600-(   )-(   ):

Property	ASTM Test Method	Typical Data
Water Absorption, %	D570	.35
Specific Gravity	----	1.94
UL 94 Flame Classification - UL File E-27875	----	94V-O
Tensile Strength, psi	D638	6,000
Flexural Strength, psi	D790	15,000
Compressive Strength, flatwise, psi	D695	15,000
Impact Strength, Izod, edgewise ft-lbs/in. notch	D256	3.5
Arc Resistance, seconds	D495	190
Track Resistance, minutes	D2303	>600
Dielectric Strength, perpendicular, short time, VPM in Oil	D149	375
Dielectric Constant @ 60 Hz	D150	5.8
Dissipation Factor @ 60 Hz	D150	295
Glass Content, percent for data shown	----	15
Range of Glass Content available %	----	5-35
Mold Shrinkage, in/in.	----	.002

HAYSITE REINFORCED PLASTICS: Bulk Molding Compounds (Continued):

Product Name:

U. L. Recognized Electrical 14100-(   )-(   ):*

Property	ASTM Test Method	Typical Data
Water Absorption, %	D570	.1
Specific Gravity	----	1.89
UL 94 Flame Classification - UL File E-27875	----	94V-O
Tensile Strength, psi	D638	6,000
Flexural Strength, psi	D790	12,500
Compressive Strength, flatwise, psi	D695	13,000
Impact Strength, Izod, edgewise ft-lbs/in. notch	D256	3.5
Arc Resistance, seconds	D495	192
Track Resistance, minutes	D2303	>600
Dielectric Strength, perpendicular, short time, VPM in Oil	D149	400
Dielectric Constant @ 60 Hz	D150	5.0
Dissipation Factor @ 60 Hz	D150	3.25
Glass Content, percent for data shown	----	15
Range of Glass Content available %	----	5-35
Mold Shrinkage, in/in.	----	.0001

* Can be furnished specially formulated for wear resistance.

HAYSITE REINFORCED PLASTICS: Bulk Molding Compounds(Continued):

Product Name:

U. L. Recognized Electrical 17000-( )-( ):

Property	ASTM Test Method	Typical Data
Water Absorption, %	D570	.10
Specific Gravity	----	1.9
U.L. Flame Classification - UL File E-27875	----	94V-O, 945-V
Tensile Strength, psi	D638	7,000
Flexural Strength, psi	D790	15,000
Compressive Strength, flatwise, psi	D695	20,000
Impact Strength, Izod, edgewise ft-lbs/in. notch	D256	5.0
Arc Resistance, seconds	D495	180
Track Resistance, minutes	D2303	600
Dielectic Strength, perpendicular, short time, VPM in Oil	D149	400
Dielectric Constant @ 60 Hz	D150	5.0
Dissipation Factor @ 60 Hz	D150	3.0
Glass Content, percent for data shown	----	20
Range of Glass Content available %	----	20
Mold Shrinkage, in/in.	----	.002

HAYSITE REINFORCED PLASTICS: Bulk Molding Compounds(Continued):

Product Name:

U. L. Recognized Electrical 20000-(   )-(   )*:

Property	ASTM Test Method	Typical Data
Water Absorption, %	D570	.1
Specific Gravity	----	1.95
UL 94 Flame Classification - UL File E-27875	----	94V-O
Tensile Strength, psi	D638	6,500
Flexural Strength, psi	D790	15,000
Compressive Strength, flatwise, psi	D695	13,000
Impact Strength, Izod, edgewise ft-lbs/in. notch	D256	4.0
Arc Resistance, seconds	D495	190
Track Resistance, minutes	D2303	>600
Dielectric Strength, perpendicular, short time, VPM in Oil	D149	400
Dielectric Constant @ 60 Hz	D150	5.5
Dissipation Factor @ 60 Hz	D150	2.85
Glass Content, percent for data shown	----	15
Range of Glass Content available %	----	5-35
Mold Shrinkage, in/in.	----	.0025

* Can be furnished specially formulated for wear resistance.

HAYSITE REINFORCED PLASTICS: Compound - BMC 14105-24-RE:

Grade 14105-24-RE is a fiberglass reinforced thermoset polyester bulk molding compound that is recognized by U.L. R under file E27875 and exhibits high compressive strength and excellent shrinkage control. It can be molded by the compression, transfer or injection process.

Property	ASTM Test	Value
Specific Gravity	D-792	1.82
Water Absorption - %	D-229	.164
UL94 Flame Resistance	UL94	94V-0
Flame Resistance/Seconds	D-229	
Ignition Time		101
Burning Time		18
Temperature Index Elec./Mech.	U.L.	130C/130C
Thermal Conductivity BTU/Hr./Sq. Ft./In./F	C-177	.657 x 10 -3
Shrink - In./In.	D-955	.0001"
Glass Content %		24%
Tensile Strength, PSI	D-638	7,000
Flexural Strength, PSI	D-790	20,000
Compressive Strength, Flatwise PSI	D-695	32,000
Impact Strength, Izod Ft./Lbs./In. Notch Edgewise	D-256	10
Impact Strength, Izod	D-256	10
Arc Resistance, Seconds	D-495	>180 Sec.
Track Resistance/Minutes	D-2303	>600
Dielectric Strength, Perpendicular VPM, Short time in oil	D-149	350
Dielectric Constant @ 60 Hz	D-150	4.9
Dissipation Factor @ 60 HZ	D-150	2.2

HAYSITE REINFORCED PLASTICS: Compound - BMC 14113-05:

Grade 14113-05 is a fiberglass reinforced thermoset polyester bulk molding compound that is recognized as a 94-VO material by U.L. under file number E27875. It can be molded by the compression, transfer or injection process.

Property	ASTM Test	Value
Specific Gravity	D-792	2.05
Water Absorption, %	D-229	.14
UL94 Flame Resistance	UL94	V-O
Flame Resistance/Seconds	D229	
Ignition Time		90
Burn Time		5
Temperature Index Elec./Mech.	U.L.	130C/130C
Shrink - In./In.	D-955	.002
Glass Content, %	-----	5%
Tensile Strength, PSI	D-638	4,000
Flexural Strength, PSI	D-790	8,200
Compression Strength, Flatwise, PSI	D-695	14,000
Compression Strength Edgewise, PSI	D-695	19,000
Impact Strength, Izod Ft./Lbs./In. Notch Edgewise	D-256	2-3
Arc Resistance, Second	D-495	216
Track Resistance, Minute	D-2303	600+
Dielectric Strength, Perpendicular VPM, short time in air	D-149	339
Dielectric Strength Parallel Step-by-step in oil, K.V.	D-149	60 KV
Comparative Tracking Index		600+ drops

HAYSITE REINFORCED PLASTICS: Compound - BMC 14113-15:

Grade 14113-15 is a fiberglass reinforced thermoset polyester bulk molding compound that is recognized as a 94-VO material by U. L. R under file number E27875. It can be molded by the compression, transfer or injection process.

Property	ASTM Test	Value
Specific Gravity	D-792	1.85
Water Absorption, %	D-229	.16
UL94 Flame Resistance	UL94	V-O
Flame Resistance/Seconds	D229	
Ignition Time		97
Burn Time		52
Temperature Index Elec./Mech.	U.L.	130C/130C
Heat Distortion Temperature	Fed Std. 406.2011	400F
Shrink - In./In.	D-955	.001/In.
Glass Content, %	-----	15%
Tensile Strength, PSI	D-638	6,000
Flexural Strength, PSI	D-790	20,000
Compression Strength, Flatwise, PSI	D-695	14,900
Compression Strength, Edgewise, PSI	D-695	20,000
Impact Strength, Izod Ft./Lbs./In. Notch Edgewise	D-256	8-9
Arc Resistance, Second	D-495	194
Track Resistance, Minute	D-2303	600+
Dielectric Strength, Perpendicular VPM, short time in air	D-149	412
Dielectric Strength Parallel Step-by-step in oil, K.V.	D-149	56 KV

HAYSITE REINFORCED PLASTICS: Compound - BMC 14113-26:

Grade 14113-26 is a fiberglass reinforced thermoset polyester bulk molding compound that is recognized as a 94-VO material by U. L. R under file E27875. This material has been tested for "Smoke Density" per NFPA 258 and *Radiant Panel per ASTM E 162. It can be molded by the compression, transfer or injection process.

Property	ASTM Test	Value
Specific Gravity	D-792	1.78
Water Absorption - %	D-299	.075%
UL 94 Flame Resistance	UL94	V-O
Radiant Panel	E162	3.6 average
Smoke Density	NFPA 258	Density @ 1.5 min.
Flamming		Average 1
Nonflamming		Average 1
Smoke Density	NFPA 258	Density @ 4.0 min.
Flamming		Average 43
Nonflamming		Average 1
Temperature Index Electrical/Mechanical	UL	130C/130C
Barcol Hardness	-----	50
Shrink In./In.	-----	.001-.0015
Glass Content %	-----	26
Tensile Strength, PSI	D-638	9400
Flex Strength, PSI	D-790	20,000
Compression Strength, Edgewise PSI	D-695	18,000 PSI
Impact Strength, Izod Ft./Lbs./In. Notch Edgewise	D-256	12.0
Arc Resistance, Sec.	D-495	185
Track Resistance, Min.	D-2303	600
Dielectric Strength V.P.M. short time in oil	D-149	379
Dielectric Constant at 60HZ	D-150	
Guarded		4.62
Unguarded		4.77

HAYSITE REINFORCED PLASTICS: HAYSITE Fiberglass Reinforced Polyester Pultrusion Grades:

Product Name:

EHC-P:

Properties	ASTM Test Method	Typical Values
I.E.E.E.* Temperature Class		155C
Specific Gravity	D-792	1.97
Water Absorption %	D-570	.16
Tensile Strength P.S.I.	D-638	35,000
Flexural Strength P.S.I.	D-790	94,000
Compressive Strength (Radially) P.S.I.	D-695	15,000
Impact Strength Izod, Ft. lbs/ in. notch	D-256	40
Dielectric Strength Perpendicular, V.P.M.	D-149	350
Dielectric Strength Parallel, K.V.	D-149	43
Arc Resistance Seconds	D-495	185
Standard Color		Light Brown

* Approximate Temperature Class is listed. Actual Classification depends on end use and locations of the material in a complete system. See NEMA L1-1 1971.

EHC-P:

A general purpose economy grade that offers 155C temperature performance. Typical uses include transformer dogbones and GPO-1 type electrical applications.

HAYSITE REINFORCED PLASTICS: HAYSITE Fiberglass Reinforced Polyester Pultrusion Grades (Continued):

Product Name:

ETR-FR-C-P:

Properties	ASTM Test Method	Typical Values
I.E.E.E.* Temperature Class		130C
Specific Gravity	D-792	1.82
Water Absorption %	D-570	.30
Tensile Strength P.S.I.	D-638	35,000
Flexural Strength P.S.I.	D-790	100,000
Compressive Strength (Radially) P.S.I.	D-695	16,000
Impact Strength Izod, Ft. lbs/in. notch	D-256	40
Dielectric Strength Perpendicular, V.P.M.	D-149	200
Dielectric Strength Parallel, K.V.	D-149	48
Arc Resistance Seconds	D-495	150
Flame Resistance		94V-O
Standard Color		Red

* Approximate Temperature Class is listed. Actual Classification depends on end use and location of the material in a complete system. See NEMA L1-1 1971

ETR-FR-C-P:

A flame retardant material that finds many uses in electrical applications. ETR-FR-C-P is available in channel and angle form.

HAYSITE REINFORCED PLASTICS: HAYSITE Fiberglass Reinforced Polyester Pultrusion Grades (Continued):

Product Name:

HST-II-P:

Properties	ASTM Test Method	Typical Values
I.E.E.E.* Temperature Class		220C
Specific Gravity	D-792	1.88
Water Absorption %	D-570	.28
Tensile Strength P.S.I.	D-638	40,000
Flexural Strength P.S.I.	D-790	100,000
Compressive Strength (Radially) P.S.I.	D-695	16,000
Impact Strength Izod, Ft. lbs/in. notch	D-256	40
Dielectric Strength Perpendicular, V.P.M.	D-149	270
Dielectric Strength Parallel, K.V.	D-149	45
Arc Resistance Seconds	D-495	185
Standard Color		Dark Brown

* Approximate Temperature Class is listed. Actual Classification depends on end use and location of the material in a complete system. See NEMA L1-1 1971.

HST-II-P:

A class 220C material having high retention of physical and electrical properties at elevated temperatures. HST-II-P is a cost effective replacement for NEMA G-7 or G-11. Uses include any application that requires a material with high thermal indices.

HAYSITE REINFORCED PLASTICS: HAYSITE Fiberglass Reinforced Polyester Pultrusion Grades(Continued):

Product Name:

HYR:

Properties	ASTM Test Method	Typical Values
I.E.E.E.* Temperature Class		130C
Specific Gravity	D-792	1.88
Water Absorption %	D-570	.20
Tensile Strength P.S.I.	D-638	100,000
Flexural Strength P.S.I.	D-790	120,000
Compressive Strength (Radially) P.S.I.	D-695	16,000
Impact Strength Izod, Ft. lbs/in. notch	D-256	58
Dielectric Strength Perpendicular, V.P.M.	D-149	350
Dielectic Strength Parallel K.V.	D-149	42
Arc Resistance Seconds	D-495	150
Standard Color		Natural

* Approximate Temperature Class is listed. Actual Classification depends on end use and location of the material in a complete system. See NEMA L1-1 1971.

HYR:

Primarily a structural grade, HYR exhibits the highest mechanical properties available. Applications are sucker rods, strain rods, and other high strength requirements.

HAYSITE REINFORCED PLASTICS: Sheet Molding Compounds:

Product Name:

Special Applications: Automotive 46000-(   )-(   ):

Property	ASTM Test Method	Typical Data
Water Absorption, %	D570	.15
Specific Gravity	----	1.8
Tensile Strength, psi	D638	12,000
Flexural Strength, psi	D790	30,000
Compressive Strength, flatwise, psi	D695	28,000
Impact Strength, Izod, edgewise ft-lbs/in. notch	D256	18.5
Arc Resistance, seconds	D495	140
Dielectric Strength, perpendicular, short time, VPM in Oil	D149	400
Dielectric Constant @ 60 Hz	D150	4.85
Dissipation Factor @ 60 Hz	D150	.6
Glass Content, percent for data shown	----	30
Range of Glass Content available %	----	30
Mold Shrinkage, in/in.	----	.0007

HAYSITE REINFORCED PLASTICS: Sheet Molding Compounds (Continued):

Product Name:

Special Applications: Corrosion Resistant 44000-(   )-(   ):

Property	ASTM Test Method	Test Method
Water Absorption, %	D570	.23
Specific Gravity	----	1.66
Tensile Strength, psi	D638	11,000
Flexural Strength, psi	D790	30,000
Compressive Strength, flatwise, psi	D695	21,000
Impact Strength, Izod, edgewise ft-lbs/in. notch	D256	16.0
Arc Resistance, seconds	D495	140
Track Resistance, minutes	D2303	200
Dielectric Strength, perpendicular, short time, VPM in oil	D149	450
Dielectric Constant @ 60 Hz	D150	4.2
Dissipation Factor @ 60 Hz	D150	2.3
Glass Content, percent for data shown	----	22
Range of Glass Content Available %	----	15-35
Mold Shrinkage, in/in.	----	.002

HAYSITE REINFORCED PLASTICS: Sheet Molding Compounds (Continued):

Product Name:

Special Applications: General Purpose 48000-(   )-(   ):*

Property	ASTM Test Method	Typical Data
Water Absoprtion	D570	.23
Specific Gravity	----	1.72
Tensile Strength, psi	D638	13,500
Flexural Strength, psi	D790	26,500
Compressive Strength, flatwise, psi	D695	27,000
Impact Strength, Izod, edgewise ft-lbs/in. notch	D256	20.0
Arc Resistance, seconds	D495	170
Track Resistance, minutes	D2303	>600
Dielectric Strength, Perpendicular, short time, VPM in Oil	D149	400
Dielectric Constant @ 60 Hz	D150	4.4
Dissipation Factor @ 60 Hz	D150	1.1
Glass Content, percent for data shown	----	22
Range of Glass Content available, %	----	15-35
Mold Shrinkage, in/in.	----	.002

* Can be furnished specially formulated for wear resistance.

HAYSITE REINFORCED PLASTICS: Sheet Molding Compounds (Continued):

Product Name:

Special Applications: High Glass 60000-( )-( ):

Property	ASTM Test Method	Typical Data
Water Absorption, %	D570	.3
Specific Gravity	----	1.75
Tensile Strength, psi	D638	30,000
Flexural Strength, psi	D790	40,000
Compressive Strength, flatwise, psi	D695	35,000
Impact Strength, Izod, edgewise ft-lbs/in. notch	D256	25.0
Arc Resistance, seconds	D495	130
Glass Content, percent for data shown	----	60
Range of Glass Content available %	----	60
Mold Shrinkage, in/in.	----	.002

HAYSITE REINFORCED PLASTICS: Sheet Molding Compounds (Continued):

Product Name:

Special Applications: High Strength Pigmentable 53000-( )-( ):

Property	ASTM Test Method	Typical Data
Water Absorption, %	D570	.23
Specific Gravity	----	1.56
Tensile Strength, psi	D638	13,000
Flexural Strength, psi	D790	20,000
Compressive Strength, flatwise, psi	D695	18,000
Impact Strength, Izod, edgewise ft-lbs/in. notch	D256	20.0
Arc Resistance, seconds	D495	140
Dielectric Strength, perpendicular, short time, VPM in Oil	D149	400
Glass Content, percent for data shown	----	35
Range of Glass Content available, %	----	35
Mold Shrinkage, in/in.	----	.002

HAYSITE REINFORCED PLASTICS: Sheet Molding Compounds (Continued):

Product Name:

Special Applications: Low Pressure 70000-(   )-(   ):

Property	ASTM Test Method	Typical Data
Water Absorption, %	D570	.2
Specific Gravity	----	1.85
Tensile Strength, psi	D638	10,000
Flexural Strength, psi	D790	21,000
Compressive Strength, Flatwise, psi	D695	21,000
Impact Strength, Izod, edgewise ft-lbs/in. notch	D256	16.0
Arc Resistance, seconds	D495	180
Track Resistance, minutes	D2303	400
Dielectric Strength, perpendicular, short time, VPM in Oil	D149	400
Dielectric Constant @ 60 Hz	D150	4.7
Dissipation Factor @ 60 Hz	D150	.8
Glass Content, percent for data shown	----	22
Range of Glass Content available %	----	15-35
Mold Shrinkage, in/in.	----	.0005

HAYSITE REINFORCED PLASTICS: Sheet Molding Compounds (Continued):

Product Name:

Special Applications: Low Profile 43000-(   )-(   ):

Property	ASTM Test Method	Typical Data
Water Absorption, %	D570	.2
Specific Gravity	----	1.78
Tensile Strength, psi	D638	10,000
Flexural Strength, psi	D790	25,000
Compressive Strength, flatwise, psi	D695	15,000
Impact Strength, Izod, edgewise ft-lbs/in. notch	D256	15.0
Arc Resistance, seconds	D495	140
Dielectric Strength, perpendicular, short time, VPM in oil	D149	380
Dielectric Constant @ 60 Hz	D150	4.6
Dissipation Factor @ 60 Hz	D150	.3
Glass Content, percent for data shown	----	22
Range of Glass Content available %	----	15-35
Mold Shrinkage, in/in.	----	.0015

HAYSITE REINFORCED PLASTICS: Sheet Molding Compounds (Continued):

Product Name:

U. L. Recognized Electrical 40000-(   )-(   ):

Property	ASTM Test Method	Typical Data
Water Absorption, %	D570	.2
Specific Gravity	----	1.85
UL 94 Flame Classification - UL File E-27875	----	94V-0
Tensile Strength, psi	D638	14,000
Flexural Strength, psi	D790	30,000
Compressive Strength, flatwise, psi	D695	30,000
Impact Strength, Izod, edgewise ft-lbs/in. notch	D256	13.5
Arc Resistance, seconds	D495	180
Track Resistance, minutes	D2303	450
Dielectric Strength, perpendicular, short time, VPM in Oil	D149	420
Dielectric Constant @ 60 Hz	D150	4.7
Dissipation Factor @ 60 Hz	D150	.96
Glass Content, percent for data shown	----	22
Range of Glass Content available, %	----	15-35
Mold Shrinkage, in/in.	----	.002

HAYSITE REINFORCED PLASTICS: Sheet Molding Compounds (Continued):

Product Name:

U. L. Recognized Electrical 42000-( )-( ):

Property	ASTM Test Method	Typical Data
Water Absorption, %	D570	.18
Specific Gravity	----	1.75
UL 94 Flame Classification - UL File E-27875	----	94V-O
Tensile Strength, psi	D638	10,000
Flexural Strength, psi	D790	25,000
Compressive Strength, flatwise, psi	D695	28,000
Impact Strength, Izod, edgewise ft-lbs/in. notch	D256	18.5
Arc Resistance, seconds	D495	180
Track Resistance, minutes	D2303	>600
Dielectric Strength, perpendicular, short time, VPM in Oil	D149	425
Dielectric Constant @ 60 Hz	D150	4.4
Dissipation Factor @ 60 Hz	D150	.96
Glass Content, percent for data shown	----	22
Range of Glass Content available,%	----	15-35
Mold Shrinkage, in/in.	----	.001

HAYSITE REINFORCED PLASTICS: Sheet Molding Compounds (Continued):

Product Name:

U. L. Recognized Electrical 47000-( )-( ):

Property	ASTM Test Method	Typical Data
Water Absorption, %	D570	.15
Specific Gravity	----	1.85
UL 94 Flame Classification - UL File E-27875	----	94V-O, 945-V
Tensile Strength, psi	D638	15,000
Flexural Strength, psi	D790	30,000
Compressive Strength, flatwise, psi	D695	30,000
Impact Strength, Izod, edgewise ft-lbs/in. notch	D256	20.0
Arc Resistance, seconds	D495	180
Track Resistance, minutes	D2303	500
Dielectric Strength, perpendicular, short time, VPM in Oil	D149	420
Dielectric Constant @ 60 Hz	D150	4.4
Dissipation Factor @ 60 Hz	D150	.95
Glass Content, percent for data shown	----	33
Range of Glass Content available %	----	33
Mold Shrinkage, in/in.	----	.002

HEXCEL: Resin Matrix Selector:

<u>Resin Systems:</u>

EPOXY:

F155:
- Fiberglass: Woven Fabric, Uni-Directional Tape
- KEVLAR: Woven Fabric, Uni-Directional Tape
- Graphite: Woven Fabric, Uni-Directional Tape
- Other: THORSTRAND Woven Fabric

F185:
- Fiberglass: Woven Fabric
- KEVLAR: Woven Fabric
- Graphite: Uni-Directional Tape

F161:
- Fiberglass: Woven Fabric, Uni-Directional Tape
- KEVLAR: Woven Fabric, Uni-Directional Tape
- Other: THORSTRAND Woven Fabric, Quartz Woven Tape

F263:
- Fiberglass: Woven Fabric, Uni-Directional Tape
- KEVLAR: Woven Fabric, Uni-Directional Tape
- Other: THORSTRAND Woven Fabric

F584:
- Graphite: Woven Fabric, Uni-Directional Tape

F593:
- Fiberglass: Woven Fabric
- Graphite: Woven Fabric

POLYESTER:

F141:
- Fiberglass: Woven Fabric
- KEVLAR: Woven Fabric

F148:
- Fiberglass: Woven Fabric

F549:
- Graphite: Uni-Directional Tape

HEXCEL: Resin Matrix Selector(Continued):

Resin Systems:

PHENOLIC:

F120:
Fiberglass: Woven Fabric
Other: NEXTEL Quartz Fabric

F886:
Fiberglass: Woven Fabric
KEVLAR: Woven Fabric

POLYIMIDE:

F174:
Fiberglass: Woven Fabric
Other: NEXTEL Woven Fabric, Quartz Woven Fabric

F178:
Fiberglass: Woven Fabric
Graphite: Woven Fabric, Uni-Directional Tape
Other: Quartz Woven Fabric

LARC-160:
Fiberglass: Woven Fabric
Graphite: Woven Fabric, Uni-Directional Tape

PMR-15:
Fiberglass: Woven Fabric
Graphite: Woven Fabric, Uni-Directional Tape
Other: NEXTEL Woven Fabric

F650:
Fiberglass: Woven Fabric
Graphite: Woven Fabric, Uni-Directional Tape
Other: NEXTEL Woven Fabric, Quartz Woven Fabric

HUNTSMAN CHEMICAL CORP.: Crystal Polystyrene:

Huntsman Chemical offers a range of hard, clear and glossy resins for applications where impact toughness is not required, but where thermal properties and a wide range of flow rates allow for economical processing. Grades 201 and 202 are high heat, low flow and low monomer materials for extrusion and molding of critical packaging. Grade 203 offers a higher flow material with excellent thermal properties for general molding of packaging, cassettes, medical wares, and quality cutlery and tumblers. Grades 206 and 208 offer increasingly high flows for general molding and compounding.

Typical Properties:
(All samples injection molded 1/8" thickness)

Grade Numbers:

201:
Melt Flow (condition G, g/10 min.): 1.7
Specific Gravity (73/73F): 1.05
Vicat Softening Point (F) (C): 225 (107)
Heat Distortion (264 psi, F) (1820 KPa, C): 200 (93)
Tensile Strength (psi) (MPa): 7400 (51)

202:
Melt Flow (condition G, g/10 min.): 3
Specific Gravity (73/73F): 1.05
Vicat Softening Point (F) (C): 220 (103)
Heat Distortion (264 psi, F) (1820 KPa, C): 200 (93)
Tensile Strength (psi) (MPa): 7000 (48)

203:
Melt Flow (condition G, g/10 min.): 8
Specific Gravity (73/73F): 1.05
Vicat Softening Point (F) (C): 217 (102)
Heat Distortion (264 psi, F) (1820 KPa, C): 200 (93)
Tensile Strength (psi) (MPa): 6000 (42)

206:
Melt Flow (condition G, g/10 min.): 14
Specific Gravity (73/73F): 1.05
Vicat Softening Point (F) (C): 206 (97)
Heat Distortion (264 psi, F) (1820 KPa, C): 193 (89)
Tensile Strength (psi) (MPa): 5700 (39)

208:
Melt Flow (condition G, g/10 min.): 18
Specific Gravity (73/73F): 1.05
Vicat Softening Point (F) (C): 200 (93)
Heat Distortion (264 psi, F) (1820 Kpa, C): 185 (85)
Tensile Strength (psi) (MPa): 5600 (35)

HUNTSMAN CHEMICAL CORP.: High Impact Polystyrene:

Huntsman Chemical high impact polystyrenes are available in a broad range of physical properties.

Grade 333 is a versatile high impact polystyrene with a good balance of properties for injection molding applications. This grade has found wide acceptance in a number of appliance, furniture, and toy applications.

Grade 331 is a high melt flow, high impact polystyrene especially designed for injection molding difficult to fill parts such as air conditioner fronts, housewares, and toys. Other difficult to mold parts involving weld line strength and appearance deficiencies have been successfully molded using this Huntsman grade.

Grade 338 is a versatile high impact polystyrene with excellent processing characteristics for both extrusion, thermoforming, and injection molding applications. It is accepted for a variety of both molding and extrusion applications such as toys, automotive parts, appliance parts and has found wide acceptance in the sheet extrusion market.

Typical Properties:
(All samples injection molded 1/8" thickness)
Grade Numbers:

331:
Melt Flow (condition G, g/10 min.): 1.5
Specific Gravity (73/73F): 1.05
Vicat Softening Point (F) (C): 190 (88)
Heat Distortion (264 psi, F) (1820 KPa, C): 175 (79)
Tensile Strength (psi) (MPa): 2900 (20)

333:
Melt Flow (condition G, g/10 min.): 8
Specific Gravity (73/73F): 1.05
Vicat Softening Point (F) (C): 202 (94)
Heat Distortion (264 psi, F) (1820 KPa, C): 185 (85)
Tensile Strength (psi) (MPa): 3400 (23)

338:
Melt Flow (condition G, g/10 min.): 4
Specific Gravity (73/73F): 1.05
Vicat Softening Point (F) (C): 207 (97)
Heat Distortion (264 psi, F) (1820 KPa, C): 190 (88)
Tensile Strength (psi) (MPa): 4200 (29)

350:
Melt Flow (condition G, g/10 min): 3
Specific Gravity (73/73F): 1.05
Vicat Softening Point (F) (C): 212 (100)
Heat Distortion (264 psi, F) (1820 KPa, C): 194 (90)
Tensile Strength (psi) (MPa): 4200 (29)

HUNSTMAN CHEMICAL CORP.: Medium Impact Polystyrene:

Huntsman Chemical medium impact polystyrenes are available in a wide range of physical properties.

Grade 312 is a low flow medium impact polystyrene with an excellent balance of thermal and impact properties. It has been well accepted in both extrusion and molding applications. Grade 314 offers high temperature properties and good practical toughness often needed for cartridge or cassette applications, small appliances, and housewares. Grade 316 is a moderately higher flow grade with good balance of impact and thermal properties. Grade 318 is a high flow grade for use in thin wall or long flow path moldings in the packaging, toy, and houseware markets.

Typical Properties:
(All samples injection molded 1/8" thickness)

Grade Numbers:

312:
Melt Flow (condition G, g/10 min.): 3
Specific Gravity (73/73F): 1.05
Vicat Softening Point (F) (C): 217 (103)
Heat Distortion (264 psi, F) (1820 KPa, C): 190 (88)
Tensile Strength (psi) (MPa): 5000 (34)

314:
Melt Flow (condition G, g/10 min.): 6
Specific Gravity (73/73F): 1.05
Vicat Softening Point (F) (C): 214 (101)
Heat Distortion (264 psi, F) (1820 KPa, C): 4800 (33)
Tensile Strength (psi) (MPa): 4800 (33)

316:
Melt Flow (condition G, g/10 min.): 10
Specific Gravity (73/73F): 1.05
Vicat Softening Point (F) (C): 202 (94)
Heat Distortion (264 psi, F) (1820 KPa, C): 4400 (30)
Tensile Strength (psi) (MPa): 4400 (30)

318:
Melt Flow (condition G, g/10 min.): 18
Specific Gravity (73/73F): 1.05
Vicat Softening Point (F) (C): 190 (88)
Heat Distortion (264 psi, F) (1820 KPa, C): 178 (81)
Tensile Strength (psi) (MPa): 4000 (28)

HUNTSMAN CHEMICAL CORP.: Polystyrene:

Type - 271
High Heat Medium Impact Polystyrene

HUNTSMAN Type 271 is a rubber-modified polystyrene which combines outstanding thermal resistance with exceptionally high gloss. Type 271 is well suited for designing furniture, parts, utensils, and other items in a variety of shapes where exceptionally fast set up times are required. Type 271 is formulated for easy processing on existing ram and screw injection molding equipment.

Physical Properties
- Impact Strength, Notched (ASTM D 256)
  - Injection Molded: 60 J/m (1.1 ft.-lb./in.)
  - Compression Molded: 43 J/m (0.8 ft.-lb./in.)
- Tensile Strength (D638): 41 MPa (6000 psi)
- Elongation (D638): 30%
- Elastic Modulus (D638): 2410 MPa (350,000 psi)
- Deflection Temperature (Under Load, 264 psi) (D648): 94 C (200F)
- Specific Gravity (D792): 1.04
- Rockwell Hardness (D785): M45
- Vicat Softening Point (D1525): 104C (219F)
- Melt Flow Point (D1238 (G)): 6 g./10 min.

Type XCI-2000
High Gloss Impact
Injection Molding Grade Polystyrene

Mechanical Properties:

- Impact Strength, Notched (D256)
  - Injection Molded: 1.3 ft.-lb./in.
  - Compression Molded: 1.1 ft.-lb./in.
- Tensile Strength (D638): 4900 psi
- Elongation (D638): 45%
- Elastic Modulus (D638): 330,000 psi
- Deflection Temperature (D648): 181F Under Load, 264 psi
- Specific Gravity (D792): 1.04
- Rockwell Hardness (D785): L-70
- Vicat Softening Point (D1525): 216F (102C)
- Melt Flow Rate: (D1238 (G)) (5.0 g/10 minutes)

HUNTSMAN CHEMICAL CORP.: Polystyrene(Continued):

Type 335
High Impact High Heat Polystyrene

HUNTSMAN Type 335 is a high impact, high heat polystyrene with an excellent balance flow and impact properties. It has been well accepted in both extrusion and molding applications.

Typical applications include toys, small appliances, packaging, displays, and household products.

Physical Properties:
(All samples injection molded 1/8" thickness)
Melt Flow (condition G, g/10 min.): 3.5 (D1238)
Specific Gravity (73/73F): 1.04 (D792)
Vicat Softening Point (F): 215 (D1525)
Heat Distortion (264 psi, F): 190 (D648)
Tensile Strength (psi): 2600 (D638)
Elongation (%): 40 (D638)
Flexural Strength (psi x 1000): 6700 (D790)
Hardness (Rockwell L): 50 (D785)
Izod Impact (ft.-lb./in. of notch): 1.6 (D256)

Type - 730
High-Impact Extrusion Grade Polystyrene

HUNTSMAN type 730 is a high-impact material specifically designed for extrusion and thermoforming applications. Its balance of rheological properties permits well controlled and uniform extrusion of sheet or profiles with minimum machine adjustments.

Its excellent appearance, low monomer content, and favorable stress crack resistance make it the resin of choice for food and dairy packaging applications. The material exhibits unusual consistency during fabrication, resulting in products characterized by a high degree of uniformity of the end-use properties.

Typical applications include profiles, lamp shades, containers, refrigerator door liners and trays, hot and cold drink cups and lids, and other packaging applications.

Physical Properties:
Impact Strength, Notched (D256)
Injection Molded: 92 J/m (1.7 ft.-lb./in. notch)
Compression Molded: 75 J/m (1.4 ft.-lb../in. notch)
Tensile Strength (D638): 25 MPa (4000 psi)
Elongation (D638): 50%
Elastic Modulus (D638): 1,800 MPa (260,000 psi)
Deflection Temperature (D648)
Under Load, 264 psi (1/4" thick): 88C (190F)
Specific Gravity (D792): 1.04
Rockwell Hardness (D785): L56

HUNTSMAN CHEMICAL CORP.: Polystyrene(Continued):

Type 731
High-Impact High Heat Polystyrene

Type 731 is a high-impact material specifically designed for compatibility with all thermoplastics processes.

Its high gloss, low monomer content, and favorable stress crack resistance are combined with good elevated temperature performance. This material exhibits unusual consistency during processing resulting in products characterized by a high degree of uniformity.

Typical applications include profiles, lamp shades, containers, refrigerator door liners and trays, hot and cold drink cups and lids, cameras, appliances and other packaging applications.

Physical Properties:
- Impact Strength, Notched (D256)
  - Injection Molded: 96 J/m (1.8 ft.-lb./in. notch)
  - Compression Molded: 80 J/m (1.5 ft.-lb./in. notch)
- Tensile Strength (D638): 28 MPa (4000 psi)
- Elongation (D638): 60%
- Elastic Modulus (D638): 1,800 MPa (260,000 psi)
- Deflection Temperature (D648): 88C (190F)
  - Under Load, 264 psi
- Specific Gravity (D792): 1.04
- Rockwell Hardness (D785): L60

HCC - 840
Extra High-Impact Polystyrene

HUNTSMAN Type 840 is an extra high-impact polystyrene. It can easily be processed under normal conditions using all conventional processes.

The smooth surface and deep draw characteristics make it superior for critical thermoforming applications. Typical applications include internal packaging articles and cottage cheese tubs, toys and wire spools.

This resin has proven itself tough enough to compete with costly engineering resins in many applications.

Physical Properties:
- Impact Strength, Notched (D256)
  - Injection Molded: 161 J/m (3.0 ft.-lb./in.)
  - Comprestion Molded: 107 J/m (2.0 ft.-lb./in.)
- Tensile Strength (D638): 25 MPa (3600 psi)
- Elongation @ Break (D638): 50%
- Elastic Modulus (D638): 50%
- Deflection Temperature (D648)
  - Under Load, 264 psi: 88C (190F)
- Specific Gravity (D792): 1.04
- Rockwell Hardness (D785): L50

HUNTSMAN CHEMICAL CORP.: Polystyrene(Continued):

HCC - 702
Modified High Impact Polystyrene

HUNTSMAN Type 702 is a high impact grade polystyrene containing chemicals to increase resistance to ignition in laboratory flame tests. This material has an Underwriters Laboratories flammability classification of 94 V-O at thicknesses of 0.058 in. and greater (File No. E39792C).

Typical applications for Type 702 modified polystyrene include TV cabinets, appliances, calculators, business machine housings, smoke detectors and various electrical appliances. Type 702 modified polystyrene can be structurally foam molded utilizing suitable physical or chemical blowing agents. Type 702 is also available in a renge of suitable colors. Its natural color is a cream white.

Physical Properties:
- Izod Impact Strength, Notched (D256)
  - Injection Molded: 102 J/m (1.9 ft.-lb./in)
  - Compression Molded: 80 J/m (1.5 ft.-lb./in)
- Tensile Strength
  - at Yield (D638): 25 MPa (3600 psi)
  - at Break (D638): 22 MPa (3200 psi)
- Elongation at Break (D638): 40%
- Elastic Modulus (D638): 1660 MPa (240,000 psi)
- Deflection Temperature
  - Under Load, 264 psi (D648): 84C (183F)
- Specific Gravity (D792): 1.17 g./cc
- Flexural Strength (D790): 31 MPa (4500 psi)
- Flexural Modulus (D790): 1930 MPa (280,000 psi)

Type 703
Modified High-Impact Polystyrene

Type 703 is a high-impact grade of polystyrene containing chemicals to increase resistance to ignition in laboratory flame tests. This material has an Underwriters Laboratories flammability classification of 94 V-O at thicknesses of 0.058 in. and greater (File No. E39792C).

Type 703 modified high-impact polystyrene is specifically formulated for optimum performance in hard to fill mold designs. Higher flow length to part thickness ratios are achieved with Type 703 in either injection molding or low pressure structural foam molding processes. Typical applications include a wide variety of electronic enclosures such as photo copy machines, cash registers, CRT data terminals, and printers. The natural color of Type 703 is cream white. It is also available in custom and standard colors.

Physical Properties:
- Melt Flow Rate (D1238(G)): 6.0 g./10 min.
- Vicat Softening Point (D1525): 96C (205F)

HUNTSMAN CHEMICAL CORP.: Polystyrene(Continued):

Type 705
Modified High-Impact Polystyrene

Type 705 is a high-impact grade of polystyrene containing chemicals to increase resistance to ignition in laboratory flame tests. This material has an Underwriters Laboratories flammability classification of 94 V-O at thicknesses of 0.058 in. and greater (File No. E39792M).

Typical applications for Type 705 modified polystyrene include TV cabinets, appliances, calculators, business machine housings, smoke detectors and various electrical applications. Type 705 modified polystyrene can be structurally foam molded utilizing suitable physical or chemical blowing agents. Type 705 is also available in a range of suitable colors. Its natural color is a cream white.

Physical Properties:
- Melt Flow Rate (D1238 (G)): 4.0 g./10 min.
- Vicat Softening Point (D1525): 101C (214F)
- Deflection Temperature at 264 psi (D648): 92C (198F)
- Izod Impact Strength, Notched Injection (D256)
  - Compression Molded: 96 J/m (1.8 ft.-lb./in.)
  - 64 J/m (1.2 ft.-lb./in.)
- Tensile Strength at Yield (D638): 32 MPa (4000 psi)
- Tensile Strength at Break (D638): 26 MPa (3800 psi)
- Elongation at Break (D638): 35%

Type 707
Modified High-Impact Polystyrene

Type 707 is a high-impact grade of polystyrene containing chemicals to increase resistance to ignition in laboratory flame tests. This material has an Underwriters Laboratories flammability classification of 94V-O at thicknesses of 0.058 in. and greater. (File No. 39792C)

Typical applications of Type 707 modified polystyrene include TV cabinets, appliances, calculators, business machine housings, smoke detectors and various electrical applications. Type 707 modified polystyrene can be structurally foam molded utilizing suitable physical or chemical blowing agents. Its natural color is a cream-white.

Physical Properties
- Melt Flow Rate (D1238, (G)): 3.5 g./10 min.
- Vicat Softening Point (D1525): 95C (203F)
- Deflection Temperature at 264 psi (D648): 83 C (181F)
- Izod Impact Strength, Notched Injection (D256)
  - Compression Molded : 170 J/m (3.2 ft.-lb./in.)
  - 106 J/m (2.0 ft. lb./in.)
- Tensile Strength at Yield (D638): 21 MPa (3100 psi)
- Tensile Strength at Break (D638): 20 MPa (2800 psi)
- Elongation at Break (D638): 45%

HUNTSMAN CHEMICAL CORP.: Polystyrene(Continued):

Polystyrene SFP 30
High Flow Ignition Retardent Impact Polystyrene

HUNTSMAN Polystyrene Structural Foam Polymers (SFP) have been designed for structural foam applications in the consumer electronics industry which requires the UL 94 VO classification. HUNTSMAN SFP 30 has higher flow characteristics which allow easier filling of long or complex flowpaths while maintaining excellent physical properties and ease of painting.

Physical Properties
Structural Foam (1/4" Nominal Wall)
Specific Gravity, 73/73F, 23/23C: 0.97 (D792)
Relative Gravity, at 73F, 23C: 0.85 (D792)
Melt Flow, condition G, dg./min. (unfoamed resin): 9.5(D1238)
Tensile at Yield, (0.2 in./min.) psi: 2100
(5 mm/min) MPa: 14
Ultimate Elongation, % (0.2 in./min.) (5 mm/min): 14 (D638)
Tensile Modulus psi (0.05 in./min.): 220,000 (D638)
MPa (1.3 mm/min.): 1517
Flexural Modulus, (0.05 in./min.) psi: 220,000 (D 790)
(1.3 mm/min.): 1551

Type 779
Modified High-Impact V-O Polystyrene
With Improved UV Stability

Type 779 is a high-impact grade of polystyrene containing chemicals to increase resistance to ignition in laboratory flame tests. Unlike conventional V-O resins, this product offers excellent UV stability in light colors. It has been processed at normal conditions on a variety of machines.

Typical applications for Type 779 modified polystyrene include TV cabinets, appliances, calculators, business machine housings, smoke detectors, and various electrical applications. Type 779 modified polystyrene can be structurally foam molded utilizing suitable physical or chemical blowing agents. Type 779 is also available pre-colored to maintain optimum color retention within the total system.

Physical Properties:
Melt Flow Rate (D1238, (G)): 3.0 g./10 min.
Vicat Softening Point (D1525): 102C (215F)
Deflection Temperature at 264 psi (D648): 85C (185F)
Izod Impact Strength, Notched Injection (D256):
Compression Molded: 102 J/m (1.9 ft.-lb./in.)
80 J/m (1.5 ft.-lb./in.)
Tensile Strength at Yield (D638): 28 MPa (4000 psi)
Tensile Strength at Break (D638): 25 MPa (3700 psi)
Elongation at Break (D638): 35%

HUNTSMAN CHEMICAL CORP.: Polystyrenes(Continued):

Type 765
High-Impact Polystyrene
with Improved UV Stability

Type 765 is a high flow, high-impact polystyrene, specifically formulated for extended long-term color stability as demonstrated in laboratory tests; however, as pigments and other additives can affect the absorption of radiation greatly, users should verify suitability for their particular use or purpose. It is ideally suited for low voltage electronic and/or mechanical enclosures intended for interior use in the home, office, or department store. Selection of Type 765 is desirable for multiple component systems since purchases of "add-on" components to an original base system may draw attention to color shifts.

Type 765 is one of a broad family of products supplied by Huntsman Chemical Corp. that are recognized by Underwriters Laboratories.

Property:
- Impact Strength, Notched (D256)
  - Injection Molded: 97 J/m (1.8 ft.-lb./in.)
  - Compression Molded: 70 J/m (1.3 ft.-lb./in.)
- Tensile Strength (D638): 25 MPa (3600 psi)
- Elongation (D638): 60%
- Elastic Modulus (D638): 2070 MPa (300,000 psi)
- Deflection Temperature
  - Under Load, 264 psi (D648): 82C (180F)
- Specific Gravity (D792): 1.04
- Rockwell Hardness (D785): L70
- Vicat Softening Point (D1525): 97C (207F)
- Melt Flow Rate (D1238 G)): 8.5 g./10 min.

HUNTSMAN CHEMICAL CORP.: Polystyrenes(Continued):

HCC - 850
Super High Impact High Heat Polystyrene

HUNTSMAN Type 850 is a high gloss, super high impact polystyrene specifically designed for applications where improved toughness is a prime requirement.

This specialty product can be used for injection molding applications, as well as extruded into sheet or profiles. It has replaced engineering resins in many applications where chemical resistance is not required. The ease of processing combined with unusual resistance has made it a desired choice.

Typical applications include refrigerator door liners, toys, appliances, luggage, tub enclosures, packaging and household products.

Physical Properties:
- Impact Strength, Notched (D256)
  - Injection Molded: 240 J/m (4.5 ft.-lb./in. notch)
  - Compression Molded: 171 J/m (3.2 ft.-lb./in. notch)
- Tensile Strength (D638): 25 MPa (3600 psi)
- Elongation (D638): 50%
- Elastic Modulus (D638): 1590 MPa (230,000 psi)
- Flexural Strength (D790): 35 MPa (5000 psi)
- Flexural Modulus (D790): 1930 MPa (280,000 psi)

Type 851
Super High-Impact Polystyrene

Physical Properties
- Impact Strength @ Normal Molding (D256): 4.0
  Conditions ft.lbs./in. Notch
- Tensile Strength (D638): 3600 psi
- Elongation, % @ Fail (D638): 50
- Tensile Modulus, psi x 10 5 (D638): 2.3
- Deflection Temperature
  Under Load 264 psi, F (D638): 186
- Specific Gravity (D792): 1.04
- Rockwell Hardness (D785): L50
- Vicat Softening Point, F (D1525): 207
- Melt Flow Rate, g./10 min. (D1238 (G)): 4.3

HUNTSMAN CHEMICAL CORP.: Polystyrene(Continued):

Type 870
Extra High Gloss Precolored Impact Polystyrene

HUNTSMAN Type 870 is polymerized under exacting conditions for applications where opaque finished part luster is a principal requirement combined with good impact resistance. Optimum results can be expected only if this product is purchased in its precolored form and processed in conjunction with quality polished molds. Mold surface temperatures need not exceed 120F for most applications. The data listed below is typical for 870 with normal pigment loadings.

Typical applications include toys, appliances, electronic components, packaging, displays and miscellaneous office and household products.

Physical Properties:
(All samples injection molded 1/8" thickness.)
Melt Flow (condition G, g/10 min.): 5.0 (D1238)
Specific Gravity (73/73F): 1.05 (D792)
Vicat Softening Point (F): 212 (D1525)
Heat Distortion (264 psi, F): 185 (D648)
Tensile Strength (psi): 4200 (D638)
Elongation (%): 45 (D638)
Flexural Modulus (psi x 1000): 290 (D790)
Flexural Strength (psi x 1000): 5500 (D790)
Hardness (Rockwell L): 60 (D785)
Izod Impact (ft.-lb./in. of notch): 2.4 (D256)

HUNTSMAN CHEMICAL CORP.: Polystyrene(Continued):

Type 879
Light Stable Polystyrene

Type 879 light stable polystyrene is specially formulated for fluorescent lighting louvers and panels. This material offers resistance to yellowing, dimensional stability, light weight and excellent processability at a low cost.

Type 879 parts can be expected to perform satisfactorily for more than 10 years under "normal conditions." These conditions are defined as exposure to standard fluorescent illumination for no more than 50 hours per week with a lamp-to-plastic spacing of at least two inches.

Molding and Extrusion:

Type 879 is fabricated and finished in the same manner as ordinary polystyrene.

Type 879 is especially suited for intricate forming and embossing during extrusion and offers the superior strength properties and high heat resistance required for larger shapes.

Type 879 is supplied without lubricants for extrusion. Here, too, conditions used are the same as those needed to extrude general purpose or high heat polystyrenes.

Typical Properties:

Mechanical Properties:
- Tensile Strength at Break (D638): 7500 PSI (52 MPa)
- Elongation at Break (D638): 2.5%
- Tensile Modulus (D638): 450,000 PSI (3100 MPa)
- Flexural Strength (D790): 12,500 PSI (86 MPa)
- Izod Impact Strength, Notched (D256): 0.4 ft.-lb./in.(21 J/m)

Thermal Properties:
- Deflection Temperature at 264 psi (D648): 195F (91C)
- Vicat Softening Point (D1525): 223F (106C)

Physical Properties
- Specific Gravity (D792): 1.05
- Rockwell Hardness (D785): M75
- Moisture Absorption (D570): 0.04%
- Mold Shrinkage (D955): 0.004 in./in. (0.004 mm/mm)

Electrical Properties
- Dielectric Constant, 60-10 6 Hz (D150): 2.6
- Dielectric Strength, Short Time (D149): 575 Volts/mil
22.6 KV/mm
- Dissipation Factor, 60-10 6 Hz (D150): 0.0003
- Volume Resistivity (D257): >10 14 ohm-cm

Optical Properties (0.10-in. thick specimens)
- Luminous Transmittance (CIE-Source C) (D1003): 0.90
- Yellowness Index (Transmittance) (D1925): 1.0

ICI AMERICAS, INC.: MELINEX Polyester Film:

MELINEX Film No.:

All Melinex Films for Membrane Switches:

Membrane Switch Use: All Layers

General Properties:
Polyester Film - The most preferred film for all kinds of membrane switch devices because of superior flexing strength, excellent dimensional stability, resistance to environmental attack, and highest printing qualities. Users choose this tough facia and substrate material because it ensures longer life, snap-back and is less likely to tear, chip or delaminate in handling, printing or stamping. Polyester is not affected by most solvents, hamburger grease, crayons and household or industrial cleaners.

MELINEX Polyester Film for Membrane Switches:
has been specially designed for rigors of membrane switch production and end user activity. You'll find better roll formation with highest quality gauge control and better overall dimensional stability for best consistency roll to roll, sheet to sheet. MELINEX films range from very glossy to pearlescent to textured matte finishes.

MX 505:

Membrane Switch Use: Fascia or circuitry layer.
General Properties:
This film offers exceptional flex and snap-back qualities. It is pretreated both sides for excellent ink adhesion.

MX 516:

Membrane Switch Use: Fascia or circuitry layer
General Properties:
This film offers exceptional flex and snap-back qualities. It is pretreated both sides for excellent ink adhesion.

MX 226:

Membrane Switch Use: Spacer Sheet
General Properties:
An opaque film which is the thickest of the touch membrane layers. MX 226 is exceptionally strong in the Z direction.

MX 054:

Membrane Switch Use: Fascia or circuitry layer
General Properties:
The lowest shrinkage, best balanced MELINEX film for membrane switch applications. Highest dimensional stability with excellent snapback.

ICI AMERICAS, INC.: PROPAFILM OPP Film:

PROPAFILM oriented polypropylene film from ICI Americas, offers a versatile range of packaging films with exceptional combinations of properties.

Moisture Barrier
Strength and Stability
Ease of Conversion
Efficient Packaging
Sales Appeal

APPLICATIONS:

* Coffee (fractional packs)
* Cheese (natural)
* Snacks (oil fried and dry roast products)
* Confectionary (collation, bags)
* Cookies (overwraps, bags, bag liners)
* Heavy weight products (single web and laminate)
* Cigarette and cigar overwrapping
* Bakery overwrap
* Boxes and collation overwrapping (toiletries)
* Fruit and vegetable packaging
* Industrial thick films

The PROPAFILM Product Range

PROPAFILM C:
PVdC-Coated, high barrier, heat sealable.

PROPAFILM CR:
High yield, PVdC-coated, barrier, heat sealable.

PROPAFILM CB:
PVdC-coated, super barrier performance.

PROPAFILM MG:
Coextruded, heat sealable, general purpose.

PROPAFILM MT:
Coextruded, heat sealable for high speed overwrapping.

PROPAFILM MV:
Coextruded, heat sealable, metallizable base film for adhesion and extrusion lamination.

PROPAFILM OM:
OPP film for converter applied coatings.

PROPAPLY ML:
Coextruded thick films for heavy duty applications.

INDUSTRIAL DIELECTRICS, INC.: DIELECTRITE 44-DR-UL BMC:

DESCRIPTION:

A flame retardant diallyl phthalate monomer molding compound with moderate strength, excellent moldability, and good electrical properties. U. L. recognized E53587.

TECHNICAL DATA

Properties	ASTM Test Methods	Values
Impact Strength, Izod, Ft. Lbs./ In.	D256	2.5-3
Flexural Strength, PSI	D790	14,000
Tensile Strength, PSI	D638	4,500
Compressive Strength, PSI	D695	21,000
Water Absorption, % in 24 Hours	D570	0.17
Heat Distortion Temp., @ 264 PSI, F	D648	>500
Specific Gravity	D792	2.00
Mold Shrinkage, in./in.	D955	.002-.004
Dielectric Strength, S.T., Perp., VPM	D149	340
Bulk Factor App.	D1895	1
Hardness, Barcol	D2583	40-50
Flame Resistance, 1/8"	UL94	94V-O
Arc Resistance, Seconds	D495	187

INDUSTRIAL DIELECTRICS, INC.: DIELECTRITE 44-O:

DESCRIPTION:

A general purpose molding compound with moderate strength and low cost. Not U. L. recognized.

TECHNICAL DATA

Properties	ASTM Test Methods	Values
Impact Strength, Izod, Ft.Lbs./In.	D256	3-4
Flexural Strength, PSI	D790	13-17,000
Flexural Modulus, PSI	D790	1.9 x 10 6
Tensile Strength, PSI	D638	5-7,000
Compressive Strength, PSI	D695	21,000
Water Absorption, % in 24 Hours	D570	0.15
Heat Distortion Temp., @ 264 PSI, F	D648	>500
Specific Gravity	D792	2.10
Mold Shrinkage, in./in.	D955	.001-.004
Arc Resistance, Seconds	D495	185
Dielectric Strength, S.T., Perp., VPM	D149	330
Dielectric Strength, S.S., Perp., VPM	D149	270
Bulk Factor App.	D1895	1
Hardness, Barcol	D2583	30-40
Flame Resistance, 1/16"	UL94	94-HB
1/8"	UL94	94-HB
1/4"	UL94	94V-O

INDUSTRIAL DIELECTRICS, INC.: DIELECTRITE 44-1 BMC:

DESCRIPTION:

A multi-purpose molding compound with moderate strength and low cost. U. L. recognized - E 53587. Ford approved ESF-M3D-56A.

TECHNICAL DATA

Properties	ASTM Test Methods	Values
Impact Strength, Izod, Ft. Lbs./ In.	D256	3-4
Flexural Strength, PSI	D790	15-18,000
Flexural Modulus, PSI	D790	1.9 x 10 6
Tensile Strength, PSI	D638	5-7,000
Compressive Strength, PSI	D695	22,000
Water Absorption, % in 24 Hours	D570	0.15
Heat Distortion Temp., @ 264 PSI, F	D648	>500
Specific Gravity	D792	1.95
Mold Shrinkage, in./in.	D955	.001-.003
Arc Resistance, Seconds	D495	180
Dielectric Strength, S.T., Perp., VPM	D149	330
Dielectric Strength, S.T., Perp., VPM	D149	270
Bulk Factor App.	D1895	1
Hardness, Barcol	D2583	30-40
Flame Resistance, 1/16"	UL94	94-HB
1/8"	UL94	94-HB
1/4"	UL94	94V-O

INDUSTRIAL DIELECTRICS, INC.: DIELECTRITE 44-10 BMC:

DESCRIPTION:

A chemical and weather resistant molding compound with moderate strength and good corrosion resistance, good electrical properties including flame and track resistance. U.L. recognized - E53587.

TECHNICAL DATA

Properties	ASTM Test Methods	Values
Impact Strength, Izod, Ft. Lbs./In.	D256	4-5
Flexural Strength, PSI	D790	15-17,000
Tensile Strength, PSI	D638	6,000
Compressive Strength, PSI	D695	21,000
Water Absorption, % in 24 Hours	D570	0.15
Heat Distortion Temp., @ 264 PSI, F	D648	>500
Specific Gravity	D792	1.96
Mold Shrinkage, in./in.	D955	.001-.003
Arc Resistance, Seconds	D495	180+
Track Resistance, Minutes	D2303	600
Dielectric Strength, S.T., Perp., VPM	D149	300
Bulk Factor App.	D1895	2
Hardness, Barcol	D2583	40-50
Flame Resistance, 1/16"	UL94	94-HB
1/8"	UL94	94V-O and 94-5V
1/4"	UL94	94V-O and 94-5V

INDUSTRIAL DIELECTRICS, INC.: DIELECTRITE 44-10 HG:

DESCRIPTION:

A chemical and weather resistant molding compound with higher strength than 44-10 and good corrosion resistance, good electrical properties including flame and track resistance. U.L. recognized - E53587.

TECHNICAL DATA

Properties	ASTM Test Methods	Values
Impact Strength, Izod, Ft. Lbs./In.	D256	6-8
Flexural Strength, PSI	D790	18-22,000
Tensile Strength, PSI	D638	7-9,000
Compressive Strength	D695	22-24,000
Water Absorption, % in 24 hours	D570	0.15
Heat Distortion Temp., @ 264 PSI F	D648	>500
Specific Gravity	D792	1.97
Mold Shrinkage, in./in.	D955	.001-.003
Arc Resistance, Seconds	D495	180+
Dielectric Strength, S.T., Perp., VPM	D149	340
Dielectric Strength, S.S., Perp., VPM	D149	304
Bulk Factor, App.	D1895	2
Hardness, Barcol	D2583	40-50
Flame Resistance, 1/8"	UL94	94V-O and 94-5V

INDUSTRIAL DIELECTRICS, INC.: DIELECTRITE 44-10 HG-C*:

DESCRIPTION:

A special conductive BMC material for EMI shielding and static dissipation with excellent strength and appearance.

* NOTE - This is a special electrically conductive grade and should not be confused with other insulating materials.

TECHNICAL DATA

Properties	ASTM Test Methods	Values
Impact Strength, Izod, Ft. Lbs./In.	D256	6-8
Flexural Strength, PSI	D790	18-22,000
Tensile Strength, PSI	D638	7-9,000
Compressive Strength	D695	22-24,000
Water Absorption, % in 24 Hours	D570	0.15
Heat Distortion Temp., @ 264 PSI, F	D648	>500
Specific Gravity	D792	1.97
Bulk Factor App.	D1895	2
Hardness, Barcol	D2583	40-50
Flame Resistance, 1/8"	UL94	94V-O and 94-5V
Surface Resistivity (Per Requirements)		1x10 2 - 1x10 5 ohms/sq*
Mold Shrinkage, in./in.	D955	.001-.003
Shielding Performance - db		10-40

* Shielding levels adjusted per individual requirements.

INDUSTRIAL DIELECTRICS, INC.: DIELECTRITE 44-22 BMC:

DESCRIPTION:

An impact modified high strength compound with excellent mold flow characteristics for injection molding. Produces parts with "Zero" mold shrinkage and excellent surface profile. Recommended for automotive external painted parts and related applications.

TECHNICAL DATA

Properties	ASTM Test Methods	Values
Izod Impact Strength Notched Ft-Lbs./In.	D256	10-12
Flexural Strength, PSI	D790	18-20,000
Flexural Modulus, PSI	D790	1.8 x 10 6
Tensile Strength, PSI	D638	6,000
Compressive Strength, PSI	D695	20,000
Barcol Hardness	D2583	45
Specific Gravity	D792	1.80
Heat Distortion Temp. F @ 264 PSI	D648	>500
Water Absorption % Gain 24 Hrs.	D570	.20
Mold Shrinkage in./in.	D955	.0000-.0005
Bulk Factor	D1895	3
Cure Time @ 300F .100" Thick (Sec)		60

INDUSTRIAL DIELECTRICS, INC.: DIELECTRITE 46-3 BMC:

DESCRIPTION:

An encapsulating molding compound for molding around delicate components. Very soft flow and low shrinkage. U.L. recognized - E53587.

TECHNICAL DATA

Properties	ASTM Test Methods	Values
Impact Strength, Izod, Ft. Lbs./In.	D256	1.5-2.0
Flexural Strength, PSI	D790	8-10,000
Tensile Strength, PSI	D638	3-4,000
Compressive Strength, PSI	D695	13,000
Water Absorption, % in 24 Hours	D570	0.15
Heat Distortion Temp., @ 264 PSI, F	D648	>400
Specific Gravity	D792	1.8
Mold Shrinkage, in./in.	D955	.0005-.002
Arc Resistance, Seconds	D495	180+
Dielectric Strength, S.T., Perp., VPM	D149	300
Bulk Factor App.	D1895	1
Hardness, Barcol	D2583	20-30
Flame Resistance, 1/16"	UL94	94V-O
1/8"	UL94	94V-O
1/4"	UL94	94V-O

INDUSTRIAL DIELECTRICS, INC.: DIELECTRITE 46-09 BMC:

DESCRIPTION:

Abrasion, arc & track resistant. Soft flow with excellent moldability. Underwriters' Laboratories recognized file number E53587B.

TECHNICAL DATA

Properties	ASTM Test Methods	Values
Impact Strength, Izod, Ft. Lbs./In.	ASTM D256	2.7
Flexural Strength, PSI	ASTM D790	10,400
Tensile Strength, PSI	ASTM D638	3,000
Compressive Strength, PSI	ASTM D695	16,000
Water Absorption, % in 24 Hours	ASTM D570	0.15
Heat Distortion Temp., @ 264 PSI F	ASTM D648	>400
Specific Gravity	ASTM D792	1.7
Mold Shrinkage, in./in.	ASTM D955	.001-.003
Arc Resistance, Seconds	ASTM D495	180+
Track Resistance (Minutes)	ASTM D2303	>600
Dielectric Strength, S.T. Perp., VPM	ASTM D149	360
Dielectric Strength, S.S. Perp., VPM	ASTM D149	310
Bulk Factor App.		1
Hardness	Barcol	10-20
Flame Resistance, 1/16"	UL 94	94V-O

INDUSTRIAL DIELECTRICS, INC.: DIELECTRITE 46-12 BMC:

DESCRIPTION:

A flame retardant molding compound with high strength and excellent electrical properties. U.L. recognized - E53587.

TECHNICAL DATA

Properties	ASTM Test Methods	Values
Impact Strength, Izod, Ft. Lbs./In.	D256	8-10
Flexural Strength, PSI	D790	19-22,000
Tensile Strength, PSI	D638	7-9,000
Compressive Strength, PSI	D695	23,000
Water Absorption, % in 24 Hours	D570	0.15
Heat Distortion Temp., @ 264 PSI, F	D648	>500
Specific Gravity	D792	1.78
Mold Shrinkage, in./in.	D955	.001-.003
Arc Resistance, Seconds	D495	180+
Track Resistance, Minutes	D2303	>600
Dielectric Strength, S.T., Perp. VPM	D149	350
Dielectric Strength, S.S., Perp. VPM	D149	300
Bulk Factor App.	D1895	2-3
Hardness, Barcol	D2583	30-50
Flame Resistance, 1/16" & Up	UL94	94V-O

INDUSTRIAL DIELECTRICS, INC.: DIELECTRITE 46-14:

DESCRIPTION:

Highest strength crack resistant molding compound with good mold flow and electrical properties.

TECHNICAL DATA

Properties	ASTM Test Methods	Values
Impact Strength, Izod, Ft. Lbs./In.	D256	10-12
Flexural Strength, PSI	D790	24-26,000
Tensile Strength, PSI	D638	9-10,000
Compressive Strength, PSI	D695	23-24,000
Water Absorption, % in 24 Hours	D570	0.15
Heat Distortion Temp., @ 264 PSI, F	D648	>500
Specific Gravity	D792	1.82
Mold Shrinkage, in./in.	D955	.001-.003
Arc Resistance, Seconds	D495	181
Dielectric Strength, S.T., Perp., VPM	D149	335
Dielectric Strength, S.S., Perp., VPM	D149	320
Bulk Factor App.	D1895	3
Hardness, Barcol	D2583	35-45
Flame Resistance, 1/8"	UL94	94V-O

INDUSTRIAL DIELECTRICS, INC.: DIELECTRITE 46-16 BMC:

DESCRIPTION:

A premium molding compound with excellent resistance to smoke generation, flame resistance, impact strength, weatherability, and corrosion resistance.

TECHNICAL DATA

Properties	ASTM Test Methods	Values
Flexural Strength, PSI	D790	18,000
Impact Strength, Ft. Lbs./In.	D256	14
Tensile Strength, PSI	D638	8,000
Compressive Strength, PSI	D695	26,000
Liquid Absorption, % Gain	D570	.15
Heat Distortion Temp., C @ 264 PSI	D648	266
Arc Resistance, Seconds	D495	189
Dissipation Factor	D150	.0224
Dielectric Content	D150	4.72
Dielectric Strength, Volts/Mil	D149	304
Inclined Plane Tracking Resistance, Minutes @ 2,500 Volts	D2303	Did not track @ 1,000
Flame Propagation Index, Is	E162	4
Flame Resistance	D229	
Ignition Time		147
Burn Time		43
Specific Gravity	D792	1.80
Smoke Generation	E662	
Flaming Ds @ 90 seconds		0
240 seconds		5
Non-Flaming Ds @ 90 seconds		0
240 seconds		3
Inclined Plane Tracking Resistance After 1,000 Hrs. in Weather-O-Meter, Minutes @ 2,500 Volts	D2303 G-23, D1499	Did not track @ 600+
Arc Resistance After 1000 Hrs. In Weather-O-Meter, Seconds	D495 G23, D1499	187
Dielectric Strength After 1,000 Hrs. in Weather-O-Meter, Volts/Mil	D149 G23, D1499	290

INDUSTRIAL DIELECTRICS, INC.: DIELECTRITE 48-50 BMC:

DESCRIPTION:

A flame retardant molding compound with moderate strength and excellent electrical properties. U.L. recognized - E53587.

TECHNICAL DATA

Properties	ASTM Test Methods	Values
Impact Strength, Izod, Ft. Lbs./In.	D256	4-5
Flexural Strength, PSI	D790	14-16,000
Flexural Modulus, PSI	D790	2.0 x 10 6
Tensile Strength, PSI	D638	6,000
Compressive Strength, PSI	D695	24,000
Water Absorption, % in 24 Hours	D570	0.11
Heat Distortion Temp., @ 264 PSI F	D648	>500
Specific Gravity	D792	1.93
Mold Shrinkage, in./in.	D955	.001-.003
Arc Resistance, Seconds	D495	190
Track Resistance, Minutes	D2303	900
Dielectric Strength, S.T., Perp., VPM	D149	385
Dielectric Strength, S.S., Perp., VPM	D149	320
Dielectric Constant @ 60 Hertz	D150	5.65
Dielectric Constant @ 1 MH	D150	5.22
Dissipation Factor @ 60 Hertz	D150	0.013
Dissipation Factor @ 1 MH	D150	0.008
Loss Index @ 60 Hertz	D150	0.073
Loss Index @ 1 MH	D150	0.040
Bulk Factor App.	D1895	1.5
Hardness, Barcol	D2583	40-50
Flame Resistance, 1/32" & Up	UL94	94V-O

INDUSTRIAL DIELECTRICS, INC.: DIELECTRITE 48-56 BMC:

DESCRIPTION:

A higher strength flame retardant molding compound than 48-50 with excellent electrical properties. U. L. recognized-E53587.

TECHNICAL DATA

Properties	ASTM Test Methods	Values
Impact Strength, Izod, Ft. Lbs./In.	D256	5-6
Flexural Strength, PSI	D790	16-18,000
Tensile Strength, PSI	D638	6,000
Compressive Strength, PSI	D695	22,000
Water Absorption, % in 24 Hours	D570	0.12
Heat Distortion Temp., @ 264 PSI, F	D648	>500
Specific Gravity	D792	1.90
Mold Shrinkage, in./in.	D955	.001-.003
Arc Resistance, Seconds	D495	185+
Track Resistance, Minutes	D2303	800
Dielectric Strength, S.T., Perp., VPM	D149	360
Dielectric Strength, S.S., Perp., VPM	D149	315
Bulk Factor App.	D1895	1.5
Hardness, Barcol	D2583	40-50
Flame Resistance, 1/16" & Up	UL94	94V-O

INDUSTRIAL DIELECTRICS, INC.: INSULSTRUC P-1001-FR BMC:

DESCRIPTION:

High impact, flame retardant, molding compound with good electrical properties. U. L. recognized E53587.

TECHNICAL DATA

Properties	ASTM Test Methods	Values
Impact Strength, Izod, Lbs./In.	D256	5-6
Flexural Strength, PSI	D790	18-20,000
Flexural Modulus, PSI	D790	1.8×10^6
Tensile Strength, PSI	D638	6-7,000
Compressive Strength, PSI	D695	22,000
Water Absorption, % in 24 Hours	D570	0.10
Heat Distortion Temp. @ 264 PSI	D648	>550
Specific Gravity	D792	2.18
Mold Shrinkage, In./In.	D955	0.002-5
Arc Resistance, Seconds	D495	180
Dielectric Strength, S.T., Perp., VPM	D149	320
Dielectric Strength, S.S., Perp., VPM	D149	300
Bulk Factor App.	D1895	2
Hardness, Barcol	D2583	65-70
Dielectric Constant @ 60 Hertz	D150	5.95
Dielectric Constant @ 1 MH	D150	4.95
Power Factor @ 60 Hertz	D150	0.040
Power Factor @ 1 MH	D150	0.016
Volume Resistivity Ohm/cm	D257	1.6×10^{14}
Flame Resistance, 1/8"	UL94	94V-O

INDUSTRIAL DIELECTRICS, INC.: INSULSTRUC: P-5000-GP:

GLASS POLYESTER MOLDING COMPOUND

DESCRIPTION:

General purpose, flame retardant, crack resistant, fast curing. Underwriters' Laboratories recognized file number E53587D.

TECHNICAL DATA

Properties	Test Method	Values
Flexural Strength PSI	ASTM D790	13-15,000
Tensile Strength PSI	ASTM D638	5- 6,000
Compressive Strength FW PSI	ASTM D695	22,000
Izod Impact Strength Ft. Lbs./In.	ASTM D256	3-4
Water Absorption %	ASTM D570	.10
Heat Distortion Temp. @ 264 PSI - F	ASTM D648	over 500
Hardness	Barcol	35-40
Arc Resistance Secs.	ASTM D495	184
Dielectrite Strength Perp. St VPM	ASTM D149	300
Power Factor @ 1 MHZ	ASTM D150	.008
Dielectric Constant @ 1 MHZ	ASTM D150	6.25
Specific Gravity	ASTM D792	2.03
Bulk Factor		2
Mold Shrinkage in./in.	ASTM D955	.001-3
Flame Resistance @ 1/4"	UL 94	94 V1
Flame Resistance @ 1/16"	UL 94	94 HB

INDUSTRIAL DIELECTRICS, INC.: INSULSTRUC P-5003-FR:

DESCRIPTION:

Medium strength, superior flame retardance, suitable for all molding techniques. U. L. recognized E53587.

TECHNICAL DATA

PROPERTIES	ASTM Test Methods	Values
Impact Strength, Izod, Lbs./In.	D256	4-5
Flexural Strength, PSI	D790	14-16,000
Flexural Modulus, PSI	D790	2.0 x 10 6
Tensile Strength, PSI	D638	6,000
Compressive Strength, PSI	D695	24,000
Water Absorption, % in 24 Hours	D570	0.11
Heat Distortion Temp. @ 264 PSI	D648	>500
Specific Gravity	D792	1.93
Mold Shrinkage, in./in.	D955	.001-3
Arc Resistance, Seconds	D495	190
Dielectric Strength, S.T., Perp., VPM	D149	350
Dielectric Constant @ 1 MH	D150	5.22
Power Factor #1 MH	D150	0.008
Bulk Factor App.	D1895	1.5
Track Resistance, Minutes	D2303	900
Hardness, Barcol	D2583	40-50
Flame Resistance, 1/16" & Up	UL94	94V-O

INDUSTRIAL DIELECTRICS, INC.: POLY-DAPS:

Product Name:

POLY-DAP 6120:

POLY-DAP 6120 is a short glass fiber filled, general purpose diallyl phthalate (ortho) type molding compound, which is supplied in free flowing granular form which can be molded in compression, transfer or injection equipment. It is readily preformed and is available in standard black, red, green, blue, and gray colors. Other colors are available upon request.

OUTSTANDING FEATURES:

High dielectric strength, low dissipation factor, low moisture absorption, and fast cure.

APPLICATIONS:

Switches, terminals, insulators, cups, cases, etc., requiring strength. Encapsulation of electrical and electronic components.

APPLICABLE SPECIFICATIONS:

POLY-DAP 6120 is certifiable to...MIL-M-14G Type SDG

POLY-DAP 6130:

POLY-DAP 6130 is a long glass fiber filled, general purpose diallyl phthalate (ortho) molding compound. POLY-DAP 6130 is available in a wide range of colors and is supplied in nodular form suitable for compression or transfer molding. POLY-DAP 6130 can also be preformed for ease in handling.

OUTSTANDING FEATURES:

POLY-DAP 6130 is particularly suited to applications requiring resistance to elevated temperature and high humidity. High impact strength, excellent compressive and flexural strength, and very good electrical properties are additional features of this compound.

APPLICATIONS:

Structural parts, terminals, housings, and insulators requiring high strength.

APPLICABLE SPECIFICATIONS:

POLY-DAP 6130 is certifiable...MIL-M-14G Type GDI-30. This product can be modified to meet other specific requirements but is not then certified to the MIL specification.

INDUSTRIAL DIELECTRICS, INC.: POLY-DAPS(Continued):

Product Name:

POLY-DAP 6130F:

POLY-DAP 6130F is a long glass fiber filled, flame retardant, general purpose diallyl phthalate (ortho) molding compound. POLY-DAP 6130F as available in a wide range of colors and is supplied in a nodular form suitable for compression or transfer molding. POLY-DAP 6130F can laso be preformed for ease in handling.

OUTSTANDING FEATURES:

POLY-DAP 6130F is particularly suited to applications requiring resistance to elevated temperatures and high humidity. High impact strength, excellent compressive and flexural strength, and very good electrical properties are additional features of this compound.

APPLICATIONS:

Structural parts, housings, terminals, and insulators requiring strength.

APPLICABLE SPECIFICATIONS:

POLY-DAP 6130F is certifiable to...MIL-M-14G, types GDI-30 and GDI-30F. This product can be modified to meet other specific requirements but is not then certified to the MIL specification.

INDUSTRIAL DIELECTRICS, INC.: POLY-DAPS(Continued):

Product Name:

POLY-DAP 6160:

POLY-DAP 6160 is a mineral filled, special purpose diallyl phthalate (ortho) molding compound which is supplied in a free flowing granular form and can be easily molded in compression, injection, or transfer equipment. It is readily preformed and is available in standard black, blue and gray colors. Other colors are available upon request.

OUTSTANDING FEATURES:

POLY-DAP 6160 exceeds all mineral compounds available today in mechanical and electrical properties. POLY-DAP 6160 has an extremely low water absorption, high dielectric strength, and low dissipation factor along with high tensile strength, exceptional arc resistance and excellent resistance to high temperatures.

APPLICATIONS:

Switches, terminals, insulators, housings, cups and cases where exposure to high humidity is encountered. Encapsulation of electrical and electronic components.

APPLICABLE SPECIFICATIONS:

POLY-DAP 6160 is certifiable to MIL-M-14G Type MDG.

This product can be modified to meet other specific requirements but is then not certified to the MIL specification.

INDUSTRIAL DIELECTRICS, INC.: POLY-DAPs(Continued):

Product Name:

POLY-DAP 6220:

POLY-DAP 6220 is a short glass fiber filled, premium diallyl iso-phthalate (meta) molding compound, which exhibits excellent resistance to high service temperatures. POLY-DAP 6220 is easily molded in injection, compression, and transfer equipment and is available in standard black, blue, green and red colors. Other colors are available upon request.

OUTSTANDING FEATURES:

POLY-DAP 6220 has good impact strength, excellent electrical properties and excellent high heat service properties. POLY-DAP 6220 has the highest arc resistance of any glass filled allylic compound offered.

APPLICATIONS:

Switches, housings, terminals, cups, cases, and insulators requiring high impact strength.

APPLICABLE SPECIFICATIONS:

POLY-DAP 6220 is certifiable to MIL-M-14G Type SDG.

This product can be modified to meet other specific requirements but is not then certified to the MIL specification.

MOLDING INFORMATION:	TYPICAL PROPERTIES
Bulk factor	2.3
Apparent density, gm/cc	0.78
Specific gravity	1.79
Mold shrinkage (bar), in/in	0.002-0.004
Available plasticity	stiff, med., soft
Molding temperatures F	300-350
Molding pressure, psi	
compression(first inch)	500-2000
transfer(medium size gate, first inch)	1000-8000

INDUSTRIAL DIELECTRICS, INC.: POLY-DAP 6230:

POLY-DAP 6230 is a long glass fiber filled, high heat resistant, diallyl iso-phthalate (meta) molding compound. POLY-DAP 6230 is available in a wide range of colors and is available in nodular and preformed forms suitable for compression or transfer molding.

OUTSTANDING FEATURES:

POLY-DAP 6230 exhibits excellent mechanical and electrical properties. Its high dielectric strength and low dielectric constant are strong recommendations for its use. High heat resistance and excellent dimensional stability are made possible by the special properties of its base polymer.

APPLICATIONS:

Switches, terminals, housings, structural parts, and insulators requiring high strength.

APPLICABLE SPECIFICATIONS:

POLY-DAP 6230 is certifiable to MIL-M-14G Type GDI-30, and can be modified to meet other specific requirements but is not then certified to the MIL specification.

MOLDING INFORMATION:	TYPICAL PROPERTIES
Bulk factor	6-7
Apparent density, gm/cc	0.26
Specific gravity	1.72
Mold shrinkage (bar), in/in	0.002-0.004
Available plasticity	stiff, med., soft
Molding temperatures F	300-330
Molding pressure, psi	
compression(first inch)	500-2000
transfer(medium size gate, first inch)	1000-8000

INDUSTRIAL DIELECTRICS, INC.: POLY-DAP 6230F:

POLY-DAP 6230F is a long glass fiber filled, flame resistant, high heat resistant diallyl iso-phthalate (meta) molding compound. POLYDAP 6230F is available in a wide range of colors and is available in nodular and preformed forms suitable for compression or transfer molding.

OUTSTANDING FEATURES:

Parts molded from POLY-DAP 6230F exhibit excellent mechanical and electrical properties. Its high dielectric strength and low dielectric constant are strong recommendations for its use. The special properties of its base polymer make possible the high heat resistance in this compound.

APPLICATIONS:

Switches, terminals, housings, structural parts, and insulators requiring high strength.

APPLICABLE SPECIFICATIONS:

POLY-DAP 6230F is certifiable to...MIL-M-14G Types GDI-30 and GDI-30F. POLY-DAP 6230F can be modified to meet other specific requirements but is not then certified to the MIL specification.

MOLDING INFORMATION:	TYPICAL PROPERTIES
Bulk factor	6-7
Apparent density, gm/cc	0.26
Specific gravity	1.72
Mold shrinkage (bar),in/in	0.002-0.004
Available plasticity	stiff, med., soft
Molding temperatures F	300-330
Molding pressure, psi	
compression(first inch)	500-2000
transfer(medium size gate, first inch)	1000-8000

INDUSTRIAL DIELECTRICS, INC.: E-7052 Molding Compound:

E-7052 is a mineral filled, high spiral flow epoxy molding compound available in granular form. It is available in a standard black color.

OUTSTANDING FEATURES:

A high spiral flow permits very low pressure molding with E-7052 so that very delicate electronic components can be encapsulated with no damage or movement. E-7052 is also available in stiff and medium flows for use in standard compression and transfer equipment.

Molding Information	Typical Values	Test Method
Available Plasticity	soft, medium, stiff	Mesa Spiral
Bulk Factor	1.5-2.5	ASTM D-1895-65T
Specific Gravity	1.84	ASTM D-792-64T
Mold shrinkage(bar), in/in	.004-.006	ASTM D-955-51
Form	granular	
Molding Temperatures, F	280-360	
Molding pressures, psi		
compression	50-500	
transfer	50-2000	

INDUSTRIAL DIELECTRICS, INC.: E-7074 Molding Compound:

E-7074 is a mineral and glass fiber reinforced high temperature epoxy molding compound available in a granular form which can be molded in compression or transfer equipment. Standard black and brown colors are available.

OUTSTANDING FEATURES:

E-7074 moldings show a weight loss less than 3% after 100 hours at 500F. The high heat distortion and good physical properties make it excellent for high temperature applications.

Molding Information	Typical Values	Test Method
Available plasticity	6-20 (max)	Mesa Spiral
Bulk factor	1.5-2.5	ASTM D-1895-65T
Specific gravity	1.9	ASTM D-792-64T
Mold shrinkage(bar), in/in	.003-.005	ASTM D-955-51
Mold temperatures, F	270-360	
Molding pressures, psi		
compression	300-1000	
transfer	1000-4000	

Mechanical Properties:

Izod impact, ft lb/in of notch	0.37-0.42	FTMS 406-1071
Tensile Strength, psix10 3	6-9	FTMS 406-1012
Flexural strength, psi x 10 3	10-16	FTMS 406-1031
Compressive strength, psi x 10 3	30-35	FTMS 406-1021
Hardness(Barcol)	78	USP PTM-25

INDUSTRIAL DIELECTRICS, INC.: EM-7032 Epoxy Molding Compound:

EM-7032 is a long glass fiber reinforced, epoxy molding compound. EM-7032 is available in a standard amber natural color in flake form.

OUTSTANDING FEATURES:

EM-7302 compression molds easily to provide high strengths in molded parts.

Molding Information	Typical Values	Test Method
Specific gravity, gm/cc	1.84	ASTM D-792-64T
Mold shrinkage(bar), in/in	0.00019	ASTM D-995-51
Molding temperatures, F	290-330F	
Comp. molding pressures, psi	2500-5000	
Cure rate(1/4" thick sec. at 300F) min.	15-20	
Resin content(DRC), percent	37+-3%	USP/PTM-11E

Mechanical Properties:		
Barcol Hardness	60-74	USP/PTM-25
Izod impact strength, ft lbs./in. notch	25-34	FTMS 406-1071
Flexural strength, psi x 10 3 @ R.T.	50-70	FTMS 406-1031
Flexural modulus, psi x 10 6 @ R.T.	2.8-3.6	FTMS 406-1031
Tensile strength psi x 10 3 @ R.T.	18-27	FTMS 406-1012
Compressive strength, psi x 10 3 @ R.T.	30-38	FTMS 406-1021

INDUSTRIAL DIELECTRICS, INC.: M-4135 Molding Compound:

M-4135 is a long glass fiber reinforced melamine formaldehyde molding compound with good compression and transfer molding characteristics. M-4135 is available in nodular and flake form in standard natural (off-white) color.

OUTSTANDING FEATURES:

Parts molded from M-4135 have high impact strength, high arc resistance and are flame resistant. Other electrical properties are also exceptionally good. M-4135 is also certifiable to MIL-M-14G, MMI-30.

APPLICATIONS:

Parts molded from M-4135 are mainly for the electrical field, including heavy duty switch gear, terminal strips and stand-offs, sockets and coil forms.

Molding Information	Typical Values	Test Method
Specific gravity	1.94-2.00	ASTM D-792
Bulk factor	5-7	
Preforming	Hand	
Available plasticity	medium, stiff	
Molding temperature F	290-320	
Molding pressures psi		
compression	3000+	
transfer	5000+	
Mold shrinkage in./in.	0.002-0.004	ASTM D-955
Cure rate, 1/8" thick section at 300F, minutes	1.5-2.0	

MECHANICAL PROPERTIES:

Izod impact strength, ft lbs./in. notch	3.5-6.0	FTMS 406-1071
Compressive strength, psi x 10 3	20-25	FTMS 406-1021
Flexural strength, psi x 10 3	8.5-15.0	FTMS 406-1031
Tensile strength, psi x 10 3	5.5-7.0	FTMS 406-1012

INDUSTRIAL DIELECTRICS, INC.: 1D30 HIF Molding Compound:

ID30 HIF is a medium glass fiber filled, flame resistant, diallyl phthalate molding compound developed especially for transfer and compression molding of small parts for high physical characteristics. 1D30 HIF is available in nodular and mulch form in standard gray, green and red colors. 1D30 HIF in nodular form can be easily preformed to aid in handling.

OUTSTANDING FEATURES:

Parts molded from 1D30 HIF possess the low moisture absorption, thermal stability and good electrical properties characteristic of diallyl phthalate compounds. Fiber orientation along flow patterns is unusually good, accounting for exceptionally high strength in thin wall sections of molded parts.

APPLICATIONS:

Structural parts, housings, terminals, connectors, and insulators requiring strength.

Molding Information	Typical Values	Test Method
Specific gravity	1.8	ASTM D-792
Bulk factor	6-7	
Available plasticity	medium, stiff	
Molding pressures, psi		
compression	500-2000	
transfer	1000-8000	
Molding temperature F	290-340	
Curing time, thin sections, min.	1	
Mold shrinkage, in./in.	0.002-0.004	ASTM D-955
Post mold shrinkage, in./in.	0.0002	480 hrs. @ 257F

Mecnanical Properties:		
Izod impact strength, ft lbs./in. notch	4-10	FTMS 406-1071
Flexural strength, psi x 10 3	10-25	FTMS 406-1031
Flexural modulus, psi x 10 6	1.70	FTMS 406-1031
Compressive strength, psi x 10 3	20-24	FTMS 406-1021
Tensile strength, psi x 10 3	7-10	FTMS 406-1012

INDUSTRIAL DIELECTRICS, INC.: 1D40 Molding Compound:

1D40 is an orlon-filled, diallyl phthalate molding compound that molds easily in standard compression and transfer equipment. It is available in a diced form in standard green and blue colors.

OUTSTANDING FEATURES:

1D40 has good flexural, compressive and tensile strengths as well as excellent post-mold shrinkage when compared to other orlon molding compounds. From the standpoint of electrical properties, 1D40 provides excellent insulation resistance after prolonged periods of exposure to 100% humidity conditions at elevated temperatures.

APPLICATIONS:

1D40 has unusual resistance to hard or sudden blows and shocks, making it an ideal compound for use in terminal boards, printed circuit boards, connectors and many other components where good impact resistance is important. 1D40 is certifiable to military specification, MIL-M-14G, Type SDI-5.

Molding Information	Typical Values	Test Method
Specific gravity	1.34	ASTM D-792
Bulk factor	2-8	
Available plasticity	stiff, medium	
Molding pressures, psi		
compression	500-2000	
transfer	1000-8000	
Molding temperature F	290-330	
Curing time, thin sections, min.	1	
Mold shrinkage, in./in.	.009-0.012	ASTM D-955
Post mold shrinkage, in./in.	.0002	480 hrs. @ 257F

Mechanical Properties:		
Izod impact strength, ft.lbs./in. notch	0.6-0.7	FTMS 406-1071
Flexural strength, psi x 10 3	9.0-12.0	FTMS 406-1031
Flexural modulus psi x 10 6	0.72	FTMS 406-1031
Compressive strength, psi x 10 3	18.0-25.0	FTMS 406-1021
Tensile Strength, psi x 10 3	3.5-6.0	FTMS 406-1012

INDUSTRIAL DIELECTRICS, INC.: 1D50 Molding Compound:

1D50 is a dacron-fiber-filled, diallyl phthalate molding compound that is easily molded in standard compression and transfer equipment. It is available in a diced form in standard green, blue and black colors.

OUTSTANDING FEATURES:

1D50 exhibits extremely good electrical properties, as evidenced by high dielectric strength and low dielectric constant. 1D50 has an impact strength considerably higher than that of orlon-filled material, which makes thin molded sections extremely strong by comparison.

APPLICATIONS:

The inherent strength of dacron fibers makes this compound excellent for molded part applications where high physical and electrical properties are demanded.

1D50 is certifiable to military specification, MIL-M-14G, Type SDI-30.

Molding Information	Typical Values	Test Method
Specific gravity	1.39	ASTM D-792
Bulk factor	2-8	
Available plasticity	medium, hard	
Molding pressures, psi		
compression	500-2000	
transfer	1000-8000	
Molding temperature F	290-330	
Curing time, thin sections, min.	1	
Molding shrinkage, in./in.	0.009-0.012	ASTM D-955
Post mold shrinkage, in./in.	0.0002	480 hrs. @ 257F

Mechanical Properties:		
Izod impact strength, ft. lbs./in. notch	2.5-3.0	FTMS 406-1071
Flexural strength, psi x 10^3	10-12	FTMS 406-1031
Flexural modulus, psi x 10^6	0.64	FTMS 406-1031
Compressive strength, psi x 10^3	16-25	FTMS 406-1021
Tensile strength, psi x 10^3	4-7	FTMS 406-1012

INDUSTRIAL DIELECTRICS, INC.: 2S30 Molding Compound:

2S30 is a long glass fiber reinforced epoxy molding compound developed for use in high strength applications. It is suited for compression and transfer molding and should be preheated until relatively soft before placing in the mold. 2S30 is available in flake and nodular form in standard green, blue and natural colors.

OUTSTANDING FEATURES:

Parts molded from 2S30 show excellent mechanical properties and good fiber distribution. The flow characteristics are excellent and curing time at high temperature is relatively short.

Molding Information	Typical Values	Test Method
Specific gravity	1.81	
Mold shrinkage, in/in(bar)	0.002-0.004	
Molding temperatures, F	300-350	
Molding pressures, psi		
transfer	100-10000	
compression	500-2000	
Cure rate, 1/4" thin sections, min.	3-4	
Bulk factor	6-7	
Plasticities available	medium, stiff	
Mechanical & Other Properties:		
Izod impact strength, ft lbs./in. notch	8-15	FTMS 406-1071
Flexural strength, psi x 10 3	20-35	FTMS 406-1031
Compressive strength, psi x 10 3	22-33	FTMS 406-1021
Tensile strength, psi x 10 3	12-18	FTMS 406-1012
Flame resistance	Self-extinguishing	ASTM D-635
Water absorption, percent	0.05-0.12	48 hrs. at 122F

INDUSTRIAL DIELECTRICS, INC.: 3D36 Molding Compound:

3D36 is a long glass fiber reinforced akyd molding compound of exceptional physical and electrical properties. It is available in a nodular and flake form in a standard gray color which can be easily molded in compression or transfer equipment.

OUTSTANDING FEATURES:

3D36 cures rapidly to give high strength parts with excellent surface. 3D36 is certifiable to MIL-M-14G, MAI-60.

Molding Information	Typical Values	Test Method
Bulk factor	6-7	
Available plasticity	stiff, medium	
Specific gravity	2.00	
Mold shrinkage, in./in.(bar)	0.002-0.004	
Molding temperature	280-350	
Molding pressures, psi		
compression	500-2000	
transfer	1000-8000	
Curing time, thin sections, min.	1/2-1.0	

Mechanical Properties:		
Izod impact strength, ft lbs./in. notch	6-7	FTMS 406-1071
Tensile strength, psi x 10 3	4-7	FTMS 406-1011
Flexural strength, psi x 10 3	12-14	FTMS 406-1031
Compressive strength, psi x 10 3	18-22	FTMS 406-1021

KOPPERS CO., INC.: ATLAC Corrosion-Resistant Polyester and Vinyl Ester Resins:

Non Flame Retardant:

ATLAC 382:

The bisphenol-A polyester known for three decades as the standard in industry for corrosion resistance. This resin is FDA approved.

ATLAC 387:

A flexible resin for blending with ATLAC 82 to increase resistance to stress cracking or crazing, and damage from impact or thermal shock.

ATLAC 4010:

Pre-blended mixture of ATLAC 382 and ATLAC 387, offered for maximum convenience to fabricators. This resin is FDA approved.

ATLAC 400:

For FDA type service this terephthalate resin has been designed to maximize corrosion resistance relative to isophthalate resins.

ATLAC 580:

A vinyl ester developed for improved corrosion resistance, retention of strength and elevated temperatures and fabrication ease over conventional flexible vinyl esters.

ATLAC 570:

An epoxy novolac vinyl ester formulated for optimization of corrosion resistance to solvents and retention of properties in elevated temperature environments.

ATPRIME 100:

A two-component secondary bonding agent for use with polyester resins for improving adhesion of secondary layups. ATPRIME 100 can eliminate the need for grinding.

KOPPERS CO., INC.: ATLAC Corrosion-Resistant Polyester and Vinyl Ester Resins (Continued):

Flame Retardant:

ATLAC 711:

A flame retardant resin with a 15 max. flame spread rating per ASTM E-84 when used with 5% antimony trioxide, thus meeting requirements of Department of Commerce PS 15-69 and NFPA #91. Chemical resistance equivalent to ATLAC 382.

ATLAC 793:

A flame retardant resin with a 25 flame spread rating with the addition of 3% antimony trioxide and a 15 with 5% antimony trioxide, per ASTM E-84. It is particularly well suited for mild corrosion service in HVAC applications.

ATLAC 797:

A chlorendic polyester resin which provides excellent corrosion resistance in oxidizing environments. A flame retardant resin which achieves a 30 flame spread rating per ASTM E-84 with the addition of 5% antimony trioxide.

KOPPERS CO., INC.: DION Resins:

DION VER 9100:
Description: Vinyl Ester. Superior corrosion resistance in acidic & alkaline services.
Application: Filament wound pipe, tanks, contact moulded, pultruded & continuously laminated structural shapes.

DION VER 9200:
Description: High Temperature Vinyl Ester. Retains properties at elevated temperatures while offering excellent resistance to solvents and caustic.
Application: Same as VER 9100 with higher thermal stability.

DION VER 9300 FR:
Description: Flame Retardant Vinyl Ester. With the addition of 3% antimony trioxide, a 25 flame spread is combined with the exceptional physical properties of vinyl esters.
Application: Hoods, ducts, scrubbers, and structural shapes.

DION COR-RES 6694:
Description: Modified Bisphenol Fumarate. Optimum chemical resistance and high temperature resistance. Not flame retardant.
Application: Tanks, pipes, etc. where maximum resistance to corrosion and high temperatures are required. Up to 350F continuous and higher for shorter time periods.

DION COR-RES 6695 FR
Description: Flame retardant Bisphenol Fumarate. Chemical resistance equivalent to DION COR-RES 6694; flame retardant with a 15 flame spread per ASTM E-84 with antimony trioxide.
Application: Same as 6694, up to 300F.

DION COR-RES 7000A:
Description: Flexibilized Bisphenol Fumarate. Premium chemical resistance and temperature resistance with high flexibility and tensile elongation. Not flame retardant.
Application: Filament wound tanks, pipes, etc. where flexibility, impact and thermal cycling resistance must be combined with premium corrosion resistance. Up to 300F continuous and higher for shorter time periods.

DION COR-RES 6693 FR:
Description: Brominated THPA. Provides excellent corrosion resistance and flame retardancy; 15 flame spread with antimony trioxide.
Application: Hood, duct, scrubber and fume handling applications where flame retardancy is also required. Often used in a composite system with COR-RES 6694. Good to 250F.

KOPPERS CO., INC.: DION Resins(Continued):

DION-ISO 6631T/6060-5:
Description: Corrosion grade Iso. Combines good corrosion resistance with excellent physical strength properties. Not flame retardant.
Application: Tanks, pipes, linings, etc. Can handle less severe conditions than those where 9100, 9200 or 6694 are used. Good to 300F.

DION-ISO 6246:
Description: Flex modified corrosion grade Iso. Good corrosion resistance. Not flame retardant.
Application: Excellent for filament winding applications. Good to 250F.

DION FR 6604T:
Description: Flame retardant/Iso; 20 flame spread with antimony trioxide & 70 flame spread unfilled.
Application: Primarily for fume handling systems. Good to 275F.

Industrial Applications of DION Resins:
* Environmental Protection Systems
* Chemical Process and Storage Tanks
* Fuel Storage Tanks
* Pickling and Plating Tanks
* Fans, Blowers, Hoods
* Fume Scrubbers, Duct Systems and Exhaust Stacks
* Chlorinators
* Caustic-Chlorine Cell Covers and Chlorine Headers
* Pollution Control Equipment
* Filament Wound Pipe for Water Treatment and Industrial Waste Disposal

MKB INDUSTRIES, INC.: BEETLE, SCARAB & MELMEX Moulding Materials:

Types:

The BIP range of thermosetting compression moulding powders includes BEETLE and SCARAB urea materials. BEETLE melamine fortified urea and MELMEX melamine materials in different flows and particle sizes (grinds). The advantages of mouldings in BEETLE and SCARAB cellulose-filled urea moulding powders include:

1. Unlimited range of translucent and opaque colours including white and pastel shades (BEETLE types only) and excellent colour retention with exposure to heat or sunlight.
2. Good electrical properties at domestic voltages, including excellent resistance to tracking.
3. Resistance to oil, grease and organic solvents.
4. Moderate resistance to weak acids and alkalis.
5. High flame retardance and good heat resistance.
6. Good flexural strength and abrasion resistance.
7. Hardness, rigidity and durability.
8. Permanent antistatic properties.
9. Freedom from odour and taste (therefore suitable for domestic articles and cosmetic containers).

Mouldings in MELMEX melamine powders have the same general characteristics as those in BEETLE urea. In addition, they have the advantage of:

1. Improved resistance to both hot and cold water.
2. Improved resistance to weak acids and alkalis.
3. Improved resistance to staining and discolouration from tea, coffee or fruit juice.
4. Improved resistance to dry heat.
5. Improved electrical properties.

Applications:

BEETLE urea:

Application: General purpose for all types of articles, particularly where good dimensional stability is required.

Grades of powder: BEETLE X/G
BEETLE GXT

Remarks: Materials with good pelleting properties. For good finish with low after-shrinkage. X/G materials are available in a wide range of colours, wherease GXT is available only in standard colours, mainly whites and creams.

MKB INDUSTRIES, INC.: BEETLE, SCARAB & MELMEX Moulding Materials(Continued):

Applications(Continued):

BEETLE Urea:

Application: Both flat and deep draw applications, particularly those involving complex shapes and thick sections.

Grades of powder: BEETLE ZXT

Remarks: For simple operation and rapid production on fully automatic presses at high mould temperatures. Available in white, cream and light grey colours only.

Application: Mouldings requiring improved heat and water resistance.

Grades of powder: BEETLE MFX

Remarks: Melamine-fortified urea powder giving substantially improved properties, particularly water resistance, but at much lower cost than MELMEX melamine materials.

SCARAB Urea:

Application: General purpose for applications where the colour is acceptable.

Grades of powder: SCARAB SB

Remarks: Economical material with good pelleting and moulding properties. Available in black and brown only.

MELMEX melamine:

Application: Particularly suitable for tableware and large flat and deep draw mouldings.

Grades of powder: MELMEX M Fine

Remarks: A fine powder normally pelleted and HF preheated, giving superior surface finish. All MELMEX mouldings have particularly low moisture absorption and good retention to acids and alkalis.

Application: General purpose including tableware and electrical mouldings.

Grades of powder: MELMEX M/G

Remarks: Gives maximum production rate with good surface finish where facilities for pelleting fine powders are not available.

Applications: All types of moulding where exceptionally good dimensional stability is required.

Grades of powder: MELMEX MP/G

Remarks: Specially plasticised grade giving very low after-shrinkage. Suitable for mouldings to be used under alternating wet and dry conditions.

MKB INDUSTRIES, INC.: BEETLE, SCARAB & MELMEX Moulding Materials (Continued):

Grades of Beetle and Scarab urea materials:

BEETLE X/G and GXT:

BEETLE X/G material is available in a wide variety of translucent, semi-translucent and opaque colours. BEETLE GXT is available only in a limited range of standard colours, mainly whites and creams. Both grades are widely used for household articles such as toilet seats and electrical accessories, bathroom cabinets and trays. Excellent light diffusion of translucent mouldings makes them ideal for lampshades. They are also suitable for buttons, closures, and similar small items. Both X/G and GXT powders have a dense granular form with a medium bulk factor and they have good moulding properties for general applications.

BEETLE ZXT:

A granular, fines-free material with particle size specially suited to fully automatic, powder-fed moulding machines. It is available in white, cream and light grey standard colours. As this material is less 'gassy' than X/G and GXT, it is particularly suited to thick section mouldings; 'breathing' of the moulds is unnecessary, thus facilitating a shorter cycle time and simpler moulding sequence.

SCARAB SB:

A relatively inexpensive, fast-curing material available in a variety of flows and grinds in black and brown shades. Suitable for a wide range of applications, especially closures of various types and electrical accessories for use at domestic voltages. Apart from colour, SCARAB material has all the advantages listed for BEETLE ureas.

BEETLE MFX:

A melamine-fortified powder which offers a solution for the application requiring properties superior to those obtainable with urea, but without the higher cost of melamine based powder.

BEETLE MFX melamine-fortified urea produces mouldings with lower water absorption, better resistance to weak acids and alkalis, and better heat resistance than BEETLE or SCARAB urea. Other characteristics are similar to BEETLE X/G. It is available in a range of standard colours.

MKB INDUSTRIES, INC.: BEETLE, SCARAB & MELMEX Moulding Materials(Continued):

Grades Of MELMEX Melamine Materials:

MELMEX M Fine and M/G:

The properties of these cellulose-filled materials make them particularly suitable for the production of high gloss tableware and heavy duty kitchen utensils. A comprehensive colour range is available.

MELMEX M Fine powder is normally used where there are facilities for pelleting and HF preheating when high production rates can be achieved. Articles made in this manner exhibit a particularly smooth and glossy finish with optimum stain resistance.

MELMEX M/G can also be pelleted and HF preheated but the lower bulk factor of this grade makes it suitable for direct moulding from cold powder. It can be supplied in a grind for fully automatic moulding if required.

MELMEX MP/G:

A plasticised grade of technical quality melamine powder. Recommended for high duty mouldings of good dimensional stability where low after shrinkage is necessary or where humidity may range from high to very low e.g. fan housings. Available in a range of standard colours.

MKB INDUSTRIES, INC.: BIP BEETLE Nylon: Special Grades for the Electrical Industry:

Special electrical grades of 'BEETLE' Nylon:

MDM51:

Electrical grade containing mineral filler and flame retardant additives.

Recommended for use where good thermal conductivity is required.

MDM52:

General purpose electrical grade containing flame retardant additives and lower filler loading than MDM51.

More suitable for applications requiring suitable higher mechanical performance, snap fits, etc.

MD81:

Flame retardant version of BEETLE grade MD3 to UL94 VO standard.

Classified as unfilled. Contains non-halogenated flame retardant additives.

Resistance to heat and fire:

In formulating these three grades, special attention has been given to the requirements of various British Standard electrical specifications including BS1363, BS5733 and BS4491.

Colours:

The natural colour of all electrical grades of BEETLE nylon is white and this permits manufacture of a full range of colours.

Processing electrical grades:

The electrical grades of BEETLE nylon can be injection moulded on conventional screw or plunger machines.

Electrical grades can be satisfactorily moulded over a range of machine settings.

Beetle electrical grades have a hard, semi-matt surface finish which complements electrical accessories manufactured in the more traditional materials

Processing characteristics:

Rapid cycling
Easy mould fill
Low shrinkage and distortion
Easy mould ejection
Bonding and welding

MKB INDUSTRIES, INC.: BIP BEETLE Nylon 6:

Grade:

BEETLE Nylon 6:
Materials for injection moulding--standard compounds.

LO3:
Special features: Fast cycling, low viscosity grade. Nucleated to give a fine crystalline structure.

MD2:
Special features: Standard melt viscosity grade for general purpose use. Most suitable for applications requiring good surface appearance.

MD2ST:
Special features: Special medium viscosity grade for thick section mouldings, eg wheels.

MD3:
Special features: Nucleated standard viscosity grade, for general purpose use. Mouldings have a highly crystalline structure and can be produced at a rapid rate.

MD3C:
Special features: Very rapid cycling, medium viscosity grade with improved moulding characteristics.

HY2ST:
Special features: Special high viscosity grade with improved toughness.

HY3:
Special features: Nucleated high viscosity grade, combining rapid cycling with crystallinity and outstanding mechanical strength.

MD4:
Special features: Similar to MD3 but containing a heat and light stabiliser.

Material for injection moulding--special toughened grade:

EN30:
Special features: High impact non-conditioning grade.

MKB INDUSTRIES, INC.: BIP 'BEETLE' Nylon 6(Continued):

Grade:

Materials for extrusion:

LO1:
Special features: Low melt viscosity grade.

MD1:
Special features: Medium melt viscosity grade.

HY1:
Special features: High melt viscosity grade.

MD5:
Special features: Heat and light stabilised extrusion grade, for secondary and primary cable sheathing.

Material for blow moulding:

AC1:
Special features: High melt viscosity grade. Nylon-acrylic copolymer.

Glass filled grade for injection moulding:

MDF1:
Special features: Medium viscosity. 30% glassfibre reinforcement.

MDF2:
Special features: Medium viscosity, 30% glassfibre reinforcement. Also contains heat and light stabilisers.

MDF4:
Special features: Medium viscosity, 15% glassfibre reinforcement.

MDS1:
Special features: Medium viscosity. 50% glass sphere filled.

MDS4:
Special features: Medium viscosity. 25% glass sphere filled.

Grades of BEETLE Nylon 6:
Grades of BEETLE Nylon 6 are manufactured in a range of viscosities.

	Type reference
Low	LO
Medium	MD
High	HY

MKB INDUSTRIES, INC.: BIP BEETLE Nylon 66:

The outstanding properties of BEETLE nylon 66 include:

High degree of toughness.
Good heat resistance.
High melting point.
Good electrical performance.
Low friction characteristics.
Good chemical resistance.
Good surface appearance.

Because of its excellent performance BEETLE nylon 66 is used extensively for applications in the automotive, electrical, domestic appliance and general engineering industries.

The good mechanical and thermal characteristics of the material can be made even better by the inclusion of fillers and fibre reinforcements. The range of BEETLE nylon 66 compounds includes many such modified grades.

BEETLE nylon 66 is supplied in granular form and is processed successfully on standard thermoplastic injection moulding machines. Grades are available in natural, black and a range of colours.

Grades:

Unfilled injection moulding grades:

A20: General purpose grade
suitable for applications requiring high strength and good shock resistance.

A25: General purpose grade
designed for ease of moulding over a wide range of applications. Available in a full range of colours.

A30: Fast cycling grade
produces mouldings with a highly crystalline structure and facilitates rapid production rates.

A40: Heat stabilised grade
for applications subjected to high temperature environments.

A60: Impact modified grade
has high impact strength and is designed to minimise the need for water treatment after moulding.

MKB INDUSTRIES, INC.: BIP BEETLE Nylon 66(Continued):

Grades:

Glassfibre reinforced and glass bead filled grades

AF307: General purpose injection moulding grade containing 33% glassfibre reinforcement. Produces mouldings with high strength, rigidity and resistance to creep.

AF407: Heat stabilised glassfibre reinforced grade containing 33% glassfibre, as AF307, but heat stabilised to give better performance at elevated temperatures.

AF403: Heat stabilised glassfibre reinforced grade, as AF407, but containing 15% glassfibre reinforcement.

AF317: Glassfibre reinforced grade; improved surface appearance.
as AF407, but containing 15% glassfibre reinforcement.

AS403: 15% glass bead filled grade.
Glass beads or spheres impart increased rigidity without the distortion problems sometimes associated with glassfibre reinforcement.

AS408: 40% glass bead filled grade.
Similar to AS403 but containing 40% glass beads. Both AS403 and AS408 are heat stabilised.

Special grades:

A81: Non halogenated flame retardant unfilled grade with UL94 VO rating.

A91: Low friction, wear resistant grade containing molybdenum disulphide.

MKB INDUSTRIES, INC.: BEETLE Thermoplastic Polyesters:

'BEETLE' PET

BEETLE PET compounds are a new generation of glassfibre reinforced polyester based on polyethylene terephthalate formulated for injection moulding.

BEETLE PET compounds possess a combination of nucleants and plasticisers which permit the production of intricate technical mouldings at mould temperatures of 85-120C: this represents a considerable advance over earlier materials. The inclusion of glass fibres allows the mechanical properties and high heat distortion temperature of the semi-crystalline material to be fully realised.

BEETLE PET compounds are characterised by outstanding mechanical, thermal and electrical properties and are ideal for applications demanding:

High stiffness and surface hardness at high temperatures.
Low tendency to "creep."
Low water absorption.
Excellent dimensional stability.
Good chemical and solvent resistance.
Reduced flammability (specialised grades).

BEETLE PET, with its excellent package of properties, offers engineers new opportunities for component design in areas such as the electrical, electronic, automotive and domestic appliance industries.

PET Grades for
extrusion and injection moulding

Grade:

Unreinforced grades for extrusion

PET100:
Description: High viscosity general purpose resin.
Type of application: Extrusion, wire covering and monofilament.

PET160:
Description: High viscosity resin.
Type of application: Extrusion, wire covering and monofilament.

PET200:
Description: Medium viscosity resin.
Easier flow characteristics than PET100.
Type of Application: Extrusion, wire covering and monofilament.

MKB INDUSTRIES, INC.: BEETLE Thermoplastic Polyesters (Continued):

PET Grades for
extrusion and injection moulding(Continued)

Unreinforced grades for injection moulding:

Grade:

PET100:
Description: High viscosity general purpose resin.
Type of application: Transparent mouldings at low mould temperatures. Good wear and impact resistance.

Glass reinforced grades for injection moulding:

PET102F:
Description: General purpose material containing 10% glass fibre and 20% filler.
Excellent surface finish and resistance to warping.
Type of application: Mouldings requiring high heat resistance, dimensional accuracy and chemical resistance eg domestic cooker parts, electric motor components automotive underbonnet parts.

PET112F:
Description: Contains mixed filler material incorporating 12 1/2% glass fibre for low warpage and good surface gloss.
Type of application: Applications requiring high heat resistance coupled with good dimensional stability and surface appearance.

PET106F:
Description: General purpose material containng 30% glassfibre.
Type of application: Mouldings requiring high heat resistance, high mechanical strength and chemical resistance--domestic appliance housings, electrical connectors, automotive parts.

PET166F:
Description: Similar to PET106F but higher mechanical strength and heat distortion temperature.
Type of application: Similar uses to PET106F but where higher mechanical and thermal performance are required--electrical lamp holders, domestic appliance components, automotive components.

PET1610F:
Description: Contains 50% glassfibre for high strength, stiffness and heat distortion temperature.
Type of application: Applications requiring maximum stiffness and thermal performance--pump housings, motor housings.

MKB INDUSTRIES, INC.: BEETLE Thermoplastic Polyesters (Continued):

PET Grades for
extrusion and injection moulding(Continued):

Glass reinforced, flame retardant grades for injection moulding

Grade:

PET804F:
Description: 20% glassfibre.
Flame retardant to UL94-VO.
Type of application: Applications requiring high thermal performance.

PET866F:
Description: 30% glassfibre.
Flame retardant to UL94-VO.
Higher strength than PET804F.
Type of application: Applications requiring high strength, heat resistance and flame retardancy.

Typical Properties:

PET100:
Special characteristics: Amorphous

PET106F:
Nominal glass content: %: 30

PET166F:
Nominal glass content:%: 30
Special characteristics: High Strength

PET1610F:
Nominal glass content: %: 50
Special characteristics: High stiffness

PET804F:
Nominal glass content: %: 20

PET866F:
Nominal glass content: %: 30
Special characteristics: High strength

PET102F:
Nominal glass content: %: 10
Nominal filler content: 20

PET112F:
Nominal glass content: %: 12.5
Norminal filler content: %: 10
Special characteristics: Good surface finish

MKB INDUSTRIES, INC.: JONYLON Nylon Compounds Modified Grades:

Lubricated (L):

These grades contain extra lubricant to improve flow and mould release in problematical tools.

Extra Nucleated (N):

These grades contain a more powerful crystallising agent which assists in the 'set-up' characteristics during moulding, with minimal effect on impact strength.

Heat and Light Stabilised (HS, LS, HL, HSX):

These grades are stabilised to combat the effects of the environment and adverse conditions. Because nylon is a high melting polymer it finds application in hot environments where other polymers would distort.

Light Stabilised (LS, HL):

The use of light stabilised grade is necessary where long term external exposure occurs.

Impact Modified (IM, T, XT):

Because nylon is a notch-sensitive material, but is otherwise tough, improvement in notched impact performance is desired and has been achieved by alloying with elastomers.

Plasticized (LP, P, PX):

The principal effect of plasticisers is to increase flexibility, and toughness, at the expense of tensile strength. Various degrees of plasticisation are available, with the following applications:

'LP' - Low plasticised: as an aid to reduce time delay in water absorption
'P' - plasticised; improved impact strength.
'PX' - extra platicised: improved impact and increased flexibility to compete with Nylon 11.

Graphite/Molybdenum Disulphide Reinforced:

These materials exhibit greater surface friction than unmodified nylons, and coupled with high viscosity polymer and increased nucleation, have exceptional surface friction and reduced wear resistance.

MKB INDUSTRIES, INC.: JONYLON Nylon Compounds Reduced Flammability Grades:

Electrical:
Mineral reinforced grades of nylon 6 and modified grades of nylon 66 are formulated to comply with the specifications relating to flammability of plastic parts used by the electrical industry.

60300:
- nylon 6 30% mineral reinforced

60030/15:
- nylon 6 15% mineral reinforced

60030/15/850:
- nylon 6 15% mineral reinforced

66100/D98:
- nylon 66 unreinforced, modified

Halogenated:
To Underwriters Laboratory Specification (UL94)

These grades are formulated with halogenated additives to yield products of excellent mouldability and good thermal stability in moulding.

60072:
- nylon 6 UL94 VO

66014:
- nylon 66 UL94 VO

60074/N54:
- nylon 6 30% glass-reinforced UL94 VO

66074/N54:
- nylon 66 30% glass-reinforced UL94 VO

MKB INDUSTRIES, INC.: JONYLON Nylon Compounds Reinforced Grades:

The JONYLON range of reinforced polyamides offer considerable advantages over unreinforced materials. These are:
* Greater stiffness (flexural modulus)
* Higher tensile strength
* Higher heat distortion temperature
* Higher impact strength
* Lower mould shrinkage
* Faster moulding cycle
* Improved dimensional stability

The principal grades available are:
- Glass-fibre reinforced (15% and 30%)
- Glass-sphere reinforced (30%)
- Mineral reinforced (30% and 40%)
- Glass-fibre/mineral reinforced (40%)

Glass-fibre Reinforced

These materials are compounded with glass-fibres which are about 300 microns long, with narrow fibre length distribution.

Glass-sphere Reinforced

These materials are compounded with glass-spheres which average 40 microns diameter, and are coated so that a chemical bond is achieved with a true reinforcing effect, resulting in an increase in both tensile strength and flexural modulus as well as surface hardness and compressive strength.

Mineral Reinforced

These grades are compounded with selected minerals which offer a true reinforcing effect and a variety of improved mechanical properties, especially the following:
* Greater stiffness and strength
* Higher temperature of deflection under load
* Greater heat resistance
* Uniform dimensions on moulding
* Easy processing and very fast cycle time

Mineral/Glass-fibre Reinforced

These grade are formulated to encompass many of the properties of glass-fibre reinforced grades, but with improved dimensional stability.

MITSUI & CO., INC.: Polymethylpentene "TPX":

Today, because of its remarkable and unique combination of transparency and resistance to heat and chemicals, TPX is used in a wide variety of applications, including medical equipment, laboratory ware, small appliances, ovenware, baking cartons, release paper, wire and cable coating, and a number of industrial uses.

Characteristics of TPX:

Excellent Heat Resistance

With a melting point of 240C (460F) and a high Vicat softening point, TPX lends itself to applications involving high service temperatures.

Outstanding Transparency

TPX has a visible light transmittance of 90% and is superior in transmittance in the ultraviolet region to glass and other transparent resins.

Excellent Electrical Properties

TPX excels in electrical insulating properties and its dielectric constant is the lowest among all known synthetic resins.

Low Specific Gravity

The density of TPX is 0.83 g/cu cm, the lowest of all commercially available thermoplastics.

Outstanding Chemical Resistance

TPX is highly resistant to most organic and inorganic commercial chemicals.

Nontoxicity

TPX conforms to established specifications of plastic containers and packaging materials for food-contact applications in Japan and other countries, including the U.S. FDA regulations and the West German BGA recommendations.

MOBAY CORP.: Commercial Resins:

BAYBLEND:

BAYBLEND resins are blends of polycarbonate and ABS that feature an excellent combination of engineering properties that include high impact strength, deflection temperatures under 264 psi load to 240F, good rigidity with flexural modulus of 320,000 to 380,000 psi and ease of processing. Available grades include general purpose, flame retardant, color stable and fast cycling types for injection molding and extrusion processing.

Physical Properties:

MC-2500/3200:
- % Filler-Type: 0
- Resin Type: GP
- Specific Gravity (Natural): 1.12
- Mold Shrinkage %: 0.5-0.7
- Water Absorption 24h Immersion, %: 0.21
- Hardness Rockwell: R 117

MH-6570/6570E:
- % Filler-Type: 0
- Resin Type: FR/CS
- Specific Gravity (Natural): 1.23
- Mold Shrinkage %: 0.5-0.7
- Water Absorption 24h Immersion, %: 0.14
- Hardness Rockwell: R 117

MH-6580:
- % Filler-Type: 0
- Resin Type: FR/CS
- Specific Gravity (Natural): 1.23
- Mold Shrinkage %: 0.5-0.7
- Water Absorption 24h Immersion, %: 0.14
- Hardness Rockwell: R 117

MH-6595:
- % Filler-Type: 0
- Resin Type: FR/CS
- Specific Gravity (Natural): 1.17
- Mold Shrinkage %: 0.5-0.7
- Water Absorption 24h Immersion, %: 0.10
- Hardness Rockwell: R 119

SYMBOLS:
- GP=General Purpose
- FR=Flame Retardant
- CS=Color Stable
- HI=High Impact
- GR=Glass Reinforced
- MF=Mineral Filled

MOBAY CORP.: Commercial Resins(Continued):

BAYBLEND PC/ABS Resin:

Physical Properties:

MJ-2500:
- % Filler-Type: 0
- Resin Type: GP/FDA
- Specific Gravity (Natural): 1.12
- Mold Shrinkage %: 0.5-0.7
- Water Absorption 24h Immersion, %: 0.24
- Hardness Rockwell: R 120

FR-1439:
- % Filler-Type: 0
- Resin Type: FR/CS
- Specific Gravity (Natural): 1.17
- Mold Shrinkage %: 0.43-0.55
- Water Absorption 24h Immersion, %: 0.15
- Hardness Rockwell: R 115

FR-1441:
- % Filler-Type: 0
- Resin Type: FR/CS
- Specific Gravity (Natural): 1.18
- Mold Shrinkage %: 0.48-0.58
- Water Absorption 24h Immersion, %: 0.15
- Hardness Rockwell: R 121

MAKROBLEND:

MAKROBLEND resins are blends of polycarbonate and other engineering resins, such as PET polyester, that offer a unique combination of properties--low temperature toughness even to -40F; superior chemical resistance to a broad range of hydrocarbons, acids, bases and solvents; and easy processability. Available grades include general purpose, flame retardant, reinforced, and FDA compliance types for injection molding and extrusion processing.

MAKROBLEND PC/Blends:

UT-1018:
- % Filler-Type: 0
- Resin Type: GP
- Specific Gravity (Natural): 1.22
- Mold Shrinkage %: 0.6-0.9
- Water Absorption 24h Immersion %: 0.16
- Hardness Rockwell: R 114

MOBAY CORP: Commercial Resins(Continued):

Physical Properties:

MAKROBLEND:

EC-900:
% Filler-Type: 0
Resin Type: GP/HI
Specific Gravity (Natural): 1.19
Mold Shrinkage %: 0.75-0.90
Water Absorption 24h Immersion, %: 0.20
Hardness Rockwell: R 125

PETLON Pet Polyester:
PETLON resins are based on modified thermoplastic PET polyester and provide excellent mechanical strength, flame retardancy and high temperature resistance with deflection temperatures under 264 psi load as high as 437F. Available grades in reinforced versions include general purpose, flame retardant and warpage control types for injection molding.

3530:
% Filler-Type: 30-GR
Resin Type: GP
Specific Gravity (Natural): 1.57
Mold Shrinkage %: Flow: 0.3
Cross Flow: 0.8
Water Absorption 24h Immersion, %: 0.05
Hardness Rockwell: R 120

4530:
% Filler-Type: 30-GR
Resin Type: FR
Specific Gravity (Natural): 1.67
Mold Shrinkage %: Flow: 0.2
Cross Flow: 0.8
Water Absorption 24h Immersion, %: 0.03
Hardness Rockwell: R 115

7530:
% Filler-Type: 35-MF/GR
Resin Type: GP
Specific Gravity (Natural): 1.64
Mold Shrinkage %: Flow: 0.4
Cross Flow: 0.7
Water Absorption 24h Immersion, %: 0.06
Hardness Rockwell: R 120

MOBAY CORP.: Commercial Resins(Continued):

POCAN PBT Polyester:

POCAN resins are thermoplastic PBT polyesters which offer good dimensional stability, excellent mechanical strength, high rigidity with a flexural modulus as high as 2,100,000 psi, good electrical properties and broad chemical resistance. Available grades include general purpose, flame retardant, unfilled, reinforced and impact modified types for injection molding and extrusion processing.

Physical Properties:

B1305/1505:
- % Filler Type: 0
- Resin Type: GP
- Specific Gravity (Natural): 1.30
- Mold Shrinkage %: Flow: 1.8
  Cross Flow: 1.8
- Water Absorption 24h Immersion, %: 0.04
- Hardness Rockwell: R 116

B3225:
- % Filler Type: 20-GR
- Resin Type: GP
- Specific Gravity (Natural): 1.46
- Mold Shrinkage %: Flow: 0.4
  Cross Flow: 1.5
- Water Absorption 24h Immersion, %: 0.05
- Hardness Rockwell: R 120

B3235:
- % Filler Type: 30-GR
- Resin Type: GP
- Specific Gravity (Natural): 1.55
- Mold Shrinkage %: Flow: 0.3
  Cross Flow: 1.4
- Water Absorption 24h Immersion, %: 0.04
- Hardness Rockwell: R 121

B4225:
- Filler Type: 20-GR
- Resin Type: FR
- Specific Gravity (Natural): 1.57
- Mold Shrinkage %: Flow: 0.4
  Cross Flow: 1.5
- Water Absorption: 24h Immersion, %: 0.05
- Hardness Rockwell: R 121

MOBAY CORP.: Commercial Resins(Continued):

POCAN PBT Polyester(Continued):

Properties:
B4235:
% Filler-Type: 30-GR
Resin Type: FR
Specific Gravity (Natural): 1.65
Mold Shrinkage %: Flow: 0.3
Cross Flow: 1.4
Water Absorption 24 h Immersion, %: 0.03
Hardness Rockwell: R 121

NYDUR Polyamide:

NYDUR Polyamide resin is a family of nylon 6 and 66 material which is characterized by good mechanical properties, excellent toughness even at low temperatures, broad chemical resistance, and easy processability. Available grades in both unfilled and reinforced versions include general purpose and impact modified types for injection molding and extrusion processing.

Properties:

BKV-130:
% Filler-Type: 30-GR
Resin Type: GP
Specific Gravity (Natural): 1.36
Mold Shrinkage %: Flow: 0.4
Cross Flow: 0.6
Water Absorption 24h Immersion, %: 0.90

BKV-140:
% Filler-Type: 40-GR
Resin Type: GP
Specific Gravity (Natural): 1.45
Mold Shrinkage %: Flow: 0.4
Cross Flow: 0.6
Water Absorption 24h Immersion, %: 0.75

BC-303:
% Filler-Type: 0
Resin Type: GP/HI
Specific Gravity (Natural): 1.07
Mold Shrinkage %: 1.6
Water Absorption 24h Immersion, %: 1.60

BC-402:
% Filler-Type: 0
Resin Type: GP/HI
Specific Gravity (Natural): 1.08
Mold Shrinkage %: 1.4
Water Absorption 24h Immersion, %: 1.70

MOBAY CORP.: MAKROLON Polycarbonate:

MAKROLON polycarbonate: The remarkable thermoplastic that stands out from the rest on its ability to meet diverse application needs.

Outstanding impact strength
Superior dimensional stability
Glass-like transparency
Remarkable thermal resistance
Good electrical properties

These key properties, along with low-temperature toughness, excellent moldability and extrudability, the availablity of flame retardant grades, and more, allow MAKROLON polycarbonate to meet tough product application requirements in a variety of markets. This alone is enough to qualify polycarbonate as an exceptional engineering thermoplastic. But what distinguishes MAKROLON resin from other engineering thermoplastics is the freedom it gives design engineers to select precise combinations of these properties to meet particular specification needs.

Summary characteristics of MAKROLON polycarbonate:

Appearance:
Naturally colorless and transparent with a smooth and glossy surface
Chemistry:
Amorphous thermoplastic of bisphenol A and carbonic acid
Form Supplied:
Cylindrical pellets
Colors:
Available in a wide variety of transparent, translucent, and opaque colors
Product Line:
Broad range of grades tailored to meet specific performance requirements
Processing:
Injection molding, extrusion, injection and extrusion blow molding, and rotational molding
Joining:
Solvent or adhesive bonding, welding, riveting, snap and press fits, and threading
Machining:
Sawing, drilling, reaming, milling, cutting, shaping, grinding, turning, and tapping
Thermoforming:
Accomplished using standard equipment and any of the current methods
Decorating:
Painting, printing, vacuum metallizing, and labeling
Applications:
Automotive and transportation, building and construction, business machines, consumer, electronics and telecommunications, lighting, medical, optical/photography, packaging, recreational

MOBIL CHEMICAL CO.: MX 5400 Super-High Impact Grade Poly-Styrene:

PRODUCT DESCRIPTION:

MOBIL MX 5400 is a super-high impact polystyrene recommended for extrusion and thermoforming appplications. It is designed to provide superior toughness while maintaining a fine balance of physical properties and excellent processability. Typical applications include appliances, luggage, coolers, automotive parts and other durable goods.

TYPICAL PROPERTIES	ASTM Method	Unit	Typical Value
Nominal Melt Flow Rate (200C, 5000g: Condition G)	D1238	g/10 min	2.5
Vicat Softening Temperature	D1525	F	214
Deflection Temp. Under Load at 264 psi (Injection Molded, Annealed)	D648	F	189
Tensile Strength at Yield	D638	psi	3,300
Break	D638	psi	3,600
Tensile Elongation at Break	D638	%	48
Tensile Modulus	D638	psi	240,000
Flexural Modulus	D790	psi	270,000
Flexural Strength at Yield	D790	psi	4,800
Izod Impact Strength at 73F (1/8" thick injection molded specimen, notched)	D256	ft.lbs/in	4.3
Hardness-Rockwell	D785	L Scale	50

MOBIL CHEMICAL CO.: PS MX7100A High Impact Extrusion Grade Polystyrene:

PRODUCT DESCRIPTION:

Mobil PS MX7100A is an extrusion grade high impact polystyrene designed specifically for use in refrigerator and freezer lines. MX7100A possesses excellent resistance to degradation by foam components associated with polyurethane insulation.

TYPICAL PROPERTIES	ASTM Method	Unit	Typical Value
Nominal Melt Flow Rate (200, 5000g: Condition G)	D1238	g/10 min	3.0
Vicat Softening Temperature	D1525	F	212
Deflection Temp. Under Load at 264 psi (Injection Molded, Annealed)	D648	F	182
Tensile Strength at Yield	D638	psi	2,800
at Break	D638	psi	4,300
Tensile Elongation at Break	D638	%	55
Tensile Modulus	D638	psi	160,000
Flexural Modulus	D790	psi	200,000
Flexural Strength	D790	psi	4,000
Izod Impact Strength at 73F (1/8" thick injection molded specimen, notched)	D256	ft.lbs/in	2.1
Hardness-Rockwell	D785	L Scale	42

MOBIL CHEMICAL CO.: PS MX7800A Translucent Impact Grade Polystyrene:

PRODUCT DESCRIPTION:

MOBIL PS MX7800A is a translucent impact polystyrene with superior gloss. This versatile material can be used in injection molding, as a gloss cap layer for co-extrusion, and extruded into films for lamination.

TYPICAL PROPERTIES	ASTM Method	Unit	Typical Value
Nominal Melt Flow Rate (200C, 5000g: Condition G)	D1238	g/10 min	4.5
Vicat Softening Temperature	D1525	F	214
Deflection Temp. Under Load at 264 psi (Injection Molded, Annealed)	D648	F	185
Tensile Strength at Yield	D638	psi	4,300
at Break	D638	psi	3,600
Tensile Elongation at Break	D638	%	25
Tensile Modulus	D638	psi	250,000
Flexural Modulus	D790	psi	280,000
Flexural Strength	D790	psi	6,200
Izod Impact Strength at 73F (1/8" thick injection molded specimen, notched)	D256	ft.lbs/in	1.0
Hardness-Rockwell	D785	L Scale	55
Gloss (at 20)	D523	Gloss Units	90

MOBIL CHEMICAL CO.: PS 8020 Ignition Resistant High Impact Grade:

PRODUCT DESCRIPTION:

MOBIL PS 8020 is a high flow, high impact flame retardant (V-O) polystyrene resin for structural foam and injection molding applicatons. Typical applications include television cabinetry, business machine enclosures, and appliances.

TYPICAL PROPERTIES	ASTM Method	Unit	Typical Value
Nominal Melt Flow Rate (200C, 5000 g: Condition G)	D1238	g/10 min	10.0
Vicat Softening Temperature	D1525	F	205
Deflection Temp. Under Load at 264 psi (Injection Molded, Annealed)	D648	F	180
Tensile Strength at Yield	D638	psi	3,500
at Break	D638	psi	3,400
Tensile Elongation at Break	D638	%	45
Tensile Modulus	D638	psi	240,000
Flexural Modulus	D790	psi	280,000
Flexural Strength at Yield	D790	psi	5400
Izod Impact Strength at 73F Notched 1/8" Injection molded specimen	D256	ft.lbs/in	2.0
Hardness-Rockwell	D785	L Scale	60
Specific Gravity, @ 73F	D792		1.16
Flammability (Solid, Black & Natural)	UL94	72 mils	V-O
(Foamed, Natural)	UL94	249 mils	V-O, 5V

MOBIL CHEMICAL CO.: PS 8120 Flame Retardant Polystyrene: An Economical Alternative to FR Engineering Resins:

Mobil's PS 8120 is a versatile flame retardant, high impact polystyrene resin with a UL94V-O rating (at 1/16 inch molded part thickness, in all colors) which is performing in critical television, computer, business machine, appliance and electrical/electronics applications.

The uniform flow, outstanding thermal properties and excellent dimensional stability of PS 8120 make it particularly well suited to thin wall injection molding and structural foam FR applications such as enclosures for business machines and computers. In addition, it is suggested for FR extrusion/thermoforming applications (such as speaker housings) because of its processability and balance of properties.

Properties of PS 8120 Structural Foam for Flame Retardant Applications (15% Density Reduction, 0.25 in. Wall Thickness)

TYPICAL PROPERTIES	ASTM Method	Unit	Typical Value
Specific Gravity (@73F)	D792		0.99
Relative Gravity (@73F)	D792		0.85
Vicat Softening Temperature	D1525	F	208
Deflection Temp. Under Load	D648		
@ 66 psi		F	185
@ 264 psi		F	170
Tensile Strength at Yield	D638	psi	2,400
Tensile Elongation at Break	D638	%	20
Tensile Modulus	D638	psi	240,000
Flexural Modulus	D790	psi	280,000
Flexural Strength at Break	D790	psi	5,900
Gardner Impact (@73F)		in.lbs.	30
Dielectric Strength	D149	v/mil	252
Mold Shrinkage	D955	in./in.	0.005-0.007
Thermal Conductivity	D177	BTU/in/h-1/ft-2/F	0.82
Flammability Rating (All colors)	UL 94	250 mils	V-O

MOBIL CHEMICAL CO.: PS 8120 Ignition Resistant High Impact Grade Polystyrene:

PRODUCT DESCRIPTION:

MOBIL PS 8120 is a high impact flame retardant (V-O) polystyrene resin for injection molding and structural foam. Typical applications include television cabinetry, business machine enclosures, and appliances.

TYPICAL PROPERTIES	ASTM Method	Unit	Typical Value
Nominal Melt Flow Rate (200C, 5000g: Condition G)	D1238	g/10 min	5.0
Vicat Softening Temperature	D1525	F	215
Deflection Temp., Under Load at 264 psi (Injection Molded, Annealed)	D648	F	188
Tensile Strength at Yield	D638	psi	3,900
at Break	D638	psi	3,700
Tensile Elongation at Break	D638	%	40
Tensile Modulus	D638	psi	260,000
Flexural Modulus	D790	psi	300,000
Flexural Strength at Break	D790	psi	5,800
Izod Impact Strength at 73F, Notched 1/8" Injection molded specimen	D256	ft.lbs/in	2.0
Hardness-Rockwell	D785	L Scale	62
Specific Gravity, @ 73F	D792		1.15
Flammability (All colors)	UL94	62 mils	V-O

MOBIL CHEMICAL CO.: PS 8720 Flame Retardant Polystyrene: A High-Performance FR Resin With Excellent Processability:

Mobil's PS 8720 is an extra-high impact flame retardant polystyrene resin with a UL94V-O rating (at 1/16 inch molded part thickness, in all colors) for extrusion/thermoforming and injection molding applications.

PS 8720's excellent melt flow characteristics at low shear rates and high stiffness make it an ideal material for thermoformed FR applications such as business machine parts and point-of-purchase displays. It also provides outstanding processability at the higher shear rates and temperatures encountered in thin-wall injection molding of parts such as television cabinetry.

Performance properties of PS 8720 include a deflection temperature under load (@264 psi) of 180F, toughness without stiffness (flexural modulus of 2.8 x 10 5 psi) and the impact resistance of more costly engineering materials (notched Izod of 3.3 ft.-lb./in.). PS 8720 is supplied in natural and custom colors.

TYPICAL PROPERTIES	ASTM METHOD	UNIT	TYPICAL VALUE
Nominal Melt Flow Rate (200C, 5000g: Condition G)	D1238	g/10 min	3.5
Vicat Softening Temperature	D1525	F	210
Deflection Temp. Under Load at 264 psi (Injection Molded, Annealed)	D648	F	180
Tensile Strength at Yield	D638	psi	3,200
at Break	D638	psi	2,800
Tensile Elongation at Break	D638	%	45
Tensile Modulus	D638	psi	270,000
Flexural Modulus	D790	psi	280,000
Flexural Strength at Break	D790	psi	4,500
Izod Impact Strength at 73F Notched 1/8" Injection molded specimen	D256	ft.lbs/in	3.3
Hardness-Rockwell	D785	L Scale	55
Specific Gravity, @ 73F	D792		1.15
Flammability (All colors)	UL 94	62 mils	V-O

MONMOUTH PLASTICS, INC.: EMPEE Flame Retardant High-Impact Polystyrene Type PS 951:

TYPICAL APPLICATIONS: Injection molded and extruded parts.

PROPERTY	TEST METHOD ASTM	UL	VALUE
Melt Flow Condition 'G'	D-1238		4.0-5.0
Density, gm/cc 23 C	D-792		1.04
*Flammability (1/32" min.)		94	V-2
Tensile Yield, psi (0.2"/min.)	D-638		4370
Ultimate Elongation, %	D-638		40
Izod Impact, Notched, ft-lbs./in.	D-256		1.7
Hardness, Rockwell M	D-1706		20
Flexural Modulus, psi x 1000	D-790		304
Heat Deflection Temperature, 66 psi, C	D-648		84
Heat Deflection Temperature, 264 psi, C	D-648		75
U.L. Temperature Index, All Properties			50C

*U.L. Yellow card No. E51346 available on request.

MONMOUTH PLASTICS, INC.: EMPEE Flame Retardant Polyethylenes:

Product Name:

Type PE 112:

TYPICAL APPLICATIONS: Injection Molded Parts (Low Density Base Resin)

PROPERTY	TEST METHOD ASTM	UL	VALUE
Melt Index	D-1238		9-12
Density	D-792		0.977
Flammability		94	V-2
Tensile At Break, psi	D-638		2050
Elongation %, Break	D-638		500
Vicat Softening Point	D-1525		(190F)
Brittle Temp.	D-746 Lower than minus		(94F)

Type PE 113:

Typical Applications: Wire insulation, i.e. coaxial cable

PROPERTY	TEST METHOD ASTM	UL	VALUE
Melt Flow, g/10 Min.	D-1238		1.0
Density, g/cc	D-792		0.98
Flammability		94	V-2
Tensile Strength at Break, psi	D-638		2000
Elongation %, Break	D-638		500
Vicat Softening Point	D-1525		88C
Brittle Temp., C	D-746 Lower than Minus		70C
Dielectric Constant: 10 6 Hz	D-150		2.25
Dissipation Factor: 10 6 Hz	D-150		0.0005

MONMOUTH PLASTICS, INC.: EMPEE Flame Retardant Polyethylenes (Continued):

Product Name:

Type PE 119:

TYPICAL APPLICATIONS: Stress Crack Tubing

PROPERTY	TEST METHOD ASTM	UL	VALUE
Melt Flow, g/10 Min.	D-1238		.4-.6
Density, g/cc	D-792		1.05
Flammability		94	V-2
Tensile Strength at Break, PSI	D-638		1800
Elongation %, Break	D-638		500
LOI	D-2863		25.7
Flexural Modulus, 1,000 PSI	D-790		80
Environmental Stress Crack Resistance Hours	D-1693		>1,000

TYPE PE 122:

TYPICAL APPLICATIONS: Wire insulation, i.e. coaxial cable

PROPERTY	TEST METHOD ASTM	UL	VALUE
Melt Flow, g/10 Min.	D-1238		3.0
Density, g/cc	D-792		0.98
Flammability		94	V-2
Oxygen Index	D-2863		26
Tensile Strength at Break, psi	D-638		2000
Elongation %, Break	D-638		500
Vicat Softening Point	D-1525		88C
Brittle Temp., C	D-746		Lower than minus 70C
Dielectric Constant: 10 6 Hz	D-150		2.34
Dissipation Factor: 10 6 Hz	D-150		0.0010
Dielectric Constant: 10 3 Hz	D-150		2.32
Dissipation Factor: 10 3 Hz	D-150		0.0003

MONMOUTH PLASTICS, INC.: EMPEE Flame Retardant Polyethylenes (Continued):

Product Name:

Type PE 125:

TYPICAL APPLICATIONS: Wire insulation, i.e. coaxial cable.

PROPERTY	TEST METHOD ASTM	UL	VALUE
Melt Flow, g/10 Min.	D-1238		3-4
Density, g/cc	D-792		1.01
Flammability		94	V-2
Oxygen Index	D-2863		27
Tensile Strength at Break, psi	D-638		2000
Elongation %, Break	D-638		500
Vicat Softening Point	D-1525		88C(190F)
Brittle Temp., C	D-746 Lower than Minus		70C(94F)
Dielectric Constant: 10 6 Hz	D-150		2.34
Dissipation Factor: 10 6 Hz	D-150		0.0010
Dielectric Constant: 10 3 Hz	D-150		2.32
Dissipation Factor: 10 3 Hz	D-150		0.0003

Type PE 129:

TYPICAL APPLICATIONS: Wire insulation i.e. coaxial cable

PROPERTY	TEST METHOD ASTM	UL	VALUE
Melt Index - g/10 Min	D-1238		.25-.50
Density - g/cc	D-792		0.97
Flammability		94	V-2
Oxygen Index %	D-2863		26
Tensile Strength @ Break, psi	D-638		2,000
Elongation @ Break %	D-638		500
Vicat Softening Point, C	D-1525		88C
Brittle Temp. C	D-746 Lower than minus		70C
Dielectric Constant: 10 6 Hz	D-150		2.34
Dissipation Factor: 10 6 Hz	D-150		0.0010
Dielectric Constant 10-Hz	D-150		2.32
Dissipation Factor 10 3 Hz	D-150		0.0003

MONMOUTH PLASTICS, INC.: EMPEE Flame Retardant Polyethylenes (Continued):

Product Name:

Type PE 152:

TYPICAL APPLICATIONS: Injected molded parts, blow molding, foaming and sheet and tube extruding (High Density Base Resin)

PROPERTY	TEST METHOD ASTM	UL	VALUE
Melt Index	D-1238		4-5
Density	D-792		1.00
Flammability (1/16", 3/32", 1/8")		94	V-2
Oxygen Index	D-2863		26.5
Tensile at Yield (0.2"/Min.) PSI	D-638		2150
Ultimate Elongation, %	D-638		650
Izod Impact Notched, Ft./Lbs./In.	D-256		1.26
Flexural Modulus, PSI	D-790		118,000
Vicat Softening Point	D-1525		121C(250F)
Low Temperature Brittleness	D-746		Lower than minus 70C (94F)

PE 152 also conforms to the following Federal Aviation Administration Tests in 1/8" thickness:

25.853 a
25.853 b
25.853 b-2
25.855 a-1

MONMOUTH PLASTICS, INC.: EMPEE Flame Retardant Polypropylene Type PP 301:

TYPICAL APPLICATIONS: Injection molded parts, extruded sheet, shapes, certain electrical applications.

PROPERTY	TEST METHOD ASTM	UL	VALUE
Melt Flow Condition 'L'	D-1238		8.0-10.0
Density, gm/cc 23 C	D-792		0.988
*Flammability (1/32", 1/16", 1/8")		94	V-O
*Flammability (0.16" - .020")		94	VTM-O
Oxygen Index	D-2863		28
Water Absorption,			
% Change in Weight	D-570		0.0119
% Soluble Matter Lost	D-570		0.1511
Tensile Yield, psi (2"/min.)	D-638		4900
Tensile Impact, ft-lbs./sq. in.	D-256		23.8
Heat Deflection Temp., 66 psi, C	D-648		106 (223F)
Dielectric Strength, volts/mil	D-149		1000
Volume Resistivity, Ohm-cm.	D-257		146 x 10 15
High Voltage Arc Resistance (1/8")	D-495		84
High Current Arc Ignition (Number Arcs to Ignite) (.062"-.125")		746	>200
High Voltage Arc Resistance (Time to Ignite, sec.) (.094"-125")		746	>300
High Voltage Arc Tracking, in./min.		746	0
Hot Wire Ignition 1/8" (Time to Ignite, Sec.)		746	35.7
Comparative Tracking Index (volts)	D-3638		600+
Weatherometer (UV)		746C	Complies
Long Term Heat Aging, C		746	105 (221F)

* U.L. Yellow card No. E-51346 available on request.

MONMOUTH PLASTICS, INC.: EMPEE Flame Retardant Polypropylene Type PP 351:

TYPICAL APPLICATIONS: Injection molded parts and certain electrical applications, which require high impact.

PROPERTY	TEST METHOD ASTM	UL	VALUE
Melt Flow Condition 'L'	D-1238		6.0-9.0
Density, gm/cc 23C	D-792		0.94
*Flammability (1/32", 1/16", 3/32" 1/8")		94	V-2
Izod Impact, Notched, ft.-lbs./in. @ 73F/23C	D-256		10
Izod Impact, Unnotched, ft.-lbs./in. @ -18C	D-256		22
Flexural Modulus, 1,000 psi	D-790		106
Tensile Impact, ft.-lbs/sq. in.	D-256		68
Tensile Yield, psi (2"/min.)	D-638		3400
Water Absorption	D-570		
% Change in Weight			0.08
% Soluble Matter Lost			0.07
Heat Deflection Temp., C	D-648		
@ 66 psi			75
@ 264 psi			49
High Current Arc Ignition (Number Arcs to Ignite)		746	
@ 0.062			84
@ 0.125			113
Hot Wire Ignition (Time to Ignite), Sec's.		746	
@ 0.062			25
@ 0.125			45
High Volt, Arc Resistance (Time to Ignite), Sec's.		746	
@ 0.062			64
@ 0.125			69
High Voltage Arc Tracking, in/min		746	0
High Voltage Arc Resistance (Time to Ignite) Sec's	D-495		
@ 0.125			136
Volume Resistivity @ 0.062, Sec's	D-257		
@ 40 hrs.			41.8 x 10 12
@ 96 hrs.			4.1 x 10 12
Dielectric Strength @ 0.062, volts/mil	D-149		978
Comparative Tracking Index, Volts	D-3638		>600
Long Term Heat Aging, C		746	100

* U.L. Yellow card No. E51346 available on request.

MONMOUTH PLASTICS, INC.: EMPEE Flame Retardant Polypropylene Type PP 401; PP 401 cs*:

TYPICAL APPLICATIONS: Injection molded parts, extruded sheet, shapes, certain electrical applications.

PROPERTY	TEST METHOD ASTM	UL	VALUE
Melt Flow Condition 'L'	D-1238		5.0-7.0
Density, gm/cc 23 C	D-792		0.938
*Flammability (1/16", 1/8", 1/4")		94	V-2
Oxygen Index	D-2863		28
Water Absorption, 24 hrs, %	D-570		0.01
Tensile Yield, psi (0.2"/min.)	D-638		4400
Ultimate Elongation, %	D-638		190
Izod Impact, Notched, ft.-lbs./in.	D-256		0.5
Hardness, Rockwell R	D-1706		100
Flexural Modulus, 1,000 psi	D-790		215
Heat Deflection Temp., 66 psi, C	D-648		122
Heat Deflection Temp., 264 psi, C	D-648		80
Dielectric Strength, volts/mil	D-149		900
Volume Resistivity, Ohm-cm.	D-257		1.5 x 10 16
High Voltage Arc Resistance (1/8")	D-495		63
Comparative Tracking Index		746	>600
High Current Arc Ignition (Number Arcs to Ignite)		746	>200
High Voltage Arc Resistance (Time to Ignite, sec.)		746	>300
High Voltage Arc Tracking, in/min		746	0.13
U.L. Temperature Index, All Properties			105C

* U.L. Yellow card No. E51346 available on request.

PP 401CS is a copper stabilized version of PP 401

MONMOUTH PLASTICS, INC.: EMPEE Flame Retardant Polypropylene Type PP 402:

TYPICAL APPLICATIONS: Injection molded parts, certain electrical applications.

PROPERTY	TEST METHOD ASTM	UL	VALUE
Melt Flow Condition 'L'	D-1238		5.0-7.0
Density, gm/cc 23C	D-792		0.938
*Flammability (1/32", 1/16", 1/8",) All Colors		94	V-2
Oxygen Index	D-2863		28
Water Absorption, 24 hrs., %	D-570		0.00
Tensile Yield, psi (2"/min.)	D-638		4338
Ultimate Elongation, %	D-638		190
Izod Impact, Notched, ft.-lbs./in.	D-256		0.5
Hardness, Rockwell R	D-1706		100
Flexural Modulus, 1,000 psi	D-790		215
Heat Deflection Temp., 66 psi C	D-648		122.
Heat Deflection Temp., 264 psi, C	D-648		80
Dielectric Strength, volts/mil	D-149		1175
Volume Resistivity, Ohm-cm	D-257		5.1x10 15
Hot Wire Ignition (1/16"; Time to Ignite, sec.)		746	33
High Current Arc Ignition (Number Arcs to Ignite)		746	>200
High Voltage Arc Resistance (Time to Ignite, sec.)		746	>300
High Voltage Arc Tracking, in./min.		746	0.10
Arc Resistance to Ignition (sec.)	D-495		140
Comparative Tracking Index, (Volts)		746	600+
U.L. Temperature Index, All Properties			115C
Dielectric Constant 10 3 HZ	D-150		2.11
Dissipation Factor 10 3 HZ	D-150		0.0008

* U.L. Yellow card No. E51346 available on request.

MONMOUTH PLASTICS, INC.: EMPEE Flame Retardant Polypropylene Type PP 554; PP 554 cs*:

TYPICAL APPLICATIONS: Injection molded parts, extruded sheet, shapes, certain electrical applications.

PROPERTY	TEST METHOD ASTM	UL	VALUE
Melt Flow Condition 'L'	D-1238		5.0-8.0
Density, gm/cc 23C	D-792		0.94
*Flammability (1/32", 1/16", 3/32", 1/8")		94	V-2
Oxygen Index	D-2863		27.6
Water Absorption, 24 hrs., %	D-570		0.01
Tensile Yield, psi (2"/min)	D-638		3800
Ultimate Elongation, %	D-638		225
Izod Impact, Notched, ft.-lbs./in.	D-256		1.5
Flexural Modulus, 1,000 psi	D-790		125
Heat Deflection temp., 66 psi, C	D-648		102
High Voltage Arc Resistance (1/8")	D-495		140
High Current Arc Ignition (Number Arcs to Ignite)		746	200+
High Voltage Arc Resistance (Time to Ignite, sec.)		746	300+
High Voltage Arc Tracking, in./min.		746	0.0
Comparative Tracking Index, Volts (1/8")		746	600+
Dielectric Strength V/MIL	D-149		900
Volume Resistivity OHM/CM	D-257		1.5×10^{16}
Dielectric Constant 10^3 Hz	D-150		2.13
Dissipation Factor 10^3 Hz	D-150		0.0007
U.L. Temperature Index, All Properties PP554			100C
U.L. Temperature Index, All Properties PP554CS			105C

* U.L. Yellow card No. 51346 available on request

* PP554CS is a copper stabilized version of PP554

MONMOUTH PLASTICS, INC.: EMPEE Flame Retardant Polypropylene Concentrates:

Product Name:

EMPEE Flame Retardant Polypropylene Type PP Conc. 4:

PP Conc. 4 is a concentrated form of EMPEE 401. It is designed that, when let-down with a natural polypropylene homopolymer resin in the ratio of 7:1, the resultant composition will have essentially the same properties as PP 401.

U.L. Yellow card No. E51530 available on request.

EMPEE Flame Retardant Polypropylene Type PP Conc. 33:

EMPEE Concentrate 33 is designed to impart good flame retardancy to most polypropylene homo polymers.

It is designed to function in let-down ratios between 10:1 up to 20:1, dependent on resin choice and retardancy required. It is particularly recommended for use in fiber and film applications. The effect on physicals will be minimal.

EMPEE Flame Retardant Polypropylene Type PP Conc. 43:

EMPEE Concentrate 43 is designed to impart good flame retardancy to most polypropylene homo and copolymers.

It is designed to function in let-down ratios between 9:1 up to 15:1, dependent on resin choice and retardancy required. It is particularly recommended for use in fibers and films and other extrusion applications. The effect on physicals will be minimal.

MONMOUTH PLASTICS, INC.: Flame Retardant Concentrates:

Product Number:

Flame Retardant Polyethylene Type PE Conc. 1:

EMPEE PE Concentrate 1 is designed to impart good flame retardancy to low and high density polyethylene.

Because there are so many varieties of polyethylene it is difficult to predict the exact let-down ratio required to meet specific flammability standards. And, this fact also precludes the submission of a generic EMPEE U.L. approval material. For starting point formulations though, Monmouth suggests:

7-13 parts by weight of Low Density polyethylene
6-10 parts by weight of High Density polyethylene
to 1 of EMPEE PE Concentrate 1. The resulting composition(s) will generally be V-2 when tested under UL 94 conditions. The effect on physicals will be minimal. Products manufactured from EMPEE PE Concentrate 1 will exhibit good heat sealability and be non-blooming.

Flame Retardant Polyethylene Type PO Conc. 61:

EMPEE PO Concentrate is designed to impart good flame retardancy to low and high density polyethylene.

Because there are so many varieties of low density polyethylene it is difficult to predict the exact let-down ratio required to meet specific flammability standards. And, this fact precludes the submission of a generic EMPEE U.L. approval material. For starting point formulations though, Monmouth suggests:

7-12 parts by weight of Low Density polyethylene
5-9 parts by weight of High Density standard and copolymer polyethylene
to 1 of EMPEE PO Concentrate 61. The resulting composition(s) will generally be V-2 when tested under UL 94 conditions. The effect on physicals will be minimal.

MONMOUTH PLASTICS, INC.: Flame Retardant Polystyrene Type PS 921:

TYPICAL APPLICATIONS: Injection molded transparent crystal parts.

PROPERTY	TEST METHOD ASTM	UL	VALUE
Melt Flow Condition 'G'	D-1238		4-5
Density, gm/cc 23C	D-792		1.05
*Flammability (1/32" min.)		94	V-2
Oxygen Index	D-2863		28.3
Water Absorption, 24 hrs, %	D-570		.035
Tensile Yield, psi (0.2"/min.)	D-638		6160
Ultimate Elongation, %	D-638		2.3
Izod Impact, Notched, ft.-lbs./in.	D-256		0.5
Hardness, Rockwell R	D-1706		85
Flexural Modulus, 1,000 psi	D-790		450
Heat Deflection Temp., 66 psi, C	D-648		97
Heat Deflection Temp., 264 psi, C	D-648		82
U.L. Temperature Index, All Properties			50C

* U.L. Yellow card No. E51346 available on request.

MONSANTO CHEMICAL CO.: CADON Engineering Thermoplastic:
Extrusion Grades:

CADON. High Heat and Chemical Resistance Without High Cost.

CADON engineering thermoplastic is a family of impact/modified styrene-maleic anhydride terpolymers developed by Monsanto. They offer an exceptional balance of processability and performance including heat, chemical and impact resistance.

CADON 152:
An Extrusion Grade With Exceptional Heat Resistance
When you want an extrusion grade that can really take the heat, pick CADON 152. It gives you high heat and medium impact resistance, plus excellent extrusion and thermoforming characteristics. Available in natural, black and custom colors.

CADON 172:
Heat Resistance & Excellent Processability
CADON 172 is an excellent blend of topnotch heat resistance plus processability. You gain the advantages of good extrusion and thermoforming characteristics, as well as medium impact strength. Available in natural, black and custom colors.

CADON FRN:
For Demanding FR Applications
For your hottest applications, flame retardant CADON FRN makes the grade. It can be used for a variety of applications requiring high heat resistance, good impact strength, low gloss and flame retardancy. All without using PVC. Available in natural, black and grey.

TYPICAL PROPERTIES:

CADON 152: Molded:
Deflection Temp. Under Load, unannealed, 66 psi, 1/2 in x 1/2 in, F(ASTM D648): 239
Izod Impact Strength, Notched, 1/2 in x 1/8 in, 73F, ft-lb/in:(ASTM D256): 3.5
Dielectric Strength, volts/mil:(ASTM D149): 440
Flammability, 0.062 in: (UL94): HB

CADON 152: Extruded:
Izod Impact Strength, Notched, 1/2 in x 1/8 in, 73F, ft-lb/in: (ASTM D256): 2.5
Dielectric Strength, volts/mil: (ASTM D149): 440
Flammability, 0.062 in: (UL94): HB

MONSANTO CHEMICAL CO.: CADON Engineering Thermoplastic Extrusion Grades(Continued):

TYPICAL PROPERTIES:

CADON 172: Molded:
Deflection Temp. Under Load, unannealed, 66 psi, 1/2 in x 1/2 in, F(ASTM D648): 230
Izod Impact Strength, Notched, 1/2 in x 1/8 in, 73F, ft-lb/in(ASTM D256): 5.2
Dielectric Strength, volts/mil (ASTM D149): 402
Flammability, 0.062 in (UL94): HB

CADON 172: Extruded:
Izod Impact Strength, Notched, 1/2 in x 1/8 in, 73F, ft-lb/in (ASTM D256): 4.5
Dielectric Strength, volts/mil (ASTM D149): 402
Flammability, 0.062 in (UL 94): HB

CADON FRN: Molded:
Deflection Temp, Under Load, unannealed, 66 psi, 1/2 in x 1/2 in, F(ASTM D648): 223
Izod Impact Strength, Notched, 1/2 in x 1/8 in, 73F, ft-lb/in(ASTM D256): 2.0
Dielectric Strength, volts/mil (ASTM D149): 690
Flammability, 0.062 in. (UL94): VO

CADON FRN: Extruded:
Izod Impact Strength, Notched, 1/2 in x 1/8 in, 73F, ft-lb/in(ASTM D256): 1.5
Dielectric Strength, volts/mil(ASTM D149): 690
Flammability, 0.062 in (UL94): VO

REGULATORY AND END-USER SPECIFICATION INFORMATION

Underwriters Laboratories
Most grades of CADON engineering thermoplastics have been evaluated and recognized by UL for up to 95C mechanical service and have been assigned flammability and temperature index ratings.

American Society for Testing and Materials
CADON engineering thermoplastic is covered by ASTM D4634.

CADON Engineering Thermoplastic in Food Contact Applications
CADON engineering thermoplastic may be used in houseware items and applications.

MONSANTO CHEMICAL CO.: CADON Engineering Thermoplastics: Injection Molding Grades:

CADON. High Heat and Chemical Resistance Without High Cost.

CADON engineering thermoplastic is a family of impact/modified styrene-maleic anhydride terpolymers developed by Monsanto. They offer an exceptional balance of processability and performance including heat, chemical and impact resistance.

In addition to superior flow for injection molding, CADON delivers high rigidity, easy colorability and good paintability.

CADON 112: The Best Impact-Resistant CADON:
For the best impact resistance in the CADON family, 112 is one tough choice. It has an excellent combination of superior impact resistance, moderate heat distortion, and good melt flow. Available in natural, black and custom colors.

CADON 127: Extra Heat Resistance:
When the going gets hot, pick the grade that can take it. CADON 127. It gives you the heat resistance of a modified PPO, plus better flow, appearance and chemical resistance. Available in natural, black and custom colors.

CADON 135PG: A Special Grade for Plating Applications:
When you want improved platability, superior thermocycle performance, plus better heat resistance than ABS, choose CADON 135PG. Use it in standard plating systems--without any special etches or longer cycles. With 135PG, you can mold parts with high plate adhesion, beautiful surface appearance, and super resistance to plate cracking and blistering. Available in grey only.

CADON 140: High Heat & Excellent Impact Resistance:
When your specs call for very high heat distortion properties combined with good impact resistance, count on CADON 140. You'll gain the heat resistance of a modified PPO, but with better flow, appearance and chemical resistance. Available in natural, black and custom colors.

CADON 160: An Alternative to High-Heat ABS:
For an alternative to high-heat ABS, try CADON 160--general purpose grade. It's specially formulated to deliver heat and impact resistance with molding performance similar to high-heat ABS. Available in natural, black and custom colors.

MONSANTO CHEMICAL CO.: CADON Engineering Thermoplastic: Injection Molding Grades(Continued):

CADON 161: New Low-Gloss CADON:

Now, go with low gloss for your automotive interior trim components. And get high impact and heat resistance to boot- with CADON 161. It delivers heat and impact resistance plus mold flow similar to high-heat ABS. Available in natural, black and custom colors.

CADON G1606: Superior Heat Resistance and High Flexural Modulus:

When you want superior heat resistance combined with excell-ent dimensional stability, pick the reinforced choice. CADON G1606. It's specially reinforced with 6% glass fiber. Available in black and grey.

CADON 2100: Super-High Heat Resistance:

With its very high heat resistance, CADON 2100 is a perfect pick for base resins in glass reinforcement. Available in black and grey.

CADON G2110: For Structural Applications:

For more consistent properties than previously available from 10% glass-reinforced products, superior heat resistance, and high flexural modulus in structural components. CADON G2110 is hard to beat. It's reinforced with 10% glass fiber for great-er dimensional stability. Available in black and grey.

CADON G2120: Reinforced for Extra Strength:

For maximum structural rigidity, choose CADON G2120. It's reinforced with 20% glass fiber for superior heat resistance and very high flexural modulus. Available in black and grey.

CADON FRX: For Demanding FR Requirements:

Even under the most demanding performance requirements at high temperatures, CADON FRX stands tough. It offers a strong combination of UL 94 V-O rating, heat distortion temperature of 115C, plus excellent processability. Parts mold quickly and easily, minimizing sink marks and providing superior weld line strength. Typical applications include electrical devices, electronic housing, mechanical parts, appliances and covers. Available in natural, black and grey.

Specification Information:

Underwriters Laboratories: Most of the non-reinforced grades of CADON engineering thermoplastic have been evaluated and recognized by UL for up to 95C mechanical service.

CADON engineering thermoplastic is covered by ASTM D4634.

CADON engineering thermoplastic is widely listed by automobile manufacturers.

CADON 127 is in complaince with the requirements of USP Class VI

CADON Engineering thermoplastic may be used in houseware items and applications.

MONSANTO CHEMICAL CO.: LUSTRAN ABS Injection Molding Resins:

For beautiful performance in virtually any application.

Superior Processability:
All grades in the line of Lustran resins have one thing in common: excellent processability.

Superior Performance:
LUSTRAN resins are also durable.

And Excellent Appearance

LUSTRAN ULTRA really makes your products shine.
For one of the brightest finishes in the business, start with one of the best looking ABS's in the business--LUSTRAN ULTRA resin.

ULTRA grades mean higher productivity, too.

Specify ULTRA MX for medium-impact resistance.
Specify ULTRA HX for high-impact resistance.

Typical applications for LUSTRAN ULTRA MX, HX.
small appliances - videocassettes - toys - office equipment - swimming pool accessories - sporting goods - power tool housings - telephones - lawn and garden equipment

Choose LUSTRAN 248 for high tensile and flexural strength.
LUSTRAN 448 combines higher impact and moldability.
LUSTRAN 337 is the low-gloss alternative.
Superior flow is a major advantage of LUSTRAN 648.
LUSTRAN 743 features super high-impact resistance.

Typical applications for LUSTRAN 248, 448, 337, 648, 743.
appliances - portable tools - shrouds - lawn and garden equipment - housewares - swimming pool accessories - photographic equipment - office equipment - toys - flashlight cases - packaging - sporting goods

When your specifications call for a flame-retardant ABS, call for LUSTRAN resins.

LUSTRAN FR-ABS molds easily--and beautifully.
Choose LUSTRAN 911 for applications with critical U.L. requirements.
LUSTRAN 921 UV withstands the test of time.
LUSTRAN 914 HM gives your product extra rigidity.

Typical applications for LUSTRAN 911, 921 UV, 914 HM.
computer terminal bases - printers - CRTs - electric appliances - modems - power tools - telecommunication work stations - business machines

MONSANTO CHEMICAL CO.: LUSTRAN SAN:

Easier to process, more cost-effective than polycarbonate or acrylic

Harder, stronger, and more chemical-resistant than polystyrene

Pick it for performance. Pick it for color.

High-performance styrenes and down-to-earth service.

SAN 31-General Purpose Grade

For Most Applications

An exceptional combination of clarity, heat resistance, chemical resistance (foodstuffs, oils, greases, acids, alkalies and petroleum derivatives), scratch resistance, strength and rigidity. Excellent base material for compounding. Available in clear, water white, transparent and opaque colors.

Typical Applications

* cosmetic packaging (lipstick tubes, compact cases, eye shadow containers, displays)
* fan blades
* toys and games
* business machines
* interior refrigerator components (crisper pans and covers, butter and cheese doors)
* medical parts (tubing connectors and valves, urine bottles)
* beverage tumblers/mugs/steins
* pitchers
* trays
* serving bowls
* food containers
* tableware
* dinnerware
* blender jars

SAN 33-Soft Flow Grade

Excellent for Thin-Wall or Intricate Parts

Same outstanding properties as LUSTRAN SAN 31, plus better chemical resistance. Easy to process due to its soft flow. Good base material for compounding. Typical applications include lipstick tubes and interior refrigerator components.

SAN 35-High-Performance Grade

For Structural Product Applications

Improved chemical and heat resistance, plus better toughnesss and rigidity. Available in clear. Excellent candidate for industrial batteries and blender jars.

MONSANTO CO.: LUSTRAN SAN(Continued):

SAN 41 & 43-Anti-Stat Grades

A Specific Grade for Anti-Stat Applications

Properties like LUSTRAN SAN 31, but with extra anti-stat protection. Typical applications include video and audio cassettes, as well as automotive applications such as indicator and interior light lenses, instrument panel lenses and electrical components.

(NOTE: The higher the number, the greater the anti-stat properties of the material.)

SAN 47 & 49-U.V.-Stabilized Grades

For U.V. Exposure Environments

For inside use where part is exposed to sunlight or U.V. light. Consider SAN 47 and 49 for inside light louvers, screen door panels, or wherever you need extra U.V. protection. LUSTRAN SAN 49 is an extrusion grade designed for those extrusion applications needing some U.V. stability.

SAN 51-Specialty Grade

For Extra Toughness and Chemical Resistance

A new grade for those jobs requiring extra toughness and chemical resistance.

LUSTRAN SAN Advantages Over PC:

* SAN exhibits greater chemical resistance than PC to common kitchen foodstuffs, fats, oils and selected solvents.
* SAN is superior to PC in hardness, flexural strength and tensile strength properties.
* SAN requires less drying time (2 hours @ 180F) than PC (4 hours @250F).
* SAN is easier to process because it offers faster cycling times and requires a lower melt temperature (450F) than PC (550F).
* SAN typically provides a 4c/cubic inch advantage over PC for a lower cost per part.

Use PC Under These Conditions

* When U.V. resistance to outdoor conditions is needed.
* When heat distortion specs call for performance in temperatures above 204F.
* When impact requirements must meet 2.0 Izod.

MONSANTO CHEMICAL CO.: LUSTRAN SAN(Continued):

LUSTRAN SAN Advantages Over Acrylic

* SAN has an average 11% lower specific gravity and a higher heat distortion temperature than acrylic.
* SAN typically provides about a 1/2c/cubic inch advantage over acrylic.
* SAN is an easier molding material than acrylic.

Use Acrylic Under These Conditions

* When a much higher U.V. Stability is needed.
* Where clarity is of extreme importance.
* When neutral coloring is required.

LUSTRAN SAN Advantages Over PS/Acrylic

* SAN exhibits greater chemical resistance than PS/Acrylic to common kitchen foodstuffs, fats, and oils.
* SAN houseware items offer significantly better thermal cycle performance and resistance to crazing in dishwashers than PS/Acrylic.
* SAN is superior to PS/Acrylic in hardness, heat distortion and all strength properties.
* SAN typically provides about 1/4c/cubic inch advantage over NAS.

Use PS/Acrylic Under These Conditions

* When a more water-clear resin is needed.

LUSTRAN SAN Advantages Over HH Crystal Polystyrene

* SAN exhibits greater chemical resistance than HH crystal to common kitchen foodstuffs, fats and oils.
* SAN houseware items offer significantly better thermal cycle performance and resistance to crazing in dishwashers than HH crystal.
* SAN is superior to HH crystal in hardness, heat distortion, and strength properties.

Use HH Crystal Polystyrene Under These Conditions

* When a more water-clear resin is needed.

MONSANTO CHEMICAL CO.: LUSTRAN SAN(Continued):

LUSTRAN SAN Advantages Over CLABS

* SAN is superior to CLABS in heat distortion, flexural strength, flow, and tensile strength properties.
* SAN has lower haze and yellowness than CLABS.
* CLABS although clear, stress whitens under load.
* SAN exhibits better chemical resistance than CLABS to common kitchen foodstuffs, fats and oils.
* SAN typically provides about a 1 1/2c/cubic inch advantage over CLABS.

Use CLABS Under These Conditions

* When Izod impact strength surpassing 0.6 ft.lb/in is needed.

PROPERTIES:

LUSTRAN SAN 31:

Tensile Stress @ Yield psi(MPa): 11,000(75.9)
Tensile Modulus psi(GPa): 490,000(3.33)
Elongation @ Fail, %: 2.7
Flexural Modulus psi(GPa): 522,000(3.60)

LUSTRAN SAN 33:

Tensile Stress @ Yield psi (MPa): 10,000(61.0)
Tensile Modulus psi (GPa): 475,000
Elongation @ Fail, %: 3.2
Flexural Modulus psi (GPa): 452,000(3.12)

LUSTRAN SAN 35:

Tensile Stress @ Yield psi (MPa): 11,500(79.4)
Tensile Modulus psi (GPa): 500,000 (3.4)
Elongation @ Fail, %: 3.0
Flexural Modulus psi (GPa): 525,000(3.62)

MONSANTO CHEMICAL CO.: TRIAX Engineering Thermoplastic:

TRIAX 1000 Series:
TRIAX 1000 engineering thermoplastics feature high impact resistance and excellent processability as a result of Monsanto's state-of-the-art alloying technology. TRIAX 1000 materials may be injection molded, blow molded, extruded and thermoformed. These semi-crystalline thermoplastics exhibit good chemical resistance, high heat performance and abrasion resistance. TRIAX 1000 products are the economical alternative for applications requiring very high toughness.

TRIAX 1125:

TRIAX 1125 is a nylon/ABS engineering thermoplastic designed for injection molding. It offers excellent heat resistance. Its combination of impact strength and processability contributes to economical production of abuse resistant parts in a wide variety of molds and surface finishes. It is an excellent material for applications requiring texturing and good appearance.

TRIAX 1125 can survive attack by many aggressive chemicals. It offers better abrasion resistance than many other engineering thermoplastics.

Parts can be finished via painting, flocking, hot stamping or sonic welding. TRIAX is available in natural or pre-colored form. It can also be colored via concentrates available from several suppliers.

Tensile Strength (ASTM Method D-638) (73F) (Kpsi) (50% R.H.):
5.9
Gardner Impact 2" Ring, 1/2" Dart 100 mils (D-3029G) (73F)
(in-lbs) (50% R.H.): >400
32F: >400
-4F: 250
-40F: 50
Minimum Molding Temperature: F: 470
Specific Gravity (ASTM D-792) (73F): 1.06

UL Data (File Number E98356)

Chemical Resistance:
TRIAX 1125 exhibits good resistance to most hydrocarbons, oils, gasoline and detergents.

Heat Resistance:
Heat resistance is one of many factors that influence the suitability of a polymer for a particular application.

Moisture and Part Growth:
Parts absorb moisture from the air until equilibrium is reached.

MONSANTO CHEMICAL CO.: VYDYNE Nylon Engineering Thermoplastic Resins:
Unreinforced VYDYNE Nylon Resins:
66 Nylon-Good Strength, Heat and Chemical Resistance
VYDYNE 21:
Description: General Purpose
Chararacteristics: Basic 66 nylon with superior color stability. Unlubricated.

VYDYNE 21X:
Description: General Purpose
Characteristics: Lubricated version of 21 resin for easy mold release.

VYDYNE 21SP:
Description: Fast Cycle
Characteristics: Highly lubricated version of 21 resin for easier mold fill and mold release.

VYDYNE 22H:
Description: Heat Stabilized
Characteristics: Good heat stability for protection against embrittlement at high temperatures. Lightly lubricated.

VYDYNE 22HSP:
Description: Heat Stabilized, Fast Cycle
Characteristics: Highly lubricated version of 22H resin for easier mold filling and mold release.

VYDYNE 24NSP:
Description: Nucleated
Characteristics: Rapid crystallization for fast cycles with slight sacrifice in ductility. Lubricated for exceptional mold release and high productivity.

VYDYNE 25W:
Description: Weather Resistant
Characteristics: Contains finely divided carbon for weather resistance.

VYDYNE 65B:
Description: Extrusion
Characteristics: Intermediate molecular weight for high melt strength.

VYDYNE 66B:
Description: Extrusion
Characteristics: High molecular weight for high melt strength.
69 Nylon--Low Moisture Absorption and Good Dimensional Stabilit
VYDYNE 60H:
Description: Heat Stabilized
Characteristics: Heat stabilized for elevated temperature service in wire jacketing applications. Lubricated for machine feed in extrusion applications.

MONSANTO CHEMICAL CO.: VYDYNE Nylon Engineering Thermoplastic Resins(Continued):

Nylon Modified with Flame Retardant Additives

VYDYNE M-340:
Description: Unreinforced
Characteristics: Rust-red injection molding resin rated UL 94 V-O and UL 94-5V.

VYDYNE M-344:
Description: Unreinforced
Characteristics: Off-White (color codeable), ductile/tough injection molding resin rated UL 94 V-O and UL 94-5V.

VYDYNE M-345:
Description: Unreinforced
Characteristics: Rust red ductile/tough injection molding resin rated UL 94 V-O and UL 94-5V.

VYDYNE 909:
Description: Glass Reinforced
Characteristics: Black injection molding resin rated UL 94 V-O and UL 94-5V. Maximum strength, stiffness and heat distortion temperature.

PROPERTIES:

MINERAL:
VYDYNE R-100: 0.2% Moisture:
Tensile Strength at Yield: psi: 73F: 14,000
psi: 170F: 9,500
Elongation at Fail: %: 73F: 9
Deflection Temperature at 66 psi: F: 460
at 264 psi: F: 360
Volume Resistivity: ohm-cm: 73F: 1.0x10 15

VYDYNE R-200: 0.2% Moisture:
Tensile Strength at Yield: psi: 73F: 14,000
Elongation at Fail: %: 9
Deflection Temperature at 66 psi: F: 460
at 264 psi: F 360
Volume Resistivity: ohm-cm: 73F: 1.5x10 15

VYDYNE R-220: 0.2% Moisture:
Tensile Strength at Yield: psi: 73F: 12,000
Elongation at Fail: %: 13
Deflection Temperature at 66 psi: F: 430
at 264 psi: F: 320
Volume Resistivity: ohm-cm: 73F: 0.77x10 15

MONSANTO CHEMICAL CO.: VYDYNE Nylon Engineering Thermoplastic Resins(Continued):

PROPERTIES:

MINERAL:

VYDYNE R-250: 0.2% Moisture:
Tensile Strength at Yield: psi: 13,000
Elongation at Fail: %: 73F: 21
Deflection Temperature at 66 psi: 408
at 264 psi: 266

MINERAL/GLASS:

VYDYNE R-400G: 0.2% Moisture:
Tensile Strength at Yield: pis: 16,500
Elongation at Fail: %: 7
Deflection Temperature at 264 psi: 425

GLASS:

VYDYNE R-513, R-513H: 0.2% Moisture:
Tensile Strength at Yield: psi: 18,000
Elongation at Fail: %: 3
Deflection Temperature at 264 psi: 480

VYDYNE R-533, R-533H: 0.2% Moisture:
Tensile Strength at Yield: psi: 28,000
Elongation at Fail: %: 5
Deflection Temperature at 264 psi: 480
Volume Resistivity: ohm-cm: 1x10 15

VYDYNE R-543, R-543H: 0.2% Moisture:
Tensile Strength at Yield: psi: 32,000
Elongation at Fail: %: 3
Deflection Temperature at 264 psi: 491

Hydrolysis Resistant VYDYNE R-538H-02: 0.2% Moisture:
Tensile Strength at Yield: psi: 29,000
Elongation at Fail: 5
Deflection Temperature at 264 psi: 480

NUODEX, INC.: HULS Plastics: VESTAMID Polyamide 12:

In non-European countries VESTAMID is marketed under the trade name "NYLON 12 huls."

VESTAMID is the registered trademark of the polyamides manufactured by huls.

The range of VESTAMID granulates is comprised of grades with different melt viscosities, containing various additives, and capable of being processed by many different methods.

Finished products of VESTAMID L....are distinguished for the following properties:

1. Low density.
2. Very low water absorption. Hence even in varied humidity conditions their dimensions remain practically unchanged and their mechanical properties are largely retained.
3. High impact and notched impact strengths in the dry state and at temperatures down to 80C.
4. Good anti-friction properties, and very good abrasion resistance even in the presence of dust and sand; gears are almost noiseless in operation.
5. Good chemical and stress-cracking resistance.
6. Many VESTAMID grades are nontoxic. They may be sterilized by steam or hot air.

RANGE OF GRADES:

Basic grades:

VESTAMID L 1600:
Processing method: injection moulding
End product: mouldings requiring easy melt flow

VESTAMID L 1700:
Processing method: injection moulding
End product: mouldings requiring easy melt flow

VESTAMID L 1700:
Processing method: extrusion
End product: coatings for plastics films, metal foils and paper

VESTAMID L 1801:
Processing method: injection moulding
End product: industrial mouldings subject to high mechanical stresses

NUODEX, INC.: HULS Plastics: VESTAMID Polyamide 12(Continued):

Basic grades

Grade:

VESTAMID L 1801:
Processing method: extrusion
End product: cable sheathings and core insulations

VESTAMID L 1801:
Processing method: extrusion
End product: flat films, coatings for plastics films, metal foils and paper, thin-walled cable sheathings and core insulations

VESTAMID L 1901:
Processing method: injection moulding
End product: technical mouldings of high impact strength

VESTAMID L 1901:
Processing method: extrusion
End product: semi-finished products, cable sheathings, core insulations

VESTAMID L 2101:
Processing method: injection moulding
End product: technical mouldings of particularly high impact strength

VESTAMID L2101:
Processing method: extrusion
End product: tubing, semi-finished products

VESTAMID L2101F:
Processing method: extrusion
End product: tubular films, sheets

VESTAMID L2101F:
Processing method: thermoforming
End product: deep-drawn articles

VESTAMID L2101F:
Processing method: extrusion blow moulding
End product: hollow mouldings

VESTAMID L2106F:
Processing method: extrusion
End product: film of high transparency and flexibility for foodstuffs packaging

NUODEX, INC.: HULS Plastics: VESTAMID Polyamide 12(Continued):

Modified grades:

Grade:

VESTAMID L 1722:
VESTAMID L 1723:
VESTAMID L 1724:
Processing method: injection moulding
End product: plasticized, heat and weathering stabilized mouldings requiring easy melt flow; with gradually increased flexibility

VESTAMID L 2121:
VESTAMID L 2122:
VESTAMID L 2124:
VESTAMID L 2128:
Processing method: injection moulding
End product: plasticized, heat and weathering stabilized mouldings having very high impact and abrasion resistance; with gradually increased flexibility from L 2121 to L 2128

VESTAMID L 2121:
VESTAMID L 2122:
VESTAMID L 2124:
VESTAMID L 2128:
Processing method: extrusion
End product: flexible tubes or hoses; profiles, sheets, with gradually increased flexibility

VESTAMID L 2123:
Processing method: extrusion
End product: tubes or hoses for air brake lines

VESTAMID L 2224:
Processing method: extrusion
End product: flexible tubes or hoses; very high viscous melt

VESTAMID L 2122 P:
VESTAMID L 2124 P:
Processing method: rotational moulding
End product: hollow mouldings with high impact strength; with gradually increased flexibility.

NUODEX, INC.: HULS Plastics: VESTAMID Polyamide 12(Continued):

Modified Grades:

Grades:

VESTAMID L 1836:
VESTAMID L 1837:
Processing method: injection moulding
End product: glass microspheres filled, heat and weathering stabilized technical mouldings requiring easy melt flow and having a very low isotropic shrinkage, great hardness, rigidity and outstanding dimensional stability

VESTAMID L 1930:
Processing method: injection moulding
End product: glass fibre-reinforced, heat and weathering stabilized mouldings of great hardness, rigidity, dimensional stability and high deflection temperature under load

VESTAMID L 1931:
Processing method: injection moulding
End product: glass fibre-reinforced industrial mouldings of very high long-term heat and weathering resistance, but unsuitable for use with hot water

VESTAMID L 1640:
Processing method: injection moulding
End product: technical mouldings having high long-term heat and weathering resistance

VESTAMID L 1640:
Processing method: extrusion
End product: very thin core insulations and cable sheathings

VESTAMID L 1940:
Processing method: injection moulding
End product: technical mouldings of high resistance to heat and weathering

VESTAMID L 1940:
Processing method: extrusion
End product: core insulations, cable sheathings, semi-finished products

VESTAMID L 2140:
Processing method: injection moulding
End product: mouldings of ultra-high impact strength and high long-term heat and weathering resistance

NUODEX, INC.: HULS Plastics: VESTAMID Polyamide 12(Continued):

Modified Grades:

Grade:

VESTAMID L 2140:
Processing method: extrusion
End product: tubing, semi-finished products

VESTAMID L 2240 F:
Processing method: extrusion
End product: films sterilizable in hot air, sheets

Processing method: thermoforming
End product: deep drawn articles

Processing method: extrusion blow moulding
End product: hollow mouldings

VESTAMID L 1941:
Processing method: injection moulding
End product: technical mouldings of ultra-high long term heat and weathering resistance, but unsuitable for use with hot water

VESTAMID L 1941:
Processing method: extrusion
End product: semi-finished products, cable sheathings

VESTAMID L 1950:
Processing method: injection moulding
End product: mouldings containing molybdenum disulphide for bearings and gear wheels with low sliding friction

VESTAMID L 1950:
Processing method: extrusion
End product: tubing, semi-finished products

VESTAMID L 1960:
Processing method: injection moulding
End product: mouldings containing graphite for bearings and gear wheels with low sliding friction

VESTAMID L 1960:
Processing method: extrusion
End product: tubing, semi-finished products

NUODEX, INC.: HULS Plastics: VESTAMID Polyamide 12(Continued):

Coating Powders:

Grade:

VESTAMID WS:
Processing method: fluidized bed coating
End product: abrasion and hot water resistant metal coatings

VESTAMID Primer WS 1-3:
Processing method: fluidized bed coating
End product: adhesion primer for heavy demands on fluidized bed coatings

Hot Melt Adhesive Grades:

Grade:

VESTAMID T 170:
Melting point: approx. 133 C
End product: adhesive filaments, industrial bondings, high viscosity

VESTAMID T 250:
Melting point: approx. 135C
End product: textile bondings, steam and laundering resistant

VESTAMID T 350:
Melting point: approx. 122C
End product: all textile bondings, also for steam press

VESTAMID T 430:
Melting point: approx. 116C
End product: hot melt adhesive fleeces, low melting point, laundering resistant

VESTAMID T 450:
Melting point: approx. 116C
End product: textile bondings, low melting point, laundering resistant

VESTAMID T 470:
Melting point: approx. 119C
End product: textile bondings only as mixtures, high viscous adhesives for industrial bondings

VEATAMID T 550:
Melting point: approx. 132 C
End product: textile bondings, good adhesion on siliconized fabrics, excellently laundering resistant

NUODEX, INC.: HULS: Polybutylene Terephthalate VESTODUR:

VESTODUR is the registered trademark of polybutylene terephthalate moulding compounds manufactured by Huls.

The key characteristics of the PBTP base polymer are:

- high heat deflection temperature under load,
- high rigidity,
- ultra-low water absorption, consequently high dimensional stability,
- great hardness,
- great strength,
- good frictional properties and low abrasion,
- good creep characteristics,
- excellent electrical properties,
- good chemical resistance,
- good weathering resistance,
- uncomplicated processing,
- no tendency towards stress cracking.

VESTODUR moulding compounds are available in a wide variety of modifications:

- A range of melt viscosities designed for processing by injection moulding and extrusion;
- Processing aids facilitate mould release;
- Polymer modifications improve flexibility and impact resistance;
- Flame retardants enable self-extinguishing moulding compounds to be produced;
- Fillers and reinforcing agents increases rigidity and the deflection temperature under load; with
  * minerals and glass microspheres also countering any tendency to warpage,
  * ground glass fibres having an excellent reinforcing effect while causing only very slight warpage,
  * chopped glass fibres causing the greatest increase in the rigidity and deflection temperature under load.

The nature of the various modifications is indicated by the abbreviations used for the grade designations.

VESTODUR Nomenclature:

Nature of Modifier	Abbreviation used
none	four-digit number
impact modifier	HI
minerals	MIN
glass microspheres	GK
ground glass fibres	MGF
chopped glass fibres	SGF
flame-retardant	SO

NUODEX, INC.: HULS: Polybutylene Terephthalate VESTODUR (Continued):

Range of Grades:

VESTODUR:

1000:
2000:
3000:
Additives: None
Modulus of elasticity: N/mm 2: 2700
2600
2600

MIN 125:
Additives: minerals
Modulus of elasticity: N/mm 2: 3000

GK 20:
GK 30:
GK 50:
Additives: Glass microspheres
Modulus of elasticity: N/mm 2: 3000
4000
5400

MGF 30:
MGF 50:
Additives: ground glass fibres
Modulus of elasticity: N/mm 2: 5500
10000

SGF 10:
SGF 15:
SGF 20:
SGF 30:
SGF 50:
Additives: chopped glass fibres
Modulus of elasticity: N/mm 2: 5000
6500
7500
10000
15000

1000 SO:
2000 SO:
3000 SO:
Additives: flame retardants
Modulus of elasticity: N/mm 2: 3000
3000
3000

NUODEX, INC.: HULS: Polybutylene Terephthalate VESTODUR (Continued):

Range of Grades:

VESTODUR:

GK 20 SO:
GK 30 SO:
GK 50 SO:
Additives: flame retardants + glass microspheres
Modulus of elasticity: N/mm 2: 3700
4500
6600

MGF 30 SO:
MGF 50 SO:
Additives: flame retardants + ground glass fibres
Modulus of elasticity: N/mm 2: 7000
14000

SGF 10 SO:
SGF 20 SO:
SGF 30 SO:
SGF 50 SO:
Additives: flame retardants + chopped glass fibres
Modulus of elasticity: N/mm 2: 5400
8500
11000
20000

HI 15:
HI 15 SO:
HI 17:
HI 19:
HI 19 SGF 30:
HI 19 SO:
HI 19 SGF 30 SO:
Additives: polymer-modified grades without and with various additives (SGF; SO)
Modulus of elasticity: N/mm 2: 1800
2300
1300
660
5400
800
6800

ORTHANE DIVISION: ORTHANE 9000 Series:

Cross-linked thermoplastic urethane compounded for applications requiring superior tensile strength, tear, abrasion and oil resistance.

Product Number:

9532:

Hardness: ASTM-D-2240: Shore A: 89
Hardness: ASTM-D-2240: Shore D: 39
Tensile: ASTM-D-412:
PSI: 7600
Modulus 100%: 1050
Modulus 200%: 1750
Modulus 300%: 3250
Elongation: ASTM-D-412: 510%
Tear Strength: ASTM-D-624 (Die C): 650
Tension Set: ASTM-D-412: 3.5%
Compression Set: ASTM-D-395: %, Method B: 24
Specific Gravity: ASTM-D-395: 1.21

9529:

Hardness: ASTM-D-2240: Shore A: 94
Hardness: ASTM-D-2240: Shore D: 46
Tensile: ASTM-D-412:
PSI: 7800
Modulus 100%: 1400
Modulus 200%: 2200
Modulus 300%: 4150
Elongation: ASTM-D-412: 470%
Tear Strength: ASTM-D-624: Die C: 710
Tension Set: ASTM-D-412: 4.0%
Compression Set: ASTM-D-395: % Method B: 28
Specific Gravity: ASTM-D-395: 1.22

9552:

Hardness: ASTM-D-2240: Shore A: 95
Hardness: ASTM-D-2240: Shore D: 52
Tensile: ASTM-D-412:
PSI: 7800
Modulus 100%: 1950
Modulus 200%: 2900
Modulus 300%: 4800
Elongation: ASTM-D-412: 430%
Tear Strength: ASTM-D-624: Die C: 790
Tension Set: ASTM-D-412: 5.0%
Compression Set: ASTM-D-395: %, Method B: 28
Specific Gravity: ASTM-D-395: 1.22

ORTHANE DIVISION: ORTHANE 9000 Series(Continued):

Product Number:

9531:

Hardness: ASTM-D-2240: Shore A: 96
Hardness: ASTM-D-2240: Shore D: 61
Tensile: ASTM-D-412:
PSI: 8000
Modulus 100%: 2300
Modulus 200%: 4100
Modulus 300%: 6400
Elongation: ASTM-D-412: 370%
Tear Strength: ASTM-D-624: Die C: 1000
Tension Set: ASTM-D-412: 6.0%
Compression Set: ASTM-D-395: %, Method B: 32
Specific Gravity: ASTM-D-395: 1.24

9528:

Hardness: ASTM-D-2240: Shore A: ---
Hardness: ASTM-D-2240: Shore D: 70
Tensile: ASTM-D-412:
PSI: 8000
Modulus 100%: 4000
Modulus 200%: 5700
Modulus 300%: 7700
Elongation: ASTM-D-412: 310%
Tear Strength: ASTM-D-624: Die C: 1150
Tension Set: ASTM-D-412: 15.5%
Compression Set: ASTM-D-395: %, Method B: 37
Specific Gravity: ASTM-D-395: 1.25

ORTHANE DIVISION: ORTHANE 12000 Series:

Cross-linked thermoplastics urethane compounded for applications requiring outstanding tensile strength, abrasion resistance and resiliency, combined with good hydrolytic stability.

Product Number:

12036:

Hardness: ASTM-D-2240: Shore A: 88
Hardness: ASTM-D-2240: Shore D: 38
Tensile: ASTM-D-412:
PSI: 6400
Modulus 100%: 990
Modulus 200%: 1300
Modulus 300%: 1870
Elongation: ASTM-D-412: 550%
Tear Strength: ASTM-D-624: Die C: 525
Tension Set: ASTM-D-412: 4.0%
Compression Set: ASTM-D-395: %, Method B: 24
Specific Gravity: ASTM-D-395: 1.12

12043:

Hardness: ASTM-D-2240: Shore A: 94
Hardness: ASTM-D-2240: Shore D: 48
Tensile: ASTM-D-412:
PSI: 6550
Modulus 100%: 1700
Modulus 200%: 2150
Modulus 300%: 3250
Elongation: ASTM-D-412: 510%
Tear Strength: ASTM-D-624: Die C: 680
Tension Set: ASTM-D-412: 6.5%
Compression Set: ASTM-D-395: %, Method B: 29
Specific Gravity: ASTM-D-395: 1.15

12057:

Hardness: ASTM-D-2240: Shore A: 96
Hardness: ASTM-D-2240: Shore D: 57
Tensile: ASTM-D-412:
PSI: 6700
Modulus 100%: 2275
Modulus 200%: 3000
Modulus 300%: 4600
Elongation: ASTM-D-412: 435%
Tear Strength: ASTM-D-624: Die C: 800
Tension Set: ASTM-D-412: 7.0%
Compression Set: ASTM-D-395: %, Method B: 32
Specific Gravity: ASTM-D-395: 1.17

ORTHANE DIVISION: ORTHANE 12000 Series(Continued):

Product Number:

12065:

Hardness: ASTM-D-2240: Shore A: ----
Hardness: ASTM-D-2240: Shore D: 62
Tensile: ASTM-D-412:
PSI: 6850
Modulus 100%: 2750
Modulus 200%: 3550
Modulus 300%: 5350
Elongation: ASTM-D-412: 390%
Tear Strength: ASTM-D-624: Die C: 890
Tension Set: ASTM-D-412: 8.5%
Compression Set: ASTM-D-395: %, Method B: 34
Specific Gravity: ASTM-D-395: 1.18

12073:

Hardness: ASTM-D-2240: Shore A: ----
Hardness: ASTM-D-2240: Shore D: 71
Tensile: ASTM-D-412:
PSI: 6650
Modulus 100%: 4700
Modulus 200%: 5690
Modulus 300%: ----
Elongation: ASTM-D-412: 240%
Tear Strength: ASTM-D-624: Die C: 1050
Tension Set: ASTM-D-412: 19%
Compression Set: ASTM-D-395: 36%
Specific Gravity: ASTM-D-395: 1.20

ORTHANE DIVISION: ORTHANE 13000 Series:

Cross-linked thermoplastic urethane compounded for applications requiring toughness, hydrolytic stability and low temperature flexibility.

Product Number:

13380:

Hardness: ASTM-D-2240: Shore A: 86
Hardness: ASTM-D-2240: Shore D: 38
Tensile: ASTM-D-412:
PSI: 6500
Modulus 100%: 850
Modulus 200%: 1400
Modulus 300%: 2300
Elongation: ASTM-D-412: 540%
Tear Strength: ASTM-D-624: Die C: PLI: 615
Tension Set: ASTM-D-412: 4.5%
Compression Set: ASTM-D-395: %, Method B: 20
Specific Gravity: ASTM-D-395: 1.16

13450:

Hardness: ASTM-D-2240: Shore A: 89
Hardness: ASTM-D-2240: Shore D: 45
Tensile: ASTM-D-412:
PSI: 7000
Modulus 100%: 1000
Modulus 200%: 1700
Modulus 300%: 2500
Elongation: ASTM-D-412: 520%
Tear Strength: ASTM-D-624: Die C: PLI: 690
Tension Set: ASTM-D-412: 5%
Compression Set: ASTM-D-395: %, Method B: 20
Specific Gravity: ASTM-D-395: 1.18

ORTHANE DIVISION: ORTHANE 13000 Series(Continued):

Product Number:

13480:

Hardness: ASTM-D-2240: Shore A: 95
Hardness: ASTM-D-2240: Shore D: 48
Tensile: ASTM-D-412:
PSI: 8000
Modulus 100%: 1600
Modulus 200%: 2300
Modulus 300%: 4000
Elongation: ASTM-D-412: 450%
Tear Strength: ASTM-D-624: Die C: PLI: 740
Tension Set: ASTM-D-412: 6%
Compression Set: ASTM-D-395: %, Method B: 23
Specific Gravity: ASTM-D-395: 1.20

13550:

Hardness: ASTM-D-2240: Shore A: 97
Hardness: ASTM-D-2240: Shore D: 55
Tensile: ASTM-D-412:
PSI: 7500
Modulus 100%: 2500
Modulus 200%: 3400
Modulus 300%: 5000
Elongation: ASTM-D-412: 410%
Tear Strength: ASTM-D-624: Die C: PLI: 845
Tension Set: ASTM-D-412: 6%
Compression Set: ASTM-D-395: %, Method B: 24
Specific Gravity: ASTM-D-395: 1.21

OWENS-CORNING FIBERGLAS CORP.: ATRYL Resin:

For Compression Molding Applications in the Automotive Industry.
The Resin Without Compromise

At last. A state-of-the-art, one component resin system that delivers a class A surface for large panel and thick part applications...without sacrificing physical properties; a resin that cures fast, bonds fast...but doesn't age fast on the shelf. A pre-engineered B-side resin that gives part-to-part and shrinkage consistency...yet is totally compatible with present molding technology.

No Compromise in Physical Properties
* Superior surface
* Excellent paint and bonding adhesion
* Excellent physical and mechanical properties
* Fast cure
* Pre-engineered B-side one component resin system
* Good shelf life and shrinkage reproducibility

Typical Liquid Resin Properties
Viscosity LVT #2 @ 60 rpm: 165 cps
Non volatiles: 51.0%
Acid value: 12.3
Water content: 0.10%
Weight/gallon: 8.80 lbs./gal.
SPI 180 cure 1.0% BPO
150-190F: 5.0 min.
150 Peak exotherm: 6.5 min.
Peak temperature: 445F

Mechanical Properties
Property (Avg. of 10)
Tensile strength (psi x 10 3): 11.0
Tensile modulus (psi x 10 6): 1.67
Tensile elongation (%): 1.12
Flexural strength (psi x 10 3): 27.9
Flexural modulus tangent (psi x 10 6): 1.77
Flexural modulus secant (psi x10 6): 1.66
Izod impact notched (ft.-lbs./in.): 17.1
Izod impact unnotched (ft.-lbs./in.): 21.5
Barcol hardness: 48.2
Specific gravity: 1.95
Water absorption (%): .3944
Coefficient of thermal expansion: 12.4
(in./in./C x 10 -6) -30 - +30C
Shrinkage (mils/in.) expansion: -.65
Interlaminar shear strength (psi) 400.0
Glass content P-980 1": 27.8%

OWENS-CORNING FIBERGLAS CORP.: PENTECH:

High Efficiency Polyester Resin for Filled Systems

PENTECH Polyester Resins Offer Increased Production at Lower System Cost.

New PENTECH polyester resins from Owens-Corning Fiberglas Corp. provide hand lay-up and spray-up fabricators of tub/shower units, recreational items and other construction units greater efficiency during manufacture than other previously available resins.

PENTECH Series Resins are offered as low viscosity, unpromoted or pre-promoted, and thixed or non-thixotropic unsaturated polyesters uniquely formulated for compounding with various filler additives to achieve laminates having low flame and smoke ratings.

PENTECH Resins: Features and Benefits

PENTECH Series Resins minimize the effects of filler and catalyst changes on the liquid and curing properties of polyester resins. The resin series is available in three formulations to meet individual fabrication/processing requirements.

PENTECH 1200
- Typical Neat Viscosity, cps: 185 cps
- Filler Range*: 40-52%

PENTECH 1300
- Typical Neat Viscosity, cps: 85 cps
- Filler Range*: 53-58%

PENTECH 1400
- Typical Neat Viscosity, cps: 50 cps
- Filler Range*: 59-63%

* Recommended filler range. Normal filler range of alumina trihydrate and calcium carbonate blends.

PENTECH Series Resins

Feature: New base resin.
Advantage: Allows higher filler loadings.
Benefit: Potential unit cost savings.

Feature: Contains a high efficiency additive.
Advantage: Lower volatilization of styrene monomer.
Benefit: Overall material savings by reducing resin loss. Improves the plant working conditions.

Feature: Optimized cure chemistry.
Advantage: Excellent cure on the mold.
Benefit: Rapid production and processing.

Feature: Wetting agent.
Advantage: Improved roll-out and wet-out of a laminate.
Benefit: Less employee discomfort, fewer defects.

Feature: Computer controlled resin manufacturing.
Advantage: Consistent performance in customer plants.
Benefit: Minimizes downtime and part rejection

OWENS-CORNING FIBERGLAS CORP.: TRUSURF LP:

Low Print Series Resins from Owens-Corning
Marine Low Print Resins Developed to Build Better Boats
Better Performing, Better Looking

New Owens-Corning Fiberglas TRUSURF LP series resins are formulated specifically to meet the performance criteria of the marine market.

A resin that offers the optimal combination of low fiber print, good laminate toughness, and secondary bonding forgiveness.

All resins offer a compromise between toughness, bonding and beauty.

Advantages of TRUSURF LP Resins:
TRUSURF LP resins offer marine manufacturers the following benefits:

* Optimized Tensile Elongation
  Provides good laminate toughness for greater crack resistance
* Excellent Secondary Bonding
  Minimal surface preparation prior to layup, even over extended time periods
* Low Fiber Print
  Better looking, lower costing boats which need less "touch up" work
* Computer Controlled Manufacturing
  Batch-to-batch consistency for greater production consistency

TRUSURF LP Resins: Typical Properties

Laminate Properties
30% Owens-Corning Fiberglas M-722 Glass Mat (3/16" laminates) data run with ASTM test methods

Flexural Strength (psi)	* 31,000
Flexural Modulus (psi)	* 1,300,000
Tensile Strength (psi)	* 14,000
Tensile Modulus (psi)	* 1,300,000

Cured Resin - data run with ASTM test methods.

Flexural Strength (psi)	* 15,500
Flexural Modulus (psi)	* 580,000
Tensile Strength (psi)	* 10,600
Tensile Elongation (%)	* 1.8
Tensile Modulus (psi)	* 470,000
Glass Transition Temp.(F)	* 190

Liquid Resin Properties

Viscosity (cps) LVT #3C60	* 500
Thix	* 2.6
Weight/Gallon (lbs.)	* 9.1
Monomer %	* 42

PENNWALT CORP.: KYNAR Polyvinylidene Fluoride:

A versatile engineering reisn with a unique balance of physical, chemical, and electrical properties for a wide range of applications.

KYNAR polyvinylidene fluoride, the homopolymer of 1,1-difluoroethene, is a tough engineering thermoplastic that offers a unique balance of properties. It has the characteristic stability fluoropolymers when exposed to harsh thermal, chemical, and ultraviolet environments, and, furthermore, the alternating CH2 and CF2 groups along the polymer chain provide a unique polarity that influence its solubility and electric properties. Thus, in addition to being readily melt-processed by standard methods of molding and extrusion, KYNAR resin, especially at elevated temperatures, can be dissolved in some polar solvents such as organic esters and amines. This selective solubility is an advantage in the preparation of corrosion-resistant coatings for chemical process equipment and long-life decorative finishes on building panels.

Important Properties of KYNAR fluoropolymers:

* Mechanical strength and toughness
* High abrasion resistance
* High thermal stability
* High dielectric strength
* High purity
* Readily melt processable
* Resistance to most chemicals and solvents
* Resistance to ultraviolet and nuclear radiation
* Resistance to weathering
* Resistance to fungi

Applications:
- Coatings
- Film
- Filter Cloth
- Instrumentation and Control Equipment Linings
- Mist Eliminators/Scrubbers
- Membranes
- Static Mixers
- Pipe & Fittings
- Pumps
- Stock Shapes
- Valves
- Electrical/Electronic
  - Plenum Cable Jacketing
- Films and Specialized Applications

PENNWALT CORP.: KYNAR Polyvinylidene Fluoride(Continued):

Typical Values or Ranges:

460:

Physical Properties:
Melting Point Range: ASTM D3418: F: 311-320
C: 155-160
Density: ASTM D792: lb/ft 3: 109-110
g/cm 3: 1.75-1.77
Specific Volume: in 3/lb: 15.5-15.8
cm 3/g: 0.56-0.57
Refractive Index: ASTM D542: n25D: 1.42
Water Absorption: ASTM D570: %: <0.05

Mechanical Properties:
Tensile Strength at Yield: ASTM D638: psi: 4000-6000
MPa: 28-41
Tensile Strength at Break: psi: 4500-7500
MPa: 31-52
Elongation at Yield: ASTM D638: %: 8-15
Elongation at Break: %: 50-250
Tensile Modulus: ASTM D882: psi: 150-200 x 10 3
MPa: 1040-1380
Flexural Strength: ASTM D790: psi: 8600-10800
MPa: 59-74
Flexural Modulus: ASTM D790: psi: 165-260 x 10 3

710-760:

Physical Properties:

Melting Point Range: ASTM D3418: F: 329-338
C: 165-170
Density: ASTM D792: lb/ft 3: 110-111
g/cm 3: 1.77-179
Specific Volume: in 3/lb: 15.5-15.8
cm 3/g: 0.56-0.57
Refractive Index: ASTM D542: n25D: 1.42
Water Absorption: ASTM D570: <0.05

Mechanical Properties:
Tensile Strength at Yield: ASTM D638: psi: 5000-7000
MPa: 34-48
Tensile Strength at Break: psi: 4500-6500
MPa: 31-45
Elongation at Yield: ASTM D638: %: 7-13
Elongation at Break: %: 50-250
Tensile Modulus: ASTM D882: psi: 200-300 x 10 3
1380-2070
Flexural Strength: ASTM D790: psi: 6500-9000
MPa: 45-62
Flexural Modulus: ASTM D790: psi: 170-325 x 10 3
MPa: 1170-2240

PERSTORP, INC.: Melamine Moulding Compounds:

Melamine products have a hard, scratch resistant surface, are stronger than urea products and also have better moisture resistance and electrical insulating properties.

Melamines can be supplied in any colour, and are easily decorated by moulding with decorative foils.

Melamines have excellent resistance to solvents and household chemicals.

Grade:

732:

Filler: Cellulose
Process method: Injection
Properties: Special compound for injection moulding, not pelletable

754:

Filler: Cellulose
Process method: Compression, special automatic grade
Properties: Good pourability, not pelletable

PERSTORP, INC.: PERSTORP Melamine Moulding Compounds:

The PERSTORP 700-Series molding compounds are based on a melamine-formaldehyde resin and an alpha cellulose filler. The product can be obtained in a broad color range.

Grades Available:

751:

This grade is considered a general purpose material because of its soft flow and medium fast cure time. It is readily adaptable to widely different molding conditions like pressure, closing speed and cure time.

752:

752 is referred to as a tropical grade. This curing is slow giving a very good flowability. When operating slowly closing presses or using a low molding pressure the 752 grade is always recommended.

When molding large, heavy parts this soft flowing material attains good surface appearance and physical properties.

This grade is also recommended for injection molding.

Molding temperature is the same as for 751.

756:

The 756 grade is a specially designed material for molding parts for which a long term excellent stain resistance is required. This molding compound is very soft flowing and fast curing in the temperature range 325-335F.

Fine Powder Grades
791, 792, 795 and 796:

All above mentioned grades can also be supplied as fine powder. The molding characteristics are the same as for the corresponding grade of granulated material.

PERSTORP, INC.: Urea Moulding Compounds:

A strong argument for urea moulding compounds is that they allow almost unlimited colourations. Urea products have a hard, scratch resistant surface and moisture resistance, and provide a high gloss finish. Ureas have good electrical insulation properties, with better resistance to tracking than phenolics.

Grade:

132:
Filler: Cellulose
Process method: Compression
Properties: Strong, low shrinkage, low water absorption

151:
Filler: Cellulose
Process method: Compression
Properties: Strong, low shrinkage, low moisture content. pelletable, not for long flows

154:
Filler: Cellulose
Process method: Compression, special automatic grade
Properties: Strong, low shrinkage, specially good pourability

161:
Filler: Cellulose
Process method: Compression
Properties: Strong, specially low shrinkage, specially low water absorption

321:
Filler: Cellulose
Process method: Compression
Properties: Good pelletability, suitable for automatic moulding of small components

332:
Filler: Cellulose
Process method: Injection
Properties: Special compound for injection, not pelletable

334:
Filler: Cellulose
Process method: Compression, special automatic grade
Properties: Specially good pourability, not pelletable

532:
Filler: Cellulose
Process method: Injection
Properties: Meets hot wire test at 500C, good dimensional stability

542:
Filler: Cellulose
Process method: Injection
Properties: Meets hot wire test at 500C, very good dimensional stability

PHILLIPS 66 CO.: RYTON PPS Resins:

RYTON A-100:

A-100 is a fiberglass reinforced compound specifically tailored for injection molding of thick-walled heavy industria parts. The PPS base resin features improved ductility to eliminate internal cracking due to differential shrinkage of parts up to one-half inch (1.27 cm) thick.

In addition to these attributes, A-100 exhibits:
* UL 94 temperature index rating of 200-220C
* UL 94 V-0/5V flammability rating
* Excellent dimensional stability
* Good moldability
* Excellent chemical resistance

Suggested applications:
Industrial items such as:
* Mechanical components
* Pump housings, impellers and seals
* Automotive under-the-hood parts
* Down-hole oil well applications

Specification Data:
* UL Recognized Component(Yellow Card)

RYTON A-200:

A-200 is a compound containing 40% glass fiber with a PPS base resin that provides improved toughness and ductility.

In addition to these attributes, A-200 exhibits:
* UL 746B temperature index rating of 200-220C depending on thickness and end use
* UL 94 V-O/5V flammability rating without additives
* Excellent dimensional stability
* Good moldability
* Excellent chemical resistance
* Good weld line strength
* Electrical and mechanical properties that are virtually unaffected by high humidity and high temperature.

Suggested Applications:
Industrial items such as:
* Mechanical components
* Pump housings, impellers and seals
* Automotive under-the-hood parts
* Electrical and electronic parts

Specification Data:
* UL Recognized Component(Yellow Card)

PHILLIPS 66 CO.: RYTON PPS Resins(Continued):

RYTON R-3 02 Black:

RYTON R3 02 is a fiberglass reinforced molding compound particularly suitable for use in thin-walled molded parts. R-3 02 has the same high temperature resistance characteristics of RYTON R-4 molding compound.

In addition to these attributes, RYTON R3 02 exhibits:

* UL 746B temperature index rating of 200-220C
* UL 94 V-O flammability rating without additives
* Excellent dimensional stability
* Good moldability
* Electrical properties that are virtually unaffected by moisture
* Easy flow

Suggested Applications
Items such as:

* Connectors
* Enscapulation cups
* Automotive and electronic parts
* Appliance components

Specification Data

* UL Recognized Component (Yellow Card)
* Mil spec Mil-P46174(MR)

RYTON R-4:

This is a unique injection molding engineering thermoplastic. This PPS compound, containing 40% glass fiber, posseses outstanding chemical resistance and mechanical properties even at elevated temperatures.

In addition to these attributes, RYTON R-4 exhibits:

* UL 746B temperature index of 200-220C depending on thickness and end use
* UL94 V-O/5V flammability rating without additives
* Excellent dimensional stability
* Good moldability
* Electrical properties that are virtually unaffected by moisture.

Suggested Applications:
Items such as:

* Automotive under-the-hood components
* Electrical and electronic parts
* Appliance components
* Pump housings and impellers
* Valves, cams, etc.

Specification Data

* UL Recognized Component (Yellow Card)
* NSF approved

PHILLIPS 66 CO.: RYTON PPS Resins(Continued):

RYTON R-4 02 Black:

This is a unique injection molding engineering thermoplastic. This PPS compound, containing 40% glass fiber, possesses outstanding chemical resistance and mechanical properties even at elevated temperatures.

In addition to these attributes, RYTON R4 02 exhibits:

* UL 746B temperature index of 200-220C depending on thickness and end use
* UL 94 V-O/5-V flammability rating without additives
* Excellent dimensional stability
* Good moldability
* Electrical properties that are virtually unaffected by moisture.

Suggested Applications
Items such as:

* Automotive under-the-hood components
* Electrical and electronic parts
* Appliance components
* Pump housings and impellers
* Valves, cams, etc.

Specification Data

* UL Recognized Component (Yellow Card)

RYTON R-4 04:

This is a unique injection molding engineering thermoplastic. This PPS compound, containing 40% glass fiber, is similar to R-4, but contains a special additive system that improves insulation in very demanding applications.

In addition to these attributes, RYTON R-4 04 exhibits:

* UL 746B temperature index of 200-220C depending on thickness and end use
* UL 94V-O/5V flammability rating without additives
* Excellent dimensional stability
* Good moldabilty
* Electrical properties that are virtually unaffected by high humidity at high temperature.
* Good hydrolytic stability.

Suggested Applications

Parts such as:

* Electrical and electronic parts used in critical insulation applications such as those requiring the properties specified in Mil Spec MIL-M-24519B(EC)

Specification Data

* UL Recognized Component (Yellow Card)
* Mil Spec MIL-M-24519B(EC), Type GST-40F

PHILLIPS 66 CO.: RYTON PPS Resins(Continued):

RYTON R-4 06:

This is a unique injection molding thermoplastic. This PPS compound, containing 40% glass fiber, is similiar to R-4, but contains a special additive system that improves insulation in very demanding applications.

In addition to these attributes, RYTON R-4 06 exhibits:
* UL 746B temperature index of 200-220C depending on thickness and end use
* UL 94V-O/5V flammability rating without additives
* Excellent dimensional stability
* Good moldability
* Electrical properties that are virtually unaffected by high humidity at high temperature.
* Good hydrolytic stability.

Suggested Applications

Parts such as:
* Electrical and electronic parts use in critical insulation applications such as those requiring the properties specified in Mil Spec MIL-M-24519B(EC)

Specification Data
* UL Recognized Component (Yellow Card)
* Mil Spec MIL-M-24519B(EC), Type GST-40F

RYTON R-5:

RYTON R-5 is a fiberglass reinforced molding compound possessing chemical resistance and mechanical properties at elevated temperatures and improved dimensional stability.

In addition to these attributes, RYTON R-5 exhibits:
* UL 746B temperature index of 200-220C depending on thickness and end use
* UL 94 V-O flammability rating without additives
* Optimum dimensional stability
* Good moldability
* Electrical properties that are virtually unaffected by moisture

Suggested Applications
Items such as:
* Connectors
* Appliance components
* Valves, cams, etc.
* Automotive under-the-hood components
* Pump housings and impellers

Specification Data
* UL Recognized Component (Yellow Card)

PHILLIPS 66 CO.: RYTON PPS Resins(Continued):

RYTON R-7:

Glass fibers and mineral fillers have been compounded with polyphenylene sulfide to produce an injection moldable high temperature engineering compound with the inherent cost saving of thermoplastics.

In addition to these attributes, RYRON R-7 exhibits:
* UL 746B temperature index of 200-220C depending on thickness and end use
* UL94 V-O flammability rating without additives
* Outstanding arc resistance
* Excellent dimensional stability
* Excellent electrical properties
* Easily moldable

Suggested Applications
Items such as:
* Electronic components
* Structural electrical parts
* Appliance components
* Automotive under-the-hood parts
* Replacements for thermosets

Specification Data
* UL Recognized Component (Yellow Card)
* UL Insulation System Recognition, Class 200(N)

RYTON R-7 02 Black:
Glass fibers and mineral fillers have been compounded with polyphenylene sulfide to produce an injection moldable high temperature engineering compound with the inherent cost savings of thermoplastics.

In addition to these attributes, RYTON R-7 02 exhibits:
* UL 746B temperature index of 200-220C depending on thickness and end use.
* UL94 V-O flammability rating without additives
* Outstanding arc resistance
* Excellent dimensional stability
* Excellent electrical properties
* Easily moldable

Suggested Applications
Items such as:
* Electronic components
* Structural electrical parts
* Appliance components
* Automotive under-the-hood parts
* Replacements for thermosets

Specification Data
* UL Recognized Component(Yellow Card)

PHILLIPS 66 CO.: RYTON PPS Resins(Continued):

RYTON R-7 03:

Glass fibers and mineral fillers have been compounded with polyphenylene sulfide to produce an injection moldable high temperature engineering compound with the inherent cost savings of thermoplastics.

In addition to these attributes, RYTON R-7 03 exhibits:

* Low DMF Extractable Chloride
* UL746B temperature index of 200-220C
* UL 94 V-O flammability rating without additives
* Excellent dimensional stability
* Excellent electrical properties
* Easily moldable

Suggested Applications

Items such as:

* Electrolytic capacitor lids

RYTON R-9 901:

RYTON R-9 901 contains a special additive to enhance encapsulation and reduce moisture penetration. Its use is recommended for encapsulating capacitors and similar devices.

In addition to these attributes, RYTON R-9 901 exhibits:

* UL 746B temperature index of 200-240C depending on thickness and end use
* UL 94V-O flammability rating without additives
* Excellent dimensional stability
* Good moldability
* Electrical properties that are virtually unaffected by moisture

Suggested Applications

Items such as:

* Radial capacitors
* Axial capacitors
* Transistors
* Resistor networks

Specification Data: UL Recognized Components (Yellow Card)

PHILLIPS 66 CO.: RYTON PPS Resins(Continued):

RYTON R-9 02:

RYTON R-9 02 contains a special additive to enhance encapsulation and reduce moisture penetration. Its use is recommended for encapsulating capacitors and similar devices.

In addition to these attributes, RYTON R9 02 exhibits:
* UL 94V-O flammability rating without additives
* Excellent dimensional stability
* Good moldability
* Electrical properties that are virtually unaffected by moisture

Suggested Applications:

* Transistors
* Resistor networks

Specification Data:
* UL Recognized Component(Yellow Card)

RYTON R-10:

Glass fibers and mineral fillers have been compounded with polyphenylene sulfide to produce a series of injection moldable high temperature engineering compounds. With the characteristic and advantages listed below, the user will accrue benefits of cost savings and consistent produce quality. In addition, R-10 compounds exhibit:

Advantages of a thermoplastic:
* Fast cycles on conventional equipment
* Regrind is useable

Thermal Stability:
* Continuous long term high temperature usage
* UL 746B temperature index of 200-240C
* Retains properties at high temperature

Excellent dimensional stability
* Low shrinkage and warpage
* Part to part reproducibility

Good electrical properties
* High arc resistance, low track rate

High affinity for fillers and pigments:
* Designer flexibility in the use of color
* Mechanical and electrical grades available

Suggested Applications:

Electronic: Bobbins, coil forms, housings, carriers, contractors, connectors and sockets
Electrical: Switch components, brush holders
Small appliances: Housings, heater bases and grills

Specification Data: UL Recognized Component(Yellow Card)

PHILLIPS 66 CO.: RYTON PPS Resins(Continued):

RYTON R-8:

Glass fibers and mineral fillers have been compounded with polyphenylene sulfide to produce an injection moldable high temperature engineering compound with the inherent cost savings of thermoplastics.

In addition to these attributes, RYTON R-8 exhibits:

* UL 94V-0/5V flammability rating without additives
* UL 746B temperature index of 200-220C depending on thickness and end use
* Outstanding arc resistance
* Excellent dimensional stability
* Excellent electrical properties
* Easily moldable

Suggested Applications:

* Electronic components
* Appliance components
* Replacements for thermosets
* Structural electrical parts
* Automotive under-the-hood parts

Specification Data:

UL Recognized Component (Yellow Card)

RYTON R-10:

Glass fibers and mineral fillers have been compounded with polyphenylene sulfide to produce a series of injection moldable high temperature engineering compounds. With the characteristics and advantages listed below, the user will accrue benefits of cost savings and consistent produce quality. In addition, R-10 compounds exhibits:

Advantages of a thermoplastic:

* Fast cycles on conventional equipment
* Regrind is useable

Thermal Stability:

* Continuous long term high temperature usage
* UL 746B temperature index of 200-240C
* Retains properties at high temperature

Excellent dimensional stability

* Low shrinkage and warpage
* Part to part reproducibility

Good electrical properties:

* High arc resistance, low track rate

High affinity for fillers and pigments

* Designer flexibility in the use of color
* Mechanical and electrical grades available

Suggested Applications:

Electronic
Electrical
Small appliances

Specification Data:

UL recognized Component(Yellow Card)

PHILLIPS 66 CO.: RYTON PPS Resins(Continued):

RYTON R-11 70:

RYTON R-11 70 is specially formulated for use in microwave cookware. The composition is essentially transparent to microwave radiation.

In addition, RYTON R-11 70 exhibits:

* Excellent resistance to staining
* UL 94 V-O flammability rating without additives
* UL 746B temperature index of 200-240C.
* Excellent dimensional stability
* Good moldability
* Excellent dishwasher durability
* Microwave transparency

Suggested Applications

* Microwave ovenware

Specification Data:

Status under Federal Food, Drug and Cosmetic Act:

* Suitable for fabricating microwave ovenware for use in the home, at temperatures not to exceed 425F (218C) by virtue of the Housewares Exemption

Meets this important requirement:

* UL Recognized Component (Yellow Card)

RYTON BRO6C:

RYTON BRO6C is an encapsulation resin compound containing a special filler system having a low ionic content. Its use is recommended for encapsulating both wire bonded and film bonded IC devices.

RYTON BR42A:

RYTON BR42A contains a special additive to reduce surface friction. It is recommended for applications requiring a low coefficient of friction and wear factor.

RYTON BR61C:

RYTON BR61C contains a special additive to enhance encapsulation and reduce moisture penetration. Its use is recommended for encapsulating capacitors and similar devices.

In addition to these attributes, RYTON BR61C exhibits:

* Excellent dimensional stability
* Good moldability
* Electrical properties that are virtually unaffected by moisture

Suggested Applications:

* Radial capacitors
* Axial capacitors
* Transistors
* Resistor networks

PHILLIPS 66 CO.: RYTON PPS Resins(Continued):

RYTON BR62B:

RYTON BR62B contains a spscial filler system to enhance arc tracking resistance. It is recommended for applications that require UL 275 volt CTI or the European VDE KC 275 volt arc tracking resistance.

RYTON BR63B:
Nominal Physical Properties:
Density, g/cm 3: ASTM D 792: 1.6
Tensile Strength, psi(MPa): ASTM D638: 17,500(121)
Elongation, %: ASTM D638: 0.9

RYTON BR70:
RYTON BR70 is a special formulation for halogen lamp applications, including both sealed and vented systems, with the inherent cost savings of thermoplastics.

RYTON BR71A:
RYTON BR71A contains a special additive to enhance encapsulation and reduce moisture penetration. Its use is recommended for encapsulating capacitors and similar devices.

RYTON BR72:
RYTON BR72 is a special formulation for halogen lamp applications, including both sealed and vented systems. It also promotes adhesion to metal and has the inherent cost savings of thermoplastics.

RYTON P-4 Powder:
P-4 polyphenylene sulfide resin is a finely divided powder produced to a controlled molecular weight so that when heated to its melting point of 545F (282C) P-4 powder melts sharply producing a medium viscosity molten polymer.

In addition to these characteristics, RYTON P-4 powder exhibits:
* Excellent processing stability
* Excellent affinity for fibrous fillers such as fiber glass and graphite
* Excellent affinity for particulate fillers such as talc, calcium carbonate clay, glass beads, and other silica type fillers.
* An affinity for blending with other resins such as Teflon, Polysulfone, etc.

Suggested Applications:
* Electrical components
* Electrical parts
* Appliance components
* Pump housings and impellers
* Automotive under-the-hood components

PLASKON ELECTRONIC MATERIALS, INC.: Long-Glass-Filled DAP:

PLASKON Diallyl Phthalate (DAP) Molding Compounds Long-Glass-Filled:

PLASKON diallyl phthalate (DAP) molding compounds provide attractive cost/performance benefits which increase sales. These benefits are derived from the outstanding properties of the DAP molding compounds, namely:

* high dimensional stability
* excellent heat resistance
* excellent chemical resistance
* high insulation resistance

In addition the DAP molding compounds retain these properties even in extreme environments of high heat and humidity.

PLASKON DAP molding compounds' outstanding properties in all areas makes DAP compounds the material of choice by design engineers when specifying materials for many electrical and electronic parts. They are part of the line of PLASKON thermosetting, high performance molding compounds. Other products offered in this line are epoxy molding compounds, phenolic molding compounds and an alkyd reinforced molding compound.

DAP molding compounds reinforced with long-glass-fibers have greater mechanical strength and considerably better impact strength than resins reinforced with short-glass-fibers.

PLASKON ELECTRONIC MATERIALS, INC.: Long-Glass-Filled DAP (Continued):

SUMMARY OF PROPERTIES

Physical

* extremely good dimensional stability
* almost no post-mold shrinkage
* chemically inert

Mechanical

* excellent strength in compression
* excellent impact resistance
* exceptional thin-wall strength

Electrical

* retains high insulation resistance at elevated temperatures
* performance virtually unaffected by high ambient humidity

Thermal

* outstanding thermal stability in normal use temperatures
* no decomposition below 350F for ortho grades. The FS(META) grades will withstand 380F continuously for most applications.
* eliminates fouling of metal contacts caused by condensation of decomposition products on contacts

Chemical

* resistant to
  * solvents
  * acids
  * alkalies
* fungus-proof

Applications

* electrical and electronic parts and components
  * switches
  * connectors
  * terminal boards
  * potting vessels
  * trim pots
  * coil forms
* structural parts
* color-fixed appliance handles

parts for:

* aircraft control systems
* aerospace control systems
* marine electronics
* telecommunications systems, etc.

PLASKON ELECTRONIC MATERIALS, INC.: Long-Glass-Filled DAP (Continued):

PLASKON Diallyl Phthalate (DAP) Molding Compounds
Typical Molded Properties and Molding Parameters

Product Number:

52-20-30:

Resin type: ortho
Physical form: flake
Bulk factor: 6
Flexural strength, psi: 15,000
Arc resistance, sec: 130
Flammability Rating: STD
Specific Gravity: 1.72
Water Absorption, %: 0.35

52-40-40:

Resin type: ortho
Physical form: flake
Bulk factor: 6
Flexural strength, psi: 15,000
Arc resistance, sec: 133
Flammability Rating: FR
Specific Gravity: 1.79
Water Absorption, %: 0.32

FS-4:

Resin type: meta
Physical form: flake
Bulk factor: 6
Flexural strength, psi: 15,000
Arc resistance, sec: 138
Flammability rating: STD
Specific Gravity: 1.64
Water Absorption:, %: 0.35

FS-80:

Resin type: meta
Physical form: flake
Bulk factor: 6
Flexural strength, psi: 15,000
Arc resistance, sec: 145
Flammability Rating: FR
Specific gravity: 1.74
Water absorption, %: 0.35

PLASKON ELECTRONIC MATERIALS, INC.: Mineral-Filled DAP:

PLASKON Diallyl Phthalate Molding Compounds
Mineral-Filled

PLASKON diallyl phthalate (DAP) molding compounds provide attractive cost/performance benefits which increase sales. These benefits are derived from the outstanding properties of the DAP molding compounds, namely:

* high dimensional stability
* excellent heat resistance
* excellent chemical resistance
* high insulation resistance

In addition the DAP molding compounds retain these properties even in extreme environments of high heat and humidity.

PLASKON DAP molding compounds' outstanding properties in all areas make DAP compounds the material of choice by design engineers when specifying materials for many electrical and electronic parts. They are part of the line of PLASKON thermosetting, high performance molding compounds. Other products offered in this line are epoxy molding compounds, phenolic molding compounds and an alkyd reinforced molding compound.

Two of these compounds are made from the ortho form of the DAP base resin. The other one is made from the meta isomer of the DAP base resin.

All the grades are readily adaptable to automatic feeding systems. These resins are suitable for compression, injection and transfer molding methods.

PLASKON ELECTRONIC MATERIALS, INC.: Mineral-Filled DAP (Continued):

SUMMARY OF PROPERTIES

PROPERTY:

Physical

* extremely good dimensional stability
* almost no post-mold shrinkage
* chemically inert - gives off no corrosive vapors

Mechanical

* excellent strength in compression
* excellent impact resistance
* exceptional thin-wall strength
* excellent for applications subjected to:
  * sudden, extreme jolts
  * severe stresses

Electrical

* retains high insulation resistance at elevated temperatures
* performs virtually unaffected by high ambient humidity

Thermal

* outstanding thermal stability in normal use temperatures
* no decomposition below ~ 300F for ortho grades. The FS (META) grades will withstand 325F continuously for most applications.

Chemical

* resistant to
  * solvents
  * acids
  * alkalies
* fungus-proof

Applications

* electrical and electronic parts and components:
  * switches
  * terminal boards
  * trim pots
  * connectors
  * potting vessels
  * coil forms
* structural parts
* color-fixed appliance handles

parts for:
* aircraft control systems
* marine electronics
* high reliance systems
* aerospace control systems
* telecommunications systems

PLASKON ELECTRONIC MATERIALS, INC.: Mineral-Filled DAP (Continued):

Product Number:

51-01 CAFR:

Resin type: ortho
Physical form: granular
Bulk factor: 2.3
Flexural strength, psi: 10,000
Arc Resistance, sec: 135
Flammability Rating: FR
Specific gravity: 1.75
Water absorption, %: 0.55

775 CAF:

Resin type: ortho
Physical form: granular
Bulk factor: 2.8
Flexural strength, psi: 10,000
Arc resistance, sec: 138
Flammability Rating: Std
Specific gravity: 1.67
Water absorption, %: 0.65

FS-6 CAF:

Resin type: meta
Physical form: granular
Bulk factor: 2.3
Flexural strength, psi: 10,500
Arc resistance, sec: 175
Flammability rating: Std
Specific gravity: 1.86
Water absorption, %: 0.50

PLASKON ELECTRONIC MATERIALS, INC.: Short-Glass-Filled DAP:

Performance-proven DAP molding compounds:

PLASKON DAP molding compounds demonstrate consistent reliability in electronic and electrical components with proven dependable performance. The cost/performance benefits of DAP molding compounds continue to make it the industry's best buy since it exhibits outstanding features and benefits such as:

* Excellent dimensional stability for greater latitude in part design.

* Superior heat resistance for minimal outgassing at elevated temperatures.

* Outstanding chemical resistance for greater latitude in type of application.

* High insulation resistance, even when wet, for superior electrical performance.

And DAP compounds continue to be the choice of design engineers who specify materials for demanding electrical and electronic parts. In addition, the long-term mechanical and electrical superiority of DAP compounds under severe environmental stress offers distinct advantages to those who write military specifications for connectors and other sophisticated electronic hardware.

Molding of PLASKON DAP compounds:

PLASKON DAP molding compounds are available in a variety of plasticities to meet most requirements for transfer, compression and injection molding.

PLASKON DAP compounds do not require special equipment They do not require fast-closing presses, special molding techniques or after-bakes. They mold as easily as most general purpose materials and will not stain or cause sticking to dies.

Curing cycle - preforming:

The use of preforms and dielectric preheating will substantially reduce cure times for both compression and transfer molding. Uniformly dense preforms having a durometer hardness in the 65-75 range are recommended. Dielectric preheating to a uniform temperature in the 190-230F range is recommended.

PLASKON ELECTRONIC MATERIALS, INC.: Short-Glass-Filled DAP (Continued):

APPLICATIONS:

PLASKON DAP molding compounds can be found in use throughout the electronics, aviation, telecommunications, medical and chemical industries, plus many others. Some typical applications include:

* Electrical and electronic parts and components. Switches, connectors, terminal boards, potting vessels, trim pots and coil forms.

* Structural parts.

* Parts for controls and other systems. Aviation controls, space controls, marine electronics, telecommunications and high-reliability systems.

* Medical instruments.

* Chemical-resistant systems. Pumps and valves.

* Waste treatment systems.

* Robotics.

SUMMARY OF PROPERTIES:

Physical
* Excellent long-term dimensional stability.
* Virtually no post-mold shrinkage.
* Chemically inert.

Mechanical
* Excellent flexural strength.
* High impact resistance.
* Exceptional thin-wall strength.
* Highly resistant to sudden, extreme jolts and severe stresses.

Electrical
* Retains high insulation resistance at elevated temperatures.
* Retains performance characteristics at high ambient humidity.

Thermal
* Outstanding thermal stability in normal-use temperatures
* No decomposition below 350F for ortho grades.
* FS (iso) grades withstand 380F continuously for most applications.

Chemical - Highly resistant to solvents, acids, alkalis, etc.

PLASKON ELECTRONIC MATERIALS, INC.: Short-Glass-Filled DAP (Continued):

Typical molded properties for DAP short-glass-filled molding compounds:

Product Number:

52-01:

Resin Type: ortho
Physical Form: granular
Bulk Factor: 2.3
Flexural Strength, psi: 12,000
UL-94 V-0: no
Specific Gravity: 1.93
Water Absorption, %, 48 hrs. @ 50C: 0.25

52-70-70VO:

Resin Type: ortho
Physical Form: granular
Bulk Factor: 2.3
Flexural Strength, psi: 14,000
UL-94 V-O: 1/16"
Specific Gravity: 1.91
Water Absorption, %, 48 hrs. @ 50C: 0.25

73-70-70c:
73-70-70VO:

Resin Type: ortho
Physical Form: granular
Bulk Factor: 2.4
Flexural Strength, psi: 13,000
UL-94 V-O: 1/16"
Specific Gravity: 1.91
Water Absorption, %, 48 hrs. @ 50C: 0.25

73-70-70R:

Resin Type: ortho
Physical Form: granular
Bulk Factor: 2.3
Flexural Strength, psi: 14,000
UL-94 V-O: 1/16"
Specific Gravity: 1.85
Water Absorption, %, 48 hrs. @ 50C: 0.30

PLASKON ELECTRONIC MATERIALS, INC.: PLASKON DAIP Molding Compounds:

Typical molded properties for DAIP short-glass-filled molding compounds:

Product Number:

FS-5:

Resin Type: iso
Physical Form: granular
Bulk Factor: 2.3
Flexural Strength, psi: 11,000
UL-94 V-O: no
Specific Gravity: 1.91
Water Absorption, %, 48 hrs. @ 50C: 0.35

FS-10:

Resin Type: iso
Physical Form: granular
Bulk Factor: 2.3
Flexural Strength, psi: 13,000
UL-94 V-O: 1/16"
Specific Gravity: 1.91
Water Absorption, %, 48 hrs. @ 50C: 0.35

PLASKON ELECTRONIC MATERIALS, INC.: PLASKON FS-10 VO:

PLASKON FS-10 VO short-glass-filled diallyl isophthalate (DAIP) molding compound demonstrates consistent reliability in electonic and electrical components with proven dependable performance. This compound exhibits excellent dimensional stability, superior heat resistance, outstanding chemical resistance and high insulation resistance.

PHYSICAL PROPERTIES:

Filler: Glass
Resin Type: Iso
Specific gravity, gm/cc: 1.91
Bulk Factor, gm/cc: 2.3
Oxygen Index: >50%
Water Absorption, % 48 hrs. @ 50C: 0.35
Flammability, UL-94, (0.0625"): VO
Flame Resistance
  Ignition Time/Burning Time, sec.: >90/<90
  Military Spec. (Mil-M-14G): SDG-F

MECHANICAL PROPERTIES:

Flexural Strength, kg/sq cm: 914
Flexural Strength, psi: 13,000
Tensile Strength, kg/sq cm: 533
Tensile Strength, psi: 8,000
Compressive Strength, psi: 28,000
Izod Impact, kg-m/cm notch: .033
Izod Impact, ft-lbs/in notch: 0.6
Barcol Hardness: 62

ELECTRICAL PROPERTIES:

Dielectric Constant, 1 MHz: 4.6
Dissipation Factor, 1 MHz: 0.016
UL Index, C: 200
Arc Resistance, sec.: 175
Dielectric Strength, v/ml: 350

PLASLOK CORP.: PLASLOK 200 Black:

PLASLOK 200 Black is a two-stage, woodflour filled, asbestos free general purpose phenolic molding compound formulated to provide maximum life for drilling or tapping tools. It has a low specific gravity, broad molding latitude, and an excellent balance of electrical and physical properties. PLASLOK 200 Black meets the requirements of ASTM D-700, Type 2.

PHYSICAL PROPERTIES	ASTM TEST	TYPICAL VALUES
Apparent Density, g/cc	D1895	0.60
Bulk Factor	D1895	2.3
Flow Grades	D731	8,10,12,16,18

MOLDED PROPERTIES		
Specific Gravity	D792	1.32-1.36
Water Absorption, %	D570	0.5
Flexural Strength, psi	D790	11,000
Tensile Strength, psi	D638	7,500
Impact Strength, ft lbs/in	D256	0.30
Compressive Strength, psi	D695	33,000
Deflection Temperature, 264 psi, F	D648	335F
Modulus in Tension, psi	D638	1.3 x 10 6
Molded Shrinkage, in/in	D955	0.007-0.010
Dielectric Strength, S/T, v/mil	D149	360
S/S, v/mil	D149	280
Dissipation Factor, 60 Hertz	D150	0.06
1 Kilohertz		0.04
1 Megahertz		0.03
Dielectric Constant, 60 Hertz	D150	5.8
1 Kilohertz		
1 Megahertz		
Volume Resistivity, ohm cm	D257	1 x 10 12

PLASLOK CORP.: PLASLOK 201 Black:

PLASLOK 201 is a two-stage, asbestos-free, woodflour and mineral-filled phenolic molding compound formulated to have an excellent balance of molded properties. This material provides good electrical insulation values and physical properties combined with versatile molding characteristics. PLASLOK 201 meets the requirements of ASTM Specification No. D700, Type 2.

PHYSICAL PROPERTIES	ASTM TEST	TYPICAL VALUES
Apparent Density, g/cc	D1895	0.60
Bulk Factor	D1895	2.3
Flow Grades	D731	10,12,14,16,18
MOLDED PROPERTIES		
Specific Gravity	D792	1.36-1.38
Water Absorption, %	D570	0.45
Flexural Strength, psi	D790	11,000
Tensile Strength, psi	D638	7,500
Impact Strength, ft lbs/in	D256	0.31
Compressive Strength, psi	D695	33,000
Deflection Temperature, 264 psi, F	D648	335F
Modulus in Tension, psi	D638	1.3×10^{6}
Molded Shrinkage, in/in	D955	0.005-0.010
Dielectric Strength, S/T, v/mil	D149	360
S/S, v/mil	D149	280
Dissipation Factor, 60 Hertz	D150	0.06
1 Kilohertz		0.04
1 Megahertz		0.03
Dielectric Constant, 60 Hertz	D150	5.8
1 Kilohertz		
1 Megahertz		
Volume Resistivity, ohm cm	D257	1×10^{12}
IEC Track Test (CTI)	U.L. 746A on .120" thickness	140
Flammability	U.L. 94 on .120" thickness	94 HB

PLASLOK CORP.: PLASLOK 202 Black:

PLASLOK 202AF Black is a two-stage, asbestos-free, woodflour and mineral-filled general purpose phenolic molding compound. It is formulated to provide improved impact and fatigue strength, good dimensional stability and excellent molded finish for automotive, electrical and mechanical applications. PLASLOK 202AF Black meets the requirements of ASTM D700,Types 2,3 or 11.

PHYSICAL PROPERTIES	ASTM TEST	TYPICAL VALUES
Apparent Density, g/cc	D1895	0.56
Bulk Factor	D1895	2.5
Flow	D731	10,12,16
MOLDED PROPERTIES		COMPRESSION
Specific Gravity	D792	1.42
Water Absorption, %	D570	0.4
Flexural Strength, psi	D790	11,000
Tensile Strength, psi	D638	8,000
Modulus in Tension, psi	D638	1.2 x 10 6
Compressive Strength, psi	D695	32,000
Impact Strength, ft lbs/in	D256	0.34
Molded Shrinkage, in/in	D955	0.006
Deflection Temperature, 264 psi, F	D648	350F
Dielectric Strength, S/T, v/mil	D149	300
S/S, v/mil	D149	275
Dissipation Factor, 60 Hertz	D150	0.10
1 Kilohertz	D150	0.06
1 Megahertz	D150	0.04
Dielectric Constant, 60 Hertz	D150	6.0
1 Kilohertz	D150	5.8
1 Megahertz	D150	4.8
Volume Resistivity, ohm cm	D257	1 x 10 12
IEC Track Test (CTI)	U.L. 746A	125 volts
	on .120" thickness	
Flammability	U.L. 94	94 V-1
	on .120" thickness	

PLASLOK CORP.: PLASLOK 203 Black:

PLASLOK 203 is a two-stage, asbestos-free, woodflour and mineral-filled phenolic molding compound formulated to have an excellent balance of properties. These provide good molded finish and dimensional stability along with other physical and electrical properties that suggests its use for appliance handles, automotive parts, wiring devices and electrical components. PLASLOK 203 Black meets the requirements of ASTM D700, Type 2.

PHYSICAL PROPERTIES	ASTM TEST	TYPICAL VALUES
Apparent Density, g/cc	D1895	0.65
Bulk Factor	D1895	2.2
Flow	D731	9,12,16
MOLDED PROPERTIES		
Specific Gravity	D792	1.42
Water Absorption, %	D570	0.50
Flexural Strength, psi	D790	11,000
Tensile Strength, psi	D638	7,000
Modulus in Tension, psi	D638	1.2 x 10 6
Compressive Strength, psi	D695	30,000
Impact Strength, psi, ft lbs/in	D256	0.28
Molded Shrinkage, in/in	D955	0.007-0.008
Deflection Temp., 264 psi, F	D648	350F
Dielectic Strength, S/T, v/mil	D149	325
S/S, v/mil	D149	280
Dissipation Factor, 60 Hertz	D150	0.09
1 Kilohertz	D150	0.06
1 Megahertz	D150	0.04
Dielectric Constant, 60 Hertz	D150	6.2
1 Kilohertz	D150	5.6
1 Megahertz	D150	4.8
Volume Resistivity, ohm cm	D257	1 x 10 12

PLASLOK CORP.: PLASLOK 203 Brown:

PLASLOK 203 Brown is a two-stage, asbestos-free, woodflour and mineral-filled phenolic molding compound formulated to have an excellent balance of properties. It provides dimensional stability and superior molded finish along with physical and electrical properties that suggests its use for appliance handles, wiring devices, and electronic parts. PLASLOK 203 Brown meets the requirements of ASTM D700, Type 2.

PHYSICAL PROPERTIES	ASTM TEST	TYPICAL VALUES
Apparent Density, g/cc	D1895	0.65
Bulk Factor	D1895	2.2
Flow	D731	9,12,16

MOLDED PROPERTIES		
Specific Gravity	D792	1.42
Water Absorption, %	D570	0.50
Flexural Strength, psi	D790	11,000
Tensile Strength, psi	D638	7,000
Modulus in Tension, psi	D638	1.2×10^{6}
Compressive Strength, psi	D695	30,000
Impact Strength, ft lbs/in	D256	0.28
Molded Shrinkage, in/in	D955	0.007-0.009
Deflection Temp. 264 psi, F	D648	340F
Dielectric Strength, S/T, v/mil	D149	325
S/S, v/mil	D149	275
Dissipation Factor, 60 Hertz	D150	0.08
1 Kilohertz	D150	0.06
1 Megahertz	D150	0.04
Dielectric Constant, 60 Hertz	D150	6.4
1 Kilohertz	D150	5.8
1 Megahertz	D150	4.7
Volume Resistivity, ohm cm	D257	1×10^{12}

PLASLOK CORP.: PLASLOK 204 Black:

PLASLOK 204 Black is a two-stage, asbestos free, woodflour and mineral filled general purpose phenolic molding compound. It is recommended for use in thin wall applications requiring U.L. 94V-O flammability ratings. PLASLOK 204 meets the requirements of ASTM D 700, Type 2.

PHYSICAL PROPERTIES	ASTM TEST	TYPICAL VALUES
Apparent Density, g/cc	D1895	0.60
Bulk Factor	D1895	2.4
Flow Grades	D731	10,12,15

MOLDED PROPERTIES		
Specific Gravity	D792	1.40-1.42
Water Absorption, %	D570	0.50
Flexural Strength, psi	D790	12,000
Modulus in Flexure, psi	D790	1 x 10 6
Tensile Strength, psi	D638	6,500
Compressive Strength, psi	D695	30,000
Impact Strength, Izod, ft lbs/in	D256	0.31
Molded Shrinkage, in/in	D955	0.004-0.006
Deflection Temperature, F, 264 psi	D648	300
Dielectric Strength, S/T, v/mil	D149	300
Dissipation Factor, 60 Hertz	D150	0.07
1 Kilohertz	D150	0.04
1 Megahertz	D150	0.03
Dielectric Constant, 60 Hertz	D150	7.0
1 Kilohertz	D150	6.2
1 Megahertz	D150	5.2
Volume Resistivity, ohm cm	D257	27 x 10 9
Flammability	U.L. 94	94V-O
	on .062" thickness	

PLASLOK CORP.: PLASLOK 301 Black:

PLASLOK 301 is a two-stage, asbestos-free, woodflour and fiber-filled phenolic molding compound. It has broad molding latitude combined with good electrical properties and improved impact strength. It is particularly suited for mechanical parts with thin sections that require wheelabrating and staking or pulling of inserts. PLASLOK 301 meets the requirements of ASTM D700, Type 12.

PHYSICAL PROPERTIES	ASTM TEST	TYPICAL VALUES
Apparent Density, g/cc	D1895	0.52
Bulk Factor	D1895	2.8
Flow Grades	D731	12,16

MOLDED PROPERTIES		
Specific Gravity	D792	1.38-1.42
Water Absorption, %	D570	1.0
Flexural Strength, psi	D790	11,000
Tensile Strength, psi	D638	6,000
Modulus in Tension, psi	D638	1.2 x 10 6
Compressive Strength, psi	D695	32,000
Impact Strength, ft lbs/in	D256	0.46
Molded Shrinkage, in/in	D955	0.006-0.008
Deflection Temperature, 264 psi, F	D648	350F
Dielectric Strength, S/T, v/mil	D149	300
S/S, v/mil	D149	275
Dissipation Factor, 60 Hertz	D150	0.09
1 Kilohertz	D150	0.09
1 Megahertz	D150	0.04
Dielectric Constant, 60 Hertz	D150	6.8
1 Kilohertz	D150	6.1
1 Megahertz	D150	4.2
Volume Resistivity, ohm cm	D257	1 x 10 12

PLASLOK CORP.: PLASLOK 302 Black:

PLASLOK 302 is a two-stage, asbestos-free, woodflour and fiber-filled phenolic molding compound. It has a low specific gravity, broad molding latitude, and an excellent balance of electrical and physical properties. Its high impact strength makes it particularly suited for thin-wall parts, or where mechanical assembly stress is experienced. It is suitable for automatic molding where toughness is desired and surface finish is not a major factor. PLASLOK 302 meets the requirements of ASTM D700, Type 12.

PHYSICAL PROPERTIES	ASTM TEST	TYPICAL VALUES
Apparent Density, g/cc	D1895	0.50
Bulk Factor	D1895	2.7
Flow Grades	D731	10,12,16
MOLDED PROPERTIES		
Specific Gravity	D792	1.31-1.34
Water Absorption, %	D570	0.8
Flexural Strength, psi	D790	11,000
Tensile Strength, psi	D638	6,000
Modulus in Tension, psi	D638	1.0×10^{6}
Compressive Strength, psi	D695	30,000
Impact Strength, ft lbs/in	D256	0.50
Molded Shrinkage, in/in	D955	0.004-0.007
Deflection Temperature, 264 psi, F	D648	330F
Dielectric Strength, S/T, v/mil	D149	310
S/S, v/mil	D149	280
Dissipation Factor, 60 Hertz	D150	0.20
1 Kilohertz	D150	0.10
1 Megahertz	D150	0.05
Dielectric Constant, 60 Hertz	D150	6.8
1 Kilohertz	D150	6.1
1 Megahertz	D150	4.2
Volume Resistivity, ohm cm	D257	1×10^{12}

PLASLOK CORP.: PLASLOK 304 Black:

PLASLOK 304 is a two-stage, asbestos-free, woodflour and cellulose-flock filled compound with improved impact strength. It is suiatble for automatic molding and recommended for thin-wall parts or where mechanical assembly stress is experienced. PLASLOK 304 meets the requirements of ASTM D700, Type 3.

PHYSICAL PROPERTIES	ASTM TEST	TYPICAL VALUES
Apparent Density, g/cc	D1895	0.50
Bulk Factor	D1895	2.6
Flow Grades	D731	10,12,16
MOLDED PROPERTIES		
Specific Gravity	D792	1.36-1.38
Water Absorption, %	D570	0.4
Flexural Strength, psi	D790	11,000
Tensile Strength, psi	D638	7,500
Modulus in Tension, psi	D638	1.2 x 10 6
Compressive Strength, psi	D695	30,000
Impact Strength, Izod, ft lbs/in	D256	0.36
Molded Shrinkage, in/in	D955	0.007-0.009
Deflection Temperature, F, 264 psi	D648	330
Dielectric Strength, S/T, v/mil	D149	340
S/S, v/mil	D149	280
Dissipation Factor, 60 Hertz	D150	0.14
1 Kilohertz	D150	0.08
1 Megahertz	D150	0.04
Dielectric Constant, 60 Hertz	D150	7.2
1 Kilohertz	D150	5.6
1 Megahertz	D150	4.8
Volume Resistivity, ohm cm	D257	1 x 10 12

PLASLOK CORP.: PLASLOK 306 Black:

PLASLOK 306 is a mineral and sisal filled two-stage compound designed to have the heat resistant characteristics of PLASLOK 504 Heat Resistant, while at the same time improved strength properties. It meets the requirements of ASTM D-700 Types 10 and 13.

PHYSICAL PROPERTIES	ASTM TEST	TYPICAL VALUES
Apparent Density, gm/cc	D1895	0.55
Bulk Factor	D1895	3.0
Flow Grades	D731	12,16
MOLDED PROPERTIES		
Specific Gravity	D792	1.52
Flexural Strength, psi	D790	6500
Impact Strength, ft lb/in	D256	.5
Molded Shrinkage, in/in	D955	.004
Dielectric Strength, v/mil	D149	225
IEC Track Index (CTI), volts	U.L.746A	180
Flammability (@ .250 in.)	U.L.94	V-O

PLASLOK CORP.: PLASLOK 401AF Black:

PLASLOK 401AF is a flock and mineral filled, asbestos free, two-stage compound suitable for use where dimensional stability and good electrical properties are desired. PLASLOK 401AF Black meets the requirements of ASTM D700, Types 13 or 22.

PHYSICAL PROPERTIES	ASTM TEST	TYPICAL VALUES
Apparent Density, g/cc	D1895	0.60-0.65
Bulk Factor	D1895	2.5-2.7
Flow Grades	D731	12,15,16,18
MOLDED PROPERTIES		COMPRESSION
Specific Gravity	D792	1.66
Water Absorption, %	D570	0.1
Flexural Strength, psi	D790	9,500
Tensile Strength, psi	D638	5,700
Modulus in Tension, psi	D638	1.6 x 10 6
Compressive Strength, psi	D695	30,000
Impact Strength, ft lbs/in	D256	0.32
Molded Shrinkage, in/in	D955	0.003
Deflection Temperature, 264 psi, F	D648	330
Dielectric Strength, S/T, v/mil	D149	375
S/S, v/mil	D149	300
Dissipation Factor, 60 Hertz	D150	0.16
1 Kilohertz	D150	0.12
1 Megahertz	D150	0.08
Dielectric Constant, 60 Hertz	D150	11.5
1 Kilohertz	D150	8.5
1 Megahertz	D150	5.5
Volume Resistivity, ohm cm	D257	1 x 10 11
IEC Track Test (CTI)	U.L. 746A	195 volts on .120" thickness
Flammability	U.L. 94	94V-O on .120" thickness

PLASLOK CORP.: PLASLOK 501 Black:

PLASLOK 501 is a two-stage, asbestos free, heat resistant, electrical molding compound. It is designed to withstand exposure to 425F for extended use and higher temperatures for shorter periods. It exhibits good dimensional stability and excellent U.L. flammability. PLASLOK 501 Black meets the requirements of ASTM D 700, Types 10, 13, and 22.

PHYSICAL PROPERTIES	ASTM TEST	TYPICAL VALUES
Apparent Density, g/cc	D1895	0.70
Bulk Factor	D1895	2.4
Flow Grades	D731	9,10,12,14,16
MOLDED PROPERTIES		
Specific Gravity	D792	1.69-1.75
Water Absorption, %	D570	0.2
Flexural Strength, psi	D790	11,000
Modulus in Flexure, psi	D790	1.3×10^6
Tensile Strength, psi	D638	6,500
Compressive Strength, psi	D695	32,000
Impact Strength, Izod, ft lbs/in	D256	0.30
Molded Shrinkage, in/in	D955	0.002-0.003
Deflection Temperature, F, 264 psi	D648	350
Dielectric Strength, S/T, v/mil	D149	330
S/S, v/mil	D149	310
Dissipation Factor, 60 Hertz	D150	0.12
1 Kilohertz	D150	0.07
1 Megahertz	D150	0.03
Dielectric Constant, 60 Hertz	D150	6.5
1 Kilohertz	D150	5.8
1 Megahertz	D150	4.8
Volume Resistivity, ohm cm	D257	1×10^{12}
IEC Track Test (CTI)	U.L. 746A	195 volts on .120" thickness
Flammability	U.L. 94	94V-O on .062" thickness

PLASLOK CORP.: PLASLOK 502 Black:

PLASLOK 502 Black is a two-stage, mineral-filled compound with improved impact strength. It is recommended for mechanical automotive components or applications where heat resistance and good dimensional stability are desired. PLASLOK 502 Black meets the requirements of ASTM D 700, Type 13.

PHYSICAL PROPERTIES	ASTM TEST	TYPICAL VALUES
Apparent Density, g/cc	D1895	0.70
Bulk Factor	D1895	2.3
Flow Grades	D731	8,10,12
MOLDED PROPERTIES		
Specific Gravity	D792	1.57-1.60
Water Absorption, %	D570	0.2
Flexural Strength, psi	D790	10,000
Modulus in Flexure, psi	D790	1.4×10^6
Tensile Strength, psi	D638	6,500
Compressive Strength, psi	D695	30,000
Impact Strength, Izod, ft lbs/in	D256	0.33
Molded Shrinkage, in/in	D955	0.003-0.004
Deflection Temperature, F, 264 psi	D648	360F
Dielectric Strength, S/T, v/mil	D149	360
Dissipation Factor, 60 Hertz	D150	0.20
1 Kilohertz	D150	0.12
1 Megahertz	D150	0.04
Dielectric Constant, 60 Hertz	D150	9.2
1 Kilohertz	D150	6.7
1 Megahertz	D150	5.1
Volume Resistivity, ohm cm	D257	3×10^{12}

PLASLOK CORP.: PLASLOK 504 Black:

PLASLOK 504 is a mineral and cellulose filled two-stage compound having better impact strength and heat resistance than woodflour filled general purpose materials. PLASLOK 504 Black is asbestos-free and meets the requirements of ASTM D700, Type 13.

PHYSICAL PROPERTIES	ASTM TEST	TYPICAL VALUES
Apparent Density, g/cc	D1895	0.60
Bulk Factor	D1895	2.5
Flow Grades	D731	9,12,15,18

MOLDED PROPERTIES		COMPRESSION
Specific Gravity	D792	1.52-1.54
Water Absorption, %	D570	0.3
Flexural Strength, psi	D790	10,500
Tensile Strength, psi	D638	6,500
Modulus in Tension, psi	D638	1.3×10^6
Compressive Strength, psi	D695	32,000
Impact Strength, ft lbs/in	D256	0.32
Molded Shrinkage, in/in	D955	0.006
Deflection Temperature, 264 psi, F	D648	345F
Dielectric Strength, S/T, v/mil	D149	400
S/S, v/mil	D149	300
Dissipation Factor, 60 Hertz	D150	0.09
1 Kilohertz	D150	0.05
1 Megahertz	D150	0.04
Dielectric Constant, 60 Hertz	D150	7.0
1 Kilohertz	D150	6.2
1 Megahertz	D150	5.0
Volume Resistivity, ohm cm	D257	1×10^{12}
IEC Track Test (CTI)	U.L. 746A	180 volts on .120" thickness
Flammability	U.L. 94	94 HB on .058" thickness

PLASLOK CORP.: PLASLOK 504 Brown:

PLASLOK 504 is a mineral and cellulose filled two-stage compound having better impact strength and heat resistance than woodflour filled general purpose materials. PLASLOK 504 Brown is asbestos-free and meets the requirements of ASTM D700, Type 13.

PHYSICAL PROPERTIES	ASTM TEST	TYPICAL VALUES
Apparent Density, g/cc	D1895	0.60
Bulk Factor	D1895	2.5
Flow Grades	D731	9,12,15,18
MOLDED PROPERTIES		
Specific Gravity	D792	1.49-1.54
Water Absorption, %	D570	0.3
Flexural Strength, psi	D790	11,000
Tensile Strength, psi	D638	7,000
Modulus in Tension, psi	D638	1.3 x 10 6
Compressive Strength, psi	D695	30,000
Impact Strength, ft lbs/in	D256	.30
Molded Shrinkage, in/in	D955	0.006
Deflection Temperature, 264 psi, F	D648	340
Dielectric Strength, S/T, v/mil	D149	350
S/S, v/mil	D149	275
Dissipation Factor, 60 Hertz	D150	0.09
1 Kilohertz	D150	0.05
1 Megahertz	D150	0.04
Dielectric Constant, 60 Hertz	D150	7.0
1 Kilohertz	D150	6.2
1 Megahertz	D150	5.0
Volume Resistivity, ohm cm	D257	1 X 10 12
IEC Track Test (CTI)	U.L. 746A on .120" thickness	180 volts
Flammability	U.L. 94 on .125" thickness	94 V-O

PLASLOK CORP.: PLASLOK 506 Black:

PLASLOK 506 Black is a two-stage, mineral-filled, heat resistant phenolic molding material. It is formulated for appliance applications such as utensil handles requiring repeated exposure to 500F. PLASLOK 506 Black meets the requirements of ASTM D 700, Type 13.

PHYSICAL PROPERTIES	ASTM TEST	TYPICAL VALUES
Apparent Density, g/cc	D1895	0.70
Bulk Factor	D1895	2.3
Flow Grades	D731	9,12,15,18
MOLDED PROPERTIES		
Specific Gravity	D792	1.44-1.47
Water Absorption, %	D570	0.30
Flexural Strength, psi	D790	10,000
Modulus in Flexure, psi	D790	1.5×10^6
Tensile Strength, psi	D638	6,000
Compressive Strength, psi	D695	30,000
Impact Strength, Izod, ft lbs/in	D256	0.30
Molded Shrinkage, in/in	D955	0.003-0.005
Deflection Temperature, F, 264 psi	D648	360F
Dielectric Strength, S/T, v/mil	D149	300
S/S, v/mil	D149	275
Dissipation Factor, 60 Hertz	D150	0.17
1 Kilohertz	D150	0.12
1 Megahertz	D150	0.06
Dielectric Constant, 60 Hertz	D150	14.0
1 Kilohertz	D150	10.0
1 Megahertz	D150	6.0
Volume Resistivity, ohm cm	D257	1×10^{12}
IEC Track Test (CTI)	U.L. 746A on .120" thickness	160
Flammability	U.L. 94 on .062" thickness	94V-O

PLASLOK CORP.: PLASLOK 506 Brown:

PLASLOK 506 Brown is a two-stage, mineral-filled, heat resistant phenolic molding material. It is formulated for appliance applications such as utensil handles requiring repeated exposure to 500F. PLASLOK 506 Brown meets the requirements of ASTM D 700, Type 13.

PHYSICAL PROPERTIES	ASTM TEST	TYPICAL VALUES
Apparent Density, g/cc	D1895	0.70
Bulk Factor	D1895	2.3
Flow Grades	D731	9,12,15,18
MOLDED PROPERTIES		
Specific Gravity	D792	1.44-1.47
Water Absorption, %	D570	0.25
Flexural Strength, psi	D790	11,000
Modulus in Flexure, psi	D790	1.5 x 10 6
Tensile Strength, psi	D638	6,000
Compressive Strength, psi	D695	30,000
Impact Strength, Izod, ft lbs/in	D256	0.30
Molded Shrinkage, in/in	D955	0.003-0.005
Deflection Temperature, F, 264 psi	D648	360F
Dielectric Strength, S/T, v/mil	D149	350
Dissipation Factor, 60 Hertz	D150	0.17
1 Kilohertz	D150	0.12
1 Megahertz	D150	0.06
Dielectric Constant, 60 Hertz	D150	14.0
1 Kilohertz	D150	10.0
1 Megahertz	D150	6.0
Volume Resistivity, ohm cm	D257	1 x 10 12
IEC Track Test (CTI)	U.L. 746A on .120" thickness	165
Flammability	U.L. 94 on .120" thickness	94V-O

PLASLOK CORP.: PLASLOK 508 Black:

PLASLOK 508 is a woodflour and mineral-filled, two-stage, heat resistant type phenolic molding compound. It will meet the requirements for ASTM Type 13, specification D700.

PHYSICAL PROPERTIES	ASTM TEST	TYPICAL VALUES
Apparent Density, g/cc	D1895	0.60
Bulk Factor	D1895	2.5
Flow Grades	D731	8,10,12,16
MOLDED PROPERTIES		COMPRESSION
Specific Gravity	D792	1.56
Water Absorption, %	D570	0.3
Flexural Strength, psi	D790	10,500
Tensile Strength, psi	D638	6,600
Modulus in Tension, psi	D638	1.6 x 10 6
Compressive Strength, psi	D695	30,000
Impact Strength, ft lbs/in	D256	0.28
Molded Shrinkage, in/in	D955	0.005-0.006
Deflection Temperature, 264 psi, F	D648	360F
Dielectric Strength, S/T, v/mil	D149	375
S/S, v/mil	D149	275
Dissipation Factor, 60 Hertz	D150	0.10
1 Kilohertz	D150	0.06
1 Megahertz	D150	0.04
Dielectric Constant, 60 Hertz	D150	7.2
1 Kilohertz	D150	6.0
1 Megahertz	D150	5.2
Volume Resistivity, ohm cm	D257	1 x 10 12

PLASLOK CORP.: PLASLOK 508 Brown:

PLASLOK 508 is a woodflour and mineral-filled, two-stage, heat resistant type phenolic molding compound. It will meet the requirements for ASTM Type 13, specification D700.

PHYSICAL PROPERTIES	ASTM TEST	TYPICAL VALUES
Apparent Density, g/cc	D1895	0.60
Bulk Factor	D1895	2.5
Flow Grades	D731	8,10,12,16
MOLDED PROPERTIES		COMPRESSION
Specific Gravity	D792	1.56
Water Absorption, %	D570	0.3
Flexural Strength, psi	D790	10,500
Tensile Strength, psi	D638	6,600
Modulus in Tension, psi	D638	1.6 x 10 6
Compressive Strength, psi	D695	30,000
Impact Strength, ft lbs/in	D256	0.28
Molded Shrinkage, in/in	D955	0.005-0.006
Deflection Temperature, 264 psi, F	D648	360F
Dielectric Strength, S/T, v/mil	D149	375
S/S, v/mil	D149	275
Dissipation Factor, 60 Hertz	D150	0.10
1 Kilohertz	D150	0.06
1 Megahertz	D150	0.04
Dielectric Constant, 60 Hertz	D150	7.2
1 Kilohertz	D150	6.0
1 Megahertz	D150	5.2
Volume Resistivity, ohm cm	D257	1 x 10 12

PLASLOK CORP.: PLASLOK 509 Black:

PLASLOK 509 is a two-stage, asbestos free, mineral and flock filled compound with medium high impact strength. It exhibits good shock resistance which makes it suitable for switch gear parts and appliance applications where addition mechanical strength is required. PLASLOK 509 meets the requirements of ASTM D-700, Types 3 or 12.

PHYSICAL PROPERTIES	ASTM TEST	TYPICAL VALUES
Apparent Density, g/cc	D1895	0.54
Bulk Factor	D1895	2.7
Flow Grades	D731	8,10,12,15,18
MOLDED PROPERTIES		
Specific Gravity	D792	1.48-1.52
Water Absorption, %	D570	0.5
Flexural Strength, psi	D790	10,500
Modulus in Flexure, psi	D790	1.0 x 10 6
Tensile Strength, psi	D638	7,200
Impact Strength, ft lbs/in	D256	0.56
Compressive Strength, psi	D695	33,000
Molded Shrinkage, in/in	D955	0.004-0.007
Deflection Temperature, F, 264 psi	D648	340
Dielectric Strength, S/T, v/mil	D149	300
S/S, v/mil	D149	275
Dissipation Factor, 60 Hertz	D150	0.12
1 Kilohertz	D150	0.08
1 Megahertz	D150	0.04
Dielectric Constant, 60 Hertz	D150	6.8
1 Kilohertz	D150	5.4
1 Megahertz	D150	4.5
Volume Resistivity, ohm cm	D257	1 x 10 11

PLASLOK CORP.: PLASLOK 678 Black:

PLASLOK 678 Black is a two-stage compound formulated to have and an excellent balance of molded properties. It is designed for use on applications where resistance to detergents, soap and water are required. PLASLOK 678 Black is asbestos-free and meets the requirements of ASTM D 700, Type 3.

PHYSICAL PROPERTIES	ASTM TEST	TYPICAL VALUES
Apparent Density, g/cc	D1895	0.52
Bulk Factor	D1895	2.8
Flow Grades	D731	8,10,12
MOLDED PROPERTIES		
Specific Gravity	D792	1.38-1.40
Water Absorption, %	D570	0.3
Flexural Strength, psi	D790	10,500
Tensile Strength, psi	D638	7,000
Modulus in Tension, psi	D638	1.2 x 10 6
Compressive Strength, psi	D695	30,000
Impact Strength, ft lbs/in	D256	0.38
Molded Shrinkage, in/in	D955	0.006-0.008
Deflection Temperature, 264 psi, F	D648	350F
Dielectric Strength, S/T, v/mil	D149	275
S/S, v/mil	D149	250
Dissipation Factor, 60 Hertz	D150	0.17
1 Kilohertz	D150	0.10
1 Megahertz	D150	0.05
Dielectric Constant, 60 Hertz	D150	7.9
1 Kilohertz	D150	6.2
1 Megahertz	D150	4.8
Volume Resistivity, ohm cm	D257	1 x 10 11

PLASLOK CORP.: PLASLOK 5000 Black:

PLASLOK 5000 is the replacement material for the former Union Carbide BMM-5000 Black. It is a low fines, dust controlled material mede from the same raw materials and the same specifications as BMM-5000. PLASLOK 5000 is a two-stage, asbestos-free woodflour filled general purpose phenolic material having versatile molding and preheating characteristics, fast cure, low specific gravity, excellent molded surface finish, and superior hot rigidity. It is designed to meet the requirements of ASTM specification 700, Type 2. PLASLOK 5000 is also available in a higher fines content equivalent to the BMG specification for BMG-5000.

PHYSICAL PROPERTIES	ASTM TEST	TYPICAL VALUES
Apparent Density, g/cc	D1895	0.62
Bulk Factor	D1895	2.2
Flow Grades	D731	10,12,14,16,18
MOLDED PROPERTIES		
Specific Gravity	D792	1.36-1.38
Water Absorption, %	D570	0.45
Flexural Strength, psi	D790	10,000
Modulus in Flexure, psi	D638	1.3 x 10 6
Tensile Strength, psi	D638	7,500
Compressive Strength, psi	D695	30,000
Impact Strength, Izod, ft lbs/in	D256	0.31
Molded Shrinkage, in/in	D955	.005-.010
Deflection Temperature, F, 264 psi	D648	360
Dielectric Strength, S/T, v/mil	D149	280
Dissipation Factor, 60 Hertz	D150	0.06
1 Kilohertz	D150	0.04
1 Megahertz	D150	0.03
Dielectric Constant, 60 Hertz	D150	5.8
1 Kilohertz	D150	5.0
1 Megahertz	D150	4.7
Volume Resistivity, ohm cm	D257	1 x 10 12

PLASLOK CORP.: PLASLOK 5303AF Black:

PLASLOK 5303AF Black is a two-stage, mineral filled material designed for use in appliance and automotive applications in which asbestos free materials are required. It will meet the requirements of ASTM D700, Types 13 and 22.

PHYSICAL PROPERTIES	ASTM TEST	TYPICAL VALUES
Apparent Density, g/cc	D1895	0.65
Bulk Factor	D1895	2.4
Flow Grades	D731	9,12,15,18
MOLDED PROPERTIES		
Specific Gravity	D792	1.56-1.60
Water Absorption, %	D570	0.18
Flexural Strength, psi	D790	10,000
Modulus in Flexure, psi	D790	1.4×10^6
Tensile Strength, psi	D638	6,500
Compressive Strength, psi	D695	30,000
Impact Strength, Izod, ft lbs/in	D256	.27
Molded Shrinkage, in/in	D955	0.002-0.004
Deflection Temperature, F, 264 psi	D648	360
Dielectric Strength, S/T, v/mil	D149	375
Dissipation Factor, 60 Hertz	D150	0.15
1 Kilohertz	D150	0.09
1 Megahertz	D150	0.04
Dielectric Constant, 60 Hertz	D150	8.0
1 Kilohertz	D150	6.4
1 Megahertz	D150	4.8
Volume Resistivity, ohm cm	D257	1×10^{12}
IEC Track Test (CTI)	U.L. 746A	185 volts on .120" thickness
Flammability	U.L. 94	94V-O on .062" thickness

PLASLOK CORP.: PLASLOK 5314 Black:

PLASLOK 5314 is the replacement material for the former Union Carbide BMMA-5314 Black. It is a two-stage, impact grade material made from the same raw materials and to the same specifications as BMMA-5314. It is suitable for compression, plunger or injection molding of parts where toughness is desired and surface finish is not a major factor. PLASLOK 5314 - 16 Flow replaces BMRS-5314. PLASLOK 5314 Black is asbestos-free and meets the requirements of ASTM D700, Types 3 or 12.

PHYSICAL PROPERTIES	ASTM TEST	TYPICAL VALUES
Apparent Density, g/cc	D1895	0.45
Bulk Factor	D1895	2.9
Flow Grades	D731	14,16
MOLDED PROPERTIES		
Specific Gravity	D792	1.34-1.38
Water Absorption, %	D570	0.8
Flexural Strength, psi	D790	9,000
Modulus in Flexure, psi	D790	1.0 x 10 6
Tensile Strength, psi	D638	6,700
Compressive Strength, psi	D695	25,000
Impact Strength, Izod, ft lbs/in	D256	0.50-.55
Molded Shrinkage, in/in	D955	0.004-0.007
Deflection Temperature, F, 264 psi	D648	320
Dielectric Strength, S/T, v/mil	D149	335
Dissipation Factor, 60 Hertz	D150	0.24
1 Kilohertz	D150	0.11
1 Megahertz	D150	0.05
Dielectric Constant, 60 Hertz	D150	8.5
1 Kilohertz	D150	7.1
1 Megahertz	D150	5.3
Volume Resistivity, ohm cm	D257	3 x 10 11

PLASLOK CORP.: PLASLOK S-5440 Black:

PLASLOK S-5440 is the replacement material for the former Union Carbide BMRS-5440 Black. It is an injection molding material made from the same raw materials and to the same specifications as BMRS-5440. It provides improved impact strength and dimensional stability with a smoother surface finish when compared to the usual woodflour filled general purpose compounds. PLASLOK S5440 Black phenolic molding compound is asbestos-free amd meets the requirements of ASTM D700, Types 2 and 11.

PHYSICAL PROPERTIES	ASTM TEST	TYPICAL VALUES
Apparent Density, g/cc	D1895	0.60
Bulk Factor	D1895	2.3
Flow Grades	D731	11,12,15,18,21
MOLDED PROPERTIES		
Specific Gravity	D792	1.38-1.40
Water Absorption, %	D570	0.30
Flexural Strength, psi	D790	8,500
Modulus in Flexure, psi	D790	1 x 10 6
Tensile Strength, psi	D638	6,500
Compressive Strength, psi	D695	30,000
Impact Strength, Izod, ft lbs/in	D256	0.31
Molded Shrinkage, in/in	D955	0.005-0.007
Deflection Temperature, F, 264 psi	D648	350F
Dissipation Factor, 60 Hertz	D150	0.20
1 Kilohertz	D150	0.10
1 Megahertz	D150	0.05
Dielectric Constant, 60 Hertz	D150	9.0
1 Kilohertz	D150	6.8
1 Megahertz	D150	5.4
Volume Resistivity, ohm cm	D257	2 x 10 12
IEC Track Test (CTI)	U.L. 746A .120" thickness	210 volts
Flammability	U.L. 94 .120" thickness	94 V-O

PLASLOK CORP.: PLASLOK 5440 Brown:

PLASLOK M-5440 Brown phenolic molding compound is asbestos-free and meets the requirements of ASTM D 700, Types 2 and 11. It provides improved impact strength and dimensional stability with a smoother surface finish when compared to the usual wood-flour-filled, general purpose phenolic compounds.

PHYSICAL PROPERTIES	ASTM TEST	TYPICAL VALUES
Apparent Density, g/cc	D1895	0.60
Bulk Factor	D1895	2.3
Flow Grades	D731	9,10,12,15,18
MOLDED PROPERTIES		
Specific Gravity	D792	1.40-1.42
Water Absorption, %	D570	0.30
Flexural Strength, psi	D790	8,500
Modulus in Flexure, psi	D790	1 x 10 6
Tensile Strength, psi	D638	6,500
Compressive Strength, psi	D695	30,000
Impact Strength, Izod, ft lbs/in	D256	0.31
Molded Shrinkage, in/in	D955	0.004-0.006
Deflection Temperature, F, 264 psi	D648	350F
Dielectric Strength, S/T, v/mil	D149	330
Dissipation Factor, 60 Hertz	D150	0.15
1 Kilohertz	D150	0.08
1 Megahertz	D150	0.04
Dielectric Constant, 60 Hertz	D150	7.2
1 Kilohertz	D150	6.4
1 Megahertz	D150	5.2
Volume Resistivity, ohm cm	D257	2 x 10 12
IEC Track Test (CTI)	U.L. 746A .120" thickness	210 volts
Flammability	U.L. 94 .120" thickness	94 V-1

PLASLOK CORP.: PLASLOK 5498 Black:

PLASLOK 5498 is a low fines, two-stage, woodflour-filled general purpose phenolic molding material.It is the replacement for Union Carbide BMM-5498 and can be supplied on request in a higher fines version to be equivalent to the former BMG-5498. It has low specific gravity, versatile molding and preheating characteristics, fast cure, rigid set, and excellent molded finish. PLASLOK 5498 Black meets the requirements of ASTM D 700, Type 2.

PHYSICAL PROPERTIES	ASTM TEST	TYPICAL VALUES
Apparent Density, g/cc	D1895	0.62
Bulk Factor	D1895	2.2
Flow Grades	D731	9,10,12,14,16,18
MOLDED PROPERTIES		
Specific Gravity	D792	1.36-1.38
Water Absorption, %	D570	0.45
Flexural Strength, psi	D790	10,000
Modulus in Flexure, psi	D790	1 x 10 6
Tensile Strength, pai	D638	7,500
Compressive Strength, psi	D695	30,000
Impact Strength, Izod, ft lbs/in	D256	0.31
Molded Shrinkage, in/in	D955	0.003-0.007
Deflection Temperature, F, 264 psi	D648	325
Dielectric Strength, S/T, v/mil	D149	315
Dissipation Factor, 60 Hertz	D150	0.07
1 Kilohertz	D150	0.04
1 Megahertz	D150	0.03
Dielectric Constant, 60 Hertz	D150	6.0
1 Kilohertz	D150	5.0
1 Megahertz	D150	4.7
Volume Resistivity, ohm cm	D257	2 x 10 12

PLASLOK CORP.: PLASLOK 7000 Black:

PLASLOK 7000 is made from the same raw materials and to the same specifications as the former Union Carbide BMM 7000 Black. It is a general purpose, asbestos-free, two-stage, woodflour-filled, phenolic molding compound with very fast cure combined with broad molding latitude. Excellent molded appearance, rigid set, low gravity, and good mold release are characteristic of this material. Predrying of the powder is recommended to achieve maximum electrical insulation properties. It is designed to meet the values for ASTM D-700 type 2.

PHYSICAL PROPERTIES	ASTM TEST	TYPICAL VALUES
Apparent Density, g/cc	D1895	0.60
Bulk Factor	D1895	2.3
Flow Grades	D731	6,8,10,12,15,18
MOLDED PROPERTIES		
Specific Gravity	D792	1.35-1.37
Water Absorption, %	D570	0.45
Flexural Strength, psi	D790	10,000
Tensile Strength, psi	D638	6,500
Modulus in Tension, psi	D638	1×10^{6}
Compressive Strength, psi	D695	34,000
Impact Strength, ft lbs/in	D256	0.30
Molded Shrinkage, in/in	D955	0.005-0.008
Deflection Temperature, 264 psi, F	D648	330F
Dielectric Strength, S/T, v/mil	D149	350
S/S, v/mil	D149	280
Dissipation Factor, 60 Hertz	D150	0.07
1 Kilohertz	D150	0.04
1 Megahertz	D150	0.04
Dielectric Constant, 60 Hertz	D150	6.0
1 Kilohertz	D150	5.0
1 Megahertz	D150	4.0
Volume Resistivity, ohm cm	D257	3×10^{12}

PLASLOK CORP.: PLASLOK 7000 Brown:

PLASLOK 7000 is made from the same raw materials and to the same specifications as the former Union Carbide BMM 7000 Black. It is a general purpose, asbestos-free, two-stage, woodflour-filled, phenolic molding compound with very fast cure combined with broad molding latitude. Excellent molded appearance, rigid set, low gravity, and good mold release are characteristic of this material. Predrying of the powder is recommended to achieve maximum electrical insulation properties. It is designed to meet the values for ASTM D-700 type 2.

PHYSICAL PROPERTIES	ASTM TEST	TYPICAL VALUES
Apparent Density, g/cc	D1895	0.60
Bulk Factor	D1895	2.3
Flow Grades	D731	6,8,10,12,15,18
MOLDED PROPERTIES		
Specific Gravity	D792	1.35-1.37
Water Absorption, %	D570	0.45
Flexural Strength, psi	D790	10,000
Tensile Strength, psi	D638	6,500
Modulus in Tension, psi	D638	1 x 10 6
Compressive Strength, psi	D695	34,000
Impact Strength, ft lbs/in	D256	0.30
Molded Shrinkage, in/in	D955	0.005-0.008
Deflection Temperature, 264 psi, F	D648	330F
Dielectric Strength, S/T, v/mil	D149	350
S/S, v/mil	D149	280
Dissipation Factor, 60 Hertz	D150	0.07
1 Kilohertz	D150	0.04
1 Megahertz	D150	0.04
Dielectric Constant, 60 Hertz	D150	6.0
1 Kilohertz	D150	5.0
1 Megahertz	D150	4.0
Volume resistivity, ohm cm	D257	3 x 10 12

PLASLOK CORP.: PLASLOK 7002 Black:

PLASLOK 7002 Black is a two-stage, asbestos-free, woodflour and mineral-filled, phenolic molding compound. It provides very fast cure, rigid set, versatile molding properties and a hard surface for the molding of cams and similar applications. PLASLOK 7002 meets the requirements of ASTM D700, Type 2. It is designed for parts requiring improved impact strength, surface smoothness and dimensional stability.

PHYSICAL PROPERTIES	ASTM TEST	TYPICAL VALUES
Apparent Density, g/cc	D1895	0.60
Bulk Factor	D1895	2.3
Flow Grades	D731	9,10,12,15,18
MOLDED PROPERTIES		
Specific Gravity	D792	1.40-1.42
Water Absorption, %	D570	0.40
Flexural Strength, psi	D790	9,000
Modulus in Flexure, psi	D790	1 x 10 6
Tensile Strength, psi	D638	6,500
Compressive Strength, psi	D695	30,000
Impact Strength, Izod, ft lbs/in	D256	0.31
Molded Shrinkage, in/in	D955	0.004-0.006
Deflection Temperature, F, 264 psi	D648	350F
Dielectric Strength, S/T, v/mil	D149	330
Dissipation Factor, 60 Hertz	D150	0.15
1 Kilohertz	D150	0.08
1 Megahertz	D150	0.04
Dielectric Constant, 60 Hertz	D150	7.2
1 Kilohertz	D150	6.4
1 Megahertz	D150	5.2
Volume Resistivity, ohm cm	D257	2 x 10 12
IEC Track Test (CTI)	U.L. 746A on .120" thickness	125
Flammability	U.L. 94 on .120" thickness	94V-1

PLASLOK CORP.: PLASLOK 7021 Black:

PLASLOK 7021 is the replacement material for the former Union Carbide BMRS-7021 Black. It is a two-stage, medium-high-impact material made from the same raw materials and to the same specifications as BMRS-7021. It is formulated for injection molding of switchgear parts or mechanical assembly requiring better impact strength. PLASLOK 7021 Black is asbestos-free and meets the requirements of ASTM D700, Types 3 or 12.

PHYSICAL PROPERTIES	ASTM TEST	TYPICAL VALUES
Apparent Density, g/cc	D1895	0.48
Bulk Factor	D1895	2.5
Flow Grades	D731	15,18,21
MOLDED PROPERTIES		
Specific Gravity	D792	1.38-1.40
Water Absorption, %	D570	0.5
Flexural Strength, psi	D790	10,800
Modulus in Flexure, psi	D790	1.0 x 10 6
Tensile Strength, psi	D638	6,000
Compressive Strength, psi	D695	33,000
Impact Strength, Izod, ft lbs/in	D256	0.45
Molded Shrinkage, in/in	D955	0.004-0.007
Deflection Temperature, F, 264 psi	D648	340F
Dielectric Strength, S/T, v/mil	D149	300
Dissipation Factor, 60 Hertz	D150	0.11
1 Kilohertz	D150	0.06
1 Megahertz	D150	0.04
Dielectric Constant, 60 Hertz	D150	6.0
1 Kilohertz	D150	5.4
1 Megahertz	D150	4.5
Volume Resistivity, ohm cm	D257	6 x 10 9

PLASTICS ENGINEERING CO.: PLENCO Alkyd Molding Compounds:

PLENCO alkyd molding compounds are dry, granular materials formulated with polyester resins.

Some PLENCO alkyd fomulations are designed to pass requirements for flammability ratings as specified by Underwriters' Laboratories and others. With mold shrinkage similar to phenolics, alkyds demonstrate excellent long-term dimensional stability.

Typical applications for alkyd molding compounds include automotive distributor caps, rotors, and coil caps, switch-gear, switch housing and circuit breakers and other electrical and electronic components. Alkyd formulations provided in a wide range of colors, including pastels, are also finding use in appliance applications.

Consideration and care must be taken when compression molding alkyd compounds because of their low melt viscosity, fast setting characteristics, and tendency to entrap air. On the other hand, PLENCO alkyds mold very well in reciprocating screw injection presses. Since the catalytic reaction of alkyd compounds begins at about 225F, material temperatures in the barrel below this point provide good "barel life."

Typical Values
PLENCO Alkyd Molding Compounds

General Purpose
- Specific Gravity (D792): 1.89
- Mold Shrinkage, in./in. (D955): 0.007
- Impact Strength (ft-lb./in of notch) (D256): 0.30
- Tensile Strength, psi (D638): 6,700
- Flexural Strength, psi (D790): 11,500
- Water Absorption (% Gain in Weight) (D570): 0.26
- Dielectric Strength (S.T.) volts/mil (D149): 340
- Dielectric Strength (18 kV at 180F): 1 hr+
- Power Factor @ 1 MHz (D150): 0.034
- Tracking Index, volts (IEC Pub. 112): 750+
- Arc Resistance, sec (D495): 188
- Deflection Temp. at 264 psi, F (D648): 360

PLASTICS ENGINEERING CO.: PLENCO Alkyd Molding Compounds (Continued):

Typical Values
PLENCO Alkyd Molding Compounds

Cellulose-Filled Colors
- Specific Gravity (D792): 1.83
- Mold Shrinkage, in/in (D955): 0.007
- Impact Strength (ft-lb./in. of notch) (D256): 0.30
- Tensile Strength, psi (D 638): 7,600
- Flexural Strength, psi (D790): 11,000
- Water Absorption (% Gain in Weight): 0.42
- Dielectric Strength (S.T.) volts/mil (D149): 340
- Dielectric Strength (18 kV at 180F): 1 hr+
- Power Factor @1 MHz (D150): 0.028
- Tracking Index, volts (IEC Pub. 112): 750+
- Arc Resistance, sec. (D495): 192
- Deflection Temp. at 264 psi, F (D648): 360

Copolymer Electrical:
- Specific Gravity: 1.83
- Mold Shrinkage, in./in. (D955): 0.0065
- Impact Strength (ft-lb/in of notch) (D256): 0.30
- Tensile Strength, psi (D638): 5,100
- Flexural Strength, psi (D790): 10,000
- Water Absorption (% Gain in Weight) (D570): 0.16
- Dielectric Strength (S.T.) volts/mil (D149): 330
- Dielectric Strength (18 kV at 180F): 1 hr+
- Power Factor @1 MHz (D150): 0.030
- Tracking Index, volts (IEC Pub. 112): 750+
- Arc Resistance, sec (D495): 191
- Deflection Temp. at 264 psi, F (D648): 300

Electrical and Electronic:
- Specific Gravity (D792): 1.96
- Mold Shrinkage, in/in (D955): 0.0045
- Impact Strength (ft-lb/in of notch) (D256): 0.45
- Tensile Strength, psi (D638): 6,000
- Flexural Strength, psi (D790): 14,500
- Water Absorption (% gain in Weight) (D570): 0.12
- Dielectric Strength (18 kV at 180F): 1 hr+
- Power Factor @ 1 MHz (D150): 0.016
- Tracking Index, volts (IEC Pub. 112): 750+
- Arc Resistance, sec. (D495): 191
- Deflection Temp. at 264 psi, F (D648): 428

Electrical:
- Specific Gravity (D792): 1.83
- Mold Shrinkage, in/in (D955): 0.006
- Impact Strength (ft-lb./in. of notch) (D256): 0.45
- Tensile Strength, psi (D638): 5,500
- Flexural Strength, psi (D790): 12,000
- Water Absorption (% Gain in Weight) (D570): 0.34
- Dielectric Strength (18 kV at 180F): 1 hr+
- Power Factor @1 MHz (D150): 0.020
- Tracking Index, volts (IEC Pub 112): 750+
- Arc Resistance, sec. (D495): 193
- Deflection Temp. at 264 psi., F (D648): 430

PLASTICS ENGINEERING CO.: PLENCO Melamine-Phenolic Molding Compounds:

Plastics Engineering Compound produces a line of colored compounds which contain melamine-phenolic resins. These resins are produced by reacting melamine, phenol, and formaldehyde and in many ways are similar to single-stage phenolics.

To produce PLENCO melamine-phenolic compounds the resin is combined with coloring agents, lubricants, modifiers, and fillers such as woodflour, cellulose, or various mineral types.

A series of standard and custom colored formulations are available both with woodflour and cellulose fillers. PLENCO mineral-phenolic compounds are also available in electrical grades for use in those applications requiring good electric arc and track resistance. These compounds serve well for switchgear and other similar applications because of their good electrical insulation and resistance to abrasion during repetitive switching cycles.

Typical Values
PLENCO Melamine-Phenolic Molding Compounds

Woodflour Filled
- Specific Gravity (D792): 1.50
- Mold Shrinkage, in/in (D955): 0.007
- Post Shrinkage, in/in (240 hrs. @ 220F): 0.012
- Impact Strength (ft-lb/in of notch) (D256): 0.26
- Tensile Strength, psi (D638): 5,500
- Flexural Strength, psi (D790): 8,000
- Water Absorption (% Gain in Weight) (D570): 0.40
- Dielectric Strength (S.T.) volts/mil (D149): 300
- Power Factor @ 1 MHz (D150): 0.07
- Heat Resistance (24-hr. exposure) F: 300

Cellulose Filled
- Specific Gravity (D792): 1.65
- Mold Shrinkage, in/in (D955): 0.0065
- Post Shrinkage, in/in. (240 hrs. @ 220F): 0.012
- Impact Strength (ft-lb/in. of notch) (D256): 0.40
- Tensile Strength, psi (D638): 6,500
- Flexural Strength, psi (D790): 10,000
- Water Absorption (% Gain in Weight) (D570): 0.40
- Power Factor @ 1 MHz (D150): 0.06
- Heat Resistance (24-hr exposure) F: 300

Electrical Grades
- Specific Gravity (D792): 1.58-1.65
- Mold Shrinkage, in/in: 0.006-0.008
- Post Shrinkage, in/in (240 hrs. @ 220F): 0.010-0.015
- Impact Strength (ft-lb/in of notch) (D256): 0.26-0.34
- Tensile Strength, psi (D638): 6,500-7,500
- Flexural Strength, psi (D790): 9,000
- Water Absorption (% Gain in Weight) (D570): 0.20-0.50
- Dielectric Strength (S.T.) volts/mil (D149): 225-250
- Power Factor @1 MHz (D150): 0.05-0.07
- Heat Resistance (24-hr exposure) F: 300

PLASTICS ENGINEERING CO.: PLENCO Phenolic Molding Compounds:

At a relatively low unit cost, phenolics provide outstanding performance in the areas of heat resistance, chemical resistance, electrical insulation; surface hardness; chemical, solvent, and detergent resistance; and dimensional stability. As there are many PLENCO phenolic molding compounds from which to choose, depending on filler composition, formulations can usually be selected which will meet the requirements of a wide range of molding techniques and end-product applications.

PLENCO phenolic molding compounds are mixtures of the phenolic resin with various fillers, coloring agents, lubricants and modifiers. This mixture is homogenized and densified on rolling mills or in extruders. The resulting compound is sold in the form of granules which have been carefully classified into mesh size ranges for optimum handling in the molding process.

Typical Values
PLENCO Phenolic Molding Compounds

General Purpose

Specific Gravity (D792): 1.36-1.45
Mold Shrinkage, in/in (D955): 0.005-0.0065
Impact Strength (ft-lb/in of notch) (D256): 0.27-0.32
Tensile Strength, psi (D638): 6,500-7,500
Flexural Strength, psi (D790): 9,000-11,000
Water Absorption (% Gain in Weight) (D 570): 0.4-0.6
Dielectric Strength (S.T.) volts/mil (D149): 200-350
Power Factor @1 Mhz (D150): 0.035-0.06
Heat Resistance (24-hr exposure) F: 300-400

Impact Resistant

Specific Gravity (D792): 1.36-1.48
Mold Shrinkage, in./in. (D955): 0.005-0.007
Impact Strength (ft-lb/in of notch) (D256): 0.33-0.50
Tensile Strength, psi (D638): 6,000-7,500
Flexural Strength, psi (D790): 9,000-10,500
Water Absorption (% Gain in Weight) (D570): 0.4-0.8
Dielectric Strength (S.T.) volts/mil (D149): 200-325
Power Factor @ 1 MHz (D150): 0.04-0.08
Heat Resistance (24-hr exposure) F: 300-350

PLASTICS ENGINEERING CO.: PLENCO Phenolic Molding Compounds (Continued):

Typical Values
PLENCO Phenolic Molding Compounds

Heat Resistant

Specific Gravity (D792): 1.45-1.85
Mold Shrinkage, in/in (D955): 0.003-0.005
Impact Strength (ft-lb/in of notch) (D256): 0.26-0.37
Tensile Strength, psi (D638): 5,000-7,000
Flexural Strength, psi (D790): 8,000-10,000
Water Absorption (% Gain in Weight) (D570): 0.04-0.4
Dielectric Strength (S.T.) volts/mil (D149): 225-350
Power Factor @1 MHz (D150): 0.05-0.03
Heat Resistance (24-hr exposure) F: 400-500

Low Loss Electrical

Specific Gravity (D792): 1.40-1.82
Mold Shrinkage, in/in. (D955): 0.003-0.005
Impact Strength (ft-lb/in of notch) (D256): 0.28-0.30
Tensile Strength (psi) (D638): 5,000-7,000
Flexural Strength, psi (D790): 7,000-10,000
Water Absorption (% Gain in Weight) (D570): 0.03-0.4
Dielectric Strength (S.T.) volts/mil (D149): 300-375
Power Factor @ 1 MHz (D150): 0.013-0.03
Heat Resistance (24-hr. exposure) F: 300-450

RTP CO.: Structural Materials:

Product Name:

RTP 100 Series:
Polypropylene - Glass Fiber - Heat Stabilized

RTP 100 Series:
Polypropylene - Glass Fiber - FDA Type

RTP 100ccA Series:
Polypropylene - Glass Fiber - Structural - Chemically Combined - Heat Stabilized

RTP 100cc Series:
Polypropylene - Structural - Glass Fiber - Chemically Combined - Heat Stabilized

RTP 100 GB Series:
Polypropylene - Glass Bead - Heat Stabilized

RTP 100 Series:
Polypropylenes - Structural - Glass Fiber & Mineral - Heat Stabilized

RTP 100 Series:
Polypropylene - Structural - Talc Filled - Heat Stabilized

RTP 128 HI:
Polypropylene - Talc - Impact Modified - Heat Stabilized

RTP 100 Series:
Polypropylene - Mineral Filled - Heat Stabilized

RTP 140 Series:
Polypropylene - Calcium Carbonate - Heat Stabilized

RTP 141 HI:
Polypropylene - Mineral Filled - Impact Modified - Structural

RTP 100 Series:
Mica - Heat Stabilized

RTP 170 Series:
Polypropylene Combinations of Glass Fiber, Glass Beads & Minerals - Heat Stabilized

RTP 100 Series:
RTP 199x50507 - Polypropylene - Structural - Mineral Filled-Heat Stabilized

RTP CO.: Structural Materials(Continued):

Product Name:

RTP 200 Series:
Nylon 6/6 - Glass Fiber

RTP 200 Series:
Nylon 6/6 - Glass Fiber

RTP 200 Series:
Nylon 6/6 - Glass Bead

RTP 200 Series:
Nylon 6/6 - Glass Bead/Glass Fiber

RTP 200 Series:
Nylon 6/6 - Heat Stabilized - Glass Fiber

RTP 200 Series:
Nylon 6/6 - Combinations of Glass Fiber, Mineral, Glass Beads

RTP 220 Series:
Nylon 6/6 - Mineral, Talc

RTP 200 Series:
Nylon 6/6 - Impact Modified - Glass Fiber - Structural

RTP 200A Series:
Nylon 6 - Glass Fiber

RTP-227A:
Nylon 6 - Mineral

RTP 200A Series:
Nylon 6 - Glass Fiber & Mineral

RTP 200B Series:
Nylon 6/10 - Glass Fiber

RTP 200C Series:
Nylon 11 - Glass Fiber

RTP 200D Series:
Nylon 6/12 - Glass Fiber

RTP 299D x 29674B:
Nylon 6/12 - Glass Fiber

RTP 200E Series:
Amorphous Nylon, Amorphous - Glass Fiber

RTP-200H Series:
Nylon, High Impact - Structural - Glass Fiber

RTP CO.: Structural Materials(Continued):

Product Name:

RTP 300 Series:
Polycarbonate - Glass Fiber

RTP 300 Series:
Polycarbonate - Glass Fiber - FDA Type Glass - Structural

RTP 400 Series:
Polystyrene - Glass Fiber

RTP 400 Series:
RTP 499x50437 - Styrene Maleic Anhydride (SMA) Terpolymer - Structural

RTP 500 Series:
SAN - Glass Fiber

RTP 500 Series:
SAN - Glass Fiber & Mineral

RTP 600 Series:
ABS - Glass Fiber

RTP 700 Series:
HDPE - Glass Fiber

RTP 700 Series:
HDPE - Mica - Structural

RTP 800 Series:
Acetal, Copolymer - Glass Fiber

RTP 800 Series:
RTP 804cc - Acetal, Chemically coupled - Glass fiber reinforced

RTP 800 Series:
Acetal, Copolymer - Glass Bead

RTP 1000 Series:
PBT - Glass Fiber - Structural

RTP 1000 Series:
PBT - Glass Bead

RTP 1020 Series:
PBT - Combinations of Glass Fiber, Glass Beads, Mineral

RTP 1000 Series:
PBT - Combinations of Glass Fiber, Glass Beads, Mineral

RTP CO.: Structural Materials(Continued):

Product Name:

RTP 1100 Series:
Polyethylene-Terephthalate(PET) - Glass Fiber - Structural

RTP 1100 Series:
Polyethylene-Terephthalate(PET) - Glass Fiber/Mineral

RTP 1600 Series:
Polyarylsulfone(PAS) - Glass Fiber - RTP 1603 - Structural

RTP 1700 Series:
PPO, Modified - Glass Fiber

RTP 1707:
PPO Modified - Glass Fiber - Structural

RTP 900 Series:
Polysulfone - Glass Fiber

RTP 900 Series:
Polysulfone - Carbon Fiber(PAN) - High Temperature - Electrically Conductive

RTP 900 Series:
Polysulfone - Glass Fiber - High Temp - FDA Materials

RTP 1300 Series:
Polyphenylene Sulfide(PPS) - Glass Fiber - High Temperature

RTP 1370 Series:
Polyphenylene Sulfide(PPS) - Glass Fiber & Mineral

RTP 1300 Series:
RTP 1399 x 50268A & B - Polyphenylene Sulfide - Glass Fiber-Mineral - High Temperature

RTP 1400 Series:
Polyethersulfone(PES) - Printed Wiring Board(PWB) - PWB 14A, 14B, & 14C - High Temperature

RTP-1400 Series:
Polyethersulfone(PES) - Glass Fiber

RTP 1475:
Polyethersulfone(PES) - Glass Fiber - Electrical Grade

RTP 1400 Series:
Polyethersulfone(PES) - Carbon Fiber(PAN) - High Temperature Electrically Conductive

RTP CO.: Structural Materials(Continued):

Product Name:

RTP 2100 Series:
Polyetherimide(PEI) - Printed Wiring Board(PWB) - PWB 21A, 21B, & 21C - High Temperature

RTP 2100 Series:
Polyetherimide(PEI) - Glass Fiber - Reinforced

RTP 2100 Series:
Polyetherimide(PEI) - Carbon Fiber(PAN) - High Temperature - Electrically Conductive

RTP 2200 Series:
Polyetheretherketone - PEEK - Glass Fiber Reinforced

RTP 2200 Series:
Polyether-ether-Ketone - PEEK - Carbon Fiber(PAN) - High Temperature - Electrically Conductive

RTP 2200 Series:
Polyether-ether Ketone - PEEK - High Temperature - Wear Resistant - RTP 2299x52566B

RTP 2200 Series:
Polyether-ether Ketone - PEEK - High Temperature - Wear Resistant - RTP 2299x52566B

High Temperature LCP 3400 Series:
Liquid Crystal Polymer - Reinforced

RTP 100 FR Series:
Polypropylene - Glass Fiber - Flame Retardant

RTP 100 Series:
Polypropylene - Non-reinforced - Flame Retardant

RTP 150 HI:
Polypropylene - Co-Polymer - Flame Retardant

RTP 155:
Polypropylene - Non-reinforced - Flame Retardant

RTP 200 FR:
Nylon 6/6 - Non-Reinforced - Flame Retardant

RTP 200 FR Series:
Nylon 6/6 - Glass Fiber - Flame Retardant

RTP CO.: Structural Materials(Continued):

Product Name:

RTP 204 GB FR:
Nylon 6/6 - Glass Bead - Flame Retardant

RTP-227 FR:
Nylon 6/6 - Mineral - Flame Retardant

RTP 200 TFE FR Series:
Nylon 6/6 - Glass Fiber - Flame Retardant - Lubricated

RTP-200A Series:
Nylon - Glass Fiber - Flame Retardant

RTP 280 Series:
RTP 282 - Nylon 6/6 - Carbon Fiber(PAN) - Flame Retardant

RTP 204D FR:
RTP 205D FR - Nylon 6/12 - Glass Fiber - Flame Retardant

RTP-204D TFE 15 FR:
Nylon 6/12 - Glass Fiber - Wear Resistant/Flame Retardant

RTP-205D FR:
Nylon 6/12 - Glass Fiber - Flame Retardant

RTP-200H-FR:
Nylon, High Impact - Flame Retardant

RTP-205H FR:
Nylon, High Impact - Glass Fiber - Flame Retardant

RTP 200 FR Series:
RTP 299 x 50252 FR - Nylon Alloy - Glass Fiber - Flame Retardant

RTP 300 Series:
Polycarbonate - Glass Fiber - Structural

RTP 301-500:
Polycarbonate - Flame Retardant

RTP 300 Series:
Polycarbonate - Glass Fiber - Structural - Flame Retardant

RTP 383 FR:
Polycarbonate - Carbon Fiber(PAN) - Flame Retardant

RTP CO.: Structural Materials(Continued):

Product Name:

RTP 400 Series:
RTP 499 x 50246 - Polystyrene - Flame Retardant

RTP 500 FR Series:
SAN - Glass Fiber - Flame Retardant

RTP 600 FR Series:
ABS - Glass Fiber - Flame Retardant

RTP 600 Series:
RTP 600 FR TFE 10 - ABS - Flame Retardant - Wear Resistant

RTP 680-FR Series:
ABS - Carbon Fiber(PAN) - Flame Retardant

RTP 600 Series:
RTP 699x50496 - ABS - Electrically Conductive - Flame Retardant

RTP-700 FR:
Polyethylene - Non-Reinforced - Flame Retardant

RTP 900 Series:
Polysulfone - Glass Fiber

RTP 1000 Series:
PBT - Flame Retardant

RTP 1000 Series:
Polybutylene Terephthalate(PBT) - Flame Retardant

PBT:
Glass Fiber - PTFE Lubricated - Flame Retardant - RTP 1005 FR TFE 15

RTP-1099 x 29687:
PBT - Warp Resistance - Flame Retardant

RTP 1100 Series:
Polyethylene-Terephthalate - (PET) - Flame Retardant

RTP 1300 Series:
Polyphenylene Sulfide - (PPS) - Glass Fiber

RTP-1400 Series:
Polyethersulfone - (PES) - Glass Fiber

RTP CO.: Structural Materials(Continued):

Product Name:

ESD 100 Series:
Polypropylene - Non Reinforced - Electrically Conductive

ESD 100 FR Series:
Polypropylene - Flame Retardant - Electrically Conductive

ESD 100 Series:
Polypropylene - Glass Fiber - Electrically Conductive

ESD 100 Series:
Polypropylene - Mineral - Electrically Conductive

ESD-C-180:
Polypropylene Co-Polymer - PAN-Carbon Fiber - Electrically Conductive

ESD 200H Series:
Nylon 6/6 - High Impact - Glass Fiber Reinforced - Electrically Conductive

ESD 200H FR:
Nylon 6/6 - Non Reinforced - Electrically Conductive - Flame Retardant

ESD 200H Series:
Nylon 6/6 - High Impact - Non Reinforced, Glass Fiber - Electrically Conductive

ESD 200H Series:
Nylon 6/6 - High Impact - Glass Fiber Reinforced - Electrically Conductive

ESD-C-280H:
High Impact Nylon 6/6 - PAN-Carbon Fiber - Electrically Conductive

RTP 200 HEC Series:
Nylon 6/6 - Nickel Plated - Graphite Fiber - High Electrical Conductivity

ESD-C-380:
Polycarbonate - Pan Carbon Fiber - Electrically Conductive

RTP 300 HEC Series:
Polycarbonate - Nickel Plated Graphite Fiber - High Electrical Conductivity

RTP CO.: Structural Materials(Continued):

Product Name:

RTP 382 HEC FR:
Polycarbonate - Nickel Plated Graphite Fiber - High Electrical Conductivity

RTP 300 HEC Series:
Polycarbonate - Nickel Plated Graphite Fiber - Flame Retardant

ESD-C-680:
ABS - Pan Carbon Fiber - Electrically Conductive

RTP 600 HEC FR Series:
Flame Retardant ABS - Nickel Plated Graphite Fiber - High Electrical Conductivity

ESD 700 Series:
High Density Polyethylene(HDPE) - Non Reinforced - Electrically Conductive

ESD 700 FR:
High Density Polyethylene(HDPE) - Non Reinforced - Flame Retardant - Electrically Conductive

ESD 700A Series:
Low Density Polyethylene(LDPE) - Electrically Conductive

ESD-C-780:
High Density Polyethylene - PAN-Carbon Fiber - Electrically Conductive

ESD 901 Series:
Polysulfone - Glass Fiber Reinforced - Electrically Conductive

ESD-C-980:
Polysulfone - PAN-Carbon Fiber - Electrically Conductive

ESD 1000 Series:
Polybutylene Terephthalate(PBT) - Non Reinforced - Electrically Conductive

ESD 1000 Series:
Polybutylene Terephthalate(PBT) - Glass Fiber Reinforced - Electrically Conductive

ESD 1005 FR Series:
PBT - Glass Fiber Reinforced - Flame Retardant - Electrically Conductive

ESD-C-1080:
PBT - PAN-Carbon Fiber - Electrically Conductive

RTP CO.: Structural Materials(Continued):

Product Name:

ESD 1300 Series:
Polyphenylene Sulfide(PPS) - Glass Fiber Reinforced - Electrically Conductive

ESD-C-1380:
Polyphenylene Sulfide - PAN-Carbon Fiber - Electrically Conductive

RTP 1300 HEC Series:
Polyphenylene Sulfide(PPS) - Nickel Plated Graphite Fiber - High Electrical Conductivity

RTP 1400 HEC Series:
Polyethersulfone - (PES) - Nickel Plated Graphite Fiber - Electrically Conductive

ESD-C-2180:
Polyetherimide - PAN-Carbon Fiber - Electrically Conductive

ESD-C-2280:
PEEK - PAN-Carbon Fiber - Electrically Conductive

RTP 2280 HEC Series:
Polyetherether Ketone - PEEK - Nickel Plated Graphite Fiber

ESD 2700 Series:
Styrene Butadiene Rubber - Non Reinforced - Electrically Conductive

ESD 2700 Series:
Styrene Butadiene Rubber - Non Reinforced - Electrically Conductive

ESD 2800 Series:
Polyolefin Rubber - Non Reinforced - Electrically Conductive

ESD 3100 Series:
Perfluorolalkoxy(PFA) - Non Reinforced - Electrically Condutive

ESD 3200 Series:
Ethylene Tetrafluoroethylene(ETFE) - Non Reinforced - Electrically Conductive

RTP 181 HI:
Polypropylene Copolymer - PAN-Carbon Fiber - Carbon Fiber

RTP 280 Series:
RTP 282 - Nylon 6/6 - Carbon Fiber(PAN)

RTP CO.: Structural Materials(Continued):

Product Name:

RTP 280 Series:
Nylon 6/6 - Carbon Fiber(PAN)

RTP 181 HI:
Polypropylene Copolymer - PAN-Carbon Fiber - Carbon Fiber

RTP 280 Series:
RTP 282 - Nylon 6/6 - Carbon Fiber(PAN)

RTP 280 Series:
Nylon 6/6 - Carbon Fiber(PAN)

RTP 200 HEC Series:
Nylon 6/6 - Nickel Plated - Graphite Fiber - High Electrical Conductivity

RTP-200P Series:
Nylon 6/6 - Pitch Fiber & Glass Fiber - Conductive

RTP-280P Series:
Nylon 6/6 - Pitch Fiber - Conductive

RTP-200H-CF Series:
Nylon, High Impact - Carbon Fiber(PAN) - Glass Fiber

RTP 280H Series:
Nylon, High Impact - Carbon Fiber(PAN)

RTP 281-H:
High Impact Nylon 6/6 - PAN-Carbon Fiber

RTP 380 Series:
Polycarbonate - Carbon Fiber(PAN)

RTP 383 FR:
Polycarbonate - Carbon Fiber(PAN) - Flame Retardant

RTP 300 HEC Series:
Polycarbonate - Nickel Plated Graphite Fiber - High Electrical Conductivity

RTP 382 HEC FR:
Polycarbonate - Nickel Plated Graphite Fiber - High Electrical Conductivity

RTP 300 HEC Series:
Polycarbonate - Nickel Plated Graphite Fiber - Flame Retardant

RTP 680 Series:
ABS - Carbon Fiber(PAN)

RTP CO.: Structural Materials(Continued):

Product Name:

RTP 680-FR Series:
ABS - Carbon Fiber(PAN) - Flame Retardant

RTP 600 HEC FR Series:
Flame Retardant ABS - Nickel Plated Graphite Fiber - High Electrical Conductivity

RTP 900 Series:
Polysulfone - Carbon Fiber(PAN) - High Temperature - Electrically Conductive

RTP 1080 Series:
PBT - Carbon Fiber(PAN)

RTP 1000 Series:
RTP 1085 TFE 15:
PBT - Carbon Fiber - PTFE Lubricated - Electrically Conductive

RTP 1380 Series:
Polyphenylene Sulfide - PPS - Carbon Fiber(PAN)

RTP 200 TFE Series:
Nylon 6/6 - PTFE - Lubricated

RTP 200 Series:
Nylon 6/6 - Glass Fiber - PTFE Lubricated

RTP 200 Series:
RTP 200 Series - RTP 204 G 10 - 6/6 Nylon - Glass Fiber - Lubricated

RTP 1300 HEC Series:
Polyphenylene Sulfide(PPS) - Nickel Plated Graphite Fiber - High Electrical Conductivity

RTP 1300 Series:
Polyphenylene Sulfide - PPS - Carbon Fiber(PAN) - High Modulus

RTP 1400 Series:
Polyethersulfone(PES) - Carbon Fiber(PAN) - High Temperature Electrically Conductive

RTP 1400 HEC Series:
Polyethersulfone - (PES) - Nickel Plated Graphite Fiber - Electrically Conductive

RTP CO.: Structural Materials(Continued):

Product Number:

RTP 2100 Series:
Polyetherimide(PEI) - Carbon Fiber(PAN) - High Temperature - Electrically Conductive

RTP 2200 Series:
Polyether-ether-Ketone - (PEEK) - Carbon Fiber(PAN) - High Temperature - Electrically Conductive

RTP 2280 HEC Series:
Polyetherether Ketone - PEEK - Nickel Plated Graphite Fiber

RTP 2280-HM:
Polyether-Ether-Ketone - PEEK - Carbon Fibers - High Modulus

RTP-200 TFE Series:
Nylon 6/6 - PTFE & Glass Fiber - Lubricated

RTP 200 TFE FR Series:
Nylon 6/6 - Glass Fiber - Flame Retardant - Lubricated

RTP 200 TFE Series:
Nylon 6/6 - Glass Fiber - PTFE & Si Lubricated

RTP 200 TFE Series:
RTP 205 TFE 15 SI:
Nylon 6/6 - PTFE & Glass Fiber - Silicone Lubricated

RTP 200 Series:
Nylon 6/6 - Glass Fiber - PTFE Lubricated - FDA Materials

RTP 200 Series:
Nylon 6/6 - Heat Stabilized - Glass Fiber - Silicon Lubricated

RTP 281 TFE 20:
RTP 283 TFE 10:
Nylon 6/6 - Carbon Fiber(PAN) - PTFE - Lubricated

RTP 200 Series:
Nylon 6/6 - Carbon Fiber, PAN - Lubricated

RTP 285 TFE 13 SI 2:
Nylon 6/6 - Carbon Fiber, PTFE, Silicone - Lubricated - Wear Resistant

RTP 200 MS Series:
Nylon 6/6 - Molybdenum Disulfide - Lubricated

RTP 200A MS Series:
Nylon 6 - Molybdenum Disulfide - Lubricated

RTP CO.: Structural Materials(Continued):

Product Name:

RTP 200A Series:
Nylon 6 - Lubricated

RTP 200B Series:
Nylon 6/10 - Lubricated

RTP 200 C TFE Series:
Nylon 11 - Lubricated

RTP 200D Series:
Nylon 6/12 - Lubricated

RTP-204D TFE 15 FR:
Nylon 6/12 - Glass Fiber - Wear Resistant/Flame Retardant

RTP 200H Series:
Nylon, High Impact - Lubricated

RTP 300 TFE Series:
Polycarbonate - PTFE - Lubricated

RTP 300 Series:
Polycarbonate - PTFE-Glass Fiber - Lubricated

RTP 300 TFE Series:
Polycarbonate - PTFE, Glass Fiber - Lubricated

RTP 300 TFE Series:
Polycarbonate - Glass Fiber, PTFE, Silicone - Lubricated

RTP 385 TFE 13 S12:
Polycarbonate - Carbon Fiber - PTFE and Silicone Lubricated

RTP 600 Series:
RTP 600 FR TFE 10:
ABS - Flame Retardant - Wear Resistant

RTP 600 Series:
RTP 600 SI 2:
ABS - Silicone Fluid - Lubricated

RTP 800 S12:
Acetal - Silicone Fluid - Lubricated

RTP800 TFE Series:
Acetal - Lubricated

RTP-800 TFE Series:
Acetal - Lubricated

RTP CO.: Structural Materials(Continued):

Product Name:

RTP-800 TFE Series:
Acetal - Lubricated

RTP-800 TFE Series:
Acetal - Lubricated

RTP 800 Series:
Acetal - Lubricated - RTP 800 TFE 20B

RTP 900 Series:
Polysulfone - Lubricated

RTP 900 Series:
Polysulfone - Lubricated

RTP 1002 TFE 15:
PBT - PTFE, Glass Fiber - Lubricated

RTP 1003 TFE 15:
Polybutylene Terephthalate - Glass Reinforced - PTFE Modified

RTP 1005 TFE 15:
PBT - PTFE, Glass Fiber - Lubricated

PBT:
Glass Fiber - PTFE Lubricated - Flame Retardant - RTP 1005 FR TFE 15

RTP 1107 TFE10:
Polyethylene-Terephthalate(PET) - Glass Fiber - PTFE Lubricated

RTP-1300 TFE Series:
Polyphenylene Sulfide - Glass Fiber, PTFE - Lubricated

RTP 1376 and RTP 1377:
Polyphenylene Sulfide - Wear Resistant

RTP 1378:
Polyphenylene Sulfide(PPS) - Lubricated

RTP 1300 TFE Series:
Polyphenylene Sulfide - PPS - Carbon Fiber(PAN) - Lubricated

RTP CO.: Structural Materials(Continued):

Product Name:

RTP 2101 TFE 15:
Polyetherimide - Glass Reinforced - PTFE Modified

RTP 100SP Series:
Polypropylene - Glass Fiber - Foam Grade

RTP 103 AV:
Polypropylene - Glass Fiber - Foam Grade

RTP 400SP Series:
Polystyrene - Glass Fiber - Foam Grade

RTP 1200-80D:
Polyurethane - 80-D Durometer - Glass Fiber

RTP 1200-90A Series:
Polyurethane - 90A Durometer - Glass Fiber - Elastomeric

RTP 1500-55D Series:
Polyester Elastomer - 55 Durometer - Glass Fiber

RTP 1500-40D Series:
Polyester Elastomer - 40 Durometer - Glass Fiber - Elastomeric

RESINOID ENGINEERING CORP.: Compound No. 1310:

RESINOID 1310 is a glass-mineral reinforced, two-step phenolic molding compound. It offers excellent impact strength, electrical properties and heat resistance. Available in powder form. Recommended for transfer or compression molding.

MATERIAL PROPERTIES	TYPICAL VALUE	ASTM METHOD
Form	Fibrous	
Bulk factor	5	D1895
Shrinkage-molded in/in(positive mold)	0.0015-0.0025	D955
Molding pressure-compression, PSI	1,500-3,000	
Molding Temperature, F	300-350	D957
Available plasticity	Medium	
Color	Black	
MECHANICAL AND PHYSICAL PROPERTIES		
Specific Gravity	1.88	D792
Water absorption % 24 hr. R.T.	0.09	D570
Tensile Strength, PSI	7,700	D651
Flexural Strength, PSI	12,900	D790
Modulus in Flex 10 6 PSI	2.2	D790
Impact (Izod) Ft.-lbs./in.	2.0	D256
Compressive Strength, PSI	25,300	D695
Hardness, Rockwell	E75/M110	D895
ELECTRICAL PROPERTIES		
Dielectric strength V/ml(S.T.) Dry	210	D149
Arc Resistance, Sec	192	D495
THERMAL PROPERTIES		
Deflection temperature @ 264 PSI, F	600	D648
Coefficient of Linear Thermal Expansion (in/in/Cx10 -5)	1.4	D696 (Mod)
Flame resistance	94V-O	U.L.

RESINOID ENGINEERING CORP.: Compound No. 1312:

RESINOID Compound 1312 is a glass-mineral reinforced, two-step phenolic molding compound. Formulated for improved mechanical properties. It does not contain asbestos. It is available in pelletized form that may be used for injection molding. It may also be preformed for transfer and compression molding.

MATERIAL PROPERTIES	TYPICAL VALUE	ASTM METHOD
Form	Pellets	
Bulk factor	2-4	D1895
Shrinkage-molded in/in(positive mold)	.003-.004	D955
Molding pressure-compression, PSI	2,000-5,000	
Molding Temperature, F	300-350	D957
Color	Black	
MECHANICAL AND PHYSICAL PROPERTIES		
Specific Gravity	1.9	D792
Water absorption % 24 hr. R.T.	0.1	D570
Tensile Strength, PSI	7,000	D651
Flexural Strength, PSI	12,200	D790
Modulus in Flex 10 6 PSI	2.0	D790
Impact (Izod) Ft.-lbs./in	0.85	D256
Compressive Strength, PSI	24,700	D695
Hardness, Rockwell		D785
ELECTRICAL PROPERTIES		
Dielectric strength V/mil(S.T.)Dry		D149
Arc Resistance, Sec		D495
THERMAL PROPERTIES		
Deflection temperature @ 264 PSI, F		D648
Coefficient of Linear Thermal expansion (in/in/C x 10 -5)		D696(Mod)
Flame resistance	94V-O	U.L.

RESINOID ENGINEERING CORP.: Compound No. 1320:

RESINOID Compound 1320 is an asbestos free glass reinforced, two-step phenolic molding compound. It exhibits outstanding dimensional stability, impact, electrical and thermal properties. This compound is in pelletized form for use in injection, transfer and compression molding. It can be preformed on automatic molding equipment.

MATERIAL PROPERTIES	TYPICAL VALUE	ASTM METHOD
Form	Pellets	
Bulk factor	2-3	D1895
Shrinkage-molded in/in(positive mold)	.002	D955
Molding pressure-compression, PSI	2,500	
Molding Temperature, F	300-350	D957
Color	Black	

MECHANICAL AND PHYSICAL PROPERTIES		
Specific Gravity	2.0	D792
Water absorption % 24 hr. R.T.	<0.1	D570
Tensile Strength, PSI	7,700	D651
Flexural Strength, PSI	11,100	D790
Modulus in Flex 10 6 PSI	2.3	D790
Impact(Izod) Ft.-lbs./in	0.7	D256
Compressive Strength, PSI	25,400	D695
Hardness, Rockwell	E87/M110	D785

ELECTRICAL PROPERTIES		
Dielectric strength V/mil(S.T.)Dry	230	D149
Arc Resistance, Sec	190	D495

THERMAL PROPERTIES		
Deflection temperature @ 264 psi, F	600	D648
Coefficient of Linear Thermal expansion (in/inCx10 5)		D696(Mod)
Flame resistance	94V-O	U.L.

RESINOID ENGINEERING CORP.: Compound No. 1321:

RESINOID Compound 1321 is a glass-mineral reinforced, two-step phenolic molding compound. Formulated without asbestos. It exhibits excellent dimensional stability, electrical and thermal properties. It has improved impact properties for those applications requiring greater impact strength. It is available in pelletized form for use in injection, transfer and compression molding. It can be preformed on automatic equipment.

MATERIAL PROPERTIES	TYPICAL VALUE	ASTM METHOD
Form	Pellets	
Bulk factor	2-3	D1895
Shrinkage-molded in/in(positive mold)	.002	D955
Molding pressure-compression, PSI	2,500	
Molding Temperature, F	300-350	D957
Color	Black	
MECHANICAL AND PHYSICAL PROPERTIES		
Specific Gravity	1.95	D792
Water absorption % 24 hr. R.T.	<0.1	D570
Tensile Strength, PSI	7,400	D651
Flexural Strength, PSI	11,900	D790
Modulus in Flex 10 6 PSI	2.2	D790
Impact (Izod) Ft.-lbs./in	0.8	D256
Compressive Strength, PSI	25,800	D695
Hardness, Rockwell	E90/M110	D785
ELECTRICAL PROPERTIES		
Dielectric strength V/mil(S.T.)Dry	250	D149
Arc Resistance, Sec	190	D495
THERMAL PROPERTIES		
Deflection temperature @ 264 PSI, F	550	D648
Coefficient of Linear Thermal expansion (in/in/Cx10 -5)		D696 (Mod)
Flame resistance	94V-O	U.L.

RESINOID ENGINEERING CORP.: Compound No. 1322:

RESINOID Compound 1322 is an asbestos free glass reinforced, two-step phenolic molding compound. It exhibits excellent dimensional stability, electrical, and thermal properties. It has improved impact properties for those applications requiring greater impact. This compound is in pelletized form for use in injection, transfer and compression molding. It can be preformed on automatic equipment.

MATERIAL PROPERTIES	TYPICAL VALUE	ASTM METHOD
Form	Pellets	
Bulk factor	2-3	D1895
Shrinkage-molded in/in(positive mold)	.002	D955
Molding pressure-compression, PSI	2,500	
Molding Temperature, F	300-350	D957
Color	Black	

MECHANICAL AND PHYSICAL PROPERTIES

Specific Gravity	2.0	D792
Water absorption % 24 hr. R.T.	0.1	D570
Tensile Strength, PSI	9,300	D651
Flexural Strength, PSI	12,800	D790
Modulus in Flex, 10 6 PSI	2.2	D790
Impact (Izod) Ft.-lbs./in.	0.9	D256
Compressive Strength, PSI	25,500	D695
Hardness, Rockwell	E90/M110	D785

ELECTRICAL PROPERTIES

Dielectric strength V/mil(S.T.) Dry	240	D149
Arc Resistance, Sec	190	D495

THERMAL PROPERTIES

Deflection temperature @ 264 PSI, F	550	D648
Flame resistance	94V-O	U.L.

RESINOID ENGINEERING CORP.: Compound No. 1323:

RESINOID Compound 1323 is an asbestos free, glass reinforced, two step phenolic molding compound. It exhibits excellent dimensional stability, electrical and thermal properties. It has improved impact properties for those applications requiring greater impact. This compound is in pelletized form for use in injection, transfer, or compression molding. It can be preformed on automatic equipment.

MATERIAL PROPERTIES	TYPICAL VALUE	ASTM METHOD
Form	Pellet	
Bulk Factor	2-3	D1895
Shrinkage-molded in/in (positive mold)	.002	D955
Molding pressure-compression, PSI	2,500	
Molding Temperature, F	300-350	D957
Color	Black	
MECHANICAL & PHYSICAL PROPERTIES		
Specific Gravity	2.0	D792
Water Absorption % 24 hr. R.T.	<0.1	D570
Tensile Strength, PSI	9,000	D651
Flexural Strength, PSI	12,900	D790
Modulus in Flex 10 6 PSI	2.8	D790
Impact (IZOD) ft.-lbs./in.	1.3	D256
Compressive Strength, PSI	28,700	D695
Hardness, Rockwell	E100/M115	D785
ELECTRICAL PROPERTIES		
Dielectric strength V/mil(S.T.)dry	295	D149
Arc Resistance, Sec.	190	D495
THERMAL PROPERTIES		
Deflection temperature @ 264 PSI,F	>600	D648
Coefficient of Linear Thermal expansion(in/in/c x 10 -5)	1.4	D696(mod)
Flame resistance	94 V-O	U.L.

RESINOID ENGINEERING CORPORATION: Compound No. 1324:

RESINOID Compound 1324 is an asbestos free, glass reinforced, two step phenolic molding compound. It exhibits excellent dimensional stability, electrical and thermal properties. It has improved impact properties for those applications requiring greater impact. This compound is in pelletized form for use in injection, transfer, or compression molding. It can be preformed on automatic equipment.

MATERIAL PROPERTIES	TYPICAL VALUE	ASTM METHOD
Form	Pellet	
Bulk Factor	2-3	D1895
Shrinkage-molded in/in(positive mold)	.002	D955
Molding pressure-compression, PSI	2,500	
Molding Temperature, F	300-350	D957
Color	Black	
MECHANICAL AND PHYSICAL PROPERTIES		
Specific Gravity	2.0	D792
Water Absorption % 24 hr. R.T.	<0.1	D570
Tensile Strength, PSI	9,000	D651
Flexural Strength, PSI	12,900	D790
Modulus in Flex 10 6 PSI	2.8	D790
Impact (IZOD) ft-lbs./in	1.0	D256
Compressive Strength, PSI	34,700	D695
Hardness, Rockwell	E100/M115	D785
ELECTRICAL PROPERTIES		
Dielectric Strength V/mil(S.T.)Dry	295	D149
Arc Resistance, sec.	190	D495
THERMAL PROPERTIES		
Deflection temperature @ 264 PSI,F	>600	D648
Coefficient of Linear Thermal expansion (in/in/C x 10 -5)	1.4	D696(mod)
Flame resistance	94 V-O	U.L.

RESINOID ENGINEERING CORP.: Compounds:

Product Name:

RESINOID 1340:

RESINOID 1340 is an asbestos-free, two step phenolic molding compound offering good mechanical properties and excellent thermal stability. This product is supplied in pelletized form suitable for injection, transfer or compression molding.

Form	Pellet
Color	Black
Bulk Factor	2-3
Specific Gravity	2.0
Tensile Strength	9,000 PSI
Flexural Strength	12,000 PSI
Flexural Modulus	2.8 x 10 6 PSI
Impact Strength	1.3 ft-lb./in.
Compressive Strength	28,000 PSI
Dielectric Strength S/T	340 VPM
S/S	300 VPM
Water Absorption	0.1%
Heat Deflection Temperature	>600F

RESINOID Compound No. 1360:

RESINOID Compound 1360 is an asbestos free, two step, glass reinforced phenolic molding compound offering excellent mechanical strength. This compound is in pelletized form suitable for compression, transfer or injection molding.

Form	Pellet
Color	Black
Specific Gravity	1.73
Shrinkage	0.001 in/in.
Impact Strength	4.1 ft-lb/in.
Flexural Strength	25,700 PSI
Flexural Modulus	2.6 X 10 6 PSI
Tensile Strength	7,000 PSI
Compressive Strength	34,800 PSI
Dielectric Strength	330 VPM
Rockwell Hardness	E106/M109
Water Absorption	0.1%
Arc Resistance	180 sec.
Deflection Temperature	>600F

RESINOID ENGINEERING CORP.: Compound No. 1370:

RESINOID 1370 is a two step, glass reinforced asbestos free phenolic molding compound recommended for applications requiring prolonged exposure to elevated temperatures. This material can be supplied in grades suitable for compression, transfer or injection molding.

Flexural Strength	18,000 PSI
Flexural Modulus	2.3 X 10 6 PSI
Compressive Strength	25,000 PSI
Specific Gravity	1.9
Impact Strength	1.1 ft-lb/in.
Water Absorption	0.1%
Tensile Strength	7,500 PSI
Heat Deflection Temperature	>600F
Dielectric Strength	
Short Time	340 VPM
Step by Step	275 VPM

RESINOID ENGINEERING CORP.: Compound No. 2002-C5:

RESINOID Compound 2002-C5 is a fabric reinforced two-step phenolic molding compound. It has improved heat distortion and dimensional stability. It has been formulated to exhibit excellent wear resistant properties. It meets CFI-10 property requirements. It is available in bulk form. It is recommended for transfer or compression molding.

MATERIAL PROPERTIES	TYPICAL VALUE	ASTM METHOD
Form	Chopped Fabric	
Bulk factor	8-10	D1895
Shrinkage-molded in/in (positive mold)	.002-.003	D955
Molding pressure-compression, PSI	2,000-4,000	
Molding Temperature, F	300-350	D957
Available plasticity	20-40 sec.	
Color	Black	

MECHANICAL AND PHYSICAL PROPERTIES		
Specific Gravity	1.45	D792
Water absorption % 24 hr. R.T.	.7	D570
Tensile Strength, PSI	7,000	D651
Flexural Strength, PSI	14,000	D790
Modulus in Flex 10 6 PSI		D790
Impact (Izod) Ft.-lbs./in	2.75	D256
Compressive Strength, PSI	22,000	D695
Hardness, Rockwell	M98	D785

THERMAL PROPERTIES		
Deflection temperature @ 264 PSI, F	400	D648
Coefficient of Linear Thermal Expansion (in/in/C x 10 -5)		D696(Mod)
Flame resistance		U.L.

RESINOID ENGINEERING CORP.: Compound No. 2002-2:

RESINOID Compound 2002-2 is a fabric reinforced, two-step phenolic molding compound. It is formulated to meet gear requirements and other stressed applications. It exhibits good dimensional stability, excellent impact and flexural properties. It is available in bulk form. It is recommended for compression or transfer molding.

MATERIAL PROPERTIES	TYPICAL VALUE	ASTM METHOD
Form	Chopped Fabric	
Bulk factor	10-12	D1895
Shrinkage-molded in/in (positive mold)	.002-.004	D955
Molding pressure-compression, PSI	2,000-4,000	
Molding Temperature, F	300-340	D957
Available plasticity	20-40 sec.	
Color	Black	

MECHANICAL AND PHYSICAL PROPERTIES		
Specific Gravity	1.41	D792
Water absorption 24 % hr. R.T.	0.3	D570
Tensile Strength, PSI	8,000	D651
Flexural Strength, PSI	14,800	D790
Modulus in Flex 10 6 PSI	1.6	D790
Impact (Izod) Ft.-lbs./in	1.3	D256
Compressive Strength, PSI	20,000	D695
Hardness, Rockwell		D785

THERMAL PROPERTIES		
Deflection temperature @ 264 PSI,F	500	D648
Coefficient of Linear Thermal expansion (in/in/C x 10 -5)		D696(Mod)
Flame resistance		U.L.

RESINOID ENGINEERING CORP.: Compound No. 2002-4:

RESINOID Compound 2002-4 is a fabric reinforced, two-step phenolic molding compound. It has been modified to meet gear requirements and other stressed applications. It possesses excellent impact and flexural strength. It is available in bulk form. It meets CFI-10 property requirements. It is recommended for compressive or transfer molding.

MATERIAL PROPERTIES	TYPICAL VALUE	ASTM METHOD
Form	Chopped Fabric	
Bulk factor	10-12	D1895
Shrinkage-molded in/in (positive mold)	.002-.004	D955
Molding pressure-compression, PSI	2,000-4,000	
Molding Temperature, F	300-350	D957
Color	Black	

MECHANICAL AND PHYSICAL PROPERTIES

Specific Gravity	1.37	D792
Water absorption % 24 hr. R.T.	0.6	D570
Tensile Strength, PSI	8,000	D651
Flexural Strength, PSI	14,800	D790
Modulus in Flex 10 6 PSI	1.6	D790
Impact (Izod) Ft.-lbs./in	2.0	D256
Compressive Strength, PSI	21,700	D695
Hardness, Rockwell		D785

ELECTRICAL PROPERTIES

Dielectric strength V/mil(S.T.) Dry	325	D149
Arc Resistance, Sec		D495

THERMAL PROPERTIES

Deflection temperature @ 264 PSI,F	350	D648
Coefficient of Linear Thermal Expansion (in/in/C x 10 -5)	1.0	D696(Mod)
Flame resistance		U.L.

RESINOID ENGINEERING CORP.: Compound No. 2002-4P:

RESINOID Compound 2002-4P is a fabric reinforced, two-step phenolic molding compound. It has been formulated for use in conduit busing and gear applications. It can readily be preformed on automatic preforming equipment. It performs well in screw preplasticizing equipment. Available in pellet form. It is recommended for compression and transfer molding.

MATERIAL PROPERTIES	TYPICAL VALUE	ASTM METHOD
Form	Pellet	
Bulk factor	2-4	D1895
Shrinkage-molded in/in (positive mold)	.003-.004	D955
Molding pressure-compression, PSI	2,000-4,000	
Molding Temperature, F	300-350	D957
Color	Black	

MECHANICAL AND PHYSICAL PROPERTIES		
Specific Gravity	1.4	D792
Water absorption % 24 hr. R.T.	0.7	D570
Tensile Strength, PSI	5,100	D651
Flexural Strength, PSI	11,000	D790
Modulus in Flex 10 6 PSI	1.4	D790
Impact (Izod) Ft-lbs./in	0.9	D256
Compressive Strength, PSI	22,500	D695
Hardness, Rockwell	E71/M106	D785

ELECTRICAL PROPERTIES		
Dielectric strength V/mil(S.T.) Dry	275	D149
Arc Resistance, Sec		D495

THERMAL PROPERTIES		
Deflection temperature @ 264 PSI,F	445	D648
Coefficient of Linear Thermal expansion (in/in/C x 10 -5)	1.0	D696(Mod)
Flame Resistance		U.L.

RESINOID ENGINEERING CORP.: Compound No. 2010:

RESINOID Compound 2010 is a chopped fabric reinforced, two-step phenolic molding compound. This material has properties equivalent to those required for CFI-10 Military Specification MIL-M-14G. It can be preformed automatically due to its lower bulk density. It offers excellent impact strength and improved surface finish. This compound is available in pelletized form for compression, transfer or injection molding.

MATERIAL PROPERTIES	TYPICAL VALUE	ASTM METHOD
Form	Pellets	
Bulk factor	2-4	D1895
Shrinkage-molded in/in (positive mold)	.002-.004	D955
Molding pressure-compression, PSI	2,000-4,000	
Molding Temperature, F	300-350	D957
Color	Black	

MECHANICAL AND PHYSICAL PROPERTIES

Specific Gravity	1.37-1.40	D792
Water absorption % 24 hr. R.T.	1.1	D570
Tensile Strength, PSI	6,000	D651
Flexural Strength, PSI	10,000	D790
Modulus in Flex 10 6 PSI	1.6	D790
Impact (Izod) Ft.-lbs./in	1.3	D256
Compressive Strength, PSI	21,000	D695
Hardness, Rockwell	E86/M112	D785

ELECTRICAL PROPERTIES

Dielectric strength V/mil(S.T.)Dry	225	D149

THERMAL PROPERTIES

Deflection temperature @ 264 PSI,F	250	D648
Flame resistance	V-O	U.L.

RESINOID ENGINEERING CORP.: Compound No. 2011E:

RESINOID Compound 2011E is a fabric reinforced specially reinforced phenolic molding compound. Designed for faster curing in gear applications. It shows excellent performance in applications requiring exposure to hot oil. Available in bulk form. It is recommended for transfer or compression molding.

MATERIAL PROPERTIES	TYPICAL VALUE	ASTM METHOD
Form	Chopped Fabric	
Bulk factor	8	D1895
Shrinkage-molded in/in (positive mold)	0.002-0.003	D955
Molding pressure-compression, PSI	2,000-4,000	
Molding Temperature, F	300-350	D957
Available plasticity	Medium	
Color	Black	

MECHANICAL AND PHYSICAL PROPERTIES

Specific Gravity	1.43	D792
Water absorption % 24 hr. R.T.	0.5	D570
Tensile Strength, PSI	7,700	D651
Flexural Strength, PSI	11,000	D790
Modulus in Flex 10 6 PSI	--	D790
Impact (Izod) Ft.-lbs./in	1.8	D256
Compressive Strength, PSI	20,000	D695
Hardness, Rockwell		D785

THERMAL PROPERTIES

Deflection temperature @ 264 PSI,F	300	D648
Coefficient of Linear Thermal expansion (in/in/C x 10 -5)		D696(Mod)
Flame Resistance		U.L.

SPECIAL PROPERTIES

Hot oil exposure performance

RESINOID ENGINEERING CORP.: Compound No. 2016:

RESINOID Compound 2016 is a two-step, fabric reinforced phenolic molding compound. Its properties meet MIL-M-14G type CFI-20 requirements. It is available in bulk form or as pellets for use in injection, transfer or compression molding.

MATERIAL PROPERTIES	TYPICAL VALUE	ASTM METHOD
Form	Chopped Fabric	
Bulk factor	8-10	D1895
Shrinkage-molded in/in (positive mold)	0.002-0.003	D955
Molding pressure-compression, PSI	2,000-4,000	
Molding Temperature, F	300-350	D957
Color	Black	
MECHANICAL AND PHYSICAL PROPERTIES		
Specific Gravity	1.37	D792
Water absorption % 24 hr. R.T.	0.8	D570
Tensile Strength, PSI	8,400	D651
Flexural Strength, PSI	12,000	D790
Modulus in Flex 10 6 PSI	----	D790
Impact (Izod) Ft.-lbs./in	2.4	D256
Compressive Strength, PSI	24,500	D695
Hardness, Rockwell	M108-E76	D785
ELECTRICAL PROPERTIES		
Dielectric strength V/mil(S.T.)Dry	245	D149
Arc Resistance, Sec	115	D495
Thermal Properties		
Deflection temperature @ 264 PSI,F	460	D648
Coefficient of Linear Thermal expansion (in/In/C x 10 -5)	2.2	D696(Mod)
Flame resistance	V-O	U.L.

RESINOID ENGINEERING CORP.: Compound No. 2017:

RESINOID Compound 2017 is a fabric reinforced, two-step phenolic molding compound. Formulated to have excellent impact strength, this compound meets the requirements for CFI-20 Specifications. It is available in bulk form. It is recommended for transfer or compression molding.

MATERIAL PROPERTIES	TYPICAL VALUE	ASTM METHOD
Form	Chopped Fabric	
Bulk factor	8-10	D1895
Shrinkage-molded in/in (positive mold)	0.004-0.005	D955
Molding pressure-compression, PSI	2,000-4,000	
Molding Temperature, F	300-350	D957
Color	Black	

MECHANICAL AND PHYSICAL PROPERTIES		
Specific Gravity	1.35	D792
Water absorption % 24 hr. R.T.	0.8	D570
Tensile Strength, PSI	7,600	D651
Flexural Strength, PSI	12,900	D790
Modulus in Flex 10 6 PSI	1.0	D790
Impact (Izod) Ft.-lbs./in	3.3	D256
Compressive Strength, PSI	20,800	D695
Hardness, Rockwell	M105/E75	D785

ELECTRICAL PROPERTIES		
Dielectric strength V/mil(S.T.)Dry	245	D149
Arc Resistance, Sec	115	D495

THERMAL PROPERTIES		
Deflection temperature @ 264 PSI,F	435	D648
Coefficient of Linear Thermal expansion (in/in/C x 10 -5)		D696 (Mod)
Flame resistance		U.L.

RESINOID ENGINEERING CORP.: Compound No. 2018:

RESINOID Compound 2018 is a fabric reinforced, two-step phenolic molding compound. It exhibits excellent moldability in transfer molding. It meets certain automotive specifications. It is available in bulk form. It is recommended for transfer or compression molding.

MATERIAL PROPERTIES	TYPICAL VALUE	ASTM METHOD
Form	Chopped Fabric	
Bulk factor	8	D1895
Shrinkage-molded in/in (positive mold)	.002-.003	D955
Molding pressure-compression, PSI	2,000-4,000	
Molding Temperature, F	300-350	D957
Color	Black	

MECHANICAL AND PHYSICAL PROPERTIES		
Specific Gravity	1.36	D792
Water absorption % 24 hr. R.T.	0.77	D570
Tensile Strength, PSI	6,350	D651
Flexural Strength, PSI	18,500	D790
Modulus in Flex 10 6 PSI	----	D790
Impact (Izod) Ft-lbs./in	1.8	D256
Compressive Strength, PSI	22,000	D695
Hardness, Rockwell	----	D785

ELECTRICAL PROPERTIES		
Dielectric strength V/mil(S.T.)Dry	265	D149
Arc Resistance, Sec	----	D495

THERMAL PROPERTIES		
Deflection temperature @ 264 PSI,F	375	D648
Coefficient of Linear Thermal expansion (in/in/C x 10 -5)		D696 (Mod)
Flame resistance		U.L.

RESINOID ENGINEERING CORP.: Compound No. 2020:

RESINOID Compound 2020 is a fabric reinforced, two-step phenolic molding compound. It exhibits excellent dimensional stability and impact strength. It is available in bulk form and is recommended for compression or transfer molding.

MATERIAL PROPERTIES	TYPICAL VALUE	ASTM METHOD
Form	Chopped Fabric	
Bulk factor	10-12	D1895
Shrinkage-molded in/in (positive mold)	0.0025-0.0035	D955
Molding pressure-compression, PSI	2,000-4,000	
Molding Temperature, F	300-340F	D957
Color	Black	

MECHANICAL AND PHYSICAL PROPERTIES		
Specific Gravity	1.40	D792
Water absorption % 24 hr. R.T.	0.5	D570
Tensile Strength, PSI	7,000	D651
Flexural Strength, PSI		D790
Modulus in Flex 10 6 PSI		D790
Impact (Izod) Ft.-lbs./in	2.1	D256
Compressive Strength, PSI	23,000	D695
Hardness, Rockwell		D785

ELECTRICAL PROPERTIES		
Dielectric strength V/mil(S.T.)Dry	220	D149
Arc Resistance, Sec		D495

THERMAL PROPERTIES		
Deflection temperature @ 264 PSI,F	400	D648
Coefficient of Linear Thermal expansion (in/in/C x 10 -5)		D696 (Mod)
Flame resistance		U.L.

RESINOID ENGINEERING CORP.: Compound No. 2205:

RESINOID Compound 2205 is a pelletized, two step fabric reinforced phenolic molding compound that exhibits good gloss. It is designed to meet Military Specification MIL-M-14G, CFI 5 type applications. It is a free flowing pellet which is suitable for preforming on most automatic equipment. The product can be injection, compression, or transfer molded.

MATERIAL PROPERTIES	TYPICAL VALUE	ASTM METHOD
Form	Pellets	
Bulk factor	2-4	D1895
Shrinkage-molded in/in (positive mold)	.003-.005	D955
Molding pressure-compression, PSI	2000-4000	
Molding Temperature, F	300-350	D957
Color	Black	

MECHANICAL AND PHYSICAL PROPERTIES

Specific Gravity	1.40	D792
Water absorption % 24 hr. R.T.	2.8	D570
Tensile Strength, PSI	7,500	D651
Flexural Strength, PSI	12,500	D790
Modulus in Flex 10 6 PSI	0.7	D790
Impact (Izod) Ft.-lbs./in	0.6	D256
Compressive Strength, PSI	30,000	D695
Hardness, Rockwell		D785

ELECTRICAL PROPERTIES

Dielectric strength V/mil(S.T.)Dry	370	D149
Arc Resistance, Sec	120	D495

THERMAL PROPERTIES

Deflection temperature @ 264 PSI,F	400	D648
Coefficient of Linear Thermal expansion (in/in/C x 10 -5)		D696 (Mod)
Flame resistance		U.L.

RESINOID ENGINEERING CORP.: Compound No. 2215:

RESINOID 2215 is an organic fiber reinforced compound offering moderate impact strength. This product is available as a small pellet suitable for compression, transfer and injection molding.

Color	Black
Specific Gravity	1.45
Impact Strength	0.7 ft - lb/in
Tensile Strength	7000 psi
Flexural Strength	11,500 psi
Flexural Modulus	1.4×10^6 psi
Compressive Strength	26,000 psi
Dielectric Strength	220 vpm
Arc Resistance	100 sec.
Water Absorption	0.5%
Heat Deflection Temperature	390F
Flame Resistance	94 V-O C 1/4 inch

RESINOID ENGINEERING CORP.: Compound No. 7003S:

RESINOID Compound 7003S is a glass reinforced two-step phenolic molding compound. The material has excellent flow characteristics, dimensional stability, and impact strength for use in parts requiring approval under Type MFH of Military Specification MIL-M-14G. It can be preformed on automatic equipment. It is available in pelletized form for use in injection, transfer and compression molding.

MATERIAL PROPERTIES	TYPICAL VALUE	ASTM METHOD
Form	Pellets	
Bulk factor	2-3	D1895
Shrinkage-molded in/in (positive mold)	0.002-0.003	D955
Molding pressure-compression, PSI	2,500	
Molding Temperature, F	300-350	D957
Color	Black	
MECHANICAL AND PHYSICAL PROPERTIES		
Specific Gravity	1.84	D792
Water Absorption % 24 hr. R.T.	0.1	D570
Tensile Strength, PSI	6,000	D651
Flexural Strength, PSI	10,700	D790
Modulus in Flex 10 6 PSI	2.0	D790
Impact (Izod) Ft.-lbs./in.	0.5	D256
Compressive Strength, PSI	24,400	D695
Hardness, Rockwell	E80/M106	D785
ELECTRICAL PROPERTIES		
Dielectric strength V/mil(S.T.)Dry	300	D149
Arc Resistance, Sec.	190	D495
THERMAL PROPERTIES		
Deflection temperature @ 264 PSI,F	390	D648
Flame resistance	94V-O	U.L.
SPECIAL PROPERTIES		
U.L. Listed	E61847	

RESINOID ENGINEERING CORP.: Compound No. 7005:

RESINOID Compound 7005 is a glass reinforced, two-step phenolic molding compound. Available in pelletized form for use in injection and compression molding. It can be preformed on automatic equipment. The material exhibits good flow characteristics and dimensional stability. It was formulated for precision molded parts, and electronic connenctors subjected to vapor phase soldering.

MATERIAL PROPERTIES	TYPICAL VALUE	ASTM METHOD
Form	Pellets	
Bulk factor	2-3	D1895
Shrinkage-molded in/in (positive mold)	.0015	D955
Molding pressure-compression, PSI	2,500	
Molding Temperature, F	300-350	D957
Color	Black	

MECHANICAL AND PHYSICAL PROPERTIES

Specific Gravity	1.93	D792
Water absorption % 24 hr. R.T.	.1	D570
Tensile Strength, PSI	7,000	D651
Flexural Strength, PSI	14,000	D790
Modulus in Flex 10 6 PSI	2.5	D790
Impact (Izod) Ft.-lbs./in	0.5	D256
Compressive Strength, PSI	30,000	D695
Hardness, Rockwell	E77/M105	D785

ELECTRICAL PROPERTIES

Dielectric strength V/mil(S.T.) Dry	>365	D149
Arc Resistance, Sec	185	D495

THERMAL PROPERTIES

Deflection Temperature @ 264 PSI,F	430	D648
Coeffiecient of Linear Thermal Expansion(in/in/Cx10 -5)	1.7	D696 (Mod)
Flame resistance	94V-O	U.L.

SPECIAL PROPERTIES

U.L. Listed	E614847

RESINOID ENGINEERING CORP.: Compound No. 7051:

RESINOID Compound 7051 is a glass reinforced, two step phenolic molding compound which has been formulated for applications requiring exceptional flexural strength. The compound contains no asbestos. It exhibits excellent dimensional stability, electrical and thermal properties. It is available in pelletized form for use in injection, transfer, and compression molding. It can be preformed on automatic equipment.

MATERIAL PROPERTIES	TYPICAL VALUE	ASTM METHOD
Form	Pellets	
Bulk factor	2-3	D1895
Shrinkage-molded in/in (positive mold)	0.001-0.002	D955
Molding pressure-compression, PSI	1,500-2,500	
Molding Temperature, F	300-360	D957
Color	Black	

MECHANICAL AND PHYSICAL PROPERTIES		
Specific Gravity	1.75	D792
Water absorption % 24 hr. R.T.	0.1	D570
Tensile Strength, PSI	9,400	D651
Flexural Strength, PSI	24,000	D790
Modulus in Flex 10 6 PSI	2.4	D790
Impact (Izod) Ft.-lbs./in	0.90	D256
Compressive Strength, PSI	40,000	D695
Hardness, Rockwell	E100/M120	D785

ELECTRICAL PROPERTIES		
Dielectric strength V/mil(S.T.)Dry	>370	D149
Arc Resistance, Sec	180	D495

THERMAL PROPERTIES		
Deflection Temperature @ 264 PSI,F	520	D648
Coefficient of Linear Thermal expansion (in/in/Cx10 -5)	2.1	D696 (Mod)
Flame Resistance	---	U.L.

RESINOID ENGINEERING CORP.: Compound No. 7051 SS:

MATERIAL PROPERTIES	TYPICAL VALUE	ASTM METHOD
Form	Pellets	
Bulk Factor	2-3	D1895
Shrinkage-molded in/in (positive mold)	0.001-0.002	D955
Molding pressure-compression, PSI	1,500-2,500	
Molding Temperature, F	300-360	D957
Color	Black	

MECHANICAL AND PHYSICAL PROPERTIES		
Specific Gravity	1.75	D792
Water absorption % 24 hr. R.T.	0.15	D570
Tensile Strength, PSI	9,400	D651
Flexural Strength, PSI	24,000	D790
Modulus in Flex 10 6 PSI	2.4	D790
Impact (Izod) Ft-lbs./in	0.90	D256
Compressive Strength, PSI	40,000	D695
Hardness, Rockwell	E100-M120	D785

ELECTRICAL PROPERTIES		
Dielectric strength V/mil(S.T.)Dry	280/290	D149
Arc Resistance, Sec	180	D495

THERMAL PROPERTIES		
Deflection temperature @ 264 PSI,F	550	D648
Coefficient of Linear Thermal expansion (in/in/Cx10 5)	2.3	D696 (Mod)
Flame resistance	94 VO @ 1/16"	U.L.

SPECIAL PROPERTIES		
Insulation Resistance, megohms	18 x 10 5	D257
Volume Resistivity, ohm-cm	1.2 x 10 11	D257

RESINOID ENGINEERING CORP.: Compound No. 7201:

Resinoid Compound 7201 is a glass reinforced, impact modified, two step phenolic molding compound. It exhibits excellent processability and molding characteristics. Lower specific gravity, impact modification and dimensional stability are outstanding properties. The material is available in pellet form for use in injection molding or transfer molding.

MATERIAL PROPERTIES	TYPICAL VALUE	ASTM METHOD
Form	Pellets	
Bulk factor	2-3	D1895
Shrinkage-molded in/in (positive mold)	0.001-0.003	D955
Molding pressure-compression, PSI	2500	
Molding Temperature, F	300-350	D957
Color	Black	
MECHANICAL AND PHYSICAL PROPERTIES		
Specific Gravity	1.6	D792
Water Absorption % 24 hr. R.T.	0.2%	D570
Tensile Strength, PSI	----	D651
Flexural Strength, PSI	12,400	D790
Modulus in Flex 10 6 PSI	1.3	D790
Impact (Izod) Ft.-lbs./in	0.65	D256
Compressive Strength, PSI	16,000	D695
Hardness, Rockwell	------	D785
ELECTRICAL PROPERTIES		
Dielectric strength V/mil(S.T.)Dry	325	D149
Arc Resistance, Sec	130	D495
THERMAL PROPERTIES		
Deflection temperature @ 264 PSI,F	525	D648
Coefficient of Linear Thermal Expansion (in/in/C x 10 -5)		D696(Mod)
Flame Resistance		U.L.

RHONE-POULENC, INC.: Program in Bismaleimides for High Temperature Applications:

KERIMID:
New options for performance and high temperature properties BMI for printed circuit substrates and advanced composites.

KINEL:
Verasility in molding for high thermo-mechanical resistance BMI molding powders for structural and friction parts.

RHODIMID:
BMI monomers.

ABLAPHENE:
Special ablative phenolics.

KERIMID 500:
High temperature adhesive.

KERIMID For Electronics:

KERIMID 601:
Bismaleimide resin provides performances, reliability and repairability in the field of high tech. multilayer printed circuits.

FE 70013 and FE 70014:
Both modified KERIMID 601 resins (glass transition temperature over 250C/482F) meet MIL 13949F and give prepregs with less than 0.1% of free MDA. Moreover FE 70014 is VO (UL 94).

FE 70016:
Diphenyl silanediol modified bismaleimide is MDA free and meets MIL 13949F.
It can be run in cheap solvents.

FE 70017, FE 70018, FE 70019:
New MDA free grades under development to meet MIL 13949F GI base material

KERIMID for Advanced Composites:

New types of modified bismaleimides are introduced for prepregging carbon fibers and fabrics.

FE 70011 and FE 70015:
are hot-melt processable. They exhibit glass transition temperatures above 360C/680F, epoxy like processing, high properties on composites (flexion, interlaminar shear, toughness...) and thermal ageing capability of polyimides.

RHONE-POULENC, INC.: Program in Bismaleimides for High Temperature Applications(Continued):

KINEL Molding Powders:

KINEL 5504:

(6mm glass fiber reinforced) gives access to high mechanical properties by compression molding. Parts obtained exhibit: high glass transition temperature (HDT over 300C/572F) and can work in long run under stress at elevated temperatures (250C/482F).

KINEL 4525:

(glass reinforced molding powder) introduces polyimide thermoset in large series through injection.

KINEL 4503 (ex. FE 74003):
PTFE filled self-lubricating molding powder affords to get themostable friction parts by injection.

RHODIMID:

BMI monomers for increasing the thermal resistance of polymers. Applications field: varnishes, matrices for composites and molding powders, adhesives...
Characteristics:
* melting point (hot plate): 155-162C
* gel-time at 175C: 6-8 minutes

ABLAPHENE RA 101 and RS 101:

RA 101 AND RS 101 are phenol formlaldehyde resol type resins in solutions. They are used associated with carbon reinforcements as blast pipe liners and heat shields for military and aerospace applications. RA 101 ABLAPHENE is mineral free.

KERIMID 500:

Polyimide-amide adhesive. It is used to bond composites and metals. Its thermal ageing capability is 2,000 hours at 250C/482F or 100 hours at 300C/572F.

RHONE-POULENC, INC.: KERIMID High Temperature Composite Polyimide Resin:

KERIMID 601 is a thermosetting polyimide resin which, when reinforced with glass fiber, silica, quartz, carbon, graphite or boron fibers, gives actually void free composite materials. This resin cures without generating water or volatiles of any kind.

Composites obtained ally lightness to high mechanical properties for long aging times at 480F (250C). They are ideally suited for high temperature uses where metals have been so far solely employed.

Their non flammability makes them a choice for Aircraft and Aerospace industries.

Their resistance to radiations is outstanding.

KERIMID 601 easy processing is reflected in a low processing cost, quite comparable to that of conventional thermosets.

Characteristics:
- Appearance: fine yellow powder
- Bulk density g/cm 3: about 0.5
- lb/cu.ft.: about 31

Properties of KERIMID 601/181 E glass cloth laminates:

Physical Properties:
- Color: brown
- Resin content (by ignition): 22 to 24%
- Specific gravity: about 1.95
- Barcol hardness: about 70

Mechanical properties:
- Flexural strength (D-790) psi x 10 3:
  - * at 77F (25C): about 70
- Flexural Modulus (D-790) psi x 10 3:
  - * at 77F (25C): about 4,000
- Tensile strength (D-638) psi x 10 3:
  - * at 77F (25C): about 50
- Compressive strength (D-695) psi x 10 3:
  - * at 77F (25C): about 50
- Delaminating strength (D-2345) psi:
  - * at 77F (25C): about 2150
- Izod impact strength (D-256):
  - * at 77F (25C)
  - - notched: ft x lb/in: about 13
  - - unnotched: ft x lb/in: about 15

RHONE-POULENC, INC.: KINEL Polyimides:

For structural components:

5504:
Specific gravity (granules): 0.25
(moulded): 1.90
Moisture absorption (24 h at 23C): 0.5
Mould shrinkage: 0.1
Glass transition temperature: 300

3514:
Specific gravity (granules): 0.36
(moulded): 1.71
Moisture absorption (24 h at 23C): 0.75
Mould shrinkage: 0.2
Glass transition temperature: 300

5514:
Specific gravity (granules): 0.36
(moulded): 1.50
Moisture absorption (24 h at 23C): 0.5
Mould shrinkage: 0.1
Glass transition temperature: 300

3515:
Specific gravity (granules): 0.5
(moulded): 1.60
Moisture absorption (24 h at 23C): 0.6
Mould shrinkage: 0.1-0.35
Glass transition temperature: 300

4525:
Specific gravity (granules): 0.575
(moulded): 1.57
Moisture absorption (24 h at 23C): 0.45
Mould shrinkage: 0.3-.0.6
Grass transition temperature: 300

For friction components:

5505:
Specific gravity (granules): 0.36
(moulded): 1.45
Moisture absorption (24 h at 23C): 0.6
Mould shrinkage: 0.6
Glass transition temperature: 300

RHONE-POULENC, INC.: KINEL Polymides(Continued):

For friction components(Continued):

5508:
Specific gravity(granules): 0.34
(moulded): 1.55
Moisture absorption (24 h at 23C): 0.6
Mould shrinkage: 0.5
Glass transition temperature: 300

4511:
Specific gravity(granules): 0.20
(moulded): 1.65
Moisture absorption (24 h at 23C): 0.6
Mould shrinkage: 0.2-0.3
Glass transition temperature: 300

5511:
Specific gravity(granules): 0.20
(moulded): 1.60
Moisture absorption (24 h at 23C): 0.4
Mould shrinkage: 0.1
Glass transition temperature: 300

5517 compression:
Specific gravity(granules): 0.35
(moulded): 1.50
Moisture absorption (24 h at 23C): 0.3
Mould shrinkage: 0.5-0.6
Glass transition temperature: 300

5517 free sintering:
Specific gravity(granules): 0.35
(moulded): 1.45
Moisture absorption (24 h at 23C): 1.2
Mould shrinkage: 0.5
Glass transition temperature: 300

4518:
Specific gravity(granules): 0.75
(moulded): 1.40
Moisture absorption (24 h at 23C): 1.25
Mould shrinkage: 1
Glass transition temperature: 300

5518:
Specific gravity(granules): 0.4
(moulded): 1.42
Moisture absorption (24 h at 23C): 0.3
Mould shrinkage: 1
Glass transition temperature: 300

RICHARDSON POLYMER CORP.: CTE Acrylic Copolymer:

CTE Styrene Methyl Methacrylate Copolymer

CTE Acrylic Copolymer is characterized by:

* Sparkling clarity
* Good scratch resistance
* Ease of processing

Typical applications include: cosmetics packaging, housewares, displays, office accessories.

CTE complies with FDA regulations for food contact applications.

PROPERTY	TEST ASTM	TYPICAL VALUE
Tensile Strength, break	D-638, 0.2"/min.	8.1 x 10 3 psi
Elongation, break	D-638	2.3%
Tensile Modulus	D-638	5.0 x 10 5 psi
Flexural Strength	D-790, 0.05"/min.	14.1 x 10 3 psi
Flexural Modulus	D-790	4.5 x 10 5 psi
Impact Strength, Izod, Notched	D-256	0.4 ft.lb./in.
Specific Gravity	D-792	1.06
Hardness, Rockwell	D-785	M-80
Deflection Temperature (annealed @ 180F) 264 psi	D-648	211F
Light Transmission	D-1003	90%
Haze	D-1003	1.5%
Refractive Index (nD)	D-542	1.58
Melt Flow Rate,	D-1238	
Cond. 190/10 (N)		2.6g/10 min.
Cond. 230/1.2 (H)		0.9g/10 min.
Cond. 230/3.8 (I)		7.4g/10 min.
Mold Shrinkage	D-955	.002-.006 in./in.
Water Absorption 24 hrs.	D-570	0.11%

RICHARDSON POLYMER CORP.: NAS Acrylic Copolymer:

NAS Styrene Methyl Methacrylate Copolymer

NAS Acrylic Copolymer is characterized by:

* Sparkling clarity
* Good scratch resistance
* Ease of processing

Typical applications include: meter covers, brush blocks, drafting instruments, bathroom accessories.

NAS complies with FDA regulations for food contact applications. Physical properties are not affected by gamma irradiation.

NAS meets USP XXI specifications for Class VI Plastics.

PROPERTY	TEST ASTM	TYPICAL VALUE
Tensile Strength, break	D-638, 0.2"/min.	8.3×10^3 psi
Elongation, break	D-638	2.0%
Tensile Modulus	D-638	5.1×10^5 psi
Flexural Strength, break	D-790, 0.05"/min.	14.1×10^3 psi
Flexural Modulus	D-790	4.5×10^5 psi
Impact Strength, Izod, Notched	D-256	0.31 ft.lb./in.
Specific Gravity	D-792	1.09
Hardness, Rockwell	D-785	M-80
Deflection Temperature (annealed @ 180F) 264 psi	D-648	208F
Light Transmission	D-1003	90%
Haze	D-1003	1.5%
Refractive Index (nD)	D-542	1.56
Melt Flow Rate,	D-1238	
Cond. 190/10 (N)		4.3g/10 min.
Cond. 230/1.2 (H)		2.0g/10 min.
Cond. 230/3.8 (I)		8.6g/10 min.
Mold Shrinkage	D-955	.002-.006 in./in.
Water Absorption 24 hrs	D-570	0.15%

RICHARDSON POLYMER CORP.: P-359 Acrylic Copolymer:

P-359 Styrene Methyl Methacrylate Copolymer

P-359 Acrylic Copolymer is characterized by:

* Excellent clarity
* Good scratch resistance
* Ease of processing
* Can be blended with K-RESIN at any level to modify flexibility, surface hardness, heat distortion temperature, etc.

P-359 complies with FDA regulations for food contact applications.

PROPERTY	TEST ASTM	TYPICAL VALUE
Tensile Strength, break	D-638, 0.2"/min	8.4×10^3 psi
Elongation, break	D-638	2.1%
Tensile Modulus	D-638	5.1×10^5 psi
Flexural Strength, break	D-790 0.05"/min.	14.2×10^3 psi
Flexural Modulus	D-790	4.6×10^5 psi
Impact Strength, Izod, Notched	D-256	0.37 ft.lb./in.
Specific Gravity	D-792	1.08
Hardness, Rockwell	D-785	M-80
Deflection Temperature (annealed @ 180F) 264 psi	D-648	210F
Light Transmission	D-1003	90%
Haze	D-1003	1.5%
Refractive Index (nD)	D-542	1.57
Melt Flow Rate,	D-1238	
Cond 190/10 (N)		2.4g/10 min.
Cond. 230/1.2 (H)		0.7g/10 min.
Cond. 230/3.8 (I)		4.3g/10 min.
Mold Shrinkage	D-955	.002-.006 in./in
Water Absorption 24 hrs.	D-570	0.11%

RICHARDSON POLYMER CORP.: RPC-440 Acrylic Copolymer:

RPC-440 Methyl Methacrylate Styrene Copolymer

RPC-440 Acrylic Copolymer is characterized by:

* Sparkling clarity
* Excellent UV-resistance and weatherability
* Excellent scratch resistance

Typical applications include: POP displays, housewares, outdoor and automotive applications.

RPC-440 complies with FDA regulations for food contact applications.

PROPERTY	TEST ASTM	TYPICAL VALUE
Tensile Strength, break	D-638, 0.2"/min.	10.1 x 10 3 psi
Elongation, break	D-638	5%
Tensile Modulus	D-638	5.2 x 10 5 psi
Flexural Strength, yield	D-790 0.05"/min.	16.0 x 10 3 psi
Flexural Modulus	D-790	4.6 x 10 5 psi
Impact Strength, Izod, Notched	D-256	.34 ft.lb./in.
Specific Gravity	D-792	1.13
Hardness, Rockwell	D-785	M-85
Deflection Temperature (annealed @ 180F) 264 psi	D-648	210F
Light Transmission	D-1003	90%
Haze	D-1003	1.5%
Refractive Index (nD)	D-542	1.53
Melt Flow Rate,	D-1238	
Cond. 190/10 (N)		0.35g/10 min.
Cond. 230/1.2 (H)		0.3g/10 min.
Cond. 230/3.8 (I)		1.1g/10 min.
Mold Shrinkage	D-955	.002-.006 in./in.
Water Absorption 24 hrs.	D-570	0.17%

RICHARDSON POLYMER CORP.: RSA-2 Clear Impact Acrylic:

RSA-2 Methyl Methacrylate Butadiene Styrene Terpolymer

RSA-2 Clear Impact Acrylic is characterized by:

* Low haze, high clarity
* Abuse/scratch resistance
* Good processability

Typical applications include: Office accessories, small appliances, printer ribbon cartridges, ice scrapers.

PROPERTY	TEST ASTM	TYPICAL VALUE
Tensile Strength, yield	D-638, 0.2"/min.	8.6×10^3 psi
Elongation, break	D-638	29%
Tensile Modulus	D-638	4.5×10^5 psi
Flexural Strength, yield	D-790, 0.05"/min.	13.5×10^3 psi
Flexural Modulus	D-790	4.0×10^5 psi
Impact Strength, Izod, Notched	D-256	0.5 ft.lb./in.
Specific Gravity	D-792	1.13
Hardness, Rockwell	D-785	M-80
Deflection Temperature (annealed @ 180F) 264 psi	D-648	208F
Light Transmission	D-1003	90%
Haze	D-1003	2.3%
Refractive Index (nD)	D-542	1.52
Melt Flow Rate,	D-1238	
Cond. 190/10 (N)		1.1g/10 min.
Cond. 230/1.2 (H)		0.9g/10 min.
Cond. 230/3.8 (I)		3.9g/10 min.
Mold Shrinkage	D-955	.004 in./in.
Water Absorption 24 hrs.	D-570	0.26%

RICHARDSON POLYMER CORP.: RSA-3G Clear Impact Acrylic:

RSA-3G Methyl Methacrylate Butadiene Styrene Terpolymer

RSA-3G Clear Impact Acrylic is characterized by:

* Excellent gamma ray recovery
* High Clarity
* Abuse/Scratch resistance

Typical applications include: medical devices, office accessories, small appliances, toys.

RSA-3G complies with FDA regulations for food contact applications. Physical properties are not affected by gamma irradiation.

RSA-3G meets USP XXI specifications for Class VI Plastics.

PROPERTY	TEST ASTM	TYPICAL VALUE
Tensile Strength, yield	D-638 0.2"/min	7.7 x 10 3 psi
Elongation, break	D-638	27%
Tensile Modulus	D-638	4.1 x 10 5 psi
Flexural Strength, yield	D-790 0.05"/min.	11.7 x 10 3 psi
Flexural Modulus	D-790	3.5 x 10 5 psi
Impact Strength, Izod, Notched	D-256	0.9 ft.lb./in.
Specific Gravity	D-792	1.12
Hardness, Rockwell	D-785	M-75
Deflection Temperature (annealed @ 160F) 264 psi	D-648	183F
Light Transmission	D-1003	87%
Haze	D-1003	3.0%
Refractive Index (nD)	D-542	1.52
Melt Flow Rate,	D-1238	
Cond. 190/10 (N)		1.5g/10 min.
Cond. 230/1.2 (H)		0.6g/10 min.
Cond. 230/3.8 (I)		3.9g/10 min.
Mold Shrinkage	D-955	.004 in./in.
Water Absorption 24 hrs.	D-570	0.24%

ROGERS CORP.: RX Engineering Thermosets: Diallyl Phthalate Molding Materials:

Grade:

RX1-503:

Reinforcement: Orlon
Resin (DAP) Form: Ortho Granular
Bulk Factor: 3.0

Specific Gravity (G/cm 3): ASTM D-792A: 1.40
Shrinkage (Comp) in/in: ASTM D-955: .010-.012
Izod Impact (Ft. Lb./In.): ASTM D-256A: 0.65
Flexural Strength (PSI): ASTM D-790: 11000
Flexural Modulus (PSI) X 10 6: ASTM D-790: 0.7
Tensile Strength (PSI): ASTM D-638: 5000
Compressive Strength (PSI): ASTM D-695: 29000
Water Absorption, %: 0.40

RX1-50N:

Reinforcement: Mineral
Resin (DAP) Form: Ortho Granular
Bulk Factor: 2.3

Specific Gravity (G/cm 3): ASTM D-792A: 1.80
Shrinkage (Comp) in/in: ASTM 955: .003-.006
Izod Impact (Ft.Lb./In.): ASTM D-256A: 0.45
Flexural Strength (PSI): ASTM D790: 13000
Flexural Modulus (PSI) X 10 6: ASTM D-790: 1.6
Tensile Strength (PSI): ASTM D-638: 6500
Compressive Strength (PSI): ASTM D-695: 24000
Water Absorption, %: 0.30

RX1-501AN:

Reinforcement: Mineral/Nylon
Resin (DAP) Form: Ortho Granular
Bulk Factor: 2.6

Specific Gravity (G/cm 3): ASTM D-792A: 1.73
Shrinkage (Comp) in/in: ASTM D-955: .007-.010
Izod Impact (Ft. Lb./In.): ASTM D-256A: 0.45
Flexural Strength (PSI): ASTM D-790: 12000
Flexural Modulus (PSI) X 10 6: ASTM D-790: 1.1
Tensile Strength (PSI): ASTM D-638: 5500
Compressive Strength (PSI): ASTM D-695: 25000
Water Absorption, %: 0.40

ROGERS CORP.: RX Engineering Thermosets: Diallyl Phthalate Molding Materials(Continued):

Grade:

RX1-50N:

Reinforcement: Mineral
Resin (DAP) Form: Ortho Granular
Bulk Factor: 2.4

Specific Gravity (G/cm 3): ASTM D-792A: 1.66
Shrinkage (Comp) in/in: ASTM D-955: .003-.006
Izod Impact (Ft.Lb./In.): ASTM D-256A: 0.45
Flexural Strength (PSI): ASTM D-790: 14000
Flexural Modulus (PSI) x 10 6: ASTM D-790: 1.3
Tensile Strength (PSI): ASTM D-790: 7500
Compressive Strength (PSI): ASTM D-695: 21000
Water Absorption, %: 0.50

RX2-501N:

Reinforcement: Mineral
Resin (DAP) Form: Iso Granular
Bulk Factor: 2.1

Specific Gravity (G/cm 3): ASTM D-792A: 1.80
Shrinkage (Comp) in/in: ASTM D-955: .003-.006
Izod Impact (Ft.Lb./In.): ASTM D-256A: 0.45
Flexural Strength (PSI): ASTM D-790: 13000
Flexural Modulus (PSI) X 10 6: ASTM D-790: 2.0
Tensile Strength (PSI): ASTM D-638: 5500
Compressive Strength (PSI): ASTM D-695: 27000
Water Absorption, %: 0.25

RX1-520:
RX1310:

Reinforcement: Short Glass
Resin (DAP) Form: Ortho Granular
Bulk Factor: 2.4

Specific Gravity (G/cm 3): ASTM D-792A: 1.83
Shrinkage (Comp) in/in: ASTM D-955: .001-.003
Izod Impact (Ft.Lb./In.): ASTM D-256A: 0.70
Flexural Strength (PSI): ASTM D-790: 18000
Flexural Modulus (PSI) x 10 6: ASTM D-790: 9000
Tensile Strength (PSI): ASTM D-638: 2.0
Compressive Strength (PSI): ASTM D-695: 25000
Water Absorption, %: 0.25

ROGERS CORP.: RX Engineering Thermosets: Diallyl Phthalate Molding Materials(Continued):

Grade:

RX1-540:

Reinforcement: Short Glass/Mineral
Resin (DAP) Form: Ortho Granular
Bulk Factor: 2.3

Specific Gravity (G/cm 3): ASTM D-792A: 1.85
Shrinkage (Comp) in/in: ASTM D-955: .001-.003
Izod Impact (Ft.Lb./In.): ASTM D-256A: 0.60
Flexural Strength (PSI): ASTM D-790: 13000
Flexural Modulus (PSI) X 10 6: ASTM D-790: 1.8
Tensile Strength (PSI): ASTM D-638: 6500
Compressive Strength (PSI): ASTM D-695: 23000
Water Absorption, %: 0.35

RX2-520:

Reinforcement: Short Glass
Resin (DAP) Form: Iso Granular
Bulk Factor: 2.5

Specific Gravity (G/cm 3): ASTM D-792A: 1.75
Shrinkage (Comp) in/in: ASTM D-955: .001-.003
Izod Impact (Ft.Lb./In.): ASTM D-256A: 0.8
Flexural Strength (PSI): ASTM D-790: 18000
Flexural Modulus (PSI) X 10 6: ASTM D-790: 1.7
Tensile Strength (PSI): ASTM D-638: 7500
Compressive Strength (PSI): ASTM D-695: 30000
Water Absorption, %: 0.25

RX3-1-501N:

Reinforcement: Mineral
Resin (DAP) Form: Ortho Granular
Bulk Factor: 2.0

Specific Gravity (G/cm 3): ASTM D-792A: 1.83
Shrinkage (Comp) in/in: ASTM D-955: .003-.006
Izod Impact (Ft.Lb./in.): ASTM D-256A: 0.60
Flexural Strength (PSI): ASTM D-790: 13000
Flexural Modulus (PSI) X 10 6: ASTM D-790: 1.7
Tensile Strength (PSI): ASTM D-638: 6500
Compressive Strength (PSI): ASTM D-695: 25000
Water Absorption, %: 0.30

ROGERS CORP.: RX Engineering Thermosets: Diallyl Phthalate Molding Materials (Continued):

Grade:

RX3-2-520F:

Reinforcement: Short Glass
Resin (DAP) Form: Iso Granular
Bulk Factor: 2.4

Specific Gravity (G/cm 3): ASTM D-792A: 1.90
Shrinkage (Comp) in/in: ASTM D-955: .001-.003
Izod Impact (Ft.Lb./In.): ASTM D-955: 0.80
Flexural Strength (PSI): ASTM D-790: 17000
Flexural Modulus (PSI) x 10 6: ASTM D-790: 1.4
Tensile Strength (PSI): ASTM D-638: 7500
Compressive Strength (PSI): ASTM D-695: 26000
Water Absorption, %: 0.25

RX3-1-525F:
RX1366FR:

Reinforcement: Short Glass
Resin (DAP) Form: Ortho Granular
Bulk Factor: 2.3

Specific Gravity (G/cm 3): ASTM D-792A: 1.87
Shrinkage (Comp) in/in: ASTM D-955: .001-.003
Izod Impact (Ft.Lb./In.): ASTM D-256A: 0.80
Flexural Strength (PSI): ASTM D-790: 18000
Flexural Modulus (PSI) X 10 6: ASTM D-790: 1.8
Tensile Strength (PSI): ASTM D-638: 8000
Compressive Strength (PSI): ASTM D-695: 25000
Water Absorption, %: 0.25

RX1-530:

Reinforcement: Long Glass
Resin (DAP) Form: Ortho Coarse Granular
Bulk Factor: 3.5

Specific Gravity (G/cm 3): ASTM D-792A: 1.74
Shrinkage (Comp) in/in: ASTM D-955: .001-.004
Izod Impact (Ft.Lb./In.): ASTM D-256A: 3.0-7.0
Flexural Strength (PSI): ASTM D-790: 18000
Flexural Modulus (PSI) X 10 6: ASTM D-790: 1.8
Tensile Strength (PSI): ASTM D-638: 8500
Compressive Strength (PSI): ASTM D-695: 27000
Water Absorption, %: 0.35

ROGERS CORP.: RX Engineering Thermosets: Diallyl Phthalate Molding Materials(Continued):

Grade:

RX2-530:

Reinforcement: Long Glass
Resin (DAP) Form: Iso Coarse Granular
Bulk Factor: 3.5

Specific Gravity (G/cm 3): ASTM D-792A: 1.73
Shrinkage (Comp) in/in: ASTM D-955: .001-.004
Izod Impact (Ft.Lb./In.): ASTM D-256A: 3.0-7.0
Flexural Strength (PSI): ASTM D-790: 18000
Flexural Modulus (PSI) X 10 6: ASTM D-790: 1.8
Tensile Strength (PSI): ASTM D-638: 7500
Compressive Strength (PSI): ASTM D-695: 25000
Water Absorption, %: 0.35

RX3-1-530:

Reinforcement: Long Glass
Resin (DAP) Form: Ortho Coarse Granular
Bulk Factor: 3.5

Specific Gravity (G/cm 3): ASTM D-792A: 1.76
Shrinkage (Comp) in/in: ASTM D-955: .001-.004
Izod Impact (Ft.Lb./In.): ASTM D-256A: 3.0-7.0
Flexural Strength (PSI): ASTM D-790: 18000
Flexural Modulus (PSI) X 10 6: ASTM D-790: 1.8
Tensile Strength (PSI): ASTM D-638: 8500
Compressive Strength (PSI): ASTM D-695: 25000
Water Absorption, %: 0.35

RX3-2-530:

Reinforcement: Long Glass
Resin (DAP) Form: Iso Coarse Granular
Bulk Factor: 3.5

Specific Gravity (G/cm 3): ASTM D-792A: 1.76
Shrinkage (Comp) in/in: ASTM D-955: .001-.004
Izod Impact (Ft. Lb./In.): ASTM D-256A: 3.0-7.0
Flexural Strength (PSI): ASTM D-790: 18000
Flexural Modulus (PSI) X 10 6: ASTM D-790: 1.7
Tensile Strength (PSI): ASTM D-638: 8500
Compressive Strength (PSI): ASTM D-695: 28000
Water Absorption, %: 0.35

ROGERS CORP.: RX Engineering Thermosets: Phenolic Molding Materials:

Grade:

Self-Lubricating:

RX340:

Reinforcement: Cell
Phenolic Resin Type: 2-STP
Form: Pellet
Bulk Factor: 2.2

Specific Gravity (G/cm 3): ASTM D-792A: 1.43
Shrinkage (Comp) in/in: ASTM D-955: .006
Izod Impact (Ft.Lb./In.): ASTM D-256A: 0.50
Flexural Strength (PSI): ASTM D-790: 10000
Flexural Modulus (PSI) X 10 6: ASTM D790: 1.2
Tensile Strength (PSI): ASTM D-638: 6000
Compressive Strength (PSI): ASTM D-695: 25000

Self-Lubricating:

RX342:

Reinforcement: Cell
Phenolic Resin Type: 2-STP
Form: Pellet
Bulk Factor: 2.3

Specific Gravity (G/cm 3): ASTM D-792A: 1.44
Shrinkage (Comp) in/in: ASTM D-955: .006
Izod Impact (Ft.Lb./In.): ASTM D-256A: 0.55
Flexural Strength (PSI): ASTM D-790: 10000
Flexural Modulus (PSI) X 10 6: ASTM D-790: 1.2
Tensile Strength (PSI): ASTM D-638: 6000
Compressive Strength (PSI): ASTM D-695: 23000

RX475:

Reinforcement: Cell
Phenolic Resin Type: 2-STP
Form: Pellet
Bulk Factor: 3.3

Specific Gravity (G/cm 3): ASTM D-792A: 1.37
Shrinkage (Comp) in/in: ASTM D-955: .006
Izod Impact (Ft.Lb./In.): ASTM D-256A: 2.0
Flexural Strength (PSI) ASTM D-790: 12000
Flexural Modulus (PSI) X 10 6: ASTM D-790: 1.3
Tensile Strength (PSI): ASTM D-638: 5500
Compressive Strength (PSI): ASTM D-695: 23000

ROGERS CORP.: RX Engineering Thermosets: Phenolic Molding Materials(Continued):

Grade:

RX431:

Reinforcement: Cell
Phenolic Resin Type: 2-STP
Form: Pellet
Bulk Factor: 3.5

Specific Gravity (G/cm 3): ASTM D-792A: 1.42
Shrinkage (Comp) in/in: ASTM D-955: .005
Izod Impact (Ft.Lb./In.): ASTM D-256A: 1.1
Flexural Strength (PSI): ASTM D-790: 12000
Flexural Modulus (PSI) X 10 6: ASTM D-790: 1.3
Tensile Strength (PSI): ASTM D638: 6300
Compressive Strength (PSI): ASTM D-695: 25500

RX448:

Reinforcement: Cell
Phenolic Resin Type: 2-STP
Form: Pellet
Bulk Factor: 2.8

Specific Gravity (G/cm 3): ASTM D-792A: 1.42
Shrinkage (Comp) in/in: ASTM D-955: .006
Izod Impact (Ft.Lb./In.): ASTM D-256A: 0.65
Flexural Strength (PSI): ASTM D-790: 11000
Flexural Modulus (PSI) X 10 6: ASTM D-790: 1.3
Tensile Strength (PSI): ASTM D-638: 6500
Compressive Strength (PSI): 25000

RX525:

Reinforcement: Cell
Phenolic Resin Type: 2-STP
Form: Pellet
Bulk Factor: 2.6

Specific Gravity (G/cm 3): ASTM D-792A: 1.43
Shrinkage (Comp) in/in: ASTM D-955: .006
Izod Impact (Ft. Lb./In.): ASTM D-256A: 0.55
Flexural Strength (PSI): ASTM D-790: 10000
Flexural Modulus (PSI) X 10 6: ASTM D-790: 1.3
Tensile Strength (PSI): ASTM D-638: 6500
Compressive Strength (PSI): ASTM D-695: 23000

ROGERS CORP.: RX Engineering Thermosets: Phenolic Molding Materials(Continued):

Grade:

RX630:

Reinforcement: Glass
Phenolic Resin Type: 2-STP
Form: Gran.
Bulk Factor: 2.1

Specific Gravity (G/cm 3): ASTM D-792A: 1.78
Shrinkage (Comp) in/in: ASTM D-955: .0015
Izod Impact (Ft.Lb./In): ASTM D-256A: 0.90
Flexural Strength (PSI): ASTM D-790: 27000
Flexural Modulus (PSI) X 10 6: ASTM D-790: 3.2
Tensile Strength (PSI): ASTM D-638: 12000
Compressive Strength (PSI): 40000

RX611:

Reinforcement: Glass
Phenolic Resin Type: 2-STP
Form: Gran.
Bulk Factor: 2.1

Specific Gravity (G/cm 3): ASTM D-792A: 1.75
Shrinkage (Comp) in/in: ASTM D-955: .002
Izod Impact (Ft.Lb./In.): ASTM D-790: 0.80
Flexural Strength (PSI): ASTM D-790: 25000
Flexural Modulus (PSI) X 10 6: ASTM D-790: 3.0
Tensile Strength (PSI): ASTM D-638: 11000
Compressive Strength (PSI): ASTM D-695: 40000

RX660:

Reinforcement: Glass
Phenolic Resin Type: 2-STP
Form: Gran.
Bulk Factor: 2.1

Specific Gravity (G/cm 3): ASTM D-792A: 1.80
Shrinkage (Comp) in/in: ASTM D-955: .002
Izod Impact (Ft.Lb./In.): ASTM D-790: 0.70
Flexural Strength (PSI): ASTM D-790: 22000
Flexural Modulus (PSI) X 10 6: ASTM D-790: 2.9
Tensile Strength (PSI): ASTM D-638: 8500
Compressive Strength (PSI): ASTM D-695: 33000

ROGERS CORP.: RX Engineering Thermosets: Phenolic Molding Materials(Continued):

Grade:

RX660C:

Reinforcement: Glass
Phenolic Resin Type: 2-STP
Form: Gran.
Bulk Factor: 2.1

Specific Gravity (G/cm 3): ASTM D-792A: 1.82
Shrinkage (Comp) in/in: ASTM D-955: .002
Izod Impact (Ft.Lb./In.): ASTM D-256A: 0.70
Flexural Strength (PSI): ASTM D-790: 22000
Flexural Modulus (PSI) X 10 6: ASTM D-790: 3.0
Tensile Strength (PSI): ASTM D-638: 8800
Compressive Strength (PSI): ASTM D-695): 33000

RX670:

Reinforcement: Glass
Phenolic Resin Type: 2-STP
Form: Gran.
Bulk Factor: 2.0

Specific Gravity (G/cm 3): ASTM D-792A: 1.90
Shrinkage (Comp) in/in: ASTM D-955: .002
Izod Impact (Ft.Lb./In.): ASTM D-256A: 0.65
Flexural Strength (PSI): ASTM D-790: 20000
Flexural Modulus (PSI) X 10 6: ASTM D-790: 20000
Tensile Strength (PSI): ASTM D-638: 8000
Compressive Strength (PSI): ASTM D-695: 37000

Ammonia Free:
RX643:

Reinforcement: Glass
Phenolic Resin Type: 1-STP
Form: Gran.
Bulk Factor: 2.1

Specific Gravity (G/cm 3): ASTM D-792A: 1.74
Shrinkage (Comp) in/in: ASTM D-955: .0015
Izod Impact (Ft.Lb./In.): ASTM D-256A: 1.0
Flexural Strength (PSI): ASTM D-790: 30000
Flexural Modulus (PSI) X 10 6: ASTM D-790: 2.7
Tensile Strength (PSI): ASTM D-638: 13000
Compressive Strength (PSI): ASTM D-695: 35000

ROGERS CORP.: RX Engineering Thermosets: Phenolic Molding Materials(Continued):

Grade:

RX640:

Reinforcment: Glass
Phenolic Resin Type: 1-STP
Form: Gran.
Bulk Factor: 2.0

Specific Gravity (G/Cm 3): ASTM D-792A: 1.74
Shrinkage (Comp) in/in: ASTM D-955: .002
Izod Impact (Ft.Lb./In.): ASTM D-256A: 0.90
Flexural Strength (PSI): ASTM D-790: 27000
Flexural Modulus (PSI) X 10 6: ASTM D-790: 2.5
Tensile Strength (PSI): ASTM D-638: 12500
Compressive Strength (PSI): ASTM D-695: 32000

RX647:

Reinforcement: Glass
Phenolic Resin Type: 1-STP
Form: Gran.
Bulk Factor: 2.0

Specific Gravity (G/cm 3): ASTM D-792A: 1.90
Shrinkage (Comp) in/in: ASTM D-955: .002
Izod Impact (Ft.Lb./In.): ASTM D-256A: 0.70
Flexural Strength (PSI): ASTM D-790: 22000
Flexural Modulus (PSI) X 10 6: ASTM D-790: 2.9
Tensile Strength (PSI): ASTM D-638: 10500
Compressive Strength (PSI): ASTM D-695: 35000

RX655:

Reinforcement: Glass
Phenolic Resin Type: 2-STP
Form: Gran.
Bulk Factor: 2.0

Specific Gravity (G/Cm 3): ASTM D-792A: 2.05
Shrinkage (Comp) in/in: ASTM D-955: .001
Izod Impact (Ft.Lb./In.): ASTM D-256A: 0.5
Flexural Strength (PSI): ASTM D-790: 14000
Flexural Modulus (PSI) X 10 6: ASTM D-790: 3.4
Tensile Strength (PSI): ASTM D-638: 7000
Compressive Strength (PSI): ASTM D-695: 32000

ROGERS CORP.: RX Engineering Thermosets: Phenolic Molding Materials(Continued):

<u>Grade:</u>

RX865:

Reinforcement: Glass
Phenolic Resin Type: 2-STP
Form: Flake
Bulk Factor: 3.0

Specific Gravity (G/cm 3): ASTM D-792A: 1.86
Shrinkage (Comp) in/in: ASTM D-955: .002
Izod Impact (Ft.Lb./In.): ASTM D-256A: 1.5
Flexural Strength (PSI): ASTM D-790: 19000
Flexural Modulus (PSI) X 10 6: ASTM D-790: 3.2
Tensile Strength (PSI): ASTM D-638: 8000
Compressive Strength (PSI): ASTM D-695: 35000

RX868:

Reinforcement: Glass
Phenolic Resin Type: 2-STP
Form: Pellet
Bulk Factor: 2.5

Specific Gravity (G/cm 3): ASTM D-792A: 1.77
Shrinkage (Comp) in/in: ASTM D-955: .002
Izod Impact (Ft.Lb./In.): ASTM D-256A: 0.70
Flexural Strength (PSI): ASTM D-790: 11000
Flexural Modulus (PSI) X 10 6: ASTM D-790: 2.1
Tensile Strength (PSI): ASTM D-638: 7500
Compressive Strength (PSI): ASTM D-695: 20000

RX867:

Reinforcement: Glass
Phenolic Resin Type: 2-STP
Form: Pellet
Bulk Factor: 2.6

Specific Gravity (G/Cm 3): ASTM D-792A: 1.77
Shrinkage (Comp) in/in: ASTM D-955: .0025
Izod Impact (Ft.Lb./In.): ASTM D-256A: 0.65
Flexural Strength (PSI): ASTM D-790: 11000
Flexural Modulus (PSI) X 10 6: ASTM D-790: 2.4
Tensile Strength (PSI): ASTM D-638: 7000
Compressive Strength (PSI): ASTM D-695: 23000

ROGERS CORP.: RX Engineering Thermosets: Phenolic Molding Materials(Continued):

Grade:

RX866:

Reinforcement: Glass
Phenolic Resin Type: 2-STP
Form: Pellet
Bulk Factor: 3.0

Specific Gravity (G/cm 3): ASTM D-792A: 1.70
Shrinkage (Comp) in/in: ASTM D-955: .002
Izod Impact (Ft.Lb./In.): ASTM D-256A: 0.65
Flexural Strength (PSI): ASTM D-790: 12500
Flexural Modulus (PSI) X 10 6: ASTM D-790: 2.3
Tensile Strength (PSI): ASTM D-638: 7700
Compressive Strength (PSI): D-695: 23000

RX853:

Reinforcement: Glass
Phenolic Resin Type: 2-STP
Form: Pellet
Bulk Factor: 3.0

Specific Gravity (G/cm 3): ASTM D-792A: 1.80
Shrinkage (Comp) in/in: ASTM D-955: .002
Izod Impact (Ft.Lb./In.): ASTM D-256A: 0.65
Flexural Strength (PSI): ASTM D-790: 14000
Flexural Modulus (PSI) X 10 6: ASTM D-790: 2.7
Tensile Strength (PSI): ASTM D-638: 7700
Compressive Strength (PSI): ASTM D-695: 26000

HT500:

Reinforcement: Glass
Phenolic Resin Type: Special thermoset alloy molding compound
Form: Gran.
Bulk Factor: 3.1

Specific Gravity (G/cm 3): ASTM D792A: 1.75
Shrinkage (Comp) in/in: ASTM D-955: .002
Izod Impact (Ft.Lb./In.): ASTM D-256A: 1.0
Flexural Strength (PSI): ASTM D-790: 13500
Flexural Modulus (PSI) X 10 6: ASTM D-790: 2.0
Tensile Strength (PSI): ASTM D-638: 8500
Compressive Strength (PSI): ASTM D-695: 25000

ROHM AND HAAS CO.: PLEXIGLAS V and Impact Modified Acrylic Molding Pellets:

Properties and Advantages:

PLEXIGLAS V Series Acrylic Pellets:
- Outstanding Environmental Stability
- Outstanding Optical Quality
- Light Weight
- Chemical Resistance
- Design Flexibility
  - Variety of Grades
  - Colors
  - Surfaces
  - Finishing and Decorating

PLEXIGLAS MI-7, MI-7G, DR, DR-G, DR-M, HFI-7, HFI-7G, HFI-10 and HFI-10G Impact Modified Acrylic Pellets:

- Excellent Impact Resistance
- Light Transmission
- Environmental Stability
- Variety of Grades

Average physical properties of PLEXIGLAS molding pellets:

PLEXIGLAS V825:
- Refractive Index ND: 1.49
- Specific Gravity: 1.19
- Tensile Strength: 10,200

PLEXIGLAS V811:
- Refractive Index ND: 1.49
- Specific Gravity: 1.19
- Tensile Strength: 10,200

PLEXIGLAS V920:
- Refractive Index ND: 1.49
- Specific Gravity: 1.19
- Tensile Strength: 10,000

PLEXIGLAS VO44/VO45/VO52:
- Refractive Index ND: 1.49
- Specific Gravity: 1.19
- Tensile Strength: 10,200

ROHM & HAAS Co.: PLEXIGLAS V and Impact Modified Acrylic Molding Pellets (Continued):

Property:

PLEXIGLAS VM:
Refractive Index ND: 1.49
Specific Gravity: 1.18
Tensile Strength: 9,600

PLEXIGLAS VMHF:
Refractive Index ND: 1.49
Specific Gravity: 1.18
Tensile Strength: 9,600

PLEXIGLAS VS:
Refractive Index ND: 1.49
Specific Gravity: 1.18
Tensile Strength: 9,400

PLEXIGLASS MI-7/MI-7G:
Refractive Index ND: 1.49
Specific Gravity: 1.17
Tensile Strength: 7,000

PLEXIGLAS DR/DR-M/DR-G:
Refractive Index ND: 1.49
Specific Gravity: 1.15
Tensile Strength: 5,500

PLEXIGLAS HFI-7/HFI-7G:
Refractive Index ND: 1.49
Specific Gravity: 1.17
Tensile Strength: 6,800

PLEXIGLAS HFI-10/HFI-10G:
Refractive Index ND: 1.49
Specific Gravity: 1.15
Tensile Strength: 5,500

ROHM AND HAAS CO.: PLEXIGLAS V and Impact Modified Acrylic Molding Pellets(Continued):

DESCRIPTION:

PLEXIGLAS V825:
Highest heat resistance of any standard acrylic molding or extrusion resin.

PLEXIGLAS V811:
Injection molding and extrusion resin. High heat resistance and medium flow.

PLEXIGLAS V920:
General purpose injection molding and extrusion resin. Well suited for decorated applications.

PLEXIGLAS VO44:
Unlubricated formulation for extrusion processing. Good heat resistance.

PLEXIGLAS VO45:
Extrusion and injection molding resin. Good heat resistance.

PLEXIGLAS VO52:
High molecular weight resin for injection molding only. Slightly less heat resistance and slightly stiffer than PLEXIGLAS V811. Superior mold release.

PLEXIGLAS VM:
Injection molding and extrusion resin. Medium heat resistance and easy flow.

PLEXIGLAS VS:
Injection molding grade only. Maximum flow and lowest heat resistance.

PLEXIGLAS DR:
High impact acrylic resin for injection molding and extrusion

PLEXIGLAS MI-7:
Medium impact acrylic resin for injection molding and extrusion.

PLEXIGLAS DR-G,
PLEXIGLAS MI-7G:
Gamma radiation resistant grades of PLEXIGLAS DR and MI-7 respectively.

PLEXIGLAS VMHF:
Injection molding resin. Medium heat resistance with excellent flow for large, thin walled light diffusers.

ROHM AND HAAS CO.: PLEXIGLAS V and Impact Modified Acrylic Molding Pellets(Continued):

DESCRIPTION(Continued):

PLEXIGLAS DR-M:
High impact acrylic resin. Low reflectance and matte surface.

PLEXIGLAS HFI-7:
Medium impact acrylic resin. Improved injection-molding characteristics.

PLEXIGLAS HFI-10:
High impact acrylic resin. Improved injection-molding characteristics.

PLEXIGLAS HFI-7G:
PLEXIGLAS HFI-10G:
Gamma radiation resistant grades of PLEXIGLAS HFI-7 and HFI-10 respectively.

Optical Grade:
Specially processed, colorless, transparent injection molding and extrusion resin. Unexcelled cleanliness for optical applications.

Ultraviolet Absorbing Grades:
Injection molding and extrusion resins that absorb UV light. Different grades available for different absporption requirements. PLEXIGLAS V811-100 UVA-5 and VO52-100 UVA-5 combine heat resistance with UV absorption for lighting applications.

PLEXIGLAS V920 UVT,
PLEXIGLAS VS UVT:
Ultraviolet transmitting resins for injection molding and extrusion. Used primarily for medical diagnostic components.

ROSTONE: 3250 BMC Series:

The 3250 series ROSITE compounds are designed for applications which require a strong, non-tracking, arc quenching and self-extinguishing material. In addition to high strength, this series of compounds have improved color.

<u>Product</u> <u>Number:</u>

ROSITE 3250A:

PROPERTIES	ASTM TEST	TYPICAL VALUES
Glass Content (%)		5
MECHANICAL		
Flexural Strength, 10 3 psi	D790	10
Impact Strength, Izod Notched, ft lb/in	D256	4
Compressive Strength, 10 3 psi	D695	20
Tensile Strength, 10 3 psi	D638	3.5
ELECTRICAL		
Arc Resistance, seconds	D495	240+
IEC Track, Comparative Track Index	*	500+
Dielectric Strength, Volts/mil	D149	300
Dielectric Constant, 10 3 Hz	D150	4
Volume Resistivity, 10 14 ohm-cm (dry)	D257	1
THERMAL		
Deflection Temperature, F, 264 psi	D648	400+
UL Thermal Index, Electrical, C	*	130
Mechanical, W/O Impact, C	*	180
Mechanical, W/Impact, C	*	150
UL Flammability Class, .060 thick	*	94V-O
Conductivity, 10^{-4} cal/sec/sq cm/ 1 (C/cm)	C177	28
OTHER		
Shrinkage, in/in		.003
Specific Gravity	D792	2.08

* Information from UL Yellow Cards: reference E23867, June 1980.

ROSTONE: 3250 BMC Series(Continued):

Product Number:

ROSITE 3250C:

PROPERTIES	ASTM TEST	TYPICAL VALUES
Glass Content (%)		15
MECHANICAL		
Flexural Strength, 10^3 psi	D790	20
Impact Strength, Izod Notched, ft lb/in	D256	7
Compressive Strength, 10^3 psi	D695	20
Tensile Strength, 10^3 psi	D638	4.5
ELECTRICAL		
Arc Resistance, seconds	D495	180+
IEC Track, Comparative Track Index	*	500+
Dielectric Strength, Volts/mil	D149	300
Dielectric Constant, 10^3 Hz	D150	4
Volume Resistivity, 10^{14} ohm-cm (dry)	D257	1
THERMAL		
Deflection Temperature, F, 264 psi	D648	400+
UL Thermal Index, Electrical, C	*	130
Mechanical, W/O Impact, C	*	180
Mechanical, W/Impact, C	*	150
UL Flammability, Class, .060 thick	*	94V-O
Conductivity, 10^{-4} Cal/sec/sq cm/1(C/cm)	C177	25
OTHER		
Shrinkage, in/in		.003
Specific Gravity	D792	1.94

ROSTONE: 3550 BMC Series:

The 3550 BMC series ROSITE Compounds are low shrink, self-extinguishing, non-tracking and arc quenching. They are designed for electrical, electro-mechanical and structural applications requiring extremely accurate dimensional control and stability. Special grades of ROSITE 3550 exhibit the wear resistant qualities useful in many designs.

<u>Product Number:</u>

ROSITE 3550A:

PROPERTIES	ASTM TEST	TYPICAL VALUES
Glass Content (%)		5
MECHANICAL		
Flexural Strength, 10^3 psi	D790	7.5
Flexural Modulus, 10^5 psi	D790	15.5
Impact Strength, Izod Notched, ft lb/in	D256	3
Compressive Strength, 10^3 psi	D695	20
Tensile Strength, 10^3 psi	D638	4
ELECTRICAL		
Arc Resistance, seconds	D495	240+
IEC Track, Comparative Track Index	*	500+
Dielectric Strength, Volts/mil	D149	300
Dielectric Constant, 10^3 Hz	D150	6
Volume Resistivity, 10^{14} ohm-cm (dry)	D257	1
THERMAL		
Deflection Temperature F, 264 psi	D648	400+
UL Thermal Index, Electrical, C	*	130
Mechanical, W/O Impact, C	*	180
Mechanical, W/Impact, C	*	150
Radiant Panel, ASTME 162	*	15
UL Flammability Class, .060 thick	*	94V-O/94-5V
.120 thick	*	94V-O/94-5V
Conductivity, 10^{-4} cal/sec/sq cm/1(C/cm)	C177	28
Coefficient of Expansion, 10^{-5} in/in/C	D696	2.0
Oxygen Index, %	D2863	62
OTHER		
Shrinkage, in/in		.0005
Specific Gravity	D792	2.08

ROSTONE: 3550 BMC Series(Continued):

Product Number:

ROSITE 3550C:

PROPERTIES	ASTM TEST	TYPICAL VALUES
Glass Content (%)		15
MECHANICAL		
Flexural Strength, 10 3 psi	D790	16.5
Flexural Modulus, 10 5 psi	D790	15.5
Impact Strength, Izod Notched, ft lb/in	D256	6
Compressive Strength, 10 3 psi	D695	20
Tensile Strength, 10 3 psi	D638	4
ELECTRICAL		
Arc Resistance, seconds	D495	180+
IEC Track, Comparative Track Index	*	500+
Dielectric Strength, Volts/mil	D149	300
Dielectric Constant, 10 3 Hz	D150	6
Volume Resistivity, 10 14 ohm-cm (dry)	D257	1
THERMAL		
Deflection Temperature, F, 264 psi	D648	400+
UL Thermal Index, Electrical, C	*	130
Mechanical, W/O Impact, C	*	180
Mechanical, W/Impact, C	*	150
Radiant Panel, ASTME 162	*	15
UL Flammability Class, .060 thick	*	94V-O/94-5V
.120 thick	*	94V-0/94-5V
Conductivity, 10 -4 Cal/sec/sq cm/ 1(C/cm)	C177	25
Coefficient of Expansion, 10 -5 in/in/C	D696	2.0
Oxygen Index, %	D2863	53
OTHER		
Shrinkage, in/in		.0005
Specific Gravity	D792	1.94

* Information from UL Yellow Cards: Reference E23867, June 1980

ROSTONE: SMC:

ROSITE 3550FM, a sheet molding compound, is designed for applications requiring high strength. This is accomplished without loss of other properties in most cases. This material is compression molded and is often used for large parts.

PROPERTIES	ASTM TEST	TYPICAL RANGES
Glass Content (%)		25-30
MECHANICAL		
Flexural Strength, 10^3 psi	D790	18-25
Impact Strength, Izod Notched, ft lb/in	D256	9-12
Compressive Strength, 10^3 psi	D695	20
Tensile Strength, 10^3 psi	D638	7-10
ELECTRICAL		
Arc Resistance, seconds	D495	180
IEC Track, Comparative Track Index	*	500+
Dielectric Strength, Volts/mil	D149	300
Volume Resistivity, 10^{14} ohm-cm (dry)	D257	1
THERMAL		
Deflection Temperature, F, 264 psi	D648	400+
UL Thermal Index, Electrical, C	*	130
Mechanical, W/O Impact, C	*	180
Mechanical, W/Impact, C	*	150
Radiant Panel, ASTM E 162	*	50
UL Flammability Class, .120 thick	*	94V-O
Conductivity, 10^{-4} Cal/sec/sq cm/1(C/cm)	C177	12
Coefficient of Expansion, 10^{-5} in/in/C	D696	1.5
Oxygen Index, %	D2863	35
OTHER		
Shrinkage, in/in		.001
Specific Gravity	D792	1.75-1.90

* Information from UL Yellow Cards: Reference E23867, June 1980

A. SCHULMAN, INC.: POLYBLAK Black Concentrates:

POLYBLAK, black concentrates, is the most efficient and most economic agent known for protecting thermoplastics against degradation from ultra violet light. The use of POLYBLAK is also the most economic means to make thermoplastics opaque. The many different grades of POLYBLAK allow the jetness of the article to be determined with considerable accuracy.

Product No.

3013:

A conventional polyethylene base concentrate containing 40% carbon black. Designed for use in very thin blown film. This product has been run in HDPE film at 0.3 mil. (7 micron).

1821:

A conventional polyethylene base concentrate containing 40% carbon black. Designed for use in blown film and can be used in film as thin as 1 mil. (25 micron).

2202:

A conventional polyethylene base concentrate containing 40% carbon black. Designed for pipe and injection molding. Under restricted conditions can be used in film greater than 3 mils. thick (75 micron).

3123:

A linear polyethylene base concentrate containing 50% carbon black. Designed for use in blown film and can be used in film as thin as 1 mil. (25 micron).

3095:

A linear low density based concentrate containing 35% Western Electric approved black for use in high performance wire and cable applications.

2821:

A conventional polyethylene base concentrate meeting NSF specifications for use in PE2306 piping system. One of a series designed for pipe applications. Other products available based upon LLDPE and HDPE.

A. SCHULMAN, INC.: POLYBLAK Black Concentrates (Continued):

Product No.:

2076:

A universal olefin base concentrate containing 40% carbon black. Generally suitable for use with polyethylene, polypropylene, PVC, polystyrene, ABS, nylon, SAN, polycarbonate, and other thermoplastics.

2112:

A universal black concentrate similar to 2076 except containing 40% medium color carbon black.

3188:

A high impact polystyrene base concentrate containing 40% carbon black. Designed for molding and extrusion applications.

2973:

A high impact polystyrene base concentrate containing 25% high color carbon black. Designed for molding and extrusion applications.

2151:

An ABS base concentrate containing 20% medium color carbon black. Designed for use in injection molding and extrusion applications requiring good gloss and property retention.

2864:

A proprietary base concentrate containing 20% carbon black. Designed to color nylon without reduction in impact strength.

3224:

A polypropylene base concentrate containing 40% carbon black. Suitable for molding and extrusion applications.

3191:

PVC based black concentrate containing 40% high color carbon black. Designed for coloring flexible, semi-rigid or rigid grades of PVC. Available in ground form only.

A. SCHULMAN, INC.: POLYFORT Polypropylenes:

Filled for High Stiffness:

Property	ASTM Method	Units
FPP 1666:		
Filler Content	----	10%
Type Filler	----	Talc
Density	D729	0.035 lb/cu in
Tensile Strength	D638	4,900 psi
Flexural Modulus	D790	275 Kpsi
Notched Izod Impact	D256	0.8 fpin
Unnotched Izod Impact	D256	15 ft-lb
Falling Weight Impact	----	12 in-lb
HDT @ 66 psi (4.6 Kg/sq cm)	D648	245F(118C)
HDT @ 264 psi (18.6 Kg/sq/cm)	D648	135F(57C)
Hardness Rockwell R	D785	100
FPP 1006:		
Filler Content	----	20%
Type Filler	----	Talc
Density	D729	0.037 lb/cu in
Tensile Strength	D638	4,800 psi
Flexural Modulus	D790	330 Kpsi
Notched Izod Impact	D256	0.7 fpin
Unnotched Izod Impact	D256	14 ft-lb
Falling Weight Impact	----	10 in-lb
HDT @ 66 psi (4.6 Kg/sq cm)	D648	257F(125C)
HDT @ 264 psi (18.6 Kg/sq cm)	D648	140F(60C)
Hardness Rockwell R	D785	98
FPP 1607:		
Filler Content	----	30
Type Filler	----	Talc
Density	D729	0.041 lb/sq in
Tensile Strength	D638	4,700 psi
Flexural Modulus	D790	400 Kpsi
Notched Izod Impact	D256	0.6 fpin
Unnotched Izod Impact	D256	10 ft-lb
Falling Weight Impact	----	8 in-lb
HDT @ 66 psi (4.6 Kg/sq cm)	D648	265F(129C)
HDT @ 264 psi (18.6 Kg/sq cm)	D648	150F(65C)
Hardness Rockwell R	D785	97

A. SCHULMAN, INC.: POLYFORT Polypropylenes(Continued):

Filled for High Stiffness(Continued):

Property	ASTM Method	Units
FPP 1007:		
Filler Content	----	40%
Type Filler	----	Talc
Density	D729	0.044
Tensile Strength	D638	4,600 psi
Flexural Modulus	D790	480 Kpsi
Notched Izod Impact	D256	0.5 fpin
Unnotched Izod Impact	D256	6 ft-lb
Falling Weight Impact	----	6 in-lb
HDT @ 66 psi (4.6 Kg/sq cm)	D648	273F(134C)
HDT @ 264 psi (18.6 Kg/sq cm)	D648	160F(71C)
Hardness Rockwell R	D785	97

A. SCHULMAN, INC.: POLYFORT Polypropylenes(Continued):

Filled for High Stiffness and Low Temperature Impact:

Property	ASTM Method	Units
FPP 1651:		
Filler Content	----	10%
Type Filler	----	Coupled Mica
Density	D729	0.035 lb/cu in
Tensile Strength	D638	4,200 psi
Flexural Modulus	D790	305 Kpsi
Notched Izod Impact	D256	0.9 fpin
Unnotched Izod Impact	D256	12 ft-lb
Falling Weight Impact	----	124 in-lb
HDT @ 66 psi (4.6 kg/sq cm)	D648	250F(121C)
HDT @ 264 psi (18.6 kg/sq cm)	D648	170F(76C)
Hardness Rockwell R	D785	90
FPP 1652:		
Filler Content	----	20%
Type Filler	----	Coupled Mica
Density	D729	0.037 lb/cu in
Tensile Strength	D638	4,465 psi
Flexural Modulus	D790	390 Kpsi
Notched Izod Impact	D256	0.7 fpin
Unnotched Izod Impact	D256	10 ft-lb
Falling Weight Impact	----	36 in-lb
HDT @ 66 psi (4.6 Kg/sq cm)	D648	270F(132C)
HDT @ 264 psi (18.6 Kg/sq cm)	D648	203F(95C)
Hardness Rockwell R	D785	90

A. SCHULMAN, INC.: POLYFORT Polypropylenes(Continued):

Filled for High Stiffness and Low Temperature Impact (Continued):

Property	ASTM Method	Units
FPP 1653:		
Filler Content	----	30%
Type Filler	----	Coupled Mica
Density	D729	0.041 lb/cu in
Tensile Strength	D638	4,851 psi
Flexural Modulus	D790	495 Kpsi
Notched Izod Impact	D256	0.5 fpin
Unnotched Izod Impact	D256	4 ft-lb
Falling Weight Impact	----	12 in-lb
HDT @ 66 psi (4.6 Kg/sq cm)	D648	275F(135C)
HDT @ 264 psi (18.6 Kg/sq cm)	D648	210F(99C)
Hardness Rockwell R	D785	89
FPP 1654:		
Filler Content	----	40%
Type Filler	----	Coupled Mica
Density	D729	0.044 lb/cu in
Tensile Strength	D638	5,005
Flexural Modulus	D790	645 Kpsi
Notched Izod Impact	D256	0.4 fpin
Unnotched Izod Impact	D256	3 ft-lb
Falling Weight Impact	----	8 in-lb
HDT @ 66 psi(4.6 Kg/sq cm)	D648	285F(140C)
HDT @ 264 psi(18.6 Kg/sq cm)	D648	220F(104C)
Hardness Rockwell R	D785	89

A. SCHULMAN, INC.: POLYFORT Polypropylenes(Continued):

Filled for Impact and High Stiffness:

Property	ASTM Method	Units
FPP 1578:		
Filler Content	----	10
Type Filler	----	Calcium Carbonate
Density	D729	0.035 cu in
Tensile Strength	D638	4,500 psi
Flexural Modulus	D790	230 Kpsi
Notched Izod Impact	D256	0.65 fpin
Unnotched Izod Impact	D256	16 ft-lb
Falling Weight Impact	----	16 in-lb
HDT @ 66 psi (4.6 Kg/sq cm)	D648	200F(93C)
HDT @ 264 psi (18.6 Kg/sq cm)	D648	135F(57C)
Hardness Rockwell R	D785	99
FPP 1237:		
Filler Content	----	20
Type Filler	----	Calcium Carbonate
Density	D729	0.037 lb/cu in
Tensile Strength	D638	3,800 psi
Flexural Modulus	D790	275 Kpsi
Notched Izod Impact	D256	0.75 fpin
Unnotched Izod Impact	D256	20 ft-lb
Falling Weight Impact	----	24 in-lb
HDT @ 66 psi (4.6 Kg/sq cm)	D648	210F(99C)
HDT @ 264 psi (18.6 Kg/sq cm)	D648	140F(60C)
Hardness Rockwell R	D785	97

A. SCHULMAN, INC.: POLYFORT Polypropylenes(Continued):

Filled for Impact and High Stiffness(Continued):

Property	ASTM Method	Units
FPP 1564:		
Filler Content	----	30%
Type Filler	----	Calcium Carbonate
Density	D729	0.041 lb/cu in
Tensile Strength	D638	3,700
Flexural Modulus	D790	330 Kpsi
Notched Izod Impact	D256	0.80 fpin
Unnotched Izod Impact	D256	23 ft-lb
Falling Weight Impact	----	64 in-lb
HDT @ 66 psi (4.6 Kg/sq cm)	D648	245F(118C)
HDT @ 264 psi (18.6 Kg/sq cm)	D648	150F(65C)
Hardness Rockwell R	D785	97
FPP 1200:		
Filler Content	----	40%
Type Filler	----	Calcium Carbonate
Density	D729	0.044 lb/cu in
Tensile Strength	D638	3,600 psi
Flexural Modulus	D790	385 Kpsi
Notched Izod Impact	D256	1.00 fpin
Unnotched Izod Impact	D256	25 ft-lb
Falling Weight Impact	----	80 in-lb
HDT @ 66 psi (4.6 Kg/sq cm)	D648	270F(132C)
HDT @ 66 psi (18.6 Kg/sq cm)	D648	160F(70C)
Hardness Rockwell R	D785	98

A. SCHULMAN, INC.: POLYFORT Polypropylenes(Continued):

Filled for Stiffness and Low Temperature Impact:

FPP 1223:

Property	ASTM Method	Units
Filler Content	----	10%
Type Filler	----	Calcium Carbonate
Density	D729	0.035 lb/cu in
Tensile Strength	D638	3,465 psi
Flexural Modulus	D790	205 Kpsi
Notched Izod Impact	D256	1.8 fpin
Unnotched Izod Impact	D256	NB
Falling Weight Impact	----	180 in-lb
HDT @ 66 psi (4.6 Kg/sq cm)	D648	178F(81C)
HDT @ 264 psi (18.6 Kg/sq cm)	D648	116F(46C)
Hardness Rockwell R	D785	89

FPP 1245:

Filler Content	----	20%
Type Filler	----	Calcium Carbonate
Density	D729	0.037 lb/cu in
Tensile Strength	D638	2,925 psi
Flexural Modulus	D790	210 Kpsi
Notched Izod Impact	D256	1.6 fpin
Unnotched Izod Impact	D256	NB
Falling Weight Impact	----	140 in-lb
HDT @ 66 psi (4.6 Kg/sq cm)	D648	175F(79C)
HDT @ 264 psi (18.6 Kg/sq cm)	D648	120F(49C)
Hardness Rockwell R	D785	86

A. SCHULMAN, INC.: POLYFORT Polypropylenes(Continued):

Filled for Stiffness and Low Temperature Impact(Continued):

Property	ASTM Method	Units
FPP 1345:		
Filler Content	----	30%
Type Filler	----	Calcium Carbonate
Density	D729	0.041 lb/cu in
Tensile Strength	D638	2,850 psi
Flexural Modulus	D790	275 Kpsi
Notched Izod Impact	D256	1.0 fpin
Unnotched Izod Impact	D256	NB
Falling Weight Impact	----	112 in-lb
HDT @ 66 psi (4.6 Kg/sq cm)	D648	200F(93C)
HDT @ 264 psi (18.6 Kg/sq cm)	D648	125F(51C)
Hardness Rockwell R	D785	85
FPP 1227:		
Filler Content	----	40%
Type Filler	----	Calcium Carbonate
Density	D729	0.044 lb/cu in
Tensile Strength	D638	2,775 psi
Flexural Modulus	D790	285 Kpsi
Notched Izod Impact	D256	0.8 fpin
Unnotched Izod Impact	D256	10 ft-lb
Falling Weight Impact	----	80 in-lb
HDT @ 66 psi (4.6 Kg/sq cm)	D648	230F(110C)
HDT @ 264 psi (18.6 Kg/sq cm)	D648	145F(62C)
Hardness Rockwell R	D785	85

A. SCHULMAN, INC.: POLYFORT Polypropylenes(Continued):

Filled for Stiffness and Low Temperature Impact(Continued):

Property	ASTM Method	Units
FPP 1609:		
Filler Content	----	10%
Type Filler	----	Mica
Density	D729	0.035 lb/cu in
Tensile Strength	D638	4,160 psi
Flexural Modulus	D790	280 Kpsi
Notched Izod Impact	D256	1.2 fpin
Unnotched Izod Impact	D256	10 ft-lb
Falling Weight Impact	----	124 in-lb
HDT @ 66 psi (4.6 Kg/sq cm)	D648	214F(101C)
HDT @ 264 psi (18.6 Kg/sq cm)	D648	134F(56C)
Hardness Rockwell R	D785	89
FPP 1488:		
Filler Content	----	20%
Type Filler	----	Mica
Density	D729	0.037 lb/cu in
Tensile Strength	D638	4,195 psi
Flexural Modulus	D790	355 Kpsi
Notched Izod Impact	D256	0.9 fpin
Unnotched Izod Impact	D256	8 ft-lb
Falling Weight Impact	----	32 in-lb
HDT @ 66 psi (4.6 Kg/sq cm)	D648	250F(121C)
HDT @ 264 psi (18.6 Kg/sq cm)	D785	89

A. SCHULMAN, INC.: POLYFORT Polypropylenes(Continued):

Filled for Stiffness and Low Temperature Impact(Continued):

Property	ASTM Method	Units
FPP 1611:		
Filler Content	----	30%
Type Filler	----	Mica
Density	D729	0.041 lb/cu in
Tensile Strength	D638	4,235 psi
Flexural Modulus	D790	450 Kpsi
Notched Izod Impact	D256	0.7 fpin
Unnotched Izod Impact	D256	3 ft-lb
Falling Weight Impact	----	8 in-lb
HDT @ 66 psi (4.6 Kg/sq cm)	D648	255F(124C)
HDT @ 264 psi (18.6 Kg/sq cm)	D648	170F(76C)
Hardness Rockwell R	D785	88
FPP 1612:		
Filler Content	----	40%
Type Filler	----	Mica
Density	D729	0.044 lb/cu in
Tensile Strength	D638	4,310 psi
Flexural Modulus	D790	495 Kpsi
Notched Izod Impact	D256	0.5 fpin
Unnotched Izod Impact	D256	2 ft-lb
Falling Weight Impact	----	6 in-lb
HDT @ 66 psi (4.6 Kg/sq cm)	D648	265F(129C)
HDT @ 264 psi (18.6 Kg/sq cm)	D648	194F(90C)
Hardness Rockwell R	D785	87

A. SCHULMAN, INC.: POLYFORT Polypropylenes(Continued):

Filled for Very High Stiffness:

Property	ASTM Method	Units
FPP 1608:		
Filler Content	----	10%
Type Filler	----	Mica
Density	D729	0.035 lb/cu in
Tensile Strength	D638	5,400 psi
Flexural Modulus	D790	370 Kpsi
Notched Izod Impact	D256	0.7 fpin
Unnotched Izod Impact	D256	12 ft-lb
Falling Weight Impact	----	10 in-lb
HDT @ 66 psi (4.6 Kg/sq cm)	D648	250F(121C)
HDT @ 264 psi (18.6 Kg/sq cm)	D648	158F(70C)
Hardness Rockwell R	D785	100
FPP 1337:		
Filler Content	----	20%
Type Filler	----	Mica
Density	D729	0.037 lb/cu in
Tensile Strength	D638	5,450 psi
Flexural Modulus	D790	470 Kpsi
Notched Izod Impact	D256	0.6 fpin
Unnotched Izod Impact	D256	10 ft-lb
Falling Weight Impact	----	8 in-lb
HDT @ 66 psi (4.6 Kg/sq cm)	D648	260F(127C)
HDT @ 264 psi (18.6 Kg/sq cm)	D648	165F(74C)
Hardness Rockwell R	D785	99

A. SCHULMAN, INC.: POLYFORT Polypropylenes(Continued):

Filled for Very High Stiffness(Continued):

Property	ASTM Method	Units
FPP1494:		
Filler Content	----	30%
Type Filler	----	Mica
Density	D729	0.041 lb/cu in
Tensile Strength	D638	5,500 psi
Flexural Modulus	D790	590 Kpsi
Notched Izod Impact	D256	0.5 fpin
Unnotched Izod Impact	D256	8 ft-lb
Falling Weight Impact	----	6
HDT @ 66 psi (4.6 Kg/sq cm)	D648	270F(132C)
HDT @ 264 psi (18.6 Kg/sq cm)	D648	180F(82C)
Hardness Rockwell R	D785	98
FPP 1257:		
Filler Content	----	40%
Type Filler	----	Mica
Density	D729	0.044 lb/cu in
Tensile Strength	D638	5,600 psi
Flexural Modulus	D790	650 Kpsi
Notched Izod Impact	D256	0.4 fpin
Unnotched Izod Impact	D256	5 ft-lb
Falling Weight Impact	----	4 in-lb
HDT @ 66 psi (4.6 Kg/sq cm)	D648	280F(138C)
HDT @ 264 psi (18.6 Kg/sq cm)	D648	215F(101C)
Hardness Rockwell R	D785	99

A. SCHULMAN, INC.: POLYFORT Polypropylenes(Continued):

Filled for Very High Stiffness(Continued):

Property	ASTM Method	Units
FPP 1309:		
Filler Content	----	10%
Type Filler	----	Coupled Mica
Density	D729	0.035 lb/cu in
Tensile Strength	D638	5,500 psi
Flexural Modulus	D790	400 Kpsi
Notched Izod Impact	D256	0.7 fpin
Unnotched Izod Impact	D256	12 ft-lb
Falling Weight Impact	----	12 in-lb
HDT @ 66 psi (4.6 Kg/sq cm)	D648	255F(124C)
HDT @ 264 psi (18.6 Kg/sq cm)	D648	155F(68C)
Hardness Rockwell R	D785	101
FPP 1310:		
Filler Content	----	20%
Type Filler	----	Coupled Mica
Density	D729	0.037 lb/cu in
Tensile Strength	D638	5,800 psi
Flexural Modulus	D790	515 Kpsi
Notched Izod Impact	D256	0.6 fpin
Unnotched Izod Impact	D256	9 ft-lb
Falling Weight Impact	----	10 in-lb
HDT @ 66 psi (4.6 Kg/sq cm)	D648	270F(132C)
HDT @ 264 psi (18.6 Kg/sq cm)	D648	175F(79C)
Hardness Rockwell R	D785	99

A. SHULMAN, INC.: POLYFORT Polypropylenes(Continued):

Filled for Very High Stiffness(Continued):

Property	ASTM Method	Units
FPP 1650:		
Filler Content	----	30%
Type Filler	----	Coupled Mica
Density	D729	0.041 lb/cu in
Tensile Strength	D638	6,300 psi
Flexural Modulus	D790	650 Kpsi
Notched Izod Impact	D256	0.5 fpin
Unnotched Izod Impact	D256	8 ft-lb
Falling Weight Impact	----	8 in-lb
HDT @ 66 psi (4.6 Kg/sq cm)	D648	280F(138C)
HDT @ 264 psi (18.6 Kg/sq cm)	D648	200F(93C)
Hardness Rockwell R	D785	99
FPP 1244:		
Filler Content	----	40%
Type Filler	----	Coupled Mica
Density	D729	0.044
Tensile Strength	D638	6,500
Flexural Modulus	D790	850 Kpsi
Notched Izod Impact	D256	0.4 fpin
Unnotched Izod Impact	D256	4 ft-lb
Falling Weight Impact	----	6 in-lb
HDT @ 66 psi(4.6 Kg/sq cm)	D648	295F(146C)
HDT @ 264 psi(18.6 Kg/sq cm)	D648	230F(110C)
Hardness Rockwell R	D785	100

A. SCHULMAN, INC.: POLYFORT Polypropylenes(Continued):

General Purpose & Impact Resistant Injection Molding Grades:

Property	ASTM Method	Units
PP 829:		
Density	D729	0.032 lb/cu in
Tensile Strength	D638	5,000 psi
Flexural Modulus	D790	225 Kpsi
Notch Izod Impact	D256	0.6 fpin
Unnotched Izod Impact(-20C)	D256	4 ft-lb
Falling Weight Impact(20C)	----	6 in-lb
HDT @ 66 psi	D648	220F(104C)
Hardness Rockwell R	D785	105
PP 828:		
Density	D729	0.032 lb/cu in
Tensile Strength	D638	4,000 psi
Flexural Modulus	D790	150 Kpsi
Notch Izod Impact	D256	2.0 fpin
Unnotched Izod Impact(-20C)	D256	17 ft-lb
Falling Weight Impact(20C)	----	165 in-lb
HDT @ 66 psi	D648	210F(99C)
Hardness Rockwell R	D785	85
PP 1549:		
Density	D729	0.032 lb/cu in
Tensile Strength	D638	3,700 psi
Flexural Modulus	D790	130 Kpsi
Notch Izod Impact	D256	8.0 fpin
Unnotched Izod Impact(-20C)	D256	25 ft-lb
Falling Weight Impact(20C)	----	400+ in-lb
HDT @ 66 psi	D648	195F(90C)
Hardness Rockwell R	D785	80
PP 1531:		
Density	D729	0.032 lb/cu in
Tensile Strength	D638	3,000 psi
Flexural Modulus	D790	120 Kpsi
Notch Izod Impact	D256	12.0 fpin
Unnotched Izod Impact(-20C)	D256	35 ft-lb
Falling Weight Impact(20C)	----	400+ in-lb
HDT @ 66 psi	D648	185F(85C)
Hardness Rockwell R	D785	70

A. SCHULMAN, INC.: POLYFORT Polypropylenes (Continued):

Glass Reinforced Injection Molding Grades:

Homopolymer:

FPP 1082	10% Glass
FPP 1080	20% Glass
FPP 1208	30% Glass

Coupled Homopolymer:

FPP 1602	10% Glass
FPP 1236	20% Glass
FPP 1239	30% Glass

High Impact:

FPP 1603	10% Glass
FPP 1285	20% Glass
FPP 1604	30% Glass

Coupled High Impact:

FPP 1605	10% Glass
FPP 1220	20% Glass
FPP 1606	30% Glass

A. SCHULMAN, INC.: POLYFORT Polypropylenes(Continued):

Hybrid Combinations, Injection Molding:

Medium Impact Copolymer:

FPP 1340	5% Glass Fiber + 30% Talc
FPP 1366	10% Glass Fiber + 20% Talc

General Purpose Homopolymer:

FPP 1640	15% Glass Fiber + 20% Talc

Medium Impact Copolymer:

FPP 1636	15% Glass Fiber + 25% Talc

General Purpose Homopolymer:

FPP 1336	20% Glass Fiber + 25% Talc

General Purpose Homopolymer:

FPP 1222	10% Glass Fiber + 30% CACO3
FPP 1580	10% Glass Fiber + 40% CACO3
FPP 1225	15% Glass Fiber + 25% CACO3

General Purpose Homopolymer:

FPP 1240	5% Glass Fiber + 15% Mica
FPP 1619	10% Glass Fiber + 25% Mica
FPP 1620	10% Glass Fiber + 30% Mica
FPP 1289	15% Glass Fiber + 25% Mica
FPP 1567	20% Glass Fiber + 20% Mica
FPP 1290	15% Glass Fiber + 25% Glass Beads
FPP 1282	30% Glass Fiber + 10% Glass Beads

Wollastonite Homopolymer:

FPP 1393	10% Talc + 10% Mica

Medium Impact Copolymer:

FPP 1537	10% Talc + 30% CACO3

General Purpose Homopolymer:

FPP 1333	10% Talc + 10% CACO3
FPP 1298	10% Talc + 20% CACO3

A. SCHULMAN, INC.: POLYFORT Polypropylenes (Continued):

Special Fillers - Injection Molding:

General Purpose Homopolymer:

FPP 1414	10% Aluminum Flake
FPP 1415	20% Aluminum Flake

Medium Impact Copolymer:

FPP 1587	15% Barium Sulfate
FPP 1588	30% Barium Sulfate
FPP 1613	60% Barium Sulfate

General Purpose Homopolymer:

FPP 1371	20% Barium Sulfate
FPP 1629	30% Barium Sulfate
FPP 1294	40% Barium Sulfate
FPP 1642	70% Barium Sulfate
FPP 1384	20% Glass Spheres
FPP 1378	40% Glass Spheres
FPP 1474	50% Glass Spheres
FPP 1370	20% Glass Beads
FPP 1420	40% Glass Beads

Medium Impact Copolymer:

FPP 1444	20% Glass Beads
FPP 1274	30% Glass Beads
FPP 1281	40% Glass Beads

General Purpose Homopolymer:

FPP 1460	20% Iron Oxide
FPP 1394	30% Iron Oxide
FPP 1273	30% Wollastonite
FPP 1183	40% Wollastonite

General Purpose Homopolymer:

FPP 1545	6% Zinc Sulfate
FPP 1459	15% Talc
FPP 1573	25% Talc
FPP 1577	35% Talc
FPP 1454	35% Woodflour

A. SCHULMAN, INC.: POLYFORT Polypropylenes(Continued):

Special Fillers-Injection Molding(Continued):

Medium Impact Copolymer:

FPP 1637	12% Talc
FPP 1513	15% Talc
FPP 1529	25% Talc
FPP 1625	35% Talc
FPP 1455	35% Woodflour

General Purpose Homopolymer:

FPP 1305	25% CACO3
FPP 1364	60% CACO3

Medium Impact Copolymer:

FPP 1590	15% CACO3
FPP 1376	60% CACO3

General Purpose Homopolymer:

FPP 1300	25% Mica

Medium Impact Copolymer:

FPP 1484	15% Mica

High Impact Copolymer:

FPP 1434	25% Mica

General Purpose Homopolymers:

FPP 1324	12% Coupled Mica
FPP 1473	15% Coupled Mica

High Impact Copolymer:

FPP 1325	12% Coupled Mica
FPP 1314	15% Coupled Mica
FPP 1385	30% Coupled Mica
FPP 1326	40% Coupled Mica

General Purpose Homopolymer:

FPP 1339	15% Coupled Glass Fiber

A. SCHULMAN, INC.: POLYFLAM Flame Retardant Thermoplastic Compounds:

Flame retardant thermoplastic compounds based on polypropylene, polyethylene and polystyrene.

* Processed under normal conditions for each polymer

* Excellent electrical properties

* Good elevated temperature stability

* Outstanding molding properties

* No mold corrosion

* Low moisture absorption

* Available in natural, black and custom colors

* Grades available for:
  Injection molding, Extrusion, Blow molding, Structural Foam (Extrusion and Blow molding grades have "E" suffix)

DESIGNATIONS:

Polypropylenes........RPP
High density
Polyethylenes.........RLP/RHD
Low density
Polyethylenes.........RCP
Polystyrenes..........RMS

A. SCHULMAN, INC.: POLYFLAM Polyethylene RHD/RCP Series:

Product Name:

RHD 260 ND/RLP 426:

A high density polyethylene homopolymer offering good low temperature impact properties with a V-O rating at a thickness of 0.125 in.

This material is suitable for injection molding but an extrusion and blow molding version is available. The extrusion/blow molding grade carries a suffix E.

The RHD260ND (E) type is particularly suitable for blow molding of institutional chairs where flame retardancy is required.

RCP 762:

A low density polyethylene material suitable for injection molding or extrusion for applications where flame retardancy and flexibility is required, e.g. snap fit closures, electrical connectors, tubing, etc.

RCP 770C:

This F.R. concentrate in low density polyethylene features good dispersion and excellent processing stability to 550F.

The material is recommended for use in both low density polyethylene and high density polyethylene. It can be used with equal success in extrusion/blow molding, injection molding, structural foam and film extrusion applications.

Flame retardant properties are achieved at let down ratios from 4:1 to 15:1.

This concentrate offers good heat sealing performance in polyethylene film applications.

A. SCHULMAN, INC.: POLYFLAM Polypropylene RPP Series:

Product Name:

RPP 932:

Non-dripping, homopolymer formulated to comply with requirements of ASTM D635. Intended primarily for injection molding applications in the automotive industry.

RPP 958:

High specification homopolymer product possessing UL94 V-O qualities and 105C temperature rating at 0.030 in. Well proven compound used for a variety of demanding injection molding applications in the radio, TV, appliance, electronic and telecommunications industries.
For extrusion applications specify RPP958E.

RPP 982:

Copolymer material with good elongation and impact strength. Can be readily molded into integral hinge designs.

RPP 985:

A copolymer material designed for sheet extrusion and subsequent thermoforming applications. Also an excellent candidate for blow molding and for injection molding into integral hinge designs.

RPP 1021:

High performance copolymer offering UL94 V-2 and 105C temperature index at thicknesses as low as 0.030 in. Good processability and excellent electrical properties.

This material meets standards for direct support of current carrying parts.

Suitable for injection molding or extrusion applications, especially as a nylon replacement material in electrical switches and connectors.

A. SCHULMAN, INC.: POLYFLAM Polypropylene RPP Series(Continued):

Product Name:

RPP 1058:

General purpose homopolymer material offering V-2 flammability and 105C temperature index at 0.030 in.

Excellent electrical properties make the material ideal for wire nuts, connectors, light sockets and many other electrical and appliance applications.

RPP 1174:

Developed as a replacement for RPP 958 as insurance against F.R. additive shortages. This material is a high performance material fully interchangeable with RPP 958 for use in a variety of electro-technical industries.

RPP 1062C:

This polypropylene F.R. concentrate when let down at a ratio of 1 part: 6 parts G.P. polypropylene offers V-2 flame retardance at 0.120 in. At this let down ratio the polypropylene homopolymer shows good physical property retention.

A. SCHULMAN, INC.: POLYFLAM Polystyrene RMS Series:

Product Name:

RMS 424:

This medium impact polystyrene compound offers high rigidity and V-O flame retardancy at 0.125 in. (V-1 at 0.060 in.)

The material is specially formulated for electrical, TV, and appliance housings. It exhibits good processability and is readily painted with conventional paint systems for polystyrene.

Also available for extrusion applications - RMS424E. This type is typically used in wall and ceiling panels, sheet and profiles etc.

RS425C3:

This FR concentraate based on polystyrene exhibits good dispersion and good processability.

When mixed with general purpose polystyrene at a ratio of 1 part concentrate: 3 parts polystyrene a UL94 V-O rating can be obtained by Underwriter's Laboratories under yellow card file No. E57641.

The composite FR material shows good retention of critical physical properties of the base polystyrene.

A. SCHULMAN, INC.: POLYMAN Family of Engineering Plastics:

Product Name:

POLYMAN 506:

A rigid polymer alloy, POLYMAN 506 is a low cost, general purpose material with high performance features. It offers high tensile impact strength of 112 ft. lb. (8 ft. lb. Izod), and combines high 1000 volt wet dielectric strength with good arc tracking resistance. It is rated UL Bulletin 94 HB. It has been successfully used for electrical terminal blocks, and is recommended for small appliance housings and automotive connectors.

POLYMAN 507:

A special outdoor rigid polymer alloy, POLYMAN 507 retains more impact strength after UV exposure than any competitive material according to UL recognition tests. It is rated UL Bulletin 94 V-O and carries a 95C continuous use temperature rating. In addition it offers the good electrical properties and tensile strength needed for wiring devices and small tool housings.

POLYMAN 508:

POLYMAN 508 is identical in properties and performance with POLYMAN 507 above, but provides improved flow characteristics when molding small parts or thinner walled large parts.

POLYMAN 509:

A rigid polymer alloy, POLYMAN 509 exhibits low shrinkage (0.0045 in./in.) and combines a 212F heat distortion temperature with a superior 430,000 flexural modulus to mold and hold strict dimensional tolerances in critical fit applications such as printed circuit card holders. Its high arc track resistance and compressive strength, along with a good balance of other properties, qualifies for applications in severe electrical service such as junction boxes and connectors.

A. SCHULMAN, INC.: POLYMAN Family of Engineering Plastics (Continued):

Product Name:

POLYMAN 510:

POLYMAN 510 offers the same properties as POLYMAN 511, but is formulated to provide improved processability for large parts requiring thinner walls.

POLYMAN 511*:

POLYMAN 511 offers the highest impact strength (83.5 tensile impact; 320 in. lb. Gardner impact) of the POLYMAN UL V-O rated rigid polymer alloys. It also has excellent UV color stability along with the exceptional stiffness and flatness required for large business machine and consumer appliance housings.

POLYMAN 526:

A more economical alloy, POLYMAN 526 combines good mechanical properties, with top flame retardancy for ordinary flow molding applications, where ultimate "cosmetics" are not required.

POLYMAN 551*:

A polymer alloy system, POLYMAN 551 combines UL 94 V-5 flame retardance at 0.100" with practical impact strength. Its high 195F heat deflection, UV stability, and excellent chemical resistance provide long lasting durability under severe environmental forces for finished part longevity. This combination of properties makes a POLYMAN 551 a cost effective choice for demanding applications where polycarbonates, FR ABS, and PPO/PS compounds were previously used.

* Indicates grades available especially formulated for extrusion viscosity; indicated by a cipher in the color designation number.

A. SCHULMAN, INC.: POLYMAN Family of Engineering Plastics (Continued):

Product Name:

POLYMAN 552*:

POLYMAN 552 offers practical impact strength, a UL 94 V-O rating at 0.060, excellent outdoor UV and weathering resistance, and an attractive balance of electrical properties. It provides good processability especially when compared to FR ABS resins.

POLYMAN 554*:

POLYMAN 554 is an economical, general purpose alloy with excellent outdoor weathering characteristics. Its good dimensional stability, high impact strength, balanced electrical properties, and resistance to environmental attack make it an excellent choice for a wide range of housing and fitting applications where more stringent UL 94 V-O and V-1 ratings are not required, including swimming pool fittings and cleaning devices and irrigation equipment.

POLYMAN 558:

An outdoor application, wide temperature range alloy with excellent impact resistance is POLYMAN 558. It is suitable for injection or extrusion in flame retardant situations of moderate requirements.

* Indicates grades available especially formulated for extrusion viscosity; indicated by a cipher in the color designation number.

A. SCHULMAN, INC.: POLYTROPE Extrusion/Blow Molding Grade Thermoplastic Elastomers:

POLYTROPE extrusion/blow molding grade thermoplastic elastomers are high performance elastolefins which provide excellent processing ideally suited to all extrusion processes. These grades fill the materials void between rubber and plastic extrusion materials with a broad range of physical properties. Extruded sheet, profiles or blow molded components of POLYTROPE can be subjected to standard post-extrusion processes such as thermoforming, solid phase forming, hot-stamping, hot-bar fusion and painting. Dual durometer, co-extruded profiles can be produced to provide varying stiffness over the cross section of the profile to optimize functionality.

A. Schulman Inc. can computer color match all POLYTROPE extrusion/blow molding grades to match any color standard and also provide a UV protection package that will withstand the most severe sunlight or accelerated weathering environment.

The low specific gravity and total recyclability of regrind makes POLYTROPE the preferred choice to other elastomeric materials with the best economics for sophisticated thermoplastic extrusion processing.

Property	ASTM Test	Units
TPP-313:		
Color	----	Natural
Density	D-792	0.96 g/cc
Shore Hardness	D-2240	65-A/14-D
Tensile Strength	D-638	600 PSI
Ultimate Elongation	D-638	550%
Melt Index	D-1238	1.0 g/10 min.
Flexural Modulus	D-790	2,000 psi
Gardner Impact	-----	>320 in-lbs
Tear Strength	D-624	150 PPI
Izod Impact Notched	D-256	No Break
TPP-319:		
Color	----	Natural
Density	D-792	0.90 g/cc
Shore Hardness	D-2240	96-A/55-D
Tensile Strength	D-638	3,200 PSI
Ultimate Elongation	D-638	850%
Melt Index	D-1238	0.5 g/10 min.
Flexural Modulus	D-790	65,000 psi
Gardner Impact	----	>320 in-lb
Tear Strength	D-624	650 PPI
Izod Impact Notched	D-256	No Break

A. SCHULMAN, INC.: POLYTROPE Extrusion/Blow Molding Grade Thermoplastic Elastomers(Continued):

Property	ASTM Test	Units
TPP-344:		
Color	----	Natural
Density	D-792	0.90 g/cc
Shore Hardness	D-2240	63-D
Tensile Strength	D-638	3,800 PSI
Ultimate Elongation	D-638	750%
Melt Index	D-1238	0.8g/10 min.
Flexual Modulus	D-790	140,000 PSI
Gardner Impact	----	>320 in-lbs
Tear Strength	D-624	800 PPI
Izod Impact Notched	D-256	No Break
TPP-402:		
Color	----	Black
Density	D-792	0.90 g/cc
Shore Hardness	D-2240	38-D
Tensile Strength	D-638	1,421 PSI
Ultimate Elongation	D-638	400%
Melt Index	D-1238	0.6 g/10 min.
Flexual Modulus	D-790	22,000 PSI
Gardner Impact	----	>320 in-lbs
Tear Strength	D-624	320 PPI
Izod Impact Notched	D-256	No Break
TPP-403:		
Color	----	Black
Density	D-792	0.90 g/cc
Shore Hardness	D-2240	45-D
Tensile Strength	D-638	2,635 PSI
Ultimate Elongation	D-638	700%
Melt Index	D-1238	0.4 g/10 min.
Flexual Modulus	D-790	30,000 PSI
Gardner Impact	----	>320 in-lbs
Tear Strength	D-624	475 PPI
Izod Impact Notched	D-256	No Break

A. SCHULMAN, INC.: POLYTROPE Extrusion/Blow Molding Grade Thermoplastic Elastomers (Continued):

Property	ASTM Test	Units
TPP 408:		
Color	----	Black
Density	D-792	0.91 g/cc
Shore Hardness	D-2240	63-D
Tensile Strength	D-638	3,400 PSI
Ultimate Elongation	D-638	640%
Melt Index	D-1238	0.7 g/10 min.
Flexual Modulus	D-790	78,000 PSI
Gardner Impact	----	>320 in-lbs
Tear Strength	D-624	700 PPI
Izod Impact Notched	D-256	No Break
TPP-425:		
Color	----	Natural
Density	D-792	0.91 g/cc
Shore Hardness	D-2240	41-D
Tensile Strength	D-638	1,900 PSI
Ultimate Elongation	D-638	450%
Melt Index	D-1238	1.5 g/10 min.
Flexual Modulus	D-790	20,000 PSI
Gardner Impact	----	>320 in-lbs
Tear Strength	D-624	500 PPI
Izod Impact Notched	D-256	No Break
TPP-442:		
Color	----	Natural
Density	D-792	0.96 g/cc
Shore Hardness	D-2240	65-A
Tensile Strength	D-638	600 PSI
Ultimate Elongation	D-638	620%
Melt Index	D-1238	1.5 g/10 min.
Flexual Modulus	D-790	1,500 PSI
Gardner Impact	----	>320 in-lbs
Tear Strength	D-624	200 PPI
Izod Impact Notched	D-256	No Break

A. SCHULMAN, INC.: POLYTROPE Extrusion/Blow Molding Grade Thermoplastic Elastomers(Continued):

Property	ASTM Test	Units
TPP-443:		
Color	----	Natural
Density	D-792	0.95 g/cc
Shore Hardness	D-2240	55-A
Tensile Strength	D-638	500 PSI
Ultimate Elongation	D-638	300%
Melt Index	D-1238	0.5 g/10 min.
Flexual Modulus	D-790	1,500 PSI
Gardner Impact	----	>320 in-lbs
Tear Strength	D-624	250 PPI
Izod Impact Notched	D-256	No Break
TPP-444:		
Color	----	Natural
Density	D-792	0.97 g/cc
Shore Hardness	D-2240	60-A
Tensile Strength	D-638	685 PSI
Ultimate Elongation	D-638	475%
Melt Index	D-1238	1.0 g/10 min.
Flexual Modulus	D-790	2,000 PSI
Gardner Impact	----	>320 in-lbs
Tear Strength	D-624	260 PPI
Izod Impact Notched	D-256	No Break
TPP-445:		
Color	----	Natural
Density	D-792	0.96 g/cc
Shore Hardness	D-2240	72-A
Tensile Strength	D-638	750 PSI
Ultimate Elongation	D-638	300%
Melt Index	D-1238	1.5 g/10 min.
Flexual Modulus	D-790	3,500 PSI
Gardner Impact	----	>320 in-lbs
Tear Strength	D-624	275 PPI
Izod Impact Notched	D-256	No Break

A. SCHULMAN, INC.: POLYTROPE Extrusion/Blow Molding Grade Thermoplastic Elastomers (Continued):

Property	ASTM Test	Units
TPP-446:		
Color	----	Natural
Density	D-792	0.92 g/cc
Shore Hardness	D-2240	88-A
Tensile Strength	D-638	1,400 PSI
Ultimate Elongation	D-638	450%
Melt Index	D-1238	3.0 g/10 min.
Flexual Modulus	D-790	8,200 PSI
Gardner Impact	----	>320 in-lbs
Tear Strength	D-624	580 PPI
Izod Impact Notched	D-256	No Break
TPP-447:		
Color	----	Natural
Density	D-792	0.95 g/cc
Shore Hardness	D-2240	80-A
Tensile Strength	D-638	850 PSI
Ultimate Elongation	D-638	500%
Melt Index	D-1238	1.0 g/10 min.
Flexual Modulus	D-790	5,000 PSI
Gardner Impact	----	>320 in-lbs
Tear Strength	D-624	300 PPI
Izod Impact Notched	D-256	No Break

A. SCHULMAN, INC.: POLYTROPE Injection Molding Grade Ethylene-Propylene Alloys:

POLYTROPE ethylene-propylene alloys (EPA) exhibit the physical characteristics of vulcanized rubber with the ease of thermoplastic processing. This combination of rubber with plastics exceeds the best performance qualities of either.

POLYTROPE EPA's offer:

* rubber-like performance
* excellent impact resistance
* low specific gravity
* ultraviolet radiation resistance
* self-releasing qualities
* non-hygroscopic characteristics
* computerized color matching
* 100% recycleability

Color matching capabilities eliminate the inconvenience of metering and mixing with color concentrates. All POLYTROPE grades are available in natural, black and colors.

POLYTROPE EPA's are excellent alternatives to flexible PVC, thermoplastic urethane, RIM, and conventionally cross-linked rubber compounds.

Property	ASTM Method	Units
TPP 336:		
Density	D-792	0.96 g/c.c.
Shore Hardness	D-2240	65A
Tensile Strength	D-638	600 PSI
Percent Elongation	D-638	550
Tear Strength	D-624	150 PPI
Flexural Modulus	D-790 @ 23C	2,000 PSI
Gardner Impact	@ -20C	>190 in-lbs
Mold Shrinkage	D-955	0.018-0.020
TPP 427:		
Density	D-792	0.92 g/c.c.
Shore Hardness	D-2240	65A
Tensile Strength	D-638	730 PSI
Percent Elongation	D-638	1500
Tear Strength	D-624	175 PPI
Flexural Modulus	D-790 @ 23C	1,600 PSI
Gardner Impact	@ -20C	>300 in-lbs
Mold Shrinkage	D-955	0.018-0.020

A. SCHULMAN, INC.: POLYTROPE Injection Molding Grade Ethylene-Propylene Alloys (Continued):

Property	ASTM Method	Units
TPP 377:		
Density	D-792	0.90 g/c.c.
Shore Hardness	D-2240	70A
Tensile Strength	D-638	975 PSI
Percent Elongation	D-638	2300
Tear Strength	D-624	224 PPI
Flexural Modulus	D-790 @ 23C	1,500 PSI
Gardner Impact	@ -20C	>320 in-lbs
Mold Shrinkage	D-955	0.022-0.025
TPP 337:		
Density	D-792	0.92 g/c.c.
Shore Hardness	D-2240	76A
Tensile Strength	D-638	1000 PSI
Percent Elongation	D-638	800
Tear Strength	D-624	300 PPI
Flexural Modulus	D-790 @ 23C	2,500 PSI
Gardner Impact	@ -20C	>320 in-lbs
Mold Shrinkage	D-955	0.016-0.018
TPP 409:		
Density	D-792	1.21 g/c.c
Shore Hardness	D-2240	82A
Tensile Strength	D-638	650 PSI
Percent Elongation	D-638	500
Tear Strength	D-624	206 PPI
Flexural Modulus	D-790 @ 23C	3,000 PSI
Gardner Impact	@ -20C	>280 in-lbs
Mold Shrinkage	D-955	0.018-0.020
TPP 318:		
Density	D-792	0.91 g/c.c.
Shore Hardness	D-2240	88A
Tensile Strength	D-638	1800 PSI
Percent Elongation	D-638	800
Tear Strength	D-624	400 PPI
Flexural Modulus	D-790 @ 23C	22,000 PSI
Gardner Impact	@ -20C	>320 in-lbs
Mold Shrinkage	D-955	0.007-0.009

A. SCHULMAN, INC.: POLYTROPE Injection Molding Grade Ethylene-Propylene Alloys (Continued):

Property	ASTM Method	Units
TPP 309:		
Density	D-792	0.90 g/c.c.
Shore Hardness	D-2240	92A
Tensile Strength	D-638	2000 PSI
Percent Elongation	D-638	800
Tear Strength	D-624	450 PPI
Flexural Modulus	D-790 @ 23C	28,000 PSI
Gardner Impact	@ -20C	>320 in-lbs
Mold Shrinkage	D-955	0.010-0.011
TPP 433:		
Density	D-792	0.91 g/c.c.
Shore Hardness	D-2240	48D
Tensile Strength	D-638	2200 PSI
Percent Elongation	D-638	1000
Tear Strength	D-624	500 PPI
Flexural Modulus	D-790 @ 23C	34,000 PSI
Gardner Impact	@ -20C	>320 in-lbs
Mold Shrinkage	D-955	0.010-0.011
TPP 393:		
Density	D-792	0.90 g/c.c.
Shore Hardness	D-2240	45D
Tensile Strength	D-638	2300 PSI
Percent Elongation	D-638	850
Tear Strength	D-624	600 PPI
Flexural Modulus	D-790 @ 23C	40,000
Gardner Impact	@ -20C	>310 in-lbs
Mold Shrinkage	D-955	0.010-0.012
TPP 320:		
Density	D-792	0.91 g/c.c.
Shore Hardness	D-2240	50D
Tensile Strength	D-638	2800 PSI
Percent Elongation	D-638	900
Tear Strength	D-624	650 PPI
Flexural Modulus	D-790 @ 23C	55,000 PSI
Gardner Impact	@ -20C	>320 in-lbs
Mold Shrinkage	D-955	0.012-0.013

A. SCHULMAN, INC.: POLYTROPE Injection Molding Grade Ethylene-Propylene Alloys (Continued):

Property	ASTM Method	Units
TPP 300:		
Density	D-792	0.91 g/c.c.
Shore Hardness	D-2240	52D
Tensile Strength	D-638	3000 PSI
Percent Elongation	D-638	800
Tear Strength	D-624	850 PPI
Flexural Modulus	D-790 @ 23C	90,000 PSI
Gardner Impact	@ -20C	>320 in-lbs
Mold Shrinkage	D-955	0.012-0.013
TPP 293:		
Density	D-792	0.94 g/c.c.
Shore Hardness	D-2240	57D
Tensile Strength	D-638	3200 PSI
Percent Elongation	D-638	700
Tear Strength	D-624	600 PPI
Flexural Modulus	D-790 @ 23C	85,000 PSI
Gardner Impact	@ -20C	>320 in-lbs
Mold Shrinkage	D-955	0.009-0.011
TPP 394:		
Density	D-792	0.98 g/c.c.
Shore Hardness	D-2240	59D
Tensile Strength	D-638	2500 PSI
Percent Elongation	D-638	500
Tear Strength	D-624	650 PPI
Flexural Modulus	D-790 @ 23C	98,000 PSI
Gardner Impact	@ -20C	>320 in-lbs
Mold Shrinkage	D-955	0.004-0.006
TPP 338:		
Density	D-792	0.90 g/c.c.
Shore Hardness	D-2240	60D
Tensile Strength	D-638	3000 PSI
Percent Elongation	D-638	600
Tear Strength	D-624	800 PPI
Flexural Modulus	D-790 @ 23C	100,000 PSI
Gardner Impact	@ -20C	>320 in-lbs
Mold Shrinkage	D-955	0.008-0.010

A. SCHULMAN, INC.: POLYTROPE Injection Molding Grade Ethylene-Propylene Alloys(Continued):

Property	ASTM Method	Units
TPP 421:		
Density	D-792	0.91 g/c.c.
Shore Hardness	D-2240	60D
Tensile Strength	D-638	2700 PSI
Percent Elongation	D-638	800
Tear Strength	D-624	900 PPI
Flexural Modulus	D-790 @ 23C	119,000
Gardner Impact	>320	0.010-0.012
TPP 317:		
Density	D-792	0.90 g/c.c.
Shore Hardness	D-2240	60D
Tensile Strength	D-638	3000 PSI
Percent Elongation	D-638	700
Tear Strength	D-624	800 PPI
Flexural Modulus	D-790 @ 23C	120,000
Gardner Impact	@ -20C	>300 in-lbs
Mold Shrinkage	D-955	0.011-0.013
TPP 414:		
Density	D-792	0.99 g/c.c.
Shore Hardness	D-2240	66D
Tensile Strength	D-638	2800 PSI
Percent Elongation	D-638	500
Tear Strength	D-624	800 PPI
Flexural Modulus	D-790 @ 23C	135,000
Gardner Impact	@ -20C	>256 in-lbs
Mold Shrinkage	D-955	0.013-0.015
TPP 420:		
Density	D-792	0.92 g/c.c.
Shore Hardness	D-2240	66D
Tensile Strength	D-638	3500 PSI
Percent Elongation	D-638	330
Tear Strength	D-624	810 PPI
Flexural Modulus	D-790 @ 23C	180,000 PSI
Gardner Impact	@ -20C	>230
Mold Shrinkage	D-955	0.009-0.011

A. SCHULMAN, INC.: POLYTROPE Injection Molding Grade Ethylene-Propylene Alloys(Continued):

Property	ASTM Method	Units
TPP 1201:		
Density	D-792	1.16 g/c.c.
Shore Hardness	D-2240	62D
Tensile Strength	D-638	2970 PSI
Percent Elongation	D-638	140
Tear Strength	D-624	550 PPI
Flexural Modulus	D-790 @ 23C	180,000 PSI
Gardner Impact	@ -20C	>174 in-lbs
Mold Shrinkage	D-955	0.013-0.015
TPP 341:		
Density	D-792	0.93 g/c.c.
Shore Hardness	D-2240	68D
Tensile Strength	D-638	4200 PSI
Percent Elongation	D-638	280
Tear Strength	D-624	1000 PPI
Flexural Modulus	D-790 @ 23C	200,000 PSI
Gardner Impact	@ -20C	>130
Mold Shrinkage	D-955	0.009-0.011

A. SCHULMAN, INC.: POLYVIN Automotive Compounds:

POLYVIN PVC automotive compounds are specially formulated to meet the requirements of today's automotive needs. They range in durometer from 30 Shore A to 65 Shore D and are available in clear, natural, and custom pre-matched colors.

POLYVIN automotive compounds are specified worldwide for applications such as steering wheels, encapsulated window glass, exterior moldings and interior trim items. Automotive companies are assured of achieving consistent color matched interior and exterior parts by designating A. Schulman, Inc. as their supplier of PVC and other thermoplastic engineering compounds. Color control is accomplished visually and through the use of a modern color computer network set up in all A. Schulman plants.

Compound	Durometer	Standard Compounds
4505	45A	Gear Shift Boots
5004	54A	Gaskets & Grommets
5310	53A	Soft Steering Wheel
5813	58A	Interior Extruded Welting
5819	58A	Interior Injection
5824	58A	Interior Injection
6006	60A	Interior Non-migrating
6820	68A	Steering Wheels
7013	70A	Steering Wheels
7419	74A	Exterior Bilaminate Extrusion
7974	79A	Exterior Encapsulated Glass
8110	81A	Interior Trim
8612	86A	Exterior Extrusions
9020	90A	Exterior Moldings
9029	90A	Interior Extruded Weltings
9074	85A	Exterior Trim
9437	37D	Exterior Extrusions
9442	37D	Windshield Locking Strips
9605	50D	Exterior Moldings

A. SCHULMAN, INC.: POLYVIN Electrical Compounds:

POLYVIN PVC electrical compounds have been developed to meet the needs of the wire and cable industry. They are available in a wide range of durometers and cover UL classifications between 60C to 105C. Jacketing compounds are available for telephone cable and automotive use.

Compound	Durometer	Specific Gravity	Application
6614	66A	1.21	Injection UL-94 Rated, Anode cups 75 & 90
7843	78A	1.35	UL105 Class 12 Wiring & Plugs
8243	82A	1.35	Extrusion - Cable Jacketing 80 Class 12
8808	88A	1.36	Extrusion 105 Wire
8817	90A	1.36	Extrusion 60 Wire
9019	39D	1.36	Extrusion 90 Wire

A. SCHULMAN, INC.: POLYVIN Food and Medical Compounds:

POLYVIN PVC food grade and medical compounds have been serving these critical industries for many years. Manufactured with high purity, FDA approved ingredients, they are available in a wide range of durometers and densities. POLYVIN PVC compounds are widely used for applications such as medical masks, intravenous solution bags, dental appliances, teething rings, bandages, tubing, meat wrappings and cosmetic bags.

Many suppliers to the food and medical industry use 9300 purging compound to keep their equipment clean and sanitary.

POLYVIN PVC food and medical grade compounds - clear choices for these critical applications.

Compound	Durometer	Specific Gravity	Application
3608	63A	1.198	Extrusion - Tubing & Profiles
3622	60A	1.21	Injection Blow Molding Teething Rings
3914	75D	1.33	Injection - Blender Tops
3816	84A	1.30	Injection - Animal Tags
8145	81A	1.34	Extrusion Blow Molding Bags & Film
9300	92A	1.32	Non-Toxic Purge for color change & clean-up

A. SCHULMAN, INC.: POLYVIN General Purpose Compounds:

POLYVIN PVC general purpose compounds are custom formulated and designed to meet a broad range of specifications. These compounds range in durometers from 35 Shore A to 65 Shore D.

POLYVIN PVC general purpose compounds can be formulated to provide excellent tensile, flexural and impact strength. Superior electrical properties, weatherability, and resistance to chemicals are basic design ingredients. Attractive cost performance ratios for the end user can be accomplished through A. Schulman's years of experience in designing POLYVIN PVC general purpose compounds. Below is a representative list of typical POLYVIN PVC compounds.

Compound	Durometer	Specific Gravity	Application
4008	40A	1.15	Injection-Gaskets-Toys
6088	60A	1.29	Injection-Protective Covers
7005	75A	1.19	Extrusion-Straps-Profiles
8629	89A	1.25	Clear tube & profiles
4527	45A	1.23	Injection-Hand Grips
5802	60A	1.34	Injection-Grommets
7705	77A	1.37	Extruded-Profiles
9108	91A	1.30	Extruded-Profiles & tubing

A. SCHULMAN, INC.: POLYVIN PVC Specialty Compounds:

POLYVIN PVC specialty compounds are formulated for specific end use applications. These include compounds providing oil and gasoline resistance, low plasticizer migration into other polymers, superior low temperature properties, x-ray shielding, superior weatherability and hot strength blow molding characteristics. Several highly specialized compounds include applications for geophysical hose, urinal pads, refrigerator gasketing and explosive devices.

POLYVIN PVC purge compound is recommended for use by industrial and specialty customers who require a fast, effective method to purge equipment and to facilitate color change.

Compound	Durometer	Specific Gravity	Application
5028	50A	1.23	Extrusion & Injection Blow Molding
7039	70A	1.22	Gasoline & Oil Resistance
6112	62A	1.24	Weather Resistant Tubing
7121	68A	1.23	Extrusion - Clear Profiles & Tubing
7709	77A	1.39	Vacuum Lines - Extrusion
8121	81A	1.34	Extrusion & Blow Molding
9025	90A	1.34	Injection - Non Migrating
9400	90A	1.36	Purging & Quick Color Change Compound

A. SCHULMAN, INC.: SCHULAMID Polyamide Reinforced:

Property	ASTM Method	Units
SCHULAMID 6 - Reinforced:		
GF 30:		
30% Glass Fiber Reinforced		
High Stiffness		
High Heat Resistance		
Specific Gravity	D792	1.35 g/cu cm
Mold Shrinkage	D955	.3%
Water Absorption, 24 hrs	D570	.9%
Dry:		
Tensile Strength	D638	23,000 psi
Elongation @ Break	D638	3.5%
Flexural Modulus	D790	1,100,000 psi
Notched Izod Impact	D256	2 Ft. Lb./In.
Cond.: Samples conditioned at 70F/50% R.H.:		
Tensile Strength	D638	14,500 psi
Elongation @ Break	D638	4.5%
Flexural Modulus	D790	800,000 psi
Notched Izod Impact	D256	3.2 Ft. Lb./In
Rockwell Hardness	D785	120 R Scale
Melting Point	D789	420F
Flammability	UL94	HB
GB 30:		
30% Glass Bead Reinforced		
Good Stiffness		
Good Heat Resistance		
Excellent Surface Finish		
Specific Gravity	D792	1.36 g/cu cm
Mold Shrinkage	D955	.3%
Water Absorption, 24 hrs	D570	.9%
Dry:		
Tensile Strength	D638	17,500 psi
Elongation at Break	D638	4%
Flexural Modulus	D790	1,000,000 psi
Notched Izod Impact	D256	1.3 Ft. Lb./In
Cond.: Samples conditioned at 70F/50% R.H.:		
Tensile Strength	D638	11,600 psi
Elongation at Break	D638	5%
Flexural Modulus	D790	600,000 psi
Notched Izod Impact	D256	3.7 Ft.Lb./In.
Rockwell Hardness	D785	117 R Scale
Melting Point	D789	420F
Flammability	UL94	HB

A. SCHULMAN, INC.: SCHULAMID Polyamide Reinforced(Continued):

Property	ASTM Method	Units
SCHULAMID 6 - reinforced:		
MF 40:		
40% Mineral Reinforced		
Low Warpage		
Specific Gravity	D792	1.47 g/cu cm
Mold Shrinkage	D955	.5%
Water Absorption, 24 hrs.	D570	1.0%
Dry:		
Tensile Strength	D638	13,000 psi
Elongation @ Break	D638	8%
Flexural Modulus	D790	800,000 psi
Notched Izod Impact	D256	.8 Ft.Lb./In.
Cond.: Samples conditioned at 70F/50% R.H.:		
Tensile Strength	D638	8,700 psi
Elongation @ Break	D638	9%
Flexural Modulus	D790	550,000 psi
Notched Izod Impact	D256	1.3 Ft.Lb./In.
Rockwell Hardness	D785	119 R Scale
Melting Point	D789	420
Flammability	UL94	HB
SCHULAMID 6.6 - reinforced:		
GF 15:		
15% Glass Fiber Reinforced		
Good Stiffness		
Good Surface		
Specific Gravity	D792	120 g/cu cm
Mold Shrinkage	D955	.8%
Water Absorption, 24 hrs.	D570	.9%
Dry:		
Tensile Strength	D638	18,000 psi
Elongation @ Break	D638	2%
Flexural Modulus	D790	740,000 psi
Notched Izod Impact	D256	.9 Ft.Lb./In.
Cond.: Samples Conditioned at 70F/50% R.H.:		
Tensile Strength	D638	16,800 psi
Elongation @ Break	D638	2%
Flexural Modulus	D790	400,000 psi
Notched Izod Impact	D256	1.8 Ft.Lb./In.
Rockwell Hardness	D785	120 R Scale
Melting Point	D789	500F
Flammability	UL94	HB

A. SCHULMAN, INC.: SCHULAMID Polyamide Reinforced(Continued):

Property	ASTM Method	Units
SCHULAMID 6.6 - reinforced(continued):		
GF 30:		
30% Glass Fiber Reinforced		
High Stiffness		
High Heat Resistance		
Specific Gravity	D792	1.35 g/cu cm
Mold Shrinkage	D955	.3%
Water Absorption, 24 hrs.	D570	.7%
Dry:		
Tensile Strength	D638	27,500 psi
Elongation @ Break	D638	2%
Flexural Modulus	D790	120,000 psi
Notched Izod Impact	D256	1.8 Ft.Lb./In.
Cond.:Samples conditioned at 70F/50% R.H.:		
Tensile Strength	D638	26,000 psi
Elongation @ Break	D638	4.0%
Flexural Modulus	D790	900,000 psi
Notched Izod Impact	D256	2.3 Ft.Lb./In.
Rockwell Hardness	D785	120 R Scale
Melting Point	D789	500F
Flammability	UL94	HB
GF 45:		
45% Glass Fiber Reinforced		
Very High Stiffness		
Good Dimensional Stability		
Specific Gravity	D792	1.50
Mold Shrinkage	D955	.4%
Water Absorption, 24 hrs.	D570	.6%
Dry:		
Tensile Strength	D638	32,000 psi
Elongation @ Break	D638	2%
Flexural Modulus	D790	1,700,000 psi
Notched Izod Impact	D256	2.2 Ft.Lb./In.
Cond.: Samples conditioned at 70F/50% R.H.:		
Tensile Strength	D638	23,000 psi
Elongation at Break	D638	3%
Flexural Modulus	D790	1,300,000 psi
Notched Izod Impact	D256	2.7 Ft.Lb/In.
Rockwell Hardness	D785	123 R Scale
Melting Point	D789	500F
Flammability	UL94	HB

A. SCHULMAN, INC.: SCHULAMID Polyamide Reinforced(Continued):

Property	ASTM Method	Units
SCHULAMID 6.6 - reinforced:		
MWG 40:		
15% Glass Fiber		
25% Mineral Reinforced		
Specific Gravity	D792	1.49g/cu cm
Mold Shrinkage	D955	.5%
Water Absorption, 24 hrs.	D570	.7%
Dry:		
Tensile Strength	D638	21,000 psi
Elongation @ Break	D638	8%
Flexural Modulus	D790	880,000 psi
Notched Izod Impact	D256	1.0 Ft.Lb./In.
Cond: Samples conditioned at 70F/50% R.H.:		
Tensile Strength	D638	20,700 psi
Elongation @ Break	D638	9%
Flexural Modulus	D790	860,000 psi
Notched Izod Impact	D256	1.1 Ft.Lb./In.
Rockwell Hardness	D785	122 R Scale
Melting Point	D789	500F
Flammability	UL94	HB
MF 40:		
40% Mineral Reinforced		
Low Warpage		
Specific Gravity	D792	1.47 g/cu cm
Mold Shrinkage	D955	.8%
Water Absorption, 24 hrs.	D570	.7%
Dry:		
Tensile Strength	D638	14,700 psi
Elongation @ Break	D638	8%
Flexural Modulus	D790	790,000 psi
Notched Izod Impact	D256	1.25 Ft.Lb./In
Cond.: Samples conditioned at 70F/50% R.H.:		
Tensile Strength	D638	14,600 psi
Elongation @ Break	D638	9%
Flexural Modulus	D790	500,000 psi
Notched Izod Impact	D256	1.60 Ft.Lb./In
Rockwell Hardness	D785	122 R Scale
Melting Point	D789	500F
Flammability	UL94	HB

A. SCHULMAN, INC.: SCHULAMID Polyamide Reinforced(Continued):

Property	ASTM Method	Units
SCHULAMID 6.6 high impact:		
N 68:		
Very High Impact		
Easy Mold Release		
Specific Gravity	D792	1.08 g/cu cm
Mold Shrinkage	D955	1.5%
Water Absorption, 24 hrs.	D570	1.2%
Dry:		
Tensile Strength	D638	7,000 psi
Elongation @ Break	D638	60%
Flexural Modulus	D790	260,000 psi
Notched Izod Impact	D256	10 Ft.Lb./In.
Cond.: Samples Conditioned at 70F/50% R.H.:		
Tensile Strength	D638	5,600 psi
Elongation @ Break	D638	170%
Flexural Modulus	D790	140,000
Notched Izod Impact	D256	14 Ft.Lb./In.
Rockwell Hardness	D785	110 R Scale
Melting Point	D789	500F
Flammability	UL94	V-2

A. SCHULMAN, INC.: SUPEROHM Electrical Compositions:

High Purity
Fine Dispersion
Consistency

Product Name:

SUPEROHM 4474 FR * 4475 FR:

Flame retardant Ethylene Propylene Terpolymer Insulations for generating station control and secondary network.

* VW - 1 flame test
* Insulation jacket (Unipass) construction
* Arrhenius plot greater than 40 years
* Radiation resistance
* Excellent heat resistance
* LOCA, IEE 383
* Superior deformation at high temperatures
* 120 Mesh Screening

SUPEROHM 4763 FR:

Flame retardant Ethylene Propylene Terpolymer Insulation for XHHW, RHH, RHW, USE, SIS

* UL tested
* VW - 1 flame test
* Ultraviolet stability
* Excellent crush and deformation
* 120 Mesh Screening

SUPEROHM 3728:

High Voltage Ethylene Propylene Terpolymer Insulation for Mining, Power, Industrial, Distribution

* Manufactured in pellet form
* Semi-Crystalline Base Polymer
* Suitable for XLPE Extrusion Line
* 200 Mesh Screening
* Low Dielectric Constant and Power Factor
* Superior 90C Water Electrical Stability
* Superior Corona Resistance
* Excellent heat resistance up through 150C
* Low deformation at emergency operating temperatures
* Tree resistant

STANDARD OIL: BAREX 210 Resin:

Total Performance.

BAREX 210 Resins are high barrier acrylonitrile copolymers, manufactured by the Standard Oil Company. No other material offers the same high barrier, clarity and physical property advantages. Combined with a wide range of processing techniques, BAREX 210 resins are the preferred material for an expanding variety of end-use packaging applications including: luncheon meats, spices, medical devices, personal care and a number of consumer and industrial chemical products.

Gas Barrier.

The gas barrier provided by BAREX 210 Resins is unparalleled. Its resistance to oxygen, carbon dioxide and other gases is unsurpassed in single layer rigid structures.

Clarity and Form.

BAREX 210 Resins are used in applications where display appearance and aesthetics are critical. For personal care products, BAREX 210 Resins exhibit the properties of glass-like clarity and feel, high gloss, scratch resistance and excellent printability.

Chemical Resistance.

BAREX 210 Resins resistance to aromatic, aliphatic and chlorinated hydrocarbons is unique. These attributes make BAREX 210 Resins the perfect choice for containing gasoline additives, agricultural chemicals, rubber cement, correction fluids and other products.

FDA Approved

BAREX 210 Resins are approved for food contact use under Food & Drug Title 21 177.1480 and are listed under FDA Drug Master File No. 2343. Both the food and medical markets have discovered BAREX 210 Resins to be an ideal packaging material based on its combination of clarity, rigidity and thermoforming detail.

Processing Versatility

BAREX 210 Resin is available in extrusion and injection molding grades, allowing conversion by virtually any process technique including: blown film, sheet and profile extrusion, co-extrusion, blow-molding and injection molding.

STANDARD OIL: EKONOL PTFE Blends:

Kennecott EKONOL polyester is a thermally stable aromatic polymer. When combined with polytetrafluoroethylene (PTFE) it produces a composite material that has excellent temperature and wear resistance. The EKONOL/PTFE blend will not wear metal mating surfaces--and it resists self-wear better than any other PTFE composition. Other EKONOL/PTFE advantages include high dielectric strength, low dissipation, high thermal conductivity, long-term thermal resistance to 600F., low thermal expansion and low moisture absorbance.

Best of all, EKONOL/PTFE is easy to fabricate. And easy to use. Applications for EKONOL/PTFE blends are varied, and include packing sets, compressor ring sets, "O" ring seals, spring-loaded seals, continuous rings, lip seals, piston rings, self-lubricating bearings; and seals, rotors or vanes of process pumps. EKONOL/PTFE works best under environmentally tough conditions where wear resistance, dimensional stability and corrosion resistance are critical.

EKONOL/PTFE is available in free-flow and standard form.

EKONOL is a homopolymer based on p-oxybenzoyl repeat units, and is a linear thermoplastic. The absence of aliphatic hydrogens contributes to its excellent thermal stability. EKONOL is a highly crystalline polymer and must be heated above its crystalline transition temperature of 625F. to obtain flow. Flow and creep are virtually nonexistent below this temperature. The molded polymer has a flexural strength of 5500 psi with a modulus of 1 x 10 6 psi and a density of 1.44 gm/cc. EKONOL has a thermal conductivity of 3.9 BTU/hr/sq ft/F/in; and coefficient of thermal expansion of 3.4 x 10 -5/F, which is approximately linear from room temperature to 572F. The dielectric constant at 10 6 Hz is 3.28 with a dissipation factor of .0025. EKONOL has excellent solvent resistance with the exception of concentrated H2SO4 and strong alkalis. It also has excellent friction and wear properties.

TYPICAL PHYSICAL PROPERTIES	25% EKONOL/PTFE
Density	1.86
Tensile Strength (psi)	1800
Strain (%)	180
Shrinkage	3
Hardness--Shore D	63
Compressive Strength (psi)	
0.2% Offset	1525
At 1% Strain	1275
At 25% Strain	4700

THERMOFIL, INC.: THERMOFIL Chemically Coupled Polypropylene:

Recent advances in reinforced polymer technology have made it possible for THERMOFIL to offer Glass Fiber Reinforced Chemically Coupled Polypropylene composites with:

* Superior Strength and Stiffness
* Improved Elevated Temperature Performance
* Better Creep Resistance
* Higher Impact Resistance

These improvements are achieved through the incorporation of grafted polymer coupling agents to enhance the adhesion between the Glass Fibers and the Polypropylene. Loading stresses are more effectively transferred to the Glass Fibers. Strength and stiffness of the Glass Fibers is more efficiently utilized.

Significantly, the highly desirable qualities of THERMOFIL Glass Fiber Reinforced Polypropylene are maintained; these include:

* Excellent Chemical Resistance
* Low Melt Viscosity (easy-flow)
* Resistance to High Temperature Water
* Mold-In Color Capability
* Low Specific Gravity

THERMOFIL Chemically Coupled Glass Fiber Reinforced Polypropylene is extremely versatile. Additional THERMOFIL products include:

* Flame Retardant Grades
* Impact Modified Formulations
* Foamable Grades

THERMOFIL Chemically Coupled Glass Fiber Reinforced Polypropylene offers:

* The Lowest Specific Gravity of any Glass Fiber Reinforced Engineering Thermoplastic
* The Lowest Cost Per Pound of any Engineering Thermoplastic.

RESULT: The Lowest Cost Per Cubic Inch of any Engineering Thermoplastic.

THERMOFIL, INC.: THERMOFIL Chemically Coupled Polypropylene (Continued):

The lowest cost per cubic inch of THERMOFIL Chemically Coupled Glass Fiber Reinforced Polypropylene combined with its characteristically high structured properties results in:

* The Highest Specific Strength of any Reinforced Thermoplastic
* The Highest Specific Flexural Modulus of any Reinforced Thermoplastic

THERMOFIL Chemically Coupled Glass Fiber Reinforced Polypropylene offers the designer:

* High Structural Properties
* Good Elevated Temperature Resistance
* Excellent Resistance to Chemicals
* Low Part Weight
* Excellent Performance to Cost Ratio

THERMOFIL Chemically Coupled Glass Fiber Reinforced Polypropylene offers the molder:

* Low Energy Requirements
* Wide Processing Latitude
* Good Surface Finish

THERMOFIL Chemically Coupled Glass Fiber Reinforced Polypropylene composites provide additional resistance to deformation in potentially elevated temperature water applications.

Relative to its cost, THERMOFIL Reinforced Chemically Coupled Polypropylene displays the highest strength and stiffness of any chemically resistant themoplastic.

THERMOFIL Reinforced Chemically Coupled Polypropylene composites often displace more traditional thermoplastics.

THERMOFIL, INC.: THERMOFIL Glass Fiber Reinforced ABS:

THERMOFIL Glass Fiber Reinforced ABS composites display the optimum combination of strength, stiffness and practical toughness in styrenic base compounds. Glass fiber reinforcement greatly improves the load bearing capacity of ABS and moderately extends the upper use temperature.

The properties of THERMOFIL Glass Fiber Reinforced ABS may be varied through the selection of the ABS and the level of reinforcement. Impact, heat and chemical resistance of THERMOFIL Glass Fiber Reinforced ABS compounds reflect the inherent characteristics of the ABS base polymer selected for compounding. Through prudent selection of the grade of ABS and the type and percentage of Glass Fiber Reinforcement, compounds can be tailored to meet specific end-use requirements.

Injection molded Glass Fiber Reinforced ABS parts are compatible with most post-finishing and joining methods traditionally utilized for unfilled ABS.

THERMOFIL Glass Fiber Reinforced ABS may be molded in color for appearance applications. A grained surface texture may be selected to mask the dull surface finish inherent in some Glass Fiber Reinforced Compounds.

THERMOFIL Glass Fiber Reinforced ABS is readily structural foam molded utilizing THERMOFOAM Blowing Agent Concentrates.

TYPICAL PROPERTY DATA:

D-10FG-0100:
10% Glass Fiber Reinforced
Impact Resistant

Reinforcement Content (ASTM D-2584): 10%
Specific Gravity (ASTM D-792): 1.10
Mold Shrinkage (ASTM D-955): 0.003 in/in
Hardness (ASTM D-785): R105
Water Absorption, 24 hr. (ASTM D-570): 0.3%

D-20FG-0100:
20% Glass Fiber Reinforced
Strength

Reinforcement Content (ASTM D-2584): 20%
Specific Gravity (ASTM D-792): 1.22
Mold Shrinkage (ASTM D-955): 0.002 in/in
Hardness (ASTM D-785): R107
Water Absorption, 24 hr. (ASTM D-570): 0.3%

THERMOFIL, INC.: THERMOFIL Glass Fiber Reinforced ABS (Continued):

TYPICAL PROPERTY DATA(Continued):

D-30FG-0100:
30% Glass Fiber Reinforced
Stiffness

Reinforcement Content (ASTM D-2584): 30%
Specific Gravity (ASTM D-792): 1.28
Mold Shrinkage (ASTM D-955): 0.002 in/in
Hardness (ASTM D-785): R110
Water Absorption, 24 hr. (ASTM D-570): 0.2%

D-9900-6500:
Flame Retardant

Specific Gravity (ASTM D-792): 1.20
Mold Shrinkage (ASTM D-955): 0.007 in/in
Hardness (ASTM D-785): R102
Water Absorption, 24 hr (ASTM D570): 0.2%

D-20FG-0500:
20% Glass Fiber Reinforced
Flame Retardant

Reinforcement Content (ASTM D-2584): 20%
Specific Gravity (ASTM D-792): 1.45
Mold Shrinkage (ASTM D-955): 0.001 in/in
Hardness (ASTM D-785): R107
Water Absorption, 24 hr. (ASTM D-570): 0.3%

THERMOFIL, INC.: THERMOFIL Glass Fiber Reinforced and Lubricated Acetal Products:

Acetal is noted for its inherently high self-lubricity. Acetal also displays good resistance to organic solvents and is unaffected by moderate temperature water. The addition of Glass Fiber Reinforcement to Acetal reduces its mold shrinkage, greatly improving its ability to meet close dimensional tolerances.

The inherently low wear and friction characteristics of Acetal are further enhanced through the addition of PTFE and other lubricants. These lubricants can be combined with reinforcements to achieve the desired combination of frictional and structural characteristics.

TYPICAL PROPERTY DATA:

G-30FG-0100:
30% Glass Fiber Reinforced
Moisture Resistance

Reinforcement Content (ASTM D-2584): 30%
Specific Gravity (ASTM D-792): 1.63
Mold Shrinkage (ASTM D-955): 0.005 in/in
Hardness (ASTM D-785): M82
Water Absorption, 24 hr. (ASTM D-570): 0.2%

G-20NF-0100:
20% Graphite Fiber Reinforced
Stiffness

Reinforcement Content (ASTM D-2584): 20%
Specific Gravity (ASTM D-792): 1.46
Mold Shrinkage (ASTM D-955): 0.006 in/in
Hardness (ASTM D-785): M81
Water Absorption, 24 hr. (ASTM D570): 0.50%

G-9900-0215:
20% PTFE Lubricated
PTFE Lubricated

Specific Gravity (ASTM D-792): 1.51
Mold Shrinkage (ASTM D-955): 0.020 in/in
Hardness (ASTM D-785): R110
Water Absorption, 24 hr. (ASTM D570): 0.013%

THERMOFIL, INC.: THERMOFIL Glass Fiber Reinforced High Heat Styrenic Copolymers:

THERMOFIL Glass Fiber Reinforced High Heat Styrenic Copolymers offer high temperature structural integrity in a relatively low cost high modulus polymer. These products maintain the excellent stiffness and dimensional stability characteristic of Glass Fiber Reinforced Styrenics at higher temperatures than traditionally attainable. THERMOFIL Glass Fiber Reinforced High Heat Styrenic Copolymers are particularly well suited for applications where good dimensional control and moderate load bearing qualities at elevated temperatures are required.

HIGH HEAT STYRENIC COPOLYMER:

TYPICAL PROPERTY DATA:

A1-20FG-0700:
20% Glass Fiber Reinforced
Impact Resistance

Reinforcement Content (ASTM D-2584): 20%
Specific Gravity (ASTM D-792): 1.19
Mold Shrinkage (ASTM D-955): 0.003 in/in
Hardness (ASTM D-785): R112
Water Absorption, 24 hr. (ASTM D-570): 0.08%

A1-20FG-0570:
20% Glass Fiber Reinforced
Flame Retardant

Reinforcement Content (ASTM D-2584): 20%
Specific Gravity (ASTM D-792): 1.40
Mold Shrinkage (ASTM D-955): 0.003 in/in
Hardness (ASTM D-785): R110
Water Absorption, 24 hr. (ASTM D-570): 0.15%

Y-20FG-0100:
20% Glass Fiber Reinforced
Heat Resistance

Reinforcement Content (ASTM D-2584): 20%
Specific Gravity (ASTM D-792): 1.22
Mold Shrinkage (ASTM D-955): 0.003 in/in
Hardness (ASTM D-785): R121
Water Absorption, 24 hr. (ASTM D570): 0.28%

THERMOFIL, INC.: THERMOFIL Glass Fiber Reinforced High Heat Styrenic Copolymers(Continued):

HIGH HEAT STYRENIC COPOLYMER(Continued):

TYPICAL PROPERTY DATA:

Y-20FG-0500:
20% Glass Fiber Reinforced
Flame Retardant

Reinforcement Content (ASTM D-2584): 20%
Specific Gravity (ASTM D-792): 1.41
Mold Shrinkage (ASTM D-955): 0.002 in/in
Hardness (ASTM D-785): R118
Water Absorption, 24 hr. (ASTM D-570): 0.3%

Y7-9900-6500:
Flame Retardant

Reinforcement Content (ASTM D-2584): 0%
Specific Gravity (ASTM D-792): 1.20
Mold Shrinkage (ASTM D-955): 0.006 in/in
Hardness (ASTM D-785): R100
Water Absorption, 24 hr. (ASTM D-570): 0.08%

THERMOFIL, INC.: THERMOFIL Glass Fiber Reinforced and Lubricated Acetal Products(Continued):

TYPICAL PROPERTY DATA:

G-20FG-0214:
20% Glass Fiber
PTFE Lubricated

Reinforcement Content (ASTM D-2584): 20%
Specific Gravity (ASTM D-792): 1.63
Mold Shrinkage (ASTM D-955): 0.003 in/in
Hardness (ASTM D-785): M80
Water Absorption, 24 hr (ASTM D-570): 0.26

G-9900-0257:
Oil/PTFE Lubricated

Specific Gravity (ASTM D792): 1.52
Mold Shrinkage (ASTM D-955): 0.015 in/in
Hardness (ASTM D-785): R100
Water Absorption, 24 hr. (ASTM D-570): 0.2

THERMOFIL, INC.: THERMOFIL Glass Fiber Reinforced Polycarbonate:

THERMOFIL Glass Fiber Reinforced Polycarbonate displays a unique combination of toughness, dimensional stability and heat resistance. At low levels of Glass Fiber Reinforcement, much of the toughness inherent in unfilled Polycarbonate is maintained, while strength and stiffness are significantly improved. At high Glass Fiber loadings, the reduction in impact resistance is offset by a significant increase in dimensional stability and creep resistance over a broad temperature and humidity range.

THERMOFIL Glass Fiber Reinforced Polycarbonate has good electrical properties and is available in flame retardant grades to meet the end-use requirements of modern electrical components.

Specific grades of THERMOFIL Glass Fiber Reinforced Polycarbonate have been formulated for structural foam molding. The dimensional stability, toughness and flame retardancy of these compounds make them particularly well suited for large electrical component housings. Excellent paintability is another advantage in appearance parts.

The melt viscosity of Glass Fiber Reinforced Polycarbonate composites is relatively high. This results in characteristics stiff flowing behavior during injection molding. Additives have been developed to reduce the melt viscosity of Polycarbonate thus facilitating the filling of thin sections and intricate parts.

TYPICAL PROPERTY DATA:

R-10FG-0100:
10% Glass Fiber Reinforcement
Dimension Stability

Reinforcement Content (ASTM D-2584): 10%
Specific Garvity (ASTM D-792): 1.26
Mold Shrinkage (ASTM D-955): 0.003 in/in
Hardness (ASTM D-785): M80
Water Absorption, 24 hr. (ASTM D-570): 0.14%

R-30FG-0100:
30% Glass Fiber Reinforcement
Strength

Reinforcement Content (ASTM D-2584): 30%
Specific Gravity (ASTM D-792): 1.43
Mold Shrinkage (ASTM D-955): 0.003 in/in
Hardness (ASTM D-785): M87
Water Absorption, 24 hr. (ASTM D-570): 0.12%

THERMOFIL, INC.: THERMOFIL Glass Fiber Reinforced Polycarbonate (Continued):

TYPICAL PROPERTY DATA(Continued):

R-40NF-0100:
10% Graphite Fiber Reinforced
Stiffness

Reinforcement Content (ASTM D-2584): 40%
Specific Gravity (ASTM D-792): 1.40
Mold Shrinkage (ASTM D-955): 0.001 in/in
Hardness (ASTM D-785): M90
Water Absorption, 24 hr. (ASTM D-570): 0.08%

R-9900-0200:
Lubricated

Specific Gravity (ASTM D-792): 1.31
Mold Shrinkage (ASTM D-955): 0.006 in/in
Hardness (ASTM D-785): M55
Water Absorption, 24 hr. (ASTM D-570): 0.10%

R-9900-0214:
PTFE Lubricated

Specific Gravity (ASTM D-792): 1.30
Mold Shrinkage (ASTM D-955): 0.006 in/in
Hardness (ASTM D-785): M55
Water Absorption, 24 hr. (ASTM D-570): 0.09%

R-30FG-0214:
30% Glass Fiber Reinforced
PTFE Lubricated

Reinforcement Content (ASTM D-2584): 30%
Specific Gravity (ASTM D-792): 1.56
Mold Shrinkage (ASTM D-955): 0.003 in/in
Hardness (ASTM D-785): M80
Water Absorption, 24 hr. (ASTM D-570): 0.10%

THERMOFIL, INC.: THERMOFIL Glass Fiber Reinforced Polycarbonate (Continued):

TYPICAL PROPERTY DATA:

R-20NF-0214:
20% Graphite Fiber Reinforced
PTFE Lubricated

Reinforcement Content (ASTM D-2584): 30%
Specific Gravity (ASTM D-792): 1.37
Mold Shrinkage (ASTM D-955): 0.001 in/in
Hardness (ASTM D-785): M83
Water Absorption, 24 hr. (ASTM D-570): 0.13%

R-10FG-0568:
10% Glass Fiber Reinforced
Flame Retardant

Reinforcement Content (ASTM D-2584): 10%
Specific Gravity (ASTM D-792): 1.34
Mold Shrinkage (ASTM D-955): 0.003 in/in
Hardness (ASTM D-785): R118
Water Absorption, 24 hr. (ASTM D-570): 0.1%

R-30FG-0568:
30% Glass Fiber Reinforced
Flame Retardant

Reinforcement Content (ASTM D-2584): 30%
Specific Gravity (ASTM D-792): 1.46
Mold Shrinkage (ASTM D-955): 0.002 in/in
Hardness (ASTM D-785): M87
Water Absorption, 24 hr. (ASTM D-570): 0.20%

THERMOFIL, INC.: THERMOFIL Glass Fiber Reinforced Polycarbonate (Continued):

P/C ALLOY:

TYPICAL PROPERTY DATA:

R2-10FG-0100:
10% Glass Fiber Reinforced
Chemical Resistance

Reinforcement Content (ASTM D-2584): 10%
Specific Graphite (ASTM D-792): 1.30
Mold Shrinkage (ASTM D-955): 0.003 in/in
Hardness (ASTM D-785): R118
Water Absorption, 24 hr. (ASTM D570): 0.08%

R2-30FG-0100:
30% Glass Fiber Reinforced
Chemical Resistance

Reinforcement Content (ASTM D-2584): 30%
Specific Gravity (ASTM D-792): 1.42
Mold Shrinkage (ASTM D-955): 0.001 in/in
Hardness (ASTM D-785): R118
Water Absorption, 24 hr. (ASTM D-570): 0.10

THERMOFIL, INC.: THERMOFIL Glass Fiber Reinforced Polystyrene:

General purpose Polystyrene offers extremely high strength and stiffness relative to its cost. THERMOFIL Glass Fiber Reinforced Polystyrene further enhances these characteristics, while improving the practical toughness of this relatively low impact material.

In addition to its high strength and stiffness, THERMOFIL Glass Fiber Reinforced Polystyrene's low mold shrinkage assures good dimensional consistency. This combined with the suppressed coefficient of thermal expansion provided by glass fiber reinforcement permits the use of these compounds in large dimensionally critical parts.

The resistance of Polystyrene to elevated temperature deformation under load is not significantly improved through Glass Fiber Reinforcement. However, Glass Fiber Reinforcement does improve Polystyrene's stiffness throughout its useful temperature range, and some improvement in elevated temperature creep resistance is achieved. Polystyrene is reasonably resistant to many chemicals, but displays low temperature resistance to aromatic and chlorinated hydrocarbons. This characteristic is not improved by Glass Fiber Reinforcement.

In application situations where the high stiffness to cost relationship of Glass Fiber Reinforced Polystyrene is attractive, but improved heat and/or chemical resistance is required, THERMOFIL Glass Fiber Reinforced SAN often meets the requirements.

THERMOFIL, INC.: THERMOFIL Glass Fiber Reinforced Polystyrene (Continued):

POLYSTYRENE:

TYPICAL PROPERTY DATA:

A-20FG-0100:
20% Glass Fiber Reinforced
Stiffness

Reinforcement Content (ASTM D-2584): 20%
Specific Gravity (ASTM D-792): 1.20
Mold Shrinkage (ASTM D-955): 0.002 in/in
Hardness (ASTM D-785): R121
Water Absorption, 24 hr. (ASTM D-570): 0.01%

A-35FG-0100:
35% Glass Fiber Reinforced
Stiffness

Reinforcement Content (ASTM D-2584): 35%
Specific Gravity (ASTM D-792): 1.33
Mold Shrinkage (ASTM D-955): 0.001 in/in
Hardness (ASTM D-785): R123
Water Absorption, 24 hr. (ASTM D-570): 0.14%

A-9900-0570:
Flame Retardant
Impact Resistant

Reinforcement Content (ASTM D-2584): 0%
Specific Gravity (ASTM D-792): 1.16
Mold Shrinkage (ASTM D-955): 0.005 in/in
Hardness (ASTM D-785): M70
Water Absorption, 24 hr. (ASTM D570): 0.4%

THERMOFIL, INC.: THERMOFIL Glass Fiber Reinforced PVC:

Rigid PVC is a unique, moderately priced Polymer offering good chemical and impact resistance, combined with inherent flame retardancy.

THERMOFIL Glass Fiber Reinforced PVC extends the usefulness of PVC through improvements in strength and stiffness. The characteristic creep resistance of rigid PVC is further enhanced when Glass Fiber reinforced. The flame retardant, structural and dimensional qualities of THERMOFIL Glass Fiber Reinforced PVC make these composites attractive for electrical and appliance housings and bases. The inherently good chemical resistance of THERMOFIL Glass Fiber Reinforced PVC has led to its use in components of chemical handling equipment.

PVC:

TYPICAL PROPERTY DATA:

V-10FG-0100:
10% Glass Fiber Reinforced
Flame Retardant

Reinforcement Content (ASTM D-2584): 10%
Specific Gravity (ASTM D-792): 1.43
Mold Shrinkage (ASTM D-955): 0.004 in/in
Hardness (ASTM D-785): D87
Water Absorption, 24 hr. (ASTM D-570): 0.10%

V-20FG-0100:
20% Glass Fiber Reinforced
Flame Retardant

Reinforcement Content (ASTM D-2584): 20%
Specific Gravity (ASTM D-792): 1.56
Mold Shrinkage (ASTM D-955): 0.002 in/in
Hardness (ASTM D-785): D90
Water Absorption, 24 hr. (ASTM D-570): 0.09%

V-30FG-0100:
30% Glass Fiber Reinforced PVC
Flame Retardant

Reinforcement Content (ASTM D-2584): 30%
Specific Gravity (ASTM D-792): 1.61
Mold Shrinkage (ASTM D-955): 0.001 in/in
Hardness (ASTM D-785): D95
Water Absorption, 24 hr. (ASTM D-570): 0.08%

THERMOFIL, INC.: THERMOFIL Glass Fiber Reinforced SAN:

The copolymerization of Styrene with Acrylonitrile to produce SAN, provides significant enhancement to the properties of general purpose Polystyrene.

THERMOFIL Glass Fiber Reinforced SAN reflects these improvements. Specifically, THERMOFIL Glass Fiber Reinforced SAN offers relatively high strength and stiffness and maintains these properties to higher use temperatures. Chemical resistance is also improved.

The low mold shrinkage and coefficient of thermal expansion of THERMOFIL Glass Fiber Reinforced SAN products accomodate close dimensional tolerances over a broad temperature range. The upper use temperature, although higher than that of Polystyrene, is limited by the softening characteristics of these amorphous polymer base compounds.

THERMOFIL Glass Fiber Reinforced SAN compounds are readily painted and hot stamped. The high modulus and low softening point of these compounds is responsible for their excellent sonic weldability. Excellent bond strength is also achieved in solvent and spin welding.

SAN:

TYPICAL PROPERTY DATA:

B-20FG-0100:
20% Glass Fiber Reinforced
Rigidity

Reinforcement Content (ASTM D-2584): 20%
Specific Gravity (ASTM D-792): 1.22
Mold Shrinkage (ASTM D-955): 0.002 in/in
Hardness (ASTM D-785): R122
Water Absorption, 24 hr. (ASTM D-570): 0.24%

B-35FG-0100:
35% Glass Fiber Reinforced
Rigidity

Reinforcement Content (ASTM D-2584): 35%
Specific Gravity (ASTM D-792): 1.35
Mold Shrinkage (ASTM D-955): 0.001 in/in
Hardness (ASTM D-785): R123
Water Absorption, 24 hr. (ASTM D-570): 0.21%

THERMOFIL, INC.: THERMOFIL Glass Fiber Reinforced SAN (Continued):

SAN(Continued):

TYPICAL PROPERTY DATA:

B-35FB-0100:
35% Glass Fiber/Glass Bead Reinforced
Surface Finish

Reinforcement Content (ASTM D-2584): 35%
Specific Gravity (ASTM D-792): 1.35
Mold Shrinkage (ASTM D-955): 0.001 in/in
Hardness (ASTM D-785): R122
Water Absorption, 24 hr. (ASTM D-570): 0.2%

THERMOFIL, INC.: Nylon (Polyamide):

THERMOFIL'S Nylon series of Reinforced Compounds encompasses a number of Thermoplastic-polyamides. This family exhibits a broad spectrum of structural, electrical, and dimensional characteristics. Properties common to all THERMOFIL Reinforced Nylon Compounds include:

* Good chemical resistance
* Good elevated temperature performance
* Low friction and wear rate

Nylons accept reinforcement well. THERMOFIL Reinforced Nylon Compounds exhibit:

* High strength and stiffness
* High elevated temperature properties
* Improved resistance to deformation under long term stress

THERMOFIL Reinforced Nylons significantly improve the dimensional stability of molded parts. This is accomplished by:

* Reduced influence of moisture on dimensional change
* Reduced coefficient of Thermal expansion
* Reduced mold shrinkage

After molding, Nylon parts absorb moisture from their surroundings until a balanced or equilibrium moisture content is achieved. This increase in moisture content is accompanied by a corresponding reduction in the strength and stiffness of the molded part.

THERMOFIL Reinforced Nylon Compounds compensate for the reduction in structural properties associated with increase in moisture content by improving the strength and stiffness of moisture conditioned Nylons.

THERMOFIL, INC.: THERMOFIL Reinforced Nylon 6 Compounds:

THERMOFIL Reinforced Nylon 6 Compounds display a good balance of structural and impact qualities. Nylon 6 is among the least expensive of the Nylons. It is readily reinforced with glass fibers, mineral reinforcements and combination of glass fibers and minerals. THERMOFIL Reinforced Nylon 6 Compounds are characterized by their:

* Elevated temperature performance
* Chemical resistance
* Inherent lubricity
* Moderate cost

TYPICAL PROPERTY DATA:

N-15FG-0100:
15% Glass Fiber Reinforced
Ductility

Reinforcement Content (ASTM D-2584): 15%
Specific Gravity (ASTM D-792): 1.25
Mold Shrinkage (ASTM D-955): 0.007 in/in
Hardness (ASTM D-785): R120
Water Absorption, 24 hr. (ASTM D-570): 1.8%

N-30FG-0100:
30% Glass Fiber Reinforced
Strength

Reinforcement Content (ASTM D-2584): 30%
Specific Gravity (ASTM D-792): 1.37
Mold Shrinkage (ASTM D-955): 0.004 in/in
Hardness (ASTM D-785): R121
Water Absorption, 24 hr. (ASTM D-570): 1.3%

N-30NF-0100:
30% Graphite Fiber Reinforced
Stiffness

Reinforcement Content (ASTM D-2584): 30%
Specific Gravity (ASTM D-792): 1.28
Mold Shrinkage (ASTM D-955): 0.002 in/in
Hardness (ASTM D-785): R127
Water Absorption, 24 hr. (ASTM D-570): 0.8%

THERMOFIL, INC.: THERMOFIL Reinforced Nylon 6 Compounds (Continued):

TYPICAL PROPERTY DATA(Continued):

N-40FM-0100:
40% Glass Fiber/Mineral Reinforced
Dimension Stability

Reinforcement Content (ASTM D-2584): 40%
Specific Gravity (ASTM D-792): 1.47
Mold Shrinkage (ASTM D-955): 0.004 in/in
Hardness (ASTM D-785): R118
Water Absorption, 24 hr. (ASTM D-570): 0.9%

N-40MF-0100:
40% Mineral Reinforced
Dimension Stability

Reinforcement Content (ASTM D-2584): 40%
Specific Gravity (ASTM D-792): 1.48
Mold Shrinkage (ASTM D-955): 0.006 in/in
Hardness (ASTM D-785): R122
Water Absorption, 24 hr. (ASTM D-570): 0.8%

N-40BG-0100:
40% Glass Bead Reinforced
Dimension Stability

Reinforcement Content (ASTM D-2584): 40%
Specific Gravity (ASTM D-792): 1.45
Mold Shrinkage (ASTM D-955): 0.010 in/in
Hardness (ASTM D-785): R118
Water Absorption, 24 hr. (ASTM D-570): 1.0%

N-30FG-0214:
30% Glass Fiber Reinforced
PTFE Lubricated

Reinforcement Content (ASTM D-2584): 30%
Specific Gravity (ASTM D-792): 1.49
Mold Shrinkage (ASTM D-955): 0.002 in/in
Hardness (ASTM D-785): R120
Water Absorption, 24 hr. (ASTM D-570): 0.8%

THERMOFIL, INC.: THERMOFIL Reinforced Nylon 6 Compounds (Continued):

TYPICAL PROPERTY DATA(Continued):

N-9900-0500:
Flame Retardant

Specific Gravity (ASTM D-792): 1.42
Mold Shrinkage (ASTM D-955): 0.008 in/in
Hardness (ASTM D-785): R118
Water Absorption, 24 hr. (ASTM D-570): 1.9%

N-15FG-0500:
15% Glass Fiber Reinforced
Flame Retardant

Reinforcement Content (ASTM D-2584): 15%
Specific Gravity (ASTM D-792): 1.51
Mold Shrinkage (ASTM D-955): 0.005 in/in
Hardness (ASTM D-785): R120
Water Absorption, 24 hr. (ASTM D-570): 0.8%

N-30FG-0500:
30% Glass Fiber Reinforced
Flame Retardant

Reinforcement Content (ASTM D-2584): 30%
Specific Gravity (ASTM D-792): 1.60
Mold Shrinkage (ASTM D-955): 0.002 in/in
Hardness (ASTM D-785): R120
Water Absorption, 24 hr. (ASTM D-570): 0.50%

NYLON 6 (HIGH IMPACT):

N11-30FG-0100:
30% Glass Fiber Reinforced
High Impact

Reinforcement Content (ASTM D-2584): 30%
Specific Gravity (ASTM D-792): 1.31
Mold Shrinkage (ASTM D-955): 0.004 in/in
Hardness (ASTM D-785): M90
Water Absorption, 24 hr. (ASTM D570): 0.8%

THERMOFIL, INC.: THERMOFIL Reinforced Nylon 6 Compounds (Continued):

TYPICAL PROPERTY DATA(Continued):

N11-30NF-0100:
30% Graphite Fiber Reinforced
High Impact

Reinforcement Content (ASTM D-2584): 30%
Specific Gravity (ASTM D-792): 1.22
Mold Shrinkage (ASTM D-955): 0.002 in/in
Hardness (ASTM D-785): R108
Water Absorption, 24 hr. (ASTM D-570): 0.80%

N11-40MF-0350:
40% Mineral Reinforced
High Impact

Reinforcement Content (ASTM D-2584): 40%
Specific Gravity (ASTM D-792): 1.40
Mold Shrinkage (ASTM D-955): 0.009 in/in
Hardness (ASTM D-785): R118
Water Absorption, 24 hr. (ASTM D-570): 0.6%

THERMOFIL, INC.: THERMOFIL Reinforced Nylon 6/6 Compounds:

THERMOFIL Reinforced Nylon 6/6 Compounds are distinguished from the rest of the Nylon family by their high melting point and superior strength and stiffness.

The load bearing capabilities of Nylon 6/6 are substantially improved by the incorporation of Glass or Carbon Fibers.

Mineral Reinforcements may supplement or replace Glass Fibers to achieve specific price/performance goals. In general, THERMOFIL Reinforced Nylon 6/6 Compounds display:

* Elevated temperature load bearing capabilities
* Good impact resistance
* Low friction and wear

Flame retardant formulations and lubricated versions of Reinforced Nylon 6/6 compounds are available to meet specific requirements. Heat stabilized grades of Reinforced Nylon 6/6 are available for applications demanding continuous elevated temperature resistance.

TYPICAL PROPERTY DATA:

N3-13FG-0100:
13% Glass Fiber Reinforced
Ductility

Reinforcement Content (ASTM D-2584): 13%
Specific Gravity (ASTM D-792): 1.22
Mold Shrinkage (ASTM D-955): 0.007 in/in
Hardness (ASTM D-785): M92
Water Absorption, 24 hr. (ASTM D-570): 1.0

N3-33FG-0100:
33% Glass Fiber Reinforced
Strength

Reinforcement Content (ASTM D-2584): 33%
Specific Gravity (ASTM D-792): 1.38
Mold Shrinkage (ASTM D-955): 0.003 in/in
Hardness (ASTM D-785): M100
Water Absorption, 24 hr. (ASTM D-570): 1.0%

THERMOFIL, INC.: THERMOFIL Reinforced Nylon 6/6 Compounds (Continued):

TYPICAL PROPERTY DATA:

N3-43FG-0100:
43% Glass Fiber Reinforced
Stiffness

Reinforcement Content (ASTM D-2584): 43%
Specific Gravity (ASTM D-792): 1.51
Mold Shrinkage (ASTM D-955): 0.002 in/in
Hardness (ASTM D-785): R121
Water Absorption, 24 hr. (ASTM D-570): 0.7%

N3-20NF-0100:
20% Graphite Fiber Reinforced
Strength

Reinforcement Content (ASTM D-2584): 20%
Specific Gravity (ASTM D-792): 1.23
Mold Shrinkage (ASTM D-955): 0.003 in/in
Hardness (ASTM D-785): R120
Water Absorption, 24 hr. (ASTM D-570): 0.6%

N3-40NF-0100:
40% Graphite Fiber Reinforced
Stiffness

Reinforcement Content (ASTM D-2584): 40%
Specific Gravity (ASTM D-792): 1.33
Mold Shrinkage (ASTM D-955): 0.002 in/in
Hardness (ASTM D-785): R120
Water Absorption, 24 hr. (ASTM D-570): 0.40%

N3-9900-0279:
40% Pitch Carbon Fiber Reinforced
Stiffness

Reinforcement Content (ASTM D-2584): 40%
Specific Gravity (ASTM D792): 1.37
Mold Shrinkage (ASTM D-955): 0.006 in/in
Hardness (ASTM D-785): R120
Water Absorption, 24 hr. (ASTM D-570): 0.08%

THERMOFIL, INC.: THERMOFIL Reinforced Nylon 6/6 Compounds (Continued):

TYPICAL PROPERTY DATA:

N3-9900-0200:
Lubricated

Specific Gravity (ASTM D-792): 1.15
Mold Shrinkage (ASTM D-955): 0.02 in/in
Hardness (ASTM D-785): M60
Water Absorption, 24 hr. (ASTM D-570): 1.2%

N3-9900-0231:
MoS2, Lubricated

Specific Gravity (ASTM D-792): 1.18
Mold Shrinkage (ASTM D-955): 0.014
Hardness (ASTM D-785): M85
Water Absorption, 24 hr. (ASTM D-570): 1.1%

N3-9900-0214:
PTFE Lubricated

Specific Gravity (ASTM D-972): 1.23
Mold Shrinkage (ASTM D-955): 0.016
Hardness (ASTM D-785): M65
Water Absorption, 24 hr. (ASTM D-570): 1.0%

N3-30FG-0214:
30% Glass Fiber Reinforced
PTFE Lubricated

Reinforcement Content (ASTM D-2584): 30%
Specific Gravity (ASTM D-792): 1.48
Mold Shrinkage (ASTM D-955): 0.004
Hardness (ASTM D-785): M95
Water Absorption, 24 hr. (ASTM D-570): 0.8%

N3-30NF-0214:
30% Graphite Fiber Reinforced
PTFE Lubricated

Reinforcement Content (ASTM D-2584): 30%
Specific Gravity (ASTM D-792): 1.38
Mold Shrinkage (ASTM D-955): 0.002
Hardness (ASTM D-785): R120
Water Absorption, 24 hr. (ASTM D-570): 0.5%

THERMOFIL, INC.: THERMOFIL Reinforced Nylon 6/6 Compounds (Continued):

TYPICAL PROPERTY DATA:

N3-30FG-0282:
30% Glass Fiber Reinforced
Lubricated

Reinforcement Content (ASTM D-2584): 30%
Specific Gravity (ASTM D-792): 1.47
Mold Shrinkage (ASTM D-955): 0.002 in/in
Hardness (ASTM D-785): R120
Water Absorption, 24 hr. (ASTM D-570): 0.45%

N3-30FG-0231:
30% Glass Fiber Reinforced
MoS2 Lubricated

Reinforcement Content (ASTM D-2584): 30%
Specific Gravity (ASTM D-792): 1.43
Mold Shrinkage (ASTM D-955): 0.005 in/in
Hardness (ASTM D-785): M95
Water Absorption, 24 hr. (ASTM D-570): 0.9%

N3-9900-0560:
Flame Retardant

Specific Gravity (ASTM D-792): 1.44
Mold Shrinkage (ASTM D-955): 0.018 in/in
Hardness (ASTM D-785): R115
Water Absorption, 24 hr. (ASTM D-570): 0.9%

N3-15FG-0560:
15% Glass Fiber Reinforced
Flame Retardant

Reinforcement Content (ASTM D-2584): 15%
Specific Gravity (ASTM D-792): 1.50
Mold Shrinkage (ASTM D-955): 0.005 in/in
Hardness (ASTM D-785): M93
Water Absorption, 24 hr. (ASTM D-570): 0.7%

THERMOFIL, INC.: THERMOFIL Reinforced Nylon 6/6 Compounds (Continued):

TYPICAL PROPERTY DATA:

N3-30FG-0560:
30% Glass Fiber Reinforced
Flame Retardant

Reinforcement Content (ASTM D-2584): 30%
Specific Gravity (ASTM D-792): 1.59
Mold Shrinkage (ASTM D-955): 0.003 in/in
Hardness (ASTM D-785): M98
Water Absorption, 24 hr. (ASTM D-570): 0.5%

N3-13FG-0700:
13% Glass Fiber Reinforced
Impact Resistant

Reinforcement Content (ASTM D-2584): 13%
Specific Gravity (ASTM D-792): 1.18
Mold Shrinkage (ASTM D-955): 0.009 in/in
Hardness (ASTM D-785): M80
Water Absorption, 24 hr. (ASTM D-570): 0.9%

N3-33FG-0700:
33% Glass Fiber Reinforced
Impact Resistant

Reinforcement Content (ASTM D-2584): 33%
Specific Gravity (ASTM D-792): 1.38
Mold Shrinkage (ASTM D-955): 0.004 in/in
Hardness (ASTM D-785): M96
Water Absorption, 24 hr. (ASTM D-570): 0.9%

N3-35FM-0310:
35% Glass Fiber/Mineral Reinforced
Impact Resistant

Reinforcemnent Content (ASTM D-2584): 35%
Specific Gravity (ASTM D-792): 1.42
Mold Shrinkage (ASTM D-955): 0.004 in/in
Hardness (ASTM D-785): R120
Water Absorption, 24 hr. (ASTM D-570): 0.7%

THERMOFIL, INC.: THERMOFIL Reinforced Nylon 6/6 Compounds (Continued):

TYPICAL PROPERTY DATA:

N3-40MF-8300:
40% Mineral Reinforced
Stiffness

Reinforcement Content (ASTM D-2584): 40%
Specific Gravity (ASTM D-792): 1.47
Mold Shrinkage (ASTM D-955): .011-.012 in/in
Hardness (ASTM D-785): R119
Water Absorption, 24 hr. (ASTM D-570): 0.8%

Nylon 6/6 (High Impact):

N3-40MF-0350:
40% Mineral Filled
Impact Modified

Reinforcement Content (ASTM D-2584): 40%
Specific Gravity (ASTM D-792): 1.47
Mold Shrinkage (ASTM D-792): 0.012
Hardness (ASTM D-785): R121
Water Absorption, 24 hr. (ASTM D-570): 0.9%

N7-13FG-0100:
13% Glass Fiber Reinforced
High Impact

Reinforcement Content (ASTM D-2584): 13%
Specific Gravity (ASTM D-792): 1.14
Mold Shrinkage (ASTM D-955): 0.010
Hardness (ASTM D-785): R105
Water Absorption, 24 hr. (ASTM D-570): 1.0%

N7-30FG-0100:
30% Glass Fiber Reinforced
High Impact

Reinforcement Content (ASTM D-2584): 30%
Specific Gravity (ASTM D-792): 1.32
Mold Shrinkage (ASTM D-955): 0.004 in/in
Hardness (ASTM D-785): R114
Water Absorption, 24 hr. (ASTM D-570): 0.6%

THERMOFIL, INC.: THERMOFIL Reinforced Nylon 6/6 Compounds (Continued):

TYPICAL PROPERTY DATA:

Nylon 6/6 (High Impact):

N7-20NF-0100:
20% Graphite Fiber Reinforced
Stiffness

Reinforcement Content (ASTM D-2584): 20%
Specific Gravity (ASTM D-792): 1.20
Mold Shrinkage (ASTM D-955): 0.001
Hardness (ASTM D-785): R113
Water Absorption, 24 hr. (ASTM D-570): 1.1%

N7-35MF-7365:
35% Mineral Reinforced
Heat Stabilized

Reinforcement Content (ASTM D-2584): 35%
Specific Gravity (ASTM D-792): 1.38
Mold Shrinkage (ASTM D-955): 0.010
Hardness (ASTM D-785): R120
Water Absorption, 24 hr. (ASTM D-570): 1.0%

N7-9900-0500:
Flame Retardant
F.R. Impact Modified

Specific Gravity (ASTM D-792): 1.35
Mold Shrinkage (ASTM D-792): 0.015
Hardness (ASTM D-785): R112
Water Absorption, 24 hr. (ASTM D-570): 1.2%

N7-20FG-0500:
20% Glass Fiber Reinforced
Flame Retardant

Reinforcement Content (ASTM D-2584): 20%
Specific Gravity (ASTM D-792): 1.53
Mold Shrinkage (ASTM D-792): 0.003
Hardness (ASTM D-785): R112
Water Absorption, 24 hr. (ASTM D-570): 0.9%

THERMOFIL, INC.: THERMOFIL Reinforced Nylon 6/6 Compounds (Continued):

Dimensionally Stable Nylon Alloy:

Most low moisture absorption Nylon composites are relatively expensive, reflecting the high price of the base Nylons utilized. To combat this cost, THERMOFIL has developed a Glass Fiber Reinforced Nylon Alloy that exhibits substantially better dimensional control than Nylon 6 or Nylon 6/6 after humid aging.

This product retains structural characteristics similar to conventional Glass Fiber Reinforced Nylon 6/6. The price of THERMOFIL Glass Fiber Reinforced Nylon Alloy is significantly lower than that of the traditional low moisture absorption Nylons.

TYPICAL PROPERTY DATA:

N5-15FG-0100:
15% Glass Fiber Reinforced
Dimension Stability

Reinforcement Content (ASTM D-2584): 15%
Specific Gravity (ASTM D-792): 1.22
Mold Shrinkage (ASTM D-955): 0.004 in/in
Hardness (ASTM D-785): R118
Water Absorption, 24 hr. (ASTM D-570): 0.4%

N5-30FG-0100:
30% Glass Fiber Reinforced
Dimension Stability

Reinforcement Content (ASTM D-2584): 30%
Specific Gravity (ASTM D-792): 1.35
Mold Shrinkage (ASTM D-955): 0.003 in/in
Hardness (ASTM D-785): M100
Water Absorption, 24 hr. (ASTM D-570): 0.5%

THERMOFIL, INC.: THERMOFIL Reinforced Nylon 6/9 Compounds:

TYPICAL PROPERTY DATA:

N12-30FG-0100:
30% Glass Fiber Reinforced
Dimension Stability

Reinforcement Content (ASTM D-2584): 30%
Specific Gravity (ASTM D-792): 1.30
Mold Shrinkage (ASTM D-955): 0.003 in/in
Hardness (ASTM D-785): M95
Water Absorption, 24 hr. (ASTM D-570): 0.2%

N12-30NF-0100:
30% Graphite Fiber Reinforced
Dimension Stability

Reinforcement Content (ASTM D-2584): 30%
Specific Gravity (ASTM D-792): 1.22
Mold Shrinkage (ASTM D-955): 0.003 in/in
Hardness (ASTM D-785): M95
Water Absorption, 24 hr. (ASTM D-570): 0.2%

N12-30FG-0282:
30% Glass Fiber Reinforced
Lubricated

Reinforcement Content (ASTM D-2584): 30%
Specific Gravity (ASTM D792): 1.42
Mold Shrinkage (ASTM D-955): 0.003 in/in
Hardness (ASTM D-785): M93
Water Absorption, 24 hr. (ASTM D-570): 0.09%

N12-30FG-0500:
30% Glass Fiber Reinforced
Flame Retardant

Reinforcement Content (ASTM D-2584): 30%
Specific Gravity (ASTM D-792): 1.57
Mold Shrinkage (ASTM D-955): 0.002 in/in
Hardness (ASTM D-785): M93
Water Absorption, 24 hr. (ASTM D-570): 0.15%

THERMOFIL, INC.: THERMOFIL Reinforced Nylon 6/12 Compounds:

TYPICAL PROPERTY DATA:

N6-30FG-0100:
30% Glass Fiber Reinforced
Dimension Stability

Reinforcement Content (ASTM D-2584): 30%
Specific Gravity (ASTM D-792): 1.30
Mold Shrinkage (ASTM D-955): 0.004 in/in
Hardness (ASTM D-785): M93
Water Absorption, 24 hr. (ASTM D-570): 0.2

N6-30NF-0100:
30% Graphite Fiber Reinforced
Dimension Stability

Reinforcement Content (ASTM D-2584): 30%
Specific Gravity (ASTM D-792): 1.22
Mold Shrinkage (ASTM D-955): 0.002 in/in
Hardness (ASTM D-785): M93
Water Absorption, 24 hr. (ASTM D-570): 0.17%

N6-30FG-0214:
30% Glass Fiber Reinforced
PTFE Lubricated

Reinforcement Content (ASTM D-2584): 30%
Specific Gravity (ASTM D-792): 1.42
Mold Shrinkage (ASTM D-955): 0.003 in/in
Hardness (ASTM D-785): R118
Water Absorption, 24 hr. (ASTM D-570): 0.2%

N6-30FG-0282:
30% Glass Fiber Reinforced
Lubricated

Reinforcement Content (ASTM D-2584): 30%
Specific Gravity (ASTM D-792): 1.41
Mold Shrinkage (ASTM D-955): 0.004
Hardness (ASTM D-785): R118
Water Absorption, 24 hr. (ASTM D-570): 0.11%

THERMOFIL, INC.: THERMOFIL Reinforced Nylon 6/12 Compounds (Continued):

TYPICAL PROPERTY DATA(Continued):

N6-9900-0500:
Flame Retardant

Specific Gravity (ASTM D-792): 1.32
Mold Shrinkage (ASTM D-955): 0.012 in/in
Hardness (ASTM D-785): R114
Water Absorption, 24 hr. (ASTM D-570): 0.8%

N6-30FG-0500:
30% Glass Fiber Reinforced
Flame Retardant

Reinforcement Content (ASTM D-2584): 30%
Specific Gravity (ASTM D-792): 1.51
Mold Shrinkage (ASTM D-955): 0.003 in/in
Hardness (ASTM D-785): R120
Water Absorption, 24 hr. (ASTM D-570): 0.6%

THERMOFIL, INC.: THERMOFIL Reinforced Nylon 11/12 Compounds:

TYPICAL PROPERTY DATA:

Nylon 11:

N8-30FG-0100:
30% Glass Fiber Reinforced
Low H2O Absorption

Reinforcement Content (ASTM D-2584): 30%
Specific Gravity (ASTM D-792): 1.28
Mold Shrinkage (ASTM D-955): 0.003 in/in
Hardness (ASTM D-785): R107
Water Absorption, 24 hr. (ASTM D-570): 0.17%

Nylon 12:

N9-30FG-0100:
30% Glass Fiber Reinforced
Low H2O Absorption

Reinforcement Content (ASTM D-2584): 30%
Specific Gravity (ASTM D-792): 1.22
Mold Shrinkage (ASTM D-955): 0.004 in/in
Hardness (ASTM D-785): R110
Water Absorption, 24 hr. (ASTM D-570): 0.6%

THERMOFIL, INC.: THERMOFIL PEEK:

TYPICAL PROPERTY DATA:

K2-30FG-0100:
30% Glass Fiber Reinforced
Stability

Reinforcement Content (ASTM D-2584): 47%
Specific Gravity (ASTM D-792): 1.70
Mold Shrinkage (ASTM D-955): 0.001 in/in
Hardness (ASTM D-785): R124
Water Absorption, 24 hr. (ASTM D-570): 0.01%

K2-30NF-0100:
30% Graphite Fiber Reinforced
Conductive

Reinforcement Content (ASTM D-2584): 30%
Specific Gravity (ASTM D-792): 1.44
Mold Shrinkage (ASTM D-955): 0.001 in/in
Hardness (ASTM D-785): R123
Water Absorption, 24 hr. (ASTM D-570): 0.1

THERMOFIL, INC.: THERMOFIL Reinforced Polyetherimide:

THERMOFIL Reinforced Polyetherimide compounds combine good Elevated Temperature Structural properties with wide processing latitude.

The Elevated Temperature characteristics of THERMOFIL Reinforced Polyetherimide include:

* High flexural modulus
* High hydrolytic stability

Outstanding processability is provided by:

* Low melt viscosity
* High melt stability
* Wide processing latitude

THERMOFIL Reinforced Polyetherimide compounds display:

* High dielectric strength
* Low dissipation factor

this combined with:

* High oxygen index and
* Low smoke generation

produces compounds that are excellent candidates for electrical and electonics applications.

POLYETHERIMIDE:

TYPICAL PROPERTY DATA:

W-10FG-0100:
10% Glass Fiber Reinforced
Heat Resistance

Reinforcement Content (ASTM D-2584): 10%
Specific Gravity (ASTM D-792): 1.34
Mold Shrinkage (ASTM D-955): 0.004 in/in
Hardness (ASTM D-785): M115
Water Absorption, 24 hr. (ASTM D570): 0.3%

THERMOFIL, INC.: THERMOFIL Reinforced Polyetherimide(Continued):

POLYETHERIMIDE(Continued):

TYPICAL PROPERTY DATA:

W-30FG-0100:
30% Glass Fiber Reinforced
Heat Resistance

Reinforcement Content (ASTM D-2584): 30%
Specific Gravity (ASTM D-792): 1.51
Mold Shrinkage (ASTM D-955): 0.002 in/in
Hardness (ASTM D-785): M125
Water Absorption, 24 hr. (ASTM D570): 0.2%

W-30NF-0100:
30% Graphite Fiber Reinforced
Stiffness

Reinforcement Content (ASTM D-2584): 30%
Specific Gravity (ASTM D-792): 1.42
Mold Shrinkage (ASTM D-955): 0.002 in/in
Hardness (ASTM D-785): M127
Water Absorption, 24 hr. (ASTM D570): 0.2%

W-30NF-0214:
30% Graphite Fiber Reinforced
PTFE Lubricated

Reinforcement Content (ASTM D-2584): 30%
Specific Gravity (ASTM D-792): 1.50
Mold Shrinkage (ASTM D-955): 0.001 in/in
Hardness (ASTM D-785): M125
Water Absorption, 24 hr. (ASTM D570): 0.17%

THERMOFIL, INC.: THERMOFIL Reinforced Polyethersulfone:

POLYETHERSULFONE:

TYPICAL PROPERTY DATA:

K-20FG-0100:
20% Glass Fiber Reinforced
Heat Resistance

Reinforcement Content (ASTM D-2584): 20%
Specific Gravity (ASTM D-792): 1.51
Mold Shrinkage (ASTM D-955): 0.003 in/in
Hardness (ASTM D-785): R122
Water Absorption, 24 hr. (ASTM D570): 0.38%

K-30FG-0100:
30% Glass Fiber Reinforced
Heat Resistance

Reinforcement Content (ASTM D-2584): 30%
Specific Gravity (ASTM D-792): 1.60
Mold Shrinkage (ASTM D-955): 0.002 in/in
Hardness (ASTM D-785): R123
Water Absorption, 24 hr. (ASTM D570): 0.35%

K-30NF-0100:
30% Graphite Fiber Reinforced
Stiffness

Reinforcement Content (ASTM D-2584): 30%
Specific Gravity (ASTM D-792): 1.48
Mold Shrinkage (ASTM D-955): 0.002 in/in
Hardness (ASTM D-785): R123
Water Absorption, 24 hr. (ASTM D570): 0.29%

K-30FG-0214:
30% Glass Fiber Reinforced
PTFE Lubricated

Reinforcement Content (ASTM D-2584): 30%
Specific Gravity (ASTM D-792): 1.69
Mold Shrinkage (ASTM D-955): 0.001 in/in
Hardness (ASTM D-785): M97
Water Absorption, 24 hr. (ASTM D570): 0.3%

THERMOFIL, INC.: THERMOFIL Reinforced Polyethylene:

THERMOFIL Reinforced Polyethylene compounds extend the usefulness of High Density Polyethylene into structural applications. Glass Fiber Reinforced Polyethylene exhibits:

* Improved stiffness
* Increased strength
* Greater resistance to long term stress

THERMOFIL Glass Fiber Reinforced Polyethylene compounds exhibit greatly improved elevated temperature load bearing capacity. In addition, these compounds provide a reasonably priced choice for load bearing applications requiring:

* Low temperature toughness
* Excellent chemical resistance
* Good electrical properties

THERMOFIL Reinforced Polyethylene compounds utilizing Mineral Reinforcements supplement the inherent qualities of Polyethylene with improved dimensional control through:

* Reduced mold shrinkage
* Reduced coefficient of expansion
* Greater crystalline uniformity

THERMOFIL Reinforced Polyethylene compounds based on Stress Crack Resistant grades and Impact Copolymers are available to meet specific application requirements.

POLYETHYLENE:

TYPICAL PROPERTY DATA:

F-10FG-0100:
10% Glass Fiber Reinforced
Impact Resistance

Reinforcement Content (ASTM D-2584): 10%
Specific Gravity (ASTM D-792): 1.04
Mold Shrinkage (ASTM D-955): 0.005 in/in
Hardness (ASTM D-785): R57
Water Absorption, 24 hr. (ASTM D570): 0.08%

F-30FG-0100:
30% Glass Fiber Reinforced
Chemical Resistance

Reinforcement Content (ASTM D-2584): 30%
Specific Gravity (ASTM D-792): 1.18
Mold Shrinkage (ASTM D-955): 0.003 in/in
Hardness (ASTM D-785): R75
Water Absorption, 24 hr. (ASTM D570): 0.06%

THERMOFIL, INC.: THERMOFIL Polyphenylene Oxide (Modified):

TYPICAL PROPERTY DATA:

L-20FG-0100:
20% Glass Fiber Reinforced
Heat Resistance

Reinforcement Content (ASTM D-2584): 20%
Specific Gravity (ASTM D-792): 1.21
Mold Shrinkage (ASTM D-955): 0.003 in/in
Hardness (ASTM D-785): R107
Water Absorption, 24 hr. (ASTM D570): 0.06%

L-30FG-0100:
30% Glass Fiber Reinforced
Stiffness

Reinforcement Content (ASTM D-2584): 30%
Specific Gravity (ASTM D-792): 1.28
Mold Shrinkage (ASTM D-955): 0.002 in/in
Hardness (ASTM D-785): M93
Water Absorption, 24 hr. (ASTM D570): 0.06%

L1-30FG-0580:
30% Glass Fiber Reinforced
Flame Retardant

Reinforcement Content (ASTM D-2584): 30%
Specific Gravity (ASTM D-792): 1.30
Mold Shrinkage (ASTM D-955): 0.001 in/in
Hardness (ASTM D-785): R116
Water Absorption, 24 hr. (ASTM D-570): 0.05%

THERMOFIL, INC.: THERMOFIL Reinforced Polyphenylene Sulfide:

THERMOFIL Reinforced Polyphenylene Sulfide compounds are noted for their excellent temperature and chemical resistance. These products exhibit:

* High thermal stability
* Excellent elevated temperature chemical resistance
* Superior structural properties at elevated temperatures

The maximum Continuous Use Temperature of THERMOFIL Reinforced Polyphenylene Sulfide compounds ranges from 450F to 500F (230C to 260C). The high Tensile Strength of THERMOFIL reinforced Polyphenylene Sulfide compounds decreases with increasing temperature up to 250F. From 250F to 500F the tensile strength remains relatively unaffected.

THERMOFIL Reinforced Polphenylene Sulfide compounds are inherently flame retardant exhibiting:

* Low flame spread
* Low smoke generation

THERMOFIL Reinforced Polyphenylene Sulfide compounds exhibit excellent performance in demanding electrical applications. These compounds provide:

* Elevated temperature resistance
* Flame retardancy
* Electrical insulating properties

THERMOLUBE Lubricated Reinforced Polyphenylene Sulfide compounds expand the use of these compounds into applications demanding low Friction and Wear Rate characteristics. The compounds exhibit a unique combination of:

* Chemical Resistance
* Strength
* Rigidity
* Wear qualities

Seals, valves and bushings in chemical conveying equipment utilize the characteristics of THERMOLUBE Lubricated Reinforced Polyphenylene Sulfide compounds.

Through the selection of specific reinforcements and combinations of reinforcements, the property characteristics of THERMOFIL Reinforced Polyphenylene Sulfide Compounds can be tailored to provide such benefits as:

* Improved dimensional stability
* Reduced product cost
* Increased electrical properties

THERMOFIL, INC.: THERMOFIL Reinforced Polyphenylene Sulfide (Continued):

THERMOFIL Reinforced Polyphenylene Sulfide compounds with Microwave Transparency and Elevated Temperature Resistance are particularly well suited for Microwave Cookware applications.

POLYPHENYLENE SULFIDE:

TYPICAL PROPERTY DATA:

T-40FG-0100:
40% Glass Fiber Reinforced
Stiffness

Reinforcement Content (ASTM D-2584): 40%
Specific Gravity (ASTM D-792): 1.64
Mold Shrinkage (ASTM D-955): 0.002 in/in
Hardness (ASTM D-785): R123
Water Absorption, 24 hr. (ASTM D-570): 0.01

T-20NF-0100:
20% Graphite Fiber Reinforced
Strength

Reinforcement Content (ASTM D-2584): 20%
Specific Gravity (ASTM D-792): 1.38
Mold Shrinkage (ASTM D-955): 0.0015 in/in
Hardness (ASTM D-785): R122
Water Absorption, 24 hr. (ASTM D-570): 0.02%

T-40NF-0100:
40% Graphite Fiber Reinforced
Stiffness

Reinforcement Content (ASTM D-2584): 40%
Specific Gravity (ASTM D-792): 1.46
Mold Shrinkage (ASTM D-955): 0.0005 in/in
Hardness (ASTM D-785): R123
Water Absorption, 24 hr. (ASTM D-5790): 0.02%

T-50FM-0100:
50% Glass Fiber/Mineral Reinforced
Economy

Reinforcement Content (ASTM D-2584): 50%
Specific Gravity (ASTM D-792): 1.67
Mold Shrinkage (ASTM D-955): 0.002 in/in
Hardness (ASTM D-785): R121
Water Absorption, 24 hr. (ASTM D-570): 0.03%

THERMOFIL, INC.: THERMOFIL Reinforced Polyphenylene Sulfide (Continued):

POLYPHENYLENE SULFIDE(Continued):

TYPICAL PROPERTY DATA:

T-50CG-0940:
50% Pitch Carbon Fiber/Glass Fiber Reinforced
Heat Resistance

Reinforcement Content (ASTM D-2584): 50%
Specific Gravity (ASTM D-792): 1.65
Mold Shrinkage (ASTM D-955): 0.001 in/in
Hardness (ASTM D-785): R121
Water Absorption, 24 hr. (ASTM D-570): 0.02%

T-40NG-0948:
40% Graphite Fiber/Glass Fiber Reinforced
Conductive

Reinforcement Content (ASTM D-2584): 40%
Specific Gravity (ASTM D-792): 1.57
Mold Shrinkage (ASTM D-955): 0.001 in/in
Hardness (ASTM D-785): R123
Water Absorption, 24 hr. (ASTM D-570): 0.02%

T-30FG-0214:
30% Glass Fiber Reinforced
PTFE Lubricated

Reinforcement Content (ASTM D-2584): 30%
Specific Gravity (ASTM D-792): 1.69
Mold Shrinkage (ASTM D-955): 0.002 in/in
Hardness (ASTM D-785): R120
Water Absorption, 24 hr. (ASTM D-570): 0.02%

T-30NF-0214:
30% Graphite Fiber Reinforced
Wear Resistance

Reinforcement Content (ASTM D-2584): 30%
Specific Gravity (ASTM D-792): 1.51
Mold Shrinkage (ASTM D-955): 0.001 in/in
Hardness (ASTM D-785): R123
Water Absorption, 24 hr. (ASTM D-570): 0.02%

THERMOFIL, INC.: THERMOFIL Reinforced Polypropylene:

THERMOFIL Reinforced Polypropylene compounds are noted for their versatility. Through prudent selection of Reinforcements and Property Modifiers, THERMOFIL Reinforced Polypropylene compounds provide:

* Structural Capacity
* Elevated Temperature Performance
* Impact Resistance
* Environmental Stability

The relatively low cost of THERMOFIL Reinforced Polypropylene compounds combined with Polypropylene's inherently low specific gravity result in:

* High Product Performance
* Low Cost Per Cubic Inch

The improved elevated temperature performance of THERMOFIL Reinforced Polypropylene compounds coupled with the inherent Chemical Resistance of these products is responsible for their continued popularity in automotive and pump applications.

THERMO-X Flame Retardant Reinforced Polypropylene compounds add another dimension to the usefulness of the compounds in Electrical and Electronics applications.

THERMOFIL Mineral Reinforced Polypropylene compounds employ naturally occurring reinforcements such as Mica, Talc, and Calcium Carbonate in specifically selected polymer formulations to provide reasonably priced composites with:

* Improved elevated temperature rigidity
* Increased dimensional control

The versatility of THERMOFIL Reinforced Polypropylene compounds is further enhanced by the incorporation of:

* THERMOFOAM - Structured Foam Concentrates
* THERMOLUBE - Friction and Wear Reducers
* Impact Modifiers
* U/V and Heat Stabilizers

THERMOFIL, INC.: THERMOFIL Reinforced Polypropylene(Continued):

POLYPROPYLENE(Continued):

TYPICAL PROPERTY DATA:

P-10FG-0100:
10% Glass Fiber Reinforced
Chemical Resistance

Reinforcement Content (ASTM D-2584): 10%
Specific Gravity (ASTM D-792): 0.98
Mold Shrinkage (ASTM D-955): 0.007 in/in
Hardness (ASTM D-785): R100
Water Absorption (ASTM D570): 0.05%

P-20FG-0100:
20% Glass Fiber Reinforced
Strength

Reinforcement Content (ASTM D-2584): 20%
Specific Gravity (ASTM D-792): 1.04
Mold Shrinkage (ASTM D-955): 0.006 in/in
Hardness (ASTM D-785): R103
Water Absorption (ASTM D570): 0.05%

P-20FG-0720:
20% Glass Fiber Reinforced
Impact Resistance

Reinforcement Content (ASTM D-2584): 20%
Specific Gravity (ASTM D-792): 1.04
Mold Shrinkage (ASTM D-955): 0.006 in/in
Hardness (ASTM D-785): R85
Water Absorption (ASTM D570): 0.01%

P-30FG-0100:
30% Glass Fiber Reinforced
Resistance

Reinforcement Content (ASTM D-2584): 30%
Specific Gravity (ASTM D-792): 1.12
Mold Shrinkage (ASTM D-955): 0.006 in/in
Hardness (ASTM D-785): R103
Water Absorption (ASTM D570): 0.04%

THERMOFIL, INC.: THERMOFIL Reinforced Polypropylene(Continued):

POLYPROPYLENE(Continued):

TYPICAL PROPERTY DATA:

P-20MI-0100:
20% Mica Reinforced
Temperature Stiffness

Reinforcement Content (ASTM D-2584): 20%
Specific Gravity (ASTM D-792): 1.05
Mold Shrinkage (ASTM D-955): 0.014 in/in
Hardness (ASTM D-785): R95
Water Absorption, 24 hr. (ASTM D-570): 0.03%

P-40MI-0100:
40% Mica Reinforced
Dimensional Stability

Reinforcement Content (ASTM D-2584): 40%
Specific Gravity (ASTM D-792): 1.25
Mold Shrinkage (ASTM D-955): 0.011 in/in
Hardness (ASTM D-785): R105
Water Absorption, 24 hr. (ASTM D-570): 0.04%

P-40TC-0100:
40% Talc Filled
Stiffness

Reinforcement Content (ASTM D-2584): 40%
Specific Gravity (ASTM D-792): 1.24
Mold Shrinkage (ASTM D-955): 0.013 in/in
Hardness (ASTM D-785): R100
Water Absorption, 24 hr. (ASTM D-570): 0.03%

P-40CC-0100:
40% Calcium Carbonate Filled
Impact Resistance

Reinforcement Content (ASTM D-2584): 40%
Specific Gravity (ASTM D-792): 1.25
Mold Shrinkage (ASTM D-955): 0.012 in/in
Hardness (ASTM D-785): R78
Water Absorption, 24 hr. (ASTM D-570): 0.05%

THERMOFIL, INC.: THERMOFIL Reinforced Polypropylene(Continued):

POLYPROPYLENE(Continued):

TYPICAL PROPERTY DATA:

P-40CC-0720:
40% Calcium Carbonate Filled
Impact Resistance

Reinforcement Content(ASTM D-2584): 40%
Specific Gravity (ASTM D-792): 1.25
Mold Shrinkage (ASTM D-955): 0.013 in/in
Hardness (ASTM D-785): R83
Water Absorption, 24 hr. (ASTM D-570): 0.1%

P-40MI-0720:
40% Mica Reinforced
Impact Resistance

Reinforcement Content (ASTM D-2584): 40%
Specific Gravity (ASTM D-792): 1.23
Mold Shrinkage (ASTM D-955): 0.007 in/in
Hardness (ASTM D-785): R85
Water Absorption, 24 hr. (ASTM D-570): 0.02%

P-9900-0500:
Flame Retardant
Economy

Specific Gravity (ASTM D-792): 1.25
Mold Shrinkage (ASTM D-955): 0.015 in/in
Hardness (ASTM D-785): R93
Water Absorption, 24 hr. (ASTM D570): 0.04%

P-9900-0570:
Flame Retardant
Impact Resistant

Specific Gravity (ASTM D-792): 1.25
Mold Shrinkage (ASTM D-955): 0.016 in/in
Hardness (ASTM D-785): R90
Water Absorption, 24 hr. (ASTM D570): 0.03%

THERMOFIL, INC.: THERMOFIL Reinforced Polypropylene(Continued):

POLYPROPYLENE(Continued):

TYPICAL PROPERTY DATA:

P-20FG-0500:
20% Glass Fiber Reinforced
Flame Retardant

Reinforcement Content (ASTM D-2584): 20%
Specific Gravity (ASTM D-792): 1.35
Mold Shrinkage (ASTM D-955): 0.010 in/in
Hardness (ASTM D-785): R103
Water Absorption, 24 hr. (ASTM D-570): 0.03%

P-20MF-0500:
20% Mineral Filled
Flame Retardant

Reinforcement Content (ASTM D-2584): 20%
Specific Gravity (ASTM D-792): 1.39
Mold Shrinkage (ASTM D-955): 0.10 in/in
Hardness (ASTM D-785): R102
Water Absorption, 24 hr. (ASTM D-570): 0.03%

THERMOFIL, INC.: THERMOFIL Reinforced Polypropylene(Continued):

POLYPROPYLENE(Chemically Coupled):

TYPICAL PROPERTY DATA:

P1-10FG-0100:
10% Glass Fiber Reinforced
Chemical Resistance

Reinforcement Content (ASTM D-2584): 10%
Specific Gravity (ASTM D-792): 0.98
Mold Shrinkage (ASTM D-955): 0.007 in/in
Hardness (ASTM D-785): R100
Water Absorption, 24 hr. (ASTM D570): 0.05%

P1-20FG-0100:
20% Glass Fiber Reinforced
Heat Resistance

Reinforcement Content (ASTM D-2584): 20%
Specific Gravity (ASTM D-792): 1.04
Mold Shrinkage (ASTM D-955): 0.006 in/in
Hardness (ASTM D-785): R103
Water Absorption, 24 hr. (ASTM D570): 0.04%

P1-30FG-0100:
30% Glass Fiber Reinforced
Strength

Reinforcement Content (ASTM D-2584): 30%
Specific Gravity (ASTM D-792): 1.12
Mold Shrinkage (ASTM D-955): 0.006 in/in
Hardness (ASTM D-785): R105
Water Absorption, 24 hr. (ASTM D570): 0.04%

P1-20FG-0720:
20% Glass Fiber Reinforced
Impact Resistance

Reinforcement Content (ASTM D-2584): 20%
Specific Gravity (ASTM D-792): 1.04
Mold Shrinkage (ASTM D-955): 0.003 in/in
Hardness (ASTM D-785): R102
Water Absorption, 24 hr. (ASTM D570): 0.04%

THERMOFIL, INC.: THERMOFIL Reinforced Polypropylene(Continued):

POLYPROPYLENE (Highly Chemically Coupled):

TYPICAL PROPERTY DATA:

P6-10FG-0100:
10% Glass Fiber Reinforced
Heat Resistance

Reinforcement Content (ASTM D-2584): 10%
Specific Gravity (ASTM D-792): 0.98
Mold Shrinkage (ASTM D-955): 0.006 in/in
Hardness (ASTM D-785): R100
Water Absorption, 24 hr. (ASTM D570): 0.05%

P6-20FG-0100:
20% Glass Fiber Reinforced
Strength

Reinforcement Content (ASTM D-2584): 20%
Specific Gravity (ASTM D-792): 1.04
Mold Shrinkage (ASTM D-955): 0.006 in/in
Hardness (ASTM D-785): R104
Water Absorption, 24 hr. (ASTM D570): 0.04%

P6-30FG-0100:
30% Glass Fiber Reinforced
Creep Resistance

Reinforcement Content (ASTM D-2584): 30%
Specific Gravity (ASTM D-792): 1.12
Mold Shrinkage (ASTM D-955): 0.006 in/in
Hardness (ASTM D-785): R106
Water Absorption, 24 hr. (ASTM D-570): 0.04%

P6-30FG-0720:
30% Glass Fiber Reinforced
Impact Resistance

Reinforcement Content (ASTM D-2584): 30%
Specific Gravity (ASTM D-792): 1.13
Mold Shrinkage (ASTM D-955): 0.002 in/in
Hardness (ASTM D-785): R102
Water Absorption, 24 hr. (ASTM D-570): 0.03%

THERMOFIL, INC.: THERMOFIL Reinforced Polypropylene(Continued):

POLYPROPYLENE (Highly Chemically Coupled) (Continued):

TYPICAL PROPERTY DATA:

P6-20FG-0500:
20% Glass Fiber Reinforced
Flame Retardant

Reinforcement Content (ASTM D-2584): 20%
Specific Gravity (ASTM D-792): 1.39
Mold Shrinkage (ASTM D-955): 0.002 in/in
Hardness (ASTM D-785): R104
Water Absorption, 24 hr. (ASTM D570): 0.04%

P6-20FG-0570:
20% Glass Fiber Reinforced
Impact F.R.

Reinforcement Content (ASTM D-2584): 20%
Specific Gravity (ASTM D-792): 1.39
Mold Shrinkage (ASTM D-955): 0.005 in/in
Hardness (ASTM D-785): R110
Water Absorption, 24 hr. (ASTM D570): 0.03%

THERMOFIL, INC.: THERMOFIL Reinforced Polysulfone:

THERMOFIL Reinforced Polysulfone compounds display excellent Elevated Temperature Resistance to Degradation.

These compounds retain Strength, Stiffness and Toughness after prolonged exposure to moisture at Elevated Temperatures. This characteristic, high temperature hydrolysis resistance, permits repeated autoclaving.

THERMOFIL Reinforced Polysulfone compounds are suitable for applications such as:

* Medical Apparatus
* Food Processing Equipment
* Microwave Ware
* Coffee Makers

THERMOFIL Reinforced Polysulfone compounds are formulated to provide:

* Flame Retardancy
* Elastic Memory
* Toughness

THERMOFIL Reinforced Polysulfone compounds display:

* Excellent Elevated Temperature Creep Resistance
* Superior Dimensional Stability
* High Continuous Use Temperature

The maximum Continuous Use Temperature of THERMOFIL Reinforced Polysulfone compounds ranges from 350F to 400F. Even at these Elevated Temperatures the compounds retain:

* High Stiffness
* Resistance to Degradation

The Thermal Stability of THERMOFIL Reinforced Polysulfone compounds are maintained at service temperatures in excess of 300F. These characteristics are substantially retained in the presence of moisture.

THERMOFIL, INC.: THERMOFIL Reinforced Polysulfone(Continued):

POLYSULFONE:

TYPICAL PROPERTY DATA:

S-20FG-0100:
20% Glass Fiber Reinforced
Heat Resistant

Reinforcement Content (ASTM D-2584): 20%
Specific Gravity (ASTM D-792): 1.38
Mold Shrinkage (ASTM D-955): 0.004 in/in
Hardness (ASTM D-785): M92
Water Absorption, 24 hr. (ASTM D-570): 0.2%

S-40FG-0100:
40% Glass Fiber Reinforced
Stiffness

Reinforcement Content (ASTM D-2584): 40%
Specific Gravity (ASTM D-792): 1.55
Mold Shrinkage (ASTM D-955): 0.002 in/in
Hardness (ASTM D-785): M92
Water Absorption, 24 hr. (ASTM D-570): 0.18%

S-30NF-0100:
30% Graphite Fiber Reinforced
Stiffness

Reinforcement Content (ASTM D-2584): 30%
Specific Gravity (ASTM D-792): 1.37
Mold Shrinkage (ASTM D-955): 0.001 in/in
Hardness (ASTM D-785): R123
Water Absorption, 24 hr. (ASTM D-570): 0.15%

S-30MF-0360:
30% Mineral Reinforced
FDA

Reinforcement Content (ASTM D-2584): 30%
Specific Gravity (ASTM D-792): 1.48
Mold Shrinkage (ASTM D-955): 0.005 in/in
Hardness (ASTM D-785): M70
Water Absorption, 24 hr. (ASTM D-570): 0.70%

THERMOFIL, INC.: THERMOFIL Reinforced Polysulfone(Continued):

POLYSULFONE(Continued):

TYPICAL PROPERTY DATA:

S-30FG-0214:
30% Glass Fiber Reinforced
PTFE

Reinforcement Content (ASTM D-2584): 30%
Specific Gravity (ASTM D-792): 1.59
Mold Shrinkage (ASTM D-955): 0.001 in/in
Hardness (ASTM D-785): R123
Water Absorption, 24 hr. (ASTM D-570): 0.10%

THERMOFIL, INC.: THERMOFIL Reinforced Polyurethanes:

THERMOFIL Reinforced Polyurethanes are noted for their superior wear and abrasion resistance. Through the utilization of specifically selected reinforcements and lubricants, THERMOFIL Reinforced Polyurethanes provide a wide range of load bearing and frictional characteristics.

The inherent vibration dampening qualities of Thermoplastic Polyurethanes can be modified through the use of reinforcements to provide the required deadening properties.

POLYURETHANE:

TYPICAL PROPERTY DATA:

U-10FG-0100:
10% Glass Fiber Reinforcement
Toughness

Reinforcement Content (ASTM D-2584): 10%
Specific Gravity (ASTM D-792): 1.22
Mold Shrinkage (ASTM D-955): 0.007 in/in
Hardness (ASTM D-785): D55
Water Absorption, 24 hr. (ASTM D-570): 0.4%

THERMOFIL, INC.: THERMOFIL Reinforced Thermoplastic Polyesters:

THERMOFIL Reinforced Thermoplastic Polyesters are characterized by their:

* Outstanding Electrical Properties
* High Mechanical Properties
* Good Dimensional Stability

Most importantly, due to their low moisture absorption, THERMOFIL Reinforced Thermoplastic Polyester compounds maintain these characteristics in High Humidity and Moderate Temperature water environments. THERMOFIL Reinforced Thermoplastic compounds exhibit good Chemical Resistance at room temperature and at slightly elevated temperatures. They are resistant to aliphatic hydrocarbons, esters, ethers, oils and greases.

THERMOFIL offers Reinforced Thermoplastic Polyester compounds available in both

* PBT (polybutylene terephalate)
  &
* PET (polyethylene terephthalate)

THERMOFIL REINFORCED PBT

THERMOFIL Reinforced PBT compounds by virtue of their chemical composition, provide specific desirable characteristics which include:

* Improved warpage control
* Reduced moisture sensitivity

They maintain excellent structural, chemical and electrical properties. The controlled crystallinity of THERMOFIL Reinforced PBT compounds is responsible for less process dependent and more predictable mold shrinkage in the molded part. The PBT based compounds are not as sensitive to the degrading influences of high pre-molding moisture content as the PET based compounds.

POLYESTER (PBT):

TYPICAL PROPERTY DATA:

E-15FG-0100:
15% Glass Fiber Reinforced
Electrical Resistance

Reinforcement Content (ASTM D-2584): 15%
Specific Gravity (ASTM D-792): 1.41
Mold Shrinkage (ASTM D-955): 0.005 in/in
Hardness (ASTM D-785): R113
Water Absorption, 24 hours (ASTM D-570): 0.07%

THERMOFIL, INC.: THERMOFIL Reinforced Thermoplastic Polyesters (Continued):

POLYESTER (PBT) (Continued):

TYPICAL PROPERTY DATA:

E-30FG-0100:
30% Glass Fiber Reinforced
Electrical Resistance

Reinforcement Content (ASTM D-2584): 30%
Specific Gravity (ASTM D-792): 1.52
Mold Shrinkage (ASTM D-955): 0.003 in/in
Hardness (ASTM D-785): R115
Water Absorption, 24 hr. (ASTM D-570): 0.06%

E-40FG-0100:
40% Glass Fiber Reinforced
Stiffness

Reinforcement Content (ASTM D-2584): 40%
Specific Gravity (ASTM D-792): 1.62
Mold Shrinkage (ASTM D-955): 0.002 in/in
Hardness (ASTM D-785): R117
Water Absorption, 24 hr. (ASTM D-570): 0.05%

E-30NF-0100:
30% Graphite Fiber Reinforced
Stiffness

Reinforcement Content (ASTM D-2584): 30%
Specific Gravity (ASTM D-792): 1.41
Mold Shrinkage (ASTM D-955): 0.002 in/in
Hardness (ASTM D-785): R118
Water Absorption, 24 hr. (ASTM D-570): 0.05%

E-40MI-0100:
40% Mica Reinforced
Dimension Stability

Reinforcement Content (ASTM D-2584): 40%
Specific Gravity (ASTM D-792): 1.65
Mold Shrinkage (ASTM D-955): 0.010 in/in
Hardness (ASTM D-785): R118
Water Absorption, 24 hr. (ASTM D-570): 0.06%

THERMOFIL, INC.: THERMOFIL Reinforced Thermoplastic Polyesters (Continued):

POLYESTER (PBT) (Continued):

TYPICAL PROPERTY DATA:

E-45FM-0319:
45% Glass Fiber/Mineral Reinforced
Stiffness

Reinforcement Content (ASTM D-2584): 45%
Specific Gravity (ASTM D-792): 1.66
Mold Shrinkage (ASTM D-955): 0.003 in/in
Hardness (ASTM D-785): M75
Water Absorption, 24 hr. (ASTM D-570): 0.05%

E-45FM-0393:
45% Glass Fiber/Mica Reinforced
Dimension Stability

Reinforcement Content (ASTM D-2584): 45%
Specific Gravity (ASTM D-792): 1.62
Mold Shrinkage (ASTM D-955): 0.003 in/in
Hardness (ASTM D-785): M76
Water Absorption, 24 hr. (ASTM D-570): 0.05%

E-30-FG-0287:
Glass Fiber Reinforced
Lubricated

Reinforcement Content (ASTM D-2584): 30%
Specific Gravity (ASTM D-792): 1.67
Mold Shrinkage (ASTM D-955): 0.002 in/in
Hardness (ASTM D-785): R116
Water Absorption, 24 hr. (ASTM D-570): 0.04%

E-9900-0590:
Flame Retardant
F.R., Non-Blooming

Specific Gravity (ASTM D-792): 1.60
Mold Shrinkage (ASTM D-955): 0.020 in/in
Hardness (ASTM D-785): R120
Water Absorption, 24 hr. (ASTM D-570): 0.08%

THERMOFIL, INC.: THERMOFIL Reinforced Thermoplastic Polyesters (Continued):

POLYESTER (PBT) (Continued):

TYPICAL PROPERTY DATA:

E-15FG-0590:
15% Glass Fiber Reinforced
F.R., Non-Blooming

Reinforcement Content (ASTM D-2584): 15%
Specific Gravity (ASTM D-792): 1.59
Mold Shrinkage (ASTM D-955): 0.005 in/in
Hardness (ASTM D-785): R120
Water Absorption, 24 hr. (ASTM D-570): 0.07%

E-30FG-0590:
30% Glass Fiber Reinforced
F.R., Non-Blooming

Reinforcement Content (ASTM D-2584): 30%
Specific Gravity (ASTM D-792): 1.63
Mold Shrinkage (ASTM D-955): 0.003 in/in
Hardness (ASTM D-785): R120
Water Absorption, 24 hr. (ASTM D-570): 0.06%

THERMOFIL, INC.: THERMOFIL Reinforced Thermoplastic Polyesters (Continued):

POLYESTER (Continued):

ALLOY:

TYPICAL PROPERTY DATA:

E1-10FG-0100:
10% Glass Fiber Reinforced
Electrical Resistance

Reinforcement Content (ASTM D-2584): 10%
Specific Gravity (ASTM D-792): 1.30
Mold Shrinkage (ASTM D-955): 0.006 in/in
Hardness (ASTM D-785): R120
Water Absorption, 24 hr. (ASTM D-570): 0.12%

E1-30FG-0100:
30% Glass Fiber Reinforced
Electrical Resistance

Reinforcement Content (ASTM D-2584): 30%
Specific Gravity (ASTM D-792): 1.50
Mold Shrinkage (ASTM D-955): 0.003 in/in
Hardness (ASTM D-785): R121
Water Absorption, 24 hr. (ASTM D-570): 0.11%

E1-30FG-0584:
30% Glass Fiber Reinforced
F. R., Non-Blooming

Reinforcement Content (ASTM D-2584): 30%
Specific Gravity (ASTM D-792): 1.64
Mold Shrinkage (ASTM D-955): 0.003 in/in
Hardness (ASTM D-785): R117
Water Absorption, 24 hr. (ASTM D-570): 0.06%

THERMOFIL, INC.: THERMOFIL Reinforced Thermoplastic Polyesters (Continued):

THERMOFIL REINFORCED PET:

THERMOFIL Reinforced PET compounds offer:

* Elevated Temperature Properties
* Creep Resistance
* Strength and Stiffness

The Maximum Continuous Use Temperature of THERMOFIL Reinforced PET compounds exceeds that of their reinforced PBT counterparts by 15-30F.

In addition, Reinforced PET compounds employing glass fiber levels in excess of 50% by weight are available for applications demanding

* Extreme Stiffness
* Electrical Insulation
* Dimensional Stability

in moist or humid environments.

POLYESTER (PET):

TYPICAL PROPERTY DATA:

E2-30FG-0100:
30% Glass Fiber Reinforced
Dimension Stability

Reinforcement Content (ASTM D-2584): 30%
Specific Gravity (ASTM D-792): 1.56
Mold Shrinkage (ASTM D-955): 0.003 in/in
Hardness (ASTM D-785): R120
Water Absorption, 24 hr. (ASTM D-570): 0.05%

E2-55FG-0100:
55% Glass Fiber Reinforced
Stiffness

Reinforcement Content (ASTM D-2584): 55%
Specific Gravity (ASTM D-792): 1.80
Mold Shrinkage (ASTM D-955): 0.002 in/in
Hardness (ASTM D-785): R122
Water Absorption, 24 hr. (ASTM D570): 0.03%

THERMOFIL, INC.: THERMOFIL Reinforced Thermoplastic Polyesters (Continued):

POLYESTER (PET) (Continued):

TYPICAL PROPERTY DATA:

E2-30NF-0100:
30% Graphite Fiber Reinforced
Stiffness

Reinforcement Content (ASTM D-2584): 30%
Specific Gravity (ASTM D-792): 1.41
Mold Shrinkage (ASTM D-955): 0.002 in/in
Hardness (ASTM D-785): R121
Water Absorption, 24 hr. (ASTM D-570): 0.05%

E2-45FM-0393:
45% Glass Fiber/Mica Reinforced
Dimension Stability

Reinforcement Content (ASTM D-2584): 45%
Specific Gravity (ASTM D-792): 1.65
Mold Shrinkage (ASTM D-955): 0.002 in/in
Hardness (ASTM D-785): R118
Water Absorption, 24 hr. (ASTM D-570): 0.05%

E2-30FG-7100:
30% Glass Fiber Reinforced
Lubricated

Reinforcement Content (ASTM D-2584): 30%
Specific Gravity (ASTM D-792): 1.56
Mold Shrinkage (ASTM D-955): 0.003 in/in
Hardness (ASTM D-785): R120
Water Absorption, 24 hr. (ASTM D-570): 0.05%

E2-30FG-0578:
30% Glass Fiber Reinforced
F.R., Non-Blooming

Reinforcement Content (ASTM D-2584): 30%
Specific Gravity (ASTM D-792): 1.69
Mold Shrinkage (ASTM D-955): 0.002 in/in
Hardness (ASTM D-785): R115
Water Absorption, 24 hr. (ASTM D-570): 0.06%

U.S.I. Chemicals: PETROTHENE Low Density Polyethylene Resins:

PETROTHENE low molecular weight polyethylene (PE) homopolymers impart strength, gloss and flexibility to a variety of adhesive and coating formulations. They are compatible with a broad range of waxes and modifiers and have flow characteristics which are suitable for many end uses.

Resin:

NA 592:

Equivalent Melt Index g/10 min. (D1238): 1.8
Density g/cc (D1505-D792): 0.918
Tensile @ Break psi (D638 Type IV Specimen): 1,900
Elongation @ Break % (D638 Type IV Specimen): 570
1% Secant Modulus psi (D638 Type IV Specimen): 95
Hardness Shore A (D2240): 95
Ring & Ball Softening Point F (E28): 380
Meets Applicable FDA Regulations for Food Packaging: Yes

NA 593:

Equivalent Melt Index g/10 min. (D1238): 22
Density g/cc (D1505/D792): 0.915
Tensile @ Break % psi (D638 Type IV Specimen): 1,360
Elongation @ Break % (D638 Type IV Specimen): 520
1% Secant Modulus psi (D638 Type IV Specimen): 19,000
Hardness Shore A (D2240): 95
Ring & Ball Softening Point F (E28): 265
Meets Applicable FDA Regulations for Food Packaging: Yes

NA 594:

Equivalent Melt Index g/10 min. (D1238): 70
Density g/cc (D1505/D792): 0.914
Tensile @ Break psi (D638 Type IV Specimen): 1,270
Elongation @ Break % (D638 Type IV Specimen): 325
1% Secant Modulus psi (D638 Type IV Specimen): 16,600
Hardness Shore A (D2240): 95
Ring & Ball Softening Point F (E28) : 250
Meets Applicable FDA Regulations for Food Packaging: Yes

U.S.I. CHEMICALS: PETROTHENE Low Density Polyethylene Resins (Continued):

Resin:

NA 596:

Equivalent Melt Index g/10 min. (D1238): 150
Density g/cc (D1505/D792): 0.913
Tensile @ Break psi (D638 Type IV Specimen): 990
Elongation @ Break % (D638 Type IV Specimen): 200
1% Secant Modulus psi (D638 Type IV Specimen): 16,000
Hardness Shore A (D2240): 94
Ring & Ball Softening Point F (E28): 230
Meets Applicable FDA Regulations for Food Packaging: Yes

NA 597:

Equivalent Melt Index g/10 min. (D1238): 250
Density g/cc (D1505/D792): 0.927
Tensile @ Break psi (D638 Type IV Specimen): 1,520
Elongation @ Break % (D638 Type IV Specimen): 80
1% Secant Modulus psi (D638 Type IV Specimen): 35,000
Hardness Shore A (D2240): 95
Ring & Ball Softening Point F: 250
Meets Applicable FDA Regulations for Food Packaging: Yes

NA 598:

Equivalent Melt Index g/10 min. (D1238): 400
Density g/cc (D1505/D792): 0.9165
Tensile @ Break psi (D638 Type IV Specimen): 1,360
Elongation @ Break % (D638 Type IV Specimen): 95
1% Secant Modulus psi (D638 Type IV Specimen): 22,600
Hardness Shore A: 92
Ring & Ball Softening Point F: 230
Meets Applicable FDA Regulations for Food Packaging: Yes

NA 601-00/04:

Equivalent Melt Index g/10 min. (D1238): 2000
Density g/cc (D1505/D792): 0.903
Tensile @ Break psi (D638 Type IV Specimen): 850
Elongation @ Break % (D638 Type IV Specimen): 90
Ring & Ball Softening Point F (E28): 102
Meets Applicable FDA Regulations for Food Packaging: Yes

U.S.I. CHEMICALS: Polyolefin Resins and Semiconductive Compounds for Wire and Cable:

Product:

PETROTHENE Low Density Polyethylene Resins:

NA301-06:
Description: Medium molecular weight, non-staining, antioxidant
Applications: Primary insulation, coaxial and control cables
Color: Natural

NA310-05:
Description: High molecular weight, non-staining, antioxidant
Applications: Primary insulation, high voltage cables
Color: Natural

NA310-06:
Description: High molecular weight, non-staining, antioxidant
Applications: Primary insulation, high voltage cables
Color: Natural

PETROTHENE Low Density Crosslinkable Polyethylene Resins:

XL3700:
Description: Unfilled homopolymer
Applications: Primary insulation high voltage cables
Color: Natural

PETROTHENE Medium Density Polyethylene Resins:

KR92562:
Descripiton: Thermoplastic, abrasion and heat deformation resistant
Applications: Jacketing for telephone and power cables
Color: Black

U.S.I. CHEMICALS: Polyolefin Resins and Semiconductive Compounds for Wire and Cable(Continued):

Product Name:

PETROTHENE High Density Polyethylene Resins:

KR21017:
Description: High molecular weight
Applications: Spacer cable insulation, track-resistant
Color: Gray

KR91028:
Description: High molecular weight
Applications: Spacer cable insulation, track-resistant
Color: Black

KR92561:
Description: Thermoplastic, high molecular weight
Applications: Spacer cable insulation, cable jacketing
Color: Black

LR580:
Description: High molecular weight
Applications: Primary insulation, telephone singles
Color: Natural

LR586:
Description: High molecular weight
Applications: Foam skin and insulation, telephone singles
Color: Natural

PETROTHENE Semiconductive Polyolefin Compounds:

YR92126:
Description: Thermoplastic, deformation resistant
Applications: Power cable shielding
Color: Black

U.S.I. CHEMICALS: ULTRATHENE Ethylene-Vinyl Acetate Copolymers:

ULTRATHENE ethylene-vinyl acetate (EVA) copolymers are available in grades containing from 9 to 33 percent vinyl acetate by weight. They are divided into three classifications: low vinyl acetate content--up to 19 percent; intermediate--25 to 30 percent, and high vinyl acetate content--30 to 33 percent. All three types can be used for both hot melt adhesives and coatings.

Resin:

UE 637-04:

Vinyl Acetate %: 9
Equivalent Melt Index g/10 min. (D1238): 3.4
Density g/cc (D1505/D792): 0.93
Tensile @ Break psi (D638 Type IV Specimen): 2,000
Elongation @ Break psi (D638 Type IV Specimen): 700
1% Secant Modulus psi (D638 Type IV Specimen): 11,500
Hardness Shore A (D2240): 89
Ring & Ball Softening Point F (E28): 335
Meets Applicable FDA Regulations for Food Packaging: Yes

UE 612-04:

Vinyl Acetate %: 18
Equivalent Melt Index g/10 min. (D1238): 150
Density g/cc (D1505/D792): 0.94
Tensile @ Break psi (D638 Type IV Specimen): 760
Elongation @ Break % (D638 Type IV Specimen): 630
1% Secant Modulus psi (D638 Type IV Specimen): 2,300
Hardness Shore A (D2240): 85
Ring & Ball Softening Point F (E28): 200
Meets Applicable FDA Regulations for Food Packaging: Yes

UE 649-04:

Vinyl Acetate %: 18
Equivalent Melt Index g/10 min. (D1238): 530
Density g/cc (D1505/D792): 0.93
Tensile @ Break psi (D638 Type IV Specimen): 450
Elongation @ Break % (D638 Type IV Specimen): 200
1% Secant Modulus psi (D638 Type IV Specimen): 2,850
Hardness Shore A (D2240): 78
Ring & Ball Softening Point F: 190
Meets Applicable FDA Regulations for Food Packaging: Yes

U.S.I. CHEMICALS: ULTRATHENE Ethylene-Vinyl Acetate Copolymers (Continued):

Resin:

UE 631-04:

Vinyl Acetate: 19%
Equivalent Melt Index g/10 min. (D1238): 2.5
Density (g/cc) (D1505/D792): 0.94
Tensile @ Break psi (D638 Type IV Specimen): 2,150
Elongation @ Break %: 710
1% Secant Modulus psi (D638 Type IV Specimen): 5,300
Hardness Shore A (D2240): 87
Ring & Ball Softening Point F (E28): 345
Meets Applicable FDA Regulations for Food Packaging: Yes

UE 652-00:

Vinyl Acetate: 19%
Equivalent Melt Index g/10 min. (D1238): 32
Density g/cc (D1505/D792): 0.938
Tensile @ Break psi (D638 Type IV Specimen): 1,170
Elongation @ Break % (D638 Type IV Specimen): 680
1% Secant Modulus psi (D638 Type IV Specimen): 4,300
Hardness Shore A (D2240): 88
Meets Applicable FDA Regulations for Food Packaging: Yes

UE 659-04:

Vinyl Acetate: 25%
Equivalent Melt Index g/10 min. (D1238): 2
Density g/cc (D1505/D792): 0.95
Tensile @ Break psi (D638 Type IV Specimen): 2,630
Elongation @ Break % (D638 Type IV Specimen): 770
1% Secant Modulus psi (D638 Type IV Specimen): 3,700
Hardness Shore A (D2240): 83
Ring & Ball Softening Point F (E28): 350
Meets Applicable FDA Regulations for Food Packaging: Yes

UE 644-04:

Vinyl Acetate: 28%
Equivalent Melt Index g/10 min. (D1238): 1.9
Density g/cc (D1505/D792): 0.951
Tensile @ Break psi (D638 Type IV Specimen): 2,820
Elongation @ Break % (D638 Type IV Specimen): 680
1% Secant Modulus psi (D638 Type IV Specimen): 2,800
Hardness Shore A (D2240): 85
Ring & Ball Softening Point F (E28): 331
Meets Applicable FDA Regulations for Food Packaging: Yes

U.S.I. CHEMICALS: ULTRATHENE Ethylene-Vinyl Acetate Copolymers (Continued):

Resin:

UE 645-04:

Vinyl Acetate: 28%
Equivalent Melt Index g/10 min. (D1238): 3
Density g/cc (D1505/D792): 0.95
Tensile @ Break psi (D638 Type IV Specimen): 2,800
Elongation @ Break % (D638 Type IV Specimen): 800
1% Secant Modulus psi (D638 Type IV Specimen): 2,500
Hardness Shore A (D2240): 76
Ring & Ball Softening Point F (E28): 330
Meets Applicable FDA Regulations for Food Packaging: Yes

UE 634-04:

Vinyl Acetate: 28%
Equivalent Melt Index g/10 min. (D1238): 6
Density g/cc (D1505/D792): 0.95
Tensile @ Break psi (D638 Type IV Specimen): 2,700
Elongation @ Break % (D638 Type IV Specimen): 770
1% Secant Modulus psi (D638 Type IV Specimen): 2,700
Hardness Shore A (D2240): 78
Ring & Ball Softening Point F: 305
Meets Applicable FDA Regulations for Food Packaging: Yes

UE 646-04:

Vinyl Acetate: 28%
Equivalent Melt Index g/10 min. (D1238): 26
Density g/cc (D1505/D792): 0.95
Tensile @ Break psi (D638 Type IV Specimen): 1,100
Elongation @ Break % (D638 Type IV Specimen): 770
1% Secant Modulus psi (D638 Type IV Specimen): 2,600
Hardness Shore A (D2240): 78
Ring & Ball Softening Point F (E28): 245
Meets Applicable FDA Regulations for Food Packaging: Yes

UE 636-04:

Vinyl Acetate: 28%
Equivalent Melt Index g/10 min. (D1238): 43
Density g/cc (D1505/D792): 0.95
Tensile @ Break psi (D638 Type IV Specimen): 1,100
Elongation @ Break % (D638 Type IV Specimen): 750
1% Secant Modulus psi (D638 Type IV Specimen): 2,500
Hardness Shore A (D2240): 76
Ring & Ball Softening Point F (E28): 225
Meets Applicable FDA Regulations for Food Packaging: Yes

U.S.I. CHEMICALS: ULTRATHENE Ethylene-Vinyl Acetate Copolymers (Continued):

Resin:

UE 639-35:

Vinyl Acetate: 28%
Equivalent Melt Index g/10 min (D1238): 150
Density g/cc (D1505/D792): 0.95
Tensile @ Break psi (D638 Type IV Specimen): 400
Elongation @ Break % (D638 Type IV Specimen): 590
1% Secant Modulus psi (D638 Type IV Specimen): 1,400
Hardness Shore A (D2240): 69
Ring & Ball Softening Point F (E28): 190
Meets Applicable FDA Regulations for Food Packaging: Yes

UE 653-35:

Vinyl Acetate: 28%
Equivalent Melt Index g/10 min. (D1238): 388
Density g/cc (D1505/D792): 0.94
Tensile @ Break psi (D638 Type IV Specimen): 330
Elongation @ Break % (D638 Type IV Specimen): 320
1% Secant Modulus psi (D638 Type IV Specimen): 1,800
Hardness Shore A (D2240): 68
Ring & Ball Softening Point F (E28): 180
Meets Applicable FDA Regulations for Food Packaging: Yes

UE 647-35:

Vinyl Acetate: 28%
Equivalent Melt Index g/10 min. (D1238): 950
Density g/cc (D1505/D792): 0.937
Tensile @ Break psi (D638 Type IV Specimen): 225
Elongation @ Break % (D638 Type IV Specimen): 300
1% Secant Modulus psi (D638 Type IV Specimen): 700
Hardness Shore A (D2240): 60
Ring & Ball Softening Point F: 171
Meets Applicable FDA Regulations for Food Packaging: Yes

UE 638-35:

Vinyl Acetate: 31%
Equivalent Melt Index g/10 min. (D1238): 24
Density g/cc (D1505/D792): 0.95
Tensile @ Break psi (D638 Type IV Specimen): 720
Elongation @ Break % (D638 Type IV Specimen): 960
1% Secant Modulus psi (D638 Type IV Specimen): 1,200
Hardness Shore A (D2240): 65
Ring & Ball Softening Point F (E28): 245
Meets Applicable FDA Regulations for Food Packaging: Yes

U.S.I. CHEMICALS: ULTRATHENE Ethylene-Vinyl Acetate Copolymers (Continued):

Resin:

UE 654-35:

Vinyl Acetate: 33%
Equivalent Melt Index g/10 min. (D1238): 43
Density g/cc (D1505/D792): 0.95
Tensile @ Break psi (D638 Type IV Specimen): 660
Elongation @ Break % (D638 Type IV Specimen): 870
1% Secant Modulus psi (D638 Type IV Specimen): 1,500
Hardness Shore A (D2240): 73
Ring & Ball Softening Point F: 225
Meets Applicable FDA Regulations for Food Packaging: Yes

U.S.I. CHEMICALS: VYNATHENE Vinyl Acetate Ethylene Copolymers:

VYNATHENE vinyl acetate-ethylene (VAE) copolymers are a series of high vinyl acetate content copolymers ranging from 40 to 51 percent vinyl acetate by weight. They offer the formulator properties heretofore unavailable in a base polymer, especially in tack and weathering characteristics.

VYNATHENE VAE copolymers are compatible with many modifiers, tackifiers, and both natural and synthetic polymers used in adhesive formulations. Adhesives with good low temperature properties, resiliency and elasticity can be prepared using these copolymers. Aging characteristics are excellent because the basic polymer structure has good resistance to ozone, ultraviolet radiation and weathering.

VYNATHENE copolymers can be used alone or in blends with other polymers to produce adhesives for product assembly, packaging and permanent and removable labels and tapes. When formulated into pressure-sensitive systems, they can eliminate environmental and safety problems of solvent evaporation, which can be a problem for user industries.

Resin:

EY 901-25:

Vinyl Acetate: 40%
Equivalent Melt Index g/10 min. (D1238): 7.5
Density g/cc (D1505/D792): 0.96
Tensile @ Break psi (D638 Type IV Specimen): 1,200
Elongation @ Break % (D638 Type IV Specimen): 700
Hardness Shore A (D2240): 63
Ring & Ball Softening Point F: 295
Meets Applicable FDA Regulations for Food Packaging: Yes

EY 902-30/35:

Vinyl Acetate: 41%
Equivalent Melt Index g/10 min. (D1238): 70
Density g/cc (D1505/D792): 0.96
Tensile @ Break psi (D638 Type IV Specimen): 800
Elongation @ Break psi (D638 Type IV Specimen): 1,000
Hardness Shore A (D2240): 56
Ring & Ball Softening Point (E28): 230
Meets Applicable FDA Regulations for Food Packaging: Yes

U.S.I. CHEMICALS: VYNATHENE Vinyl Acetate Ethylene Copolymers (Continued):

Resin:

EY 903:

Vinyl Acetate: 45%
Equivalent Melt Index g/10 min. (D1238): 7.5
Density g/cc (D1505/D792): 0.98
Hardness Shore A (D2240): 35
Ring & Ball Softening Point F (E28): 290
Meets Applicable FDA Regulations for Food Packaging: Yes

EY 904-00/25:

Vinyl Acetate: 51%
Equivalent Melt Index g/10 min. (D1238): 3.5
Density g/cc (D1505/D792): 0.98
Hardness Shore A (D2240): 35
Ring & Ball Softening Point F (E28): 310
Meets Applicable FDA Regulations for Food Packaging: Yes

EY 905:

Vinyl Acetate: 51%
Equivalent Melt Index g/10 min. (D1238): 18
Density g/cc (D1505/D792): 0.98
Hardness Shore A (D2240): 32
Ring & Ball Softening Point F (E28): 250
Meets Applicable FDA Regulations for Food Packaging: Yes

UNION CARBIDE ENGINEERING POLYMERS: RADEL A-400 Polyarylsulfone:

A Thermoplastic Engineered for High Performance in a Harsh Environment.

Outstanding A-400 Polyarylsulfone thermal resistance with high impact strength, transparency and hydrolytic stability.

RADEL Polyarylsulfone is a new high temperature material possessing a superior combination of properties for injection molding and extrusion applications. It will compete in high performance applications not only with other engineering resins, but with metals, glass and ceramics, offering significant fabrication cost advantages. The outstanding properties of RADEL Polyarylsulfone include:

* Thermal stability for extended use
* Excellent toughness
* Superior steam and boiling water resistance
* Transparency
* Exceptional creep resistance
* High heat deflection temperature
* Good stress-crack resistance
* Good electrical properties
* Injection moldable to close tolerances
* Desirable combustion characteristics

This unique combination of properties offers a host of new design possibilities. Besides giving a greater margin of performance in such areas as toughness in comparison with other plastics, RADEL Polyarylsulfones can replace metals in many applications. It is suitable for use in electrical and electronic parts, withstanding solder temperatures. Possible uses include printed circuit boards (sheet and molded), snap-apart connectors and lamp housings for severe environments. Its combustion characteristics recommend investigation in transportation components such as thermoformed housings and structural composites. Its combination of hydrolytic, chemical and impact resistance with transparency should be useful in replacing metal and glass components in the processing industries. RADEL Polyarylsulfone gives the extra margin of toughness in high temperature use, the bonuses of transparency and easier processing, and the economy of long life under stress.

TYPICAL PROPERTIES:

Property	ASTM Method	English	Standard
Melt Flow @ 752F(400C)	D1238	----	30g/10min.
Density	D1505	85.5 lbs/cu ft	1.37 Mg/cu m
Mold Shrinkage	D955	0.6%	0.6%
Tensile Strength @ Yield	D638	12,000 psi	82.7 MPa
Tensile Elongation @ Yield	D638	6.5%	6.5%

UNION CARBIDE ENGINEERING POLYMERS: ARDEL Polarylate:

The tough weatherable thermoplastic.

Tough, and heat resistant, with outstanding UV stability, easy to injection mold and extrude.

ARDEL Polyarylates are a new family of engineering polymers that exhibit outstanding performance in many critical areas. D-100 offers a unique combination of properties, enabling it to outperform metals, glass, ceramics and other plastics in many applications. Among the many excellent properties of ARDEL D-100 are:

* High heat deflection temperatures, up to 345F
* Excellent UV resistance
* Outstanding toughness
* Transparency
* Desirable combustion characteristics
* Excellent flexural recovery, high elastic limits
* Good electrical properties
* Good mechanical properties, including outstanding creep resistance

WEATHER RESISTANCE
Superior UV Resistance

ARDEL polyarylate offers outstanding weatherability. It provides the best known UV protection as a coating, laminate or coextruded film. Surpassing PET in UV resistance, it has even been added to that polymer as a UV stabilizer.

FLEXURAL CHARACTERISTICS
Snap-fit Applications

The flexural recovery or elastic rebound of ARDEL D-100, even at high temperatures, is outstanding, as is its resistance to tensile and compressive deformation.

HEAT
Exceptional thermal resistance

A key feature of ARDEL D-100 is heat resistance.

CHEMICAL RESISTANCE

COMBUSTION
Desirable performance profile

CREEP
Dimensionally stable

MECHANICAL
Tough, high modulus

EXCELLENT ELECTRICAL PROPERTIES

VALITE DIVISION: VALITE Phenolic Molding Compounds:

Molding Compound 759:

VALITE 759 is a black, two stage, asbestos free phenolic molding compound. This product exhibits good electrical and heat resistance properties, along with a good molded appearance. It is a versatile compound which can be used in electrical, automotive and appliance applications.

Material Properties:
- Specific Gravity: 1.50 (D792)
- Bulk Factor: 2.0 (D1895)
- Apparent Density, gm/cc: 0.76 (D1895)
- Mold Shrinkage, in/in: 0.007 (D955)
- Water Absorption @ 24 hr., % 0.3 (D570)

Physical Properties:
- Izod Impact, ft-lb/in: 0.30 (D256)
- Tensile Strength, psi: 7,000 (D638)
- Compressive Strength, psi: 25,000 (D695)
- Flexural Strength, psi: 10,000 (D790)
- Deflection Temp. @ 264 psi F: 365 (D648)

Electrical Properties:
- Dielectric Strength, v/mil: 350 (D149)
- Short Time Step by Step
- Volume Resistivity, ohm, cm: 1.0 X10 11 (D257)
- Surface Resistivity, ohm: 2.0 X 10 13 (D257)

Molding Compound 759-PS-2:

VALITE 759-PS-2 is a black, two stage, asbestos free phenolic molding compound. This material is heat resistant and exhibits good overall physical properties. It is primarily used in injection molding for appliance and electrical parts.

Material Properties:
- Specific Gravity: 1.56 (D792)
- Bulk Factor: 1.8 (D1895)
- Apparent Density, gm/cc: 0.87 (D1895)
- Mold Shrinkage, in/in: 0.004-0.007 (D955)
- Water Absorption, %: 0.20 (D570)

Physical Properties:
- Izod Impact 1/2" ft-lb/in: 0.33 (D256)
- Tensile Strength, psi: 7,500 (D651)
- Compressive Strength, psi: 28,000 (D695)
- Flexural Strength, psi: 11,000 (D790)
- Flexural Modulus, psi: 1.3 x 10 6 (D790)
- Deflection Temperature, F: 380 (D648)

Electrical Properties:
- Dielectric Strength, v/mil
  - Short Time: 355 (D149)
- Volume Resistivity, ohm/cm: 1.0 X 10 11 (D257)
- Surface Resistivity, ohm: 1.5 X 10 13 (D257)

VALITE DIVISION: VALITE Phenolic Molding Compounds(Continued):

Molding Compound 1121-D:

VALITE 1121-D is a black, two stage, glass-filled, phenolic molding compound. This material can be compression, transfer or injection molded. Typical applications are for molded parts in the electrical and appliance fields.

Material Properties:
- Specific Gravity: 1.41 (D792)
- Bulk Factor: 2.1 (D1895)
- Apparent Density, gm/cc: 0.56 (D1895)
- Mold Shrinkage, in/in: 0.006-0.009 (D955)
- Water Absorption, %: 0.35 (D570)

Physical Properties:
- Izod Impact, 1/2", Notched, ft-lb/in: 0.37 (D256)
- Tensile Strength, psi: 7,500 (D651)
- Compressive Strength, psi: 28,000 (D695)
- Flexural Strength, psi: 14,000 (D790)
- Flexural Modulus, psi: 1.2×10^{6} (D790)
- Deflection Temp. @ 264 psi F: 390 (D648)

Electrical Properties:
- Dielectric Strength, v/mil
  - Short Time: 360 (D149)
  - Step by Step
- Volume Resistivity, ohm/cm: 1.0×10^{11} (D257)
- Surface Resistivity, ohm: 6.0×10^{13} (D257)

Molding Compound 1248:

VALITE 1248 is a black, two stage, asbestos free phenolic molding compound. The material is heat resistant and exhibits good overall strength properties with a fine molded appearance. It can be readily molded by compression, transfer and injection molding.

Material Properties:
- Specific Gravity: 1.48 (D792)
- Bulk Factor: 2.1 (D1895)
- Apparent Density, gm/cc: 0.70 (D1895)
- Mold Shrinkage, in/in: 0.007 (D955)
- Water Absorption, %: 0.30 (D570)

Physical Properties:
- Izod Impact, ft-lb/in: 0.32 (D256)
- Tensile Strength, psi: 7,000 (D651)
- Compressive Strength, psi: 27,000 (D695)
- Flexural Strength, psi: 11,000 (D790)
- Deflection Temp. @ 264 psi F: 375 (D648)

Electrical Properties:
- Dielectric Strength, v/psi
  - Short Time: 385 (D149)
- Volume Resistivity, ohm/cm: 1.0×10^{11} (D257)
- Surface Resistivity, ohm: 4.0×10^{13} (D257)

VALITE DIVISION: VALITE Phenolic Molding Compounds(Continued):

Molding Compound 1248 BR:

VALITE 1248 BR is a brown, two stage, asbestos free phenolic molding compound. The material is heat resistant and exhibits good overall strength properties with a fine molded appearance. It can be readily molded by compression, transfer and injection molding.

Material Properties:
- Specific Gravity: 1.48 (D792)
- Bulk Factor: 2.1 (D1895)
- Appearance Density, gm/cc: 0.70 (D1895)
- Mold Shrinkage, in/in: 0.004-0.007 (D955)
- Water Absorption, % 0.30

Physical Properties:
- Izod Impact, ft-lb/in: 0.32 (D256)
- Tensile Strength, psi: 7,000 (D651)
- Compressive Strength, psi: 27,000 (D695)
- Flexural Strength, psi: 11,000 (D790)
- Deflection Temp. @ 264 psi F: 375 (D648)

Electrical Properties:
- Dielectric Strength, v/psi
  - Short Time: 385 (D149)
- Volume Resistivity, ohm/cm: 1.1 X10 11 (D257)
- Surface Resistivity, ohm: 4.0 X10 13 (D257)

Molding Compound 2300:

VALITE 2300 is a black, general purpose, medium impact phenolic molding compound. This material can be compression molded into parts where some impact resistance and/or low mold shrinkage is required.

Material Properties:
- Specific Gravity: 1.38 (D792)
- Bulk Factor: 2.7 (D1895)
- Apparent Density, gm/cc: 0.51 (D1895)
- Mold Shrinkage, in/in: 0.002-0.005 (D955)
- Water Absorption, %: 0.60 (D570)

Physical Properties:
- Izod Impact, 1/2, Notched, ft-lb/in: 0.55 (D256)
- Tensile Strength, psi: 7,000 (D651)
- Compressive Strength, psi: 25,000 (D695)
- Flexural Strength, psi: 10,000 (D790)
- Flexural Modulus, psi: 1.0 X10 6 (D790)
- Deflection Temp. @ 264 psi F: 360 (D648)

Electrical Properties:
- Dielectrical Strength, v/psi
  - Short Time: 200 (D149)
  - Step by Step
- Volume Resistivity, ohm/cm: 6.0 X 10 11 (D257)
- Surface Resistivity, ohm: 4.0 X 10 12 (D257)

VALITE DIVISION: VALITE Phenolic Molding Compounds(Continued):

Molding Compound 3100:

VALITE 3100 is a black, two stage, general purpose phenolic molding compound. It can be compresson molded giving a good surface appearance. Typical applications are for molded parts in the appliance and electrical fields.

Material Properties:
- Specific Gravity: 1.44 (D792)
- Bulk Factor: 2.4 (D1895)
- Apparent Density, gm/cc: 0.62 (D1895)
- Mold Shrinkage, in/in: 0.005-0.008 (D955)
- Water Absorption, %: 0.35 (D570)

Physical Properties:
- Izod Impact, 1/2", Notched, ft-lb/in: 0.29 (D256)
- Tensile Strength, psi: 6,500 (D651)
- Compressive Strength, psi: 28,000 (D695)
- Flexural Strength, psi: 10,000 (D790)
- Flexural Modulus, psi: 1.0 X10 6 (D790)
- Deflection Temp. @ 264 psi F: 360 (D648)

Electrical Properties:
- Dielectric Strength
  - Short Time: 330 (D149)
  - Step by Step
- Volume Resistivity, ohm/cm: 1 X 10 10 (D257)
- Surface Resistivity, ohm: 1.6 X 10 11 (D257)

Molding Compound 4100:

VALITE 4100 is a black, two stage, asbestos free, phenolic molding compound. This material is heat resistant and exhibits a good molded appearance. It can be processed by compression, transfer and injection molding. There is good application of this material in the appliance and electrical fields.

Material Properties:
- Specific Gravity: 1.45 (D792)
- Bulk Factor: 1.90 (D1895)
- Apparent Density, gm/cc: 0.75 (D1895)
- Mold Shrinkage, in/in: 0.005-0.008 (D955)
- Water Absorption, %: 0.32 (D570)

Physical Properties:
- Izod Impact, Notched, ft-lb/in: 0.33 (D256)
- Tensile Strength, psi: 6,500 (D651)
- Compressive Strength, psi: 30,000 (D695)
- Flexural Strength, psi: 11,000 (D790)
- Flexural Modulus, psi: 1.0 X 10 6 (D790)
- Deflection Temperature, F: 360 (D648)

Electrical Properties:
- Dielectrical Strength
  - Short Time: 330 (D149)
- Volume Resistivity, ohm/cm: 3.8 X 10 10 (D257)
- Surface Resitivity, ohm: 3.2 X 10 13 (D257)

VALITE DIVISION: VALITE Phenolic Molding Compounds(Continued):

Molding Compound 4134:

VALITE 4134 is a black, two stage, wood flour filled, asbestos free phenolic molding compound. It can be molded by compression, transfer and injection molding. It can be used for electrical and appliance parts.

Material Properties:
Specific Gravity: 1.34 (D792)
Bulk Factor: 2.3 (D1895)
Apparent Density, gm/cc: 0.58 (D1895)
Mold Shrinkage in/in: 0.006-0.008 (D955)
Water Absorption, %: 0.40 (D570)
Physical Properties:
Izod Impact, 1/2" ft-lb/in: 0.36 (D256)
, 1/4" ft-lb/in: 0.25 (D256)
Tensile Strength, psi: 7,000 (D651)
Flexural Strength, psi: 13,000 (D790)
Flexural Modulus, psi: 1.0 X 10 6 (D790)
Deflection Temp. @ 264 psi F: 340 (D648)
Electrical Properties:
Dielectric Strength, v/mil
Short Time: 320 (D149)
Volume Resistivity, ohm/cm: 3.0 X 10 10 (D257)
Surface Resistivity, ohm: 3.6 X 10 11 (D257)

Molding Compound 4224:

VALITE 4224 is a black, two stage, general purpose phenolic molding compound. It has good electrical properties and can be used for electrical and automotive applications. This material can be compression, transfer and injection molded giving an excellent surface finish.

Material Properties:
Specific Gravity: 1.39 (D792)
Bulk Factror: 2.3 (D1895)
Apparent Density, gm/cc: 0.60 (D1895)
Mold Shrinkage, in/in: 0.005-0.008 (D955)
Water Absorption, %: 0.40 (D570)
Physical Properties:
Izod Impact, 1/2, Notched ft-lb/in: 0.35 (D256)
Tensile Strength, psi: 7,000 (D651)
Compressive Strength, psi: 26,000 (D695)
Flexural Strength, psi: 12,000 (D790)
Flexural Modulus, psi: 1.1 X 10 6 (D790)
Deflection Temp. @ 264 psi F: 350 (D648)
Electrical Properties:
Dielectric Strength, v/psi
Short Time: 350 (D149)
Step by Step
Volume Resistivity, ohm/cm: 1.0 X 10 11 (D257)
Surface Resistivity, ohm: 2.0 X 10 13 (D257)

VALITE DIVISION: VALITE Phenolic Molding Compounds (Continued):

Molding Compound 4512:

VALITE 4512 is a black, two stage, wood flour filled, general purpose phenolic molding compound readily molded by either compression or transfer molding. This material exhibits excellent molding characteristics. This compound is used in the appliance industry and has passed the cookware test.

Material Properties:
- Specific Gravity: 1.44 (D792)
- Bulk Factor: 2.2 (D1895)
- Apparent Density, gm/cc: 0.70 (D1895)
- Mold Shrinkage, in/in: 0.005-0.008 (D955)
- Water Absorption, %: 0.30 (D570)

Physical Properties:
- Izod Impact, ft-lb/in: 0.35 (D256)
- Tensile Strength, psi: 7,000 (D638)
- Compressive Strength, psi: 30,000 (D695)
- Flexural Strength, psi: 10,000 (D790)
- Flexural Modulus, psi: 1.0 X 10 6 (D790)
- Deflection Temp. @ 264 psi F: 360 (D648)

Electrical Properties:
- Dielectric Strength, v/mil
  - Short Time: 340 (D149)
- Volume Resistivity, ohm/cm: 1.0 X 10 11 (D257)
- Surface Resistivity, ohm: 1.0 X 10 13 (D257)

Molding Compound 6100:

VALITE 6100 is a black, two stage, mineral filled, phenolic molding compound. It exhibits good heat resistance and electrical properties with a good molded surface. It can be molded by compression, transfer and injection molding.

Material Properties:
- Specific Gravity: 1.54 (D792)
- Bulk Factor: 2.1 (D1895)
- Apparent Density, gm/cc: 0.73 (D1895)
- Mold Shrinkage, in/in: 0.003-0.006 (D955)
- Water Absorption, %: 0.25 (D570)

Physical Properties:
- Izod Impact, 1/2", Notched, ft-lb/in: 0.35 (D256)
- Tensile Strength, psi: 7,000 (D651)
- Compressive Srength, psi: 28,000 (D695)
- Flexural Strength, psi: 11,000 (D790)
- Flexural Modulus, psi: 1.2 X 10 6 (D790)
- Deflection Temperature, F: 380 (D648)

Electrical Properties:
- Dielectric Strength, v/mil,
  - Short Time: 375 (D149)
- Volume Resistivity, ohm/cm: 1 X 10 12 (D257)
- Surface Resistivity, ohm: 1.8 X 10 14 (D257)

WASHINGTON PENN PLASTIC CO., INC.: Polypropylene:

Product Name:

PP-HIFR:
A V-2 flame retardant polypropylene, UL approved.
Typical Applications: Electrical appliances
Physical Properties:
Melt Flow rate, g/10 min: 1.0-1.5: D-1238 (Condition L)
Density, g/cc: 0.950: D-792

PP-H2MFQ-4:
A 40% mica filled coupled polypropylene.
Typical Applications: Automotive
Physical Properties:
Melt Flow Rate, g/10 min: 3.0-7.0: D-1238 (Condition L)
Density, g/cc: 1.23: D-792

PP-H3MFQ-5:
A 50% mica filled coupled polypropylene.
Typical Applications: Many products now using glass filled polypropylene
Physical Properties:
Melt Flow Rate, g/10 min: 6.0-10.0: D-1238 (Condition L)
Density, g/cc: 1.35: D-792

PP-H2TF-2:
A 20% talc filled polypropylene.
Typical Applications: Medical instruments, auto components, pump housings, instrument panels
Physical Properties:
Melt Flow Rate, g/10 min: 3.0-7.0: D-1238 (Condition L)
Density, g/cc: 1.04: D-792

PP-H2TF-4:
A 40% talc filled polypropylene.
Typical Applications: Air conditioning housings, fan shrouds, electrical connectors
Physical Properties:
Melt Flow Rate, g/10 min: 3.0-7.0: D-1238 (Condition L)
Density, g/cc: 1.24: D-792

PP-C2TF-2:
A 20% talc filled copolymer polypropylene.
Typical Applications: Laundry tubs, toys, housewares
Physical Properties:
Melt Flow Rate, g/10 min: 6.0-10.0: D-1238 (Condition L)
Density, g/cc: 1.03: D-792

WASHINGTON PENN PLASTIC CO., INC.: Polypropylene(Continued):

Product Name:

PP-C4IM-V:
A medium impact polypropylene.
Typical Applications: Automotive components, housewares, tote containers
Physical Properties:
Melt Flow Rate, g/10 min: 8.0-12.0: D-1238 (Condition L)
Density, g/cc: 0.905: D-792

PP-C3CC-4:
A 40% calcium carbonate filled polypropylene.
Typical Applications: Automotive components, toys, housewares
Physical Properties:
Melt Flow Rate, g/10 min: 6.0-10.0: D-1238 (Condition L)
Density, g/cc: 1.23: D-792

PP-VB-511:
A 10% talc filled high flow polypropylene.
Typical Applications: Automotive components (door panels, consoles)
Physical Properties:
Melt Flow Rate, g/10 min: 5.0-8.0: D-1238 (Condition L)
Density, g/cc: .980: D-792

PP-HFG-2:
A 20% fiberglass reinforced uncoupled polypropylene.
Typical Applications: Automotive components, appliances
Physical Properties:
Density, g/cc: 1.04: D-792

PP-HFG-3:
A 30% fiberglass reinforced polypropylene.
Typical Applications: Automotive components, appliances
Physical Properties:
Density, g/cc: 1.13: D-792

PP-HFGC-3:
A 30% fiberglass reinforced coupled polypropylene.
Typical Applications: Automotive components, appliances
Physical Properties:
Density, g/cc: 1.13: D-792

WASHINGTON PENN PLASTIC CO., INC.: Polypropylene(Continued):

Product Name:

PP-HFGC-2:
A 20% fiberglass reinforced coupled polypropylene.
Typical Applications: Automotive
Density, g/cc: 1.04: D-792

PP-G2MF-4:
A 40% mica filled general purpose polypropylene.
Typical Applications: Many products now using glass filled polypropylene.
Density, g/cc: 1.23: D-792

PP-G2MF-5:
A 50% mica filled general purpose polypropylene.
Typical Applications: Many products now using glass filled polypropylene.
Density, g/cc: 1.37: D-792

PP-H2MF-5:
A 50% mica filled polypropylene.
Typical Applications: Automotive Components
Physical Properties:
Density, g/cc: 1.33: D-792 (23C)

PP-H2MFS-4:
A 40% mica filled polypropylene.
Typical Applications: Crash panels, instrument panels
Physical Properties:
Density, g/cc: 1.23: D-792

WASHINGTON PENN PLASTIC CO., INC.: Polyethylene:

Product Name:

PE-25:
A medium density extrusion grade polyethylene.
Typical Applications: Wires, cables, conduits
Physical Properties:
Melt Flow Rate, g/10 min: 1.0: D-1238
Density, g/cc: 0.938: D-792 (23C)

PE-3CC-3:
A 30% calcium carbonate filled impact modified polyethylene.
Typical Applications: Housewares
Physical Properties:
Density, g/cc: 1.20: D-792

Foam Concentrates:

PE-8FC:
A 20% structural foam concentrate.
Typical Applications: Furniture, automotive and marine parts, soda cases, recreational vehicles, toys.

PE-3UV:
A 10% stabilizer concentrate.
Typical Applications: All thermoplastic resins.

WASHINGTON PENN PLASTIC CO., INC.: Polystyrene:

PS-H63CC-1:
A 10% calcium carbonate filled medium impact polystyrene.
Typical Applications: Interior & exterior shutters, housewares.
Physical Properties:
Melt Flow Rate, g/10 min: 7.5: D-1238 (Condition G)
Density, g/cc: 1.10: D-792 (23C)

PS-H23MF-1:
A 10% mica filled high impact polystyrene.
Typical Applications: Automotive components
Physical Properties:
Melt Flow Rate, g/10 min: 10.0: D-1238 (Condition G)
Density, g/cc: 1.08: D-792 (23C)

WASHINGTON PENN PLASTIC CO., INC.: ABS Terpolymer:

Product Name:

T-254:
Density: D-792: g/cc: 1.05
Mold Shrinkage: D-955: in/in: .04-.05

T-344:
Density: D-792: g/cc: 1.05
Mold Shrinkage: D-955: in/in: .04-.05

T-433:
Density: D-792: g/cc: 1.04
Mold Shrinkage: D-955: in/in: .04-.05

T-723:
Density: D-792: g/cc: 1.03

T-523-RE:
Density: D-792: g/cc: 1.04
Mold Shrinkage: D-955: in/in: .04-.05

T-254:
Characteristics: Medium impact.
High flow.
Excellent surface gloss.
Applications: Pieces requiring an excellent surface appearance with an adequate mechanical resistance.

T-344:
Characteristics: High impact and good fluidity. Good mechanical resistance with a great surface gloss.
Applications: Toys, telephones, radio, TV, appliances, bath fixtures, etc.

T-433:
Characteristics: Super high impact.
Good surface gloss.
Applications: Pieces requiring excellent mechanical resistance. Electrical materials, machine-tools, auto parts.

T-723:
Characteristics: Excellent mechanical properties. Specially designed for extrusion but can be injection molded. Good surface gloss.
Applications: Extrusion of sheet and profiles. Parts requiring good mechanical resistance such as shoe heels.

T-523-RE:
Characteristics: Special grade for electroplating.
Applications: Molded parts that are later electroplated. Parts for automotive, appliance fittings, etc.

WASHINGTON PENN PLASTIC CO., INC: TPO Advanced Thermoplastic Alloys:

TPO (Thermo Plastic Olefin) is a mechanically blended alloy, consisting of synthetic rubber(s), olefin type polymer(s), an engineered additive package which can include reinforcing agents, weathering stabilizers, antioxidants, or chemical modifiers.

TPOs are light-weight, cost effective, and versatile materials, to be used in both painted and non-painted applications. It's a versatility tool for the design and materials engineer, satisfying a wide range of part requirements, and competitive, not just with other families of plastics, but with other materials as well.

Product No./Description:

TPO-136:
Melt Flow Rate: g/10 min: D-1238: 0.2-3.0
Density: g/cc: .999

TPO-147:
Melt Flow Rate: g/10 min: D-1238: 0.2-3.0
Density: g/cc: .904

TPO-2261:
Melt Flow Rate: g/10 min: D-1238: 3.0-7.0
Density: g/cc: .920

TPO-226:
Melt Flow Rate: g/10 min: D-1238: 3.0-7.0
Density, g/cc: .900

TPO-236:
Melt Flow Rate: g/10 min: 3.0-7.0
Density, g/cc: .918

TPO-257:
Melt Flow Rate: g/10 min: D-1238: 3.0-7.0
Density, g/cc: .906

TPO-2674:
Melt Flow Rate: g/10 min: D-1238: 3.0-7.0
Density, g/cc: .957

TPO-266:
Melt Flow Rate: g/10 min: D-1238: 3.0-7.0
Density, g/cc: 1.07

TPO-287:
Melt Flow Rate: g/10 min: D-1238: 3.0-7.0
Density, g/cc: 1.03

TPO-297:
Melt Flow Rate: g/10 min: D-1238: 3.0-7.0
Density, g/cc: 1.13

WASHINGTON PENN PLASTICS CO., INC.: TPO Advanced Thermoplastic Alloys (Continued):

<u>Product No./Description:</u>

TPO-326:
Melt Flow Rate: g/10 min: D-1238: 6.0-10.0
Density: g/cc: .905

TPO-336:
Melt Flow Rate: g/10 min: D-1238: 6.0-10.0
Density: g/cc: .921

TPO-347:
Melt Flow Rate: g/10 min: D-1238: 6.0-10.0
Density: g/cc: .966

TPO-357:
Melt Flow Rate: g/10 min: D-1238: 6.0-10.0
Density: g/cc: .929

TPO-367:
Melt Flow Rate: g/10 min: D-1238: 6.0-10.0
Density: g/cc: .903

TPO-3671:
Melt Flow Rate: g/10 min: D-1238: 6.0-10.0
Density: g/cc: .905

TPO-457:
Melt Flow Rate: g/10 min: D-1238: 8.0-12.0
Density: g/cc: .910

TPO-467:
Melt Flow Rate: g/10 min: D-1238: 8.0-12.0
Density: g/cc: 1.16

TPO-525:
Melt Flow Rate: g/10 min: 12.0-18.0
Density: g/cc: .975

TPO-557:
Melt Flow Rate: g/10 min: 12.0-18.0
Density: g/cc: .944

TPO-567:
Melt Flow Rate: g/10 min: 12.0-18.0
Density: g/cc: .958

WASHINGTON PENN PLASTIC CO., INC.: TPO Advanced Thermoplastic Alloys(Continued):

Product No./Description:

TPO-646:
Melt Flow Rate: g/10 min: D-1238: 18.0-25.0
Density: g/cc: .970

TPO-656:
Melt Flow Rate: g/10 min: D-1238: 18.0-25.0
Density: g/cc: 1.17

TPO-667:
Melt Flow Rate: g/10 min: 18.0-25.0
Density: g/cc: 1.17

WELLMAN, INC.: WELLPET Engineering Thermoplastics PET Resins:

GF1020 20% Glass Reinforced:
- Tensile Strength (D638) psi: 17,000
- Elongation at Break (D638) %: 3
- Flexural Modulus (D790) psi 10 5: 9.0
- Flexural Strength (D790) psi: 26,000
- Izod Impact (D256) ft.lb./in. of notch:
  - Notched 1/8": 1.5
  - Unnotched 1/8": 12
- Hardness Rockwell (D785):
  - M Scale: 85
  - R Scale: 120
- Heat Deflection Temperature @ 264 psi (D648) C: 218

GF1030 30% Glass Reinforced:
- Tensile Strength (D638) psi: 22,000
- Elongation at Break (D638) %: 3
- Flexural Modulus (D790) psi 10 5: 13.0
- Flexural Strength (D790) psi: 31,000
- Izod Impact (D256) ft.lb./in. of notch:
  - Notched 1/8": 1.8
  - Unnotched 1/8": 15
- Hardness Rockwell (D785):
  - M Scale: 88
  - R Scale: 120
- Heat Deflection Temperature @ 264 psi (D648) C: 220

GF1045 45% Glass Reinforced:
- Tensile Strength (D638) psi: 26,000
- Elongation at Break (D638) %: 2
- Flexural Modulus (D790) psi 10 5: 20.0
- Flexural Strength (D790) psi: 40,000
- Izod Impact (D256) ft.lb./in. of notch:
  - Notched 1/8": 2.0
  - Unnotched 1/8": 16
- Hardness Rockwell (D785):
  - M Scale: 90
  - R Scale: 120
- Heat Deflection Temperature @ 264 psi (D648) C: 222

FRGF2030 30% Glass Reinforced UL-94 VO:
- Tensile Strength (D638) psi: 20,000
- Elongation at Break (D638) %: 2
- Flexural Modulus (D790) psi: 14.0
- Flexural Strength (D790) psi: 27,000
- Izod Impact (D256) ft.lb./in. of notch:
  - Notched 1/8": 1.5
  - Unnotched 1/8": 10
- Hardness Rockwell (D785):
  - M Scale: 88
  - R Scale: 120
- Heat Deflection Temperature @ 264 psi (D648) C: 215

WELLMAN, INC.: WELLITE Engineering Thermoplastics PBT Resins:

9000GP Unreinforced:
Tensile Strength (D638) psi: 8,000
Elongation at Break (D638) %: 150
Flexural Modulus (D790) psi 10 5: 3.6
Flexural Strength (D790) psi: 13,000
Izod Impact (D256) ft.lb./in. of notch:
Notched 1/8": .70
Unnotched 1/8": NB
Hardness Rockwell (D785):
M Scale: 80
R Scale: 118

9320GR 20% Glass Reinforced:
Tensile Strength (D638) psi: 16,000
Elongation at Break (D638) %: 3
Flexural Modulus (D790) psi 10 5: 8.25
Flexural Strength (D790) psi: 25,000
Izod Impact (D256) ft.lb./in. of notch:
Notched 1/8": 1.2
Unnotched 1/8": 12
Hardness Rockwell (D785):
M Scale: 85
R Scale: 120

9330GR 30% Glass Reinforced:
Tensile Strength (D638) psi: 18,000
Elongation at Break (D638) %: 3
Flexural Modulus (D790): psi 10 5: 11.00
Flexural Strength (D790) psi: 28,000
Izod Impact (D256) ft.lb./in. of notch:
Notched 1/8": 1.7
Unnotched 1/8": 15
Hardness Rockwell (D785):
M Scale: 88
R Scale: 120

9730GRVO 30% Glass Reinforced UL-94 VO:
Tensile Strength (D638) psi: 17,000
Elongation at Break (D638): 2
Flexural Modulus (D790) psi 10 5: 11.50
Flexural Strength (D790): 26,500
Izod Impact (D256) ft.lb./in. of notch:
Notched 1/8": 1.2
Unnotched 1/8": 12
Hardness Rockwell (D785):
M Scale: 89
R Scale: 120

WELLMAN, INC.: WELLAMID Engineering Resins Nylon 6 (Virgin):

420-XE-N:
42L-XE-N:
Tensile Strength (D638): psi: 10,500
Elongation at Break (D638): %: 90
Flexural Modulus (D790) psi 10 5: 4.0
Flexural Strength: psi: 18,000
Hardness Rockwell (D785):
M scale: 80
R scale: 119
Deflection Temp. (D648):
@ 66 psi: F: 360
@ 264 psi: F: 150
Izod Impact Strength (D256) Ft.lb./in. of notch:
Notched 1/8": 1.0
Melting Point (D789) F: 420

42LH-XE-N:
Tensile Strength (D638) psi: 10,500
Elongation at Break (D638) %: 75
Flexural Modulus (D790) psi 10 5: 4.0
Flexural Strength: psi: 18,000
Hardness Rockwell (D785):
M Scale: 80
R Scale: 119
Deflection Temp. (D648):
@ 66 psi: F: 360
@ 264 psi: F: 150
Izod Impact Strength (D256) Ft.lb./in. of notch:
Notched 1/8": 1.0
Melting Point (D789) F: 420

42LN2-XE-N:
Tensile Strength (D638) psi: 11,000
Elongation at Break (D638) %: 35
Flexural Modulus (D790): psi 10 5: 4.1
Flexural Strength: psi: 19,000
Hardness Rockwell (D785):
M scale: 80
R scale: 119
Deflection Temp. (D648):
@ 66 psi: F: 360
@ 264 psi: 155
Izod Impact Strength (D256) Ft.lb./in. of notch:
Notched 1/8": 0.9
Melting Point (D789) F: 420

WELLMAN, INC.: WELLAMID Engineering Resins Nylon 66 (Virgin):

220-N:
Tensile Strength (D638) psi: 12,000
Elongation at Break (D638) %: 50
Flexural Modulus (D790) psi 10 5: 4.1
Flexural Strength: psi: 20,000
Hardness Rockwell (D785):
M Scale: 80
R Scale: 118
Deflection Temp. (D648):
@ 66 psi: F: 455
@ 264 psi: F: 185

22L-N:
Tensile Strength (D638) psi: 12,000
Elongation at Break (D638) %: 50
Flexural Modulus (D790) psi 10 5: 4.1
Flexural Strength: psi: 20,000
Hardness Rockwell (D785):
M Scale: 80
R Scale: 118
Deflection Temp. (D648):
@ 66 psi: 455
@ 264 psi: 185

22LH-N:
Tensile Strength (D638) psi: 12,000
Elongation at Break (D638) %: 45
Flexural Modulus (D790): 4.1
Flexural Strength: psi: 20,000
Hardness Rockwell (D785):
M Scale: 80
R Scale: 118
Deflection Temp. (D648):
@ 66 psi: 455
@ 264 psi: 185

22LN2-H:
Tensile Strength (D638) psi: 12,500
Elongation at Break (D638) %: 30
Flexural Modulus (D790): 4.1
Flexural Strength: psi: 19,000
Hardness Rockwell (D785):
M Scale: 80
R Scale: 118
Deflection Temp; (D648):
@ 66 psi: F: 465
@ 264 psi: F: 190

WELLMAN, INC.: WELLAMID Glass Fiber Reinforced Engineering Resins Nylon 66 and 6 (Virgin):

GF30-60XE-N:
Tensile Strength (D638) psi: 23,000
Elongation at Break (D638) %: 5
Flexural Modulus (D790) psi 10 5: 12.0
Flexural Strength: psi: 35,000
Hardness Rockwell (D785):
M Scale: 90
R Scale: 120

GF13-66XE-N:
Tensile Strength (D638) psi: 15,000
Elongation at Break (D638) %: 5
Flexural Modulus (D790) psi 10 5: 7.0
Flexural Strength: psi: 23,000
Hardness Rockwell (D785):
M Scale: 94
R Scale: 120

GF33-66XE-N:
GF33-66XE-N1:
Tensile Strength (D638) psi: 27,000
Elongation at Break (D638) %: 5
Flexural Modulus (D790): psi 10 5: 12.5
Flexural Strength: psi: 38,000
Hardness Rockwell (D785):
M Scale: 100

GF43-66XE-N:
Tensile Strength (D638) psi: 30,000
Elongation at Break (D638) %: 3
Flexural Modulus (D790) psi 10 5: 16.0
Flexural Strength: psi: 40,000
Hardness Rockwell (D785):
M Scale: 101

GFT13-66XE-N:
Tensile Strength (D638): 13,000
Elongation at Break (D638) %: 10
Flexural Modulus (D790) psi 10 5: 5.1
Flexural Strength: psi: 21,000
Hardness Rockwell (D785):
M Scale: 77
R Scale: 116

WILSON-FIBERFIL INTERNATIONAL: PLASLUBE Internally Lubricated and Reinforced Engineering Resins:

The cost-effective alternative to metal moving parts.

PLASLUBE is Wilson-Fiberfil International's tradename for a family of internally-lubricated and reinforced engineering resin composites.

Injection-moldable thermoplastics have replaced metal parts in many applications, reducing weight, friction, vibration, corrosion, noise and maintenance and providing many other advantages over metals. However, many thermoplastic materials lack lubricity, strength, and dimensional stability and have coefficients of expansion and other characteristics which severely limits their application--partcicularly in replacing moving parts, such as gears.

The family of PLASLUBE composites was developed to provide better lubricity and resistance to wear, less thermal expansion, better fatigue endurance, and creep resistance as well as providing improved load bearing capacity, impact strength, and dimensional stability.

Reading a FIBERFIL Product Code:

J-50/20/TF/15/BK:

J-50:

Product Prefix: This letter-number combination indicates base resin and type of reinforcement. Refer to product class code lists for explanation. (50 is polycarbonate)

/20/:

Reinforcement percentage. The number immediately following the product class indicates percent of fiberglass, milled fiber or metallized glass reinforcement. If there is no number, another type of reinforcement is used. (20% fiberglass)

TF/15:

Additional reinforcement and additives. Other types of reinforcement and additives are indicated by letter. The number following gives the percent. Refer to reinforcements, additives, and lubricants lists. (PTFE fluorocarbon 15%)

BK:

Other additives. All other additives, including those for color, are indicated by letter abbreviations at the end of the product code. Refer to additive and color code lists (Black)

WILSON-FIBERFIL INTERNATIONAL: PLASLUBE Internally Lubricated and Reinforced Engineering Resins(Continued):

PRODUCT CLASS:

G - Long fiber(glass, carbon, metal)
J - Chopped fiber(glass, carbon)

RESIN CODES	RESINS	TRADENAMES
NY1, 10, 29	Nylon(type 66)	NYLAFIL
NY4	Nylon(type 612)	NYLAFIL
NY8	Nylon(Toughened)	NYLAFIL
NY16, 17, 70, 71	Nylon(Modified)	NYLAFIL
PC50	Polycarbonate	POLYCARBAFIL
AC80	Acetal(copolymer)	FORMALDAFIL
PPS1300	Polyphenylene sulfide	SUFIL

REINFORCEMENTS:

CF--Carbon fiber(PAN)
MF--Mineral fiber

LUBRICANTS:

TF--PTFE fluorocarbon
MS--Molybdenum disulfide
SI--Silicone
GR--Graphite

COLOR CODES:

BK--Black

WILSON-FIBERFIL INTERNATIONAL: PLASLUBE Internally Lubricated and Reinforced Engineering Resins (Continued):

PLASLUBE:

6-6 Nylon:

Wear Factor: 10 -10 cu. in. min: 200
Coefficient of Friction: Static: 0.20
Dynamic: 0.28
Limiting PV 100 F.P.M.: 2500

NY-1/SI/5:

Wear Factor: 10 -10 cu. in. min: 40
Coefficient of Friction: Static: 0.08
Dynamic: 0.08
Limiting PV 100 F.P.M.: 6000

NY-1/TF/15:

Wear Factor: 10 -10 cu.in. min.: 12
Coefficient of Friction: Static: 0.11
Dynamic: 0.18
Limiting PV 100 F.P.M.: 12000

NY-1/TF/10:

Wear Factor: 10 -10 cu.in. min.: 20
Coefficient of Friction: Static: 0.12
Dynamic: 0.20
Limiting PV 100 F.P.M.: 9000

NY-1/TF/30:

Wear Factor: 10 -10 cu.in. min.: 10
Coefficient of Friction: Static: 0.10
Dynamic: 0.18
Limiting PV 100 F.P.M.: 12000

NY-1/MS/5:

Wear Factor: 10 -10 cu.in. min.: 160
Coefficient of Friction: Static: 0.28
Dynamic: 0.30
Limiting PV 100 F.P.M.: 7000

WILSON-FIBERFIL INTERNATIONAL: PLASLUBE Internally Lubricated and Reinforced Engineering Resins(Continued):

PLASLUBE:

NY-1/MS/5/TF/30:

Wear Factor: 10 -10 cu.in. min: 12
Coefficient of Friction: Static: 0.12
Dynamic: 0.15
Limiting PV 100 F.P.M.: 9000

G-1/30/MS/5:

Wear Factor: 10 -10 cu.in. min: 75
Coefficient of Friction: Static: 0.24
Dynamic: 0.31
Limiting PV 100 F.P.M.: 15000

G-1/30/SI/2:

Wear Factor: 10 -10 cu.in. min: 100
Coefficient of Friction: Static: 0.14
Dynamic: 0.15

G-1/30/TF/15:

Wear Factor: 10 -10 cu.in. min: 17
Coefficient of Friction: Static: 0.20
Dynamic: 0.25
Limiting PV 100 F.P.M.: 20000

J-1/30/MS/5:

Wear Factor: 10 -10 cu.in. min: 75
Coefficient of Friction: Static: 0.24
Dynamic: 0.31
Limiting PV 100 F.P.M.: 15000

J-1/30/SI/3:

Wear Factor: 10 -10 cu.in. min: 65
Coefficient of Friction: Static: 0.14
Dynamic: 0.15
Limiting PV 100 F.P.M.: 12000

WILSON-FIBERFIL INTERNATIONAL: PLASLUBE Internally Lubricated and Reinforced Engineering Resins(Continued):

PLASLUBE:

J-1/30/TF/15:

Wear Factor: 10 -10 cu.in. min: 17
Coefficient of Friction: Static: 0.20
Dynamic: 0.25
Limiting PV 100 F.P.M.: 20000

J-1/33/TF/13/SI/2:

Wear Factor: 10 -10 cu.in. min: 9
Coefficient of Friction: Static: 0.12
Dynamic: 0.14
Limiting PV 100 F.P.M.: 20000

J-1/CF/15/TF/20:

Coefficient of Friction: Static: 0.11
Dynamic: 0.15
Limiting PV 100 F.P.M.: 30000

J-1/CF/30/TF/15:

Wear Factor: 10 -10 cu.in. min: 10
Coefficient of Friction: Static: 0.11
Dynamic: 0.15
Limiting PV 100 F.P.M.: 42000

J-1/CF/30/TF/13/SI/2:

Wear Factor: 10 -10 cu.in. min: 6
Coefficient of Friction: Static: 0.10
Dynamic: 0.11
Limiting PV 100 F.P.M.: 43000

6-12 Nylon:

Wear Factor: 10 -10 cu.in. min: 190
Coefficient of Friction: Static: 0.27
Dynamic: 0.31
Limiting PV 100 F.P.M.: 2000

WILSON-FIBERFIL INTERNATIONAL: PLASLUBE Internally Lubricated and Reinforced Engineering Resins(Continued):

PLASLUBE:

NY-4/TF/10:

Wear Factor: 10 -10 cu.in. min: 20
Coefficient of Friction: Static: 0.14
Dynamic: 0.22
Limiting PV 100 F.P.M.: 14000

J-4/30/TF/15:

Wear Factor: 10 -10 cu.in. min: 17
Coefficient of Friction: Static: 0.2
Dynamic: 0.25
Limiting PV 100 F.P.M.: 20000

J-4/CF/30/TF/10:

Wear Factor: 10 -10 cu.in. min: 12
Coefficient of Friction: Static: 0.14
Dynamic: 0.18
Limiting PV 100 F.P.M.: 36000

J-4/CF/30/TF/13/SI/2:

Wear Factor: 10 -10 cu.in. min: 6
Coefficient of Friction: Static: 0.10
Dynamic: 0.11
Limiting PV 100 F.P.M.: 43000

Polyphenylene Sulfide:

Wear Factor: 10 -10 cu.in. min: 570
Coefficient of Friction: Static: 0.22
Dynamic: 0.27
Limiting PV 100 F.P.M.: 3000

J-1300/30/TF/15:

Wear Factor: 10 -10 cu.in. min: 110
Coefficient of Friction: Static: 0.15
Dynamic: 0.17
Limiting PV 100 F.P.M.: 27000

WILSON-FIBERFIL INTERNATIONAL: PLASLUBE Internally Lubricated and Reinforced Engineering Resins(Continued):

PLASLUBE:

J-1300/CF/30/TF/15:

Wear Factor: 10 -10 cu.in. min: 75
Coefficient of Friction: Static: 0.15
Dynamic: 0.16
Limiting PV 100 F.P.M.: 28000

J-1300/CF/20/MS/10/TF/15:

Coefficient of Friction: Static: 0.53

6-Nylon:

Wear Factor: 10 -10 cu.in. min: 200
Coefficient of Friction: Static: 0.22
Dynamic: 0.26
Limiting PV 100 F.P.M.: 2000

G-3/40/MS/5:

Wear Factor: 10 -10 cu.in. min: 70
Coefficient of Friction: Static: 0.26
Dynamic: 0.33

J-3/30/MS/5:

Wear Factor: 10 -10 cu.in. min: 75
Coefficient of Friction: Static: 0.24
Dynamic: 0.31

Polycarbonate:

Wear Factor: 10 -10 cu.in. min: 2500
Coefficient of Friction: Static: 0.31
Dynamic: 0.38
Limiting PV 100 F.P.M.: 500

PC-50/TF/10:

Wear Factor: 10 -10 cu.in. min: 85
Coefficient of Friction: Static: 0.11
Dynamic: 0.17
Limiting PV 100 F.P.M.: 18000

WILSON-FIBERFIL INTERNATIONAL: PLASLUBE Internally Lubricated and Reinforced Engineering Resins(Continued):

PLASLUBE:

PC-50/TF/15:

Wear Factor: 10 -10 cu.in. min: 75
Coefficient of Friction: Static: 0.09
Dynamic: 0.15
Limiting PV 100 F.P.M.: 20000

PC-50/TF/30:

Wear Factor: 10 -10 cu.in. min: 65
Coefficient of Friction: Static: 0.08
Dynamic: 0.14

G-50/20/TF/10:

Wear Factor: 10 -10 cu.in. min: 50
Coefficient of Friction: Static: 0.20
Dynamic: 0.24

G-50/20/TF/15:

Wear Factor: 10 -10 cu.in. min: 40
Coefficient of Friction: Static: 0.19
Dynamic: 0.22
Limiting PV 100 F.P.M.: 25000

J-50/20/TF/10:

Wear Factor: 10 -10 cu.in. min: 40
Coefficient of Friction: Static: 0.20
Dynamic: 0.24
Limiting PV 100 F.P.M.: 25000

J-50/20/TF/15:

Wear Factor: 10 -10 cu.in. min: 40
Coefficient of Friction: Static: 0.19
Dynamic: 0.22
Limiting PV 100 F.P.M.: 30000

WILSON-FIBERFIL INTERNATIONAL: PLASLUBE Internally Lubricated and Reinforced Engineering Resins (Continued):

PLASLUBE:

J-50/30/TF/10:

Wear Factor: 10 - 10 cu.in. min: 40
Coefficient of Friction: Static: 0.21
Dynamic: 0.25

J-50/30/TF/15:

Wear Factor: 10 -10 cu.in. min: 30
Coefficient of Friction: Static: 0.18
Dynamic: 0.20
Limiting PV 100 F.P.M.: 30000

J-50/30/SI/2:

Tensile Strength: D638-PSI(MPa): 16000(110)
Izod Notched: D-256-ft.lb./in.(J/m): 3.5(187)

J-50/CF/10/TF/15/FR:

Tensile Strength: D638-PSI(MPa): 14000(97)
Izod Notched: D-256-ft.lb./in.(J/m): 1.5(80)

J-50/CF/20/TF/15:

Tensile Strength: D638-PSI(MPa): 20000(138)
Izod Notched: D-256-ft.lb./in.(J/m): 1.5(80)

Acetal-Copolymer:

Wear Factor: 10 -10 cu.in. min: 65
Coefficient of Friction: Static: 0.18
Dynamic: 0.21
Limiting PV 100 F.P.M.: 3500

AC-80/TF/10:

Wear Factor: 10 -10 cu.in. min: 30
Coefficient of Friction: Static: 0.10
Dynamic: 0.21

WILSON-FIBERFIL INTERNATIONAL: PLASLUBE Internally Lubricated and Reinforced Engineering Resins (Continued):

PLASLUBE:

AC-80/TF/15:

Wear Factor(10 -10 cu.in. min): 20
Coefficient of Friction: Static: 0.08
Dynamic: 0.16

AC-80/TF/20:

Wear Factor(10 -10 cu.in. min): 14
Coefficient of Friction: Static: 0.07
Dynamic: 0.15
Limiting PV 100 F.P.M.: 1250

AC-80/TF/30:

Wear Factor(10 -10 cu.in. min): 12
Coefficient of Friction: Static: 0.06
Dynamic: 0.13

J-80/20/TF/15:

Wear Factor(10 -10 cu.in. min): 200
Coefficient of Friction: Static: 0.20
Dynamic: 0.28
Limiting PV 100 F.P.M.: 17500

J-80/30/TF/15:

Wear Factor(10 -10 cu.in. min): 200
Coefficient of Friction: Static: 0.20
Dynamic: 0.28
Limiting PV 100 F.P.M.: 17500

J80/CF/10/TF/10:

Tensile Strength: D638-PSI(MPa): 12000(83)
Izod Notched: D-256-ft.lb./in.(J/m): 0.7(37.4)

Suppliers' Addresses

Allied-Signal Inc.
Columbia Road and Park Ave.
Morristown, NJ 07960

Amoco Performance Products
P. O. Box 446
Marietta, OH 45750

Applied Composites
333 North Sixth St.
St. Charles, IL 60174

Arco Chemical Co.
1500 Market St.
Philadelphia, PA 19102

Ashley Polymers, Inc.
5114 Ft. Hamilton Pkwy.
Brooklyn, NY 11219

Atochem, Inc.
Polymers Division
266 Harristown Road
P. O. Box 607
Glen Rock, NJ 07452

BASF Corp.
Engineering Plastics
510 Heron Drive
P. O. Box 405
Bridgeport, NJ 08014

Borg Warner Chemicals, Inc.
International Center
PNB Square
Parkersburg, WV 26102

The Budd Co.
Plastics Division
32055 Edward Ave.
Madison Heights, MI 48071

Celanese Engineering Resins
26 Main St.
Chatham, NJ 07928

Colorite Plastics Co.
101 Railroad Ave.
Ridgefield, NJ 07657

Continental Polymers, Inc.
2225 East Del Amo Blvd.
Compton, CA 90220

Cosmic Plastics, Inc.
12314 Gladstone Ave.
San Fernando, CA 91342

Custom Resins
P. O. Box 933
Henderson, KY 42420

Dartco Manufacturing, Inc.
3702 Clanton Road
P. O. Box 5867
Augusta, GA 30906

Dexter Plastics
294 Town Center Drive
Troy, MI 48084

Dow Chemical U.S.A.
Midland, MI 48667

DuPont Co.
1007 Market St.
Wilmington, DE 19898

Eagle-Picher Plastics Division
Eagle-Picher Industries, Inc.
Grabill, IN 46741

Eastman Chemicals Division
Eastman Kodak Co.
P. O. Box 1969
Kingsport, TN 37662

El Paso Products Co.
P. O. Box 3986
Odessa, TX 79760

Enron Chemical Co.
3100 Golf Rd.
Rolling Meadows, IL 60008

Eval Co. of America
1001 Warrenville Rd., Suite 201
Lisle, IL 60532-1359

Fiberite
501 West Third St.
Winona, MN 55987

Fina Oil and Chemical Co.
8350 N. Central Expry.
Dallas, TX 75206

Freeman Chemical Corp.
217 Freeman Drive
P. O. Box 247
Port Washington, WI 53074

General Electric Co.
Plastics Group
One Plastics Ave.
Pittsfield, MA 01201

Georgia Gulf Corp.
P. O. Box 629
Plaquemine, LA 70765-0629

B. F. Goodrich
Chemical Group
6100 Oak Tree Blvd.
Cleveland, OH 44131

Haysite Reinforced Plastics
5599 New Perry Hgwy.
Erie, PA 16509

Hexcel
11711 Dublin Blvd.
Dublin, CA 94566

Huntsman Chemical Corp.
2000 Eagle Gate Tower
Salt Lake City, Utah 84111

ICI Americas, Inc.
Wilmington, DE 19897

Industrial Dielectrics, Inc.
P. O. Box 357
Nobleboro, IN 46060

Koppers Co., Inc.
1900 Koppers Bldg.
Pittsburgh, PA 15219

MKB Industries, Inc.
111 Research Drive
Stratford, CT 06497

Mitsui Petrochemicals(America)
200 Park Ave.
Suite 4114-11
New York, NY 10166

Mobay Corp.
Plastics and Rubber Division
Mobay Road
Pittsburgh, PA 15205-9741

Mobil Chemical Co.
15600 Drummett Blvd.
Suite 900
Houston, TX 77032-2343

Monmouth Plastics, Inc.
814 Asbury Ave.
Asbury Park, NJ 07712

Monsanto Chemical Co.
800 N. Lindbergh Blvd.
St. Louis, MO 63167

Nuodex, Inc.
Turner Place
P. O. Box 365
Piscataway, NJ 08854

Orthane Division
Ohio Rubber
Denton, TX 76201

Owens-Corning Fiberglas Corp.
Resins and Coatings Division
Fiberglas Tower
Toledo, OH 43659

Pennwalt Corp.
Plastics Dept.
Three Parkway
Philadelphia, PA 19102

Perstorp, Inc.
Molding Compounds Division
238 Nonotuck St.
Florence, MA 01060

Phillips 66 Co.
Plastics Division
Bartlesville, OK 74004

Plaskon Electronic Materials
2829 Glendale Ave.
Toledo, OH 43614

Plaslock Corp.
3155 Broadway
Buffalo, NY 14227

Plastics Engineering Co.
3518 Lakeshore Road
P. O. Box 758
Sheboygan, WI 53081

RTP Co.
P. O. Box 439
580 East Front St.
Winona, MN 55987

Resinoid Engineering Corp.
7557 N. St. Louis Ave.
Skokie, IL 60076

Rhone-Poulenc, Inc.
Specialty Plastics Division
P. O. Box 125
Monmouth Junction, NJ 08852

Richardson Polymer Corp.
17 Woodland Road
Madison, CT 06443-2399

Rogers Corp.
One Technology Drive
Rogers, CT 06263

Rohm and Haas
Independence Mall West
Philadelphia, PA 19105

Rostone
P. O. Box 7497
2450 Sagamore Parkway South
Lafayette, IN 47903

A. Schulman, Inc.
3550 W. Market St.
Akron, OH 44313

Standard Oil
Engineered Materials Co.
2050 Cory Drive
Sanborn, NY 14132-9633

Thermofil, Inc.
P. O. Box 489
6150 Whitmore Lake Rd.
Brighton, MI 48116-0489

U.S.I. Chemicals
11500 Northlake Drive
Cincinnati, OH 45249

Union Carbide Corp.
Engineering Polymers
Old Ridgebury Rd.
Danbury, CT 06817-0001

Valite Division
Valentine Sugars, Inc.
Route 2, Box 625
Louisiana Hwy. #308
Lockport, LA 70374

Washington Penn Plastic Co.
2080 North Main St.
Washington, PA 15301

Wellman, Inc.
Plastics Division
P. O. Box 188
Johnsonville, SC 29555

Wilson-Fiberfil International
P. O. Box 3333
2267 West Mill Road
Evansville, IN 47732

Trade Name Index

CYCOLAC	Borg Warner Chemicals, Inc.
CYCOLOY	Borg Warner Chemicals, Inc.
CYCOVIN	Borg Warner Chemicals, Inc.
DELRIN	E.I. DuPont de Nemours & Co., Inc.
DEXCARB	Dexter Plastics
DEXLON	Dexter Plastics
DEXLOY	Dexter Plastics
DEXPRO	Dexter Plastics
DIELECTRITE	Industrial Dielectrics, Inc.
DION	Koppers Co., Inc.
DUREL	Celanese Engineering Resins, Inc.
DYLARK	Arco Chemical Co.
DYLITE	Arco Chemical Co.
DYPRO	Fina Oil and Chemical Co.
DYTHERM	Arco Chemical Co.
EKONOL	Standard Oil
EKTAR	Eastman Chemicals Division
EMPEE	Monmouth Plastics, Inc.
ESD	RTP Co.
EVAL	Eval Co. of America
FIBERFIL	Wilson-Fiberfil International
FINA	Fina Oil & Chemical Co.
FORAFLON	Atochem, Inc.
FORMALDAFIL	Wilson-Fiberfil International
FORTRON	Celanese Engineering Resins, Inc.
GELOY	General Electric Co.
GEON	B.F. Goodrich Co.
HAYSITE	Haysite Reinforced Plastics
HULS	Nuodex, Inc.
HUNTSMAN	Huntsman Chemical Corp.
HYTREL	E.I. DuPont de Nemours & Co., Inc.
INSULSTRUC	Industrial Dielectrics, Inc.
INTERPOL	Freeman Chemical Co.
JONYLON	MKB Industries, Inc.
KERIMID	Rhone-Poulenc, Inc.
KEVLAR	Hexcel
KINEL	Rhone-Poulenc, Inc.
KYNAR	Pennwalt Corp.
LEXAN	General Electric Co.
LOMOD	General Electric Co.
LUSTRAN	Monsanto Chemical Co.
MAGNUS	Dow Chemical U.S.A.
MAKROBLEND	Mobay Corp.
MAKROLON	Mobay Corp.
MELINEX	ICI Americas, Inc.
MELMEX	MKB Industries, Inc.
MINDEL	Amoco Performance Products
MINLON	E.I. DuPont de Nemours & Co., Inc.

RYTON	Phillips 66 Co.
SCARAB	MBK Industries, Inc.
SCHULAMID	A. Schulman, Inc.
SPECTRIM MM	Dow Chemical U.S.A.
SPECTRIM RIM	Dow Chemical U.S.A.
STYPOL	Freeman Chemical Co.
STYRON	Dow Chemical U.S.A.
SUFIL	Wilson-Fiberfil International
SUPEROHM	A. Schulman, Inc.
THERMOFIL	Thermofil, Inc.
THERMOFOAM	Thermofil, Inc.
THERMOLUBE	Thermofil, Inc.
THERMO-X	Thermofil, Inc.
THORSTRAND	Hexcel
TORLON	Amoco Performance Products
TOUGH X	Celanese Engineering Resins, Inc.
TPX	Mitsui & Co., Inc.
TRIAX	Monsanto Chemical Co.
TRUSURF LP	Owens-Corning Fiberglas Corp.
TYRIL	Dow Chemical U.S.A.
UDEL	Amoco Performance Products
ULTEM	General Electric Co.
ULTRA	Monsanto Chemical Co.
ULTRAFORM	BASF Corp.
ULTRATHENE	U.S.I. Chemicals
UNICHEM	Colorite Plastics Co.
UNICHEM RADIOPAQUE	Colorite Plastics Co.
VALITE	Valite Division
VALOX	General Electric Co.
VECTRA	Celanese Engineering Resins, Inc.
VESTAMID	Nuodex, Inc.
VESTODUR	Nuodex, Inc.
VOLTALEF	Atochem, Inc.
VYDYNE	Monsanto Chemical Co.
VYNATHENE	U.S.I. Chemicals
WELLAMID	Wellman, Inc.
WELLITE	Wellman, Inc.
WELLPET	Wellman, Inc.
XENOY	General Electric Co.
XYDAR	Dartco Manufacturing, Inc.
ZYTEL	E.I. DuPont de Nemours & Co., Inc.